T0396557

Tea Polyphenols,

Oxidative Stress and Health Effects

Volume 1

Tea Polyphenols,
Oxidative Stress and Health Effects
Volume 1

Edited by
Baolu Zhao
Institute of Biophysics
Chinese Academy of Sciences, China

World Scientific

NEW JERSEY · LONDON · SINGAPORE · BEIJING · SHANGHAI · HONG KONG · TAIPEI · CHENNAI · TOKYO

Published by

World Scientific Publishing Co. Pte. Ltd.

5 Toh Tuck Link, Singapore 596224

USA office: 27 Warren Street, Suite 401-402, Hackensack, NJ 07601

UK office: 57 Shelton Street, Covent Garden, London WC2H 9HE

Library of Congress Cataloging-in-Publication Data
Names: Zhao, Baolu, editor.
Title: Tea Polyphenols, Oxidative Stress and Health Effects, In 2 Volumes / edited by
    Baolu Zhao, Institute of Biophysics, Chinese Academy of Sciences, China.
Description: New Jersey : World Scientific, [2024] | Includes bibliographical references and index.
Identifiers: LCCN 2023016990 | ISBN 9789811274206 (set) (hardcover) |
    ISBN 9789811274213 (set) (ebook) | ISBN 9789811274220 (set) (ebook for individuals) |
    ISBN 9789811285301 Volume 1 | ISBN 9789811285318 Volume 2
Subjects: LCSH: Tea--Health aspects. | Tea--Therapeutic use. | Polyphenols--Therapeutic use. |
    Oxidative stress.
Classification: LCC RM251 .T44 2024 | DDC 615.3/21--dc23/eng/20230823
LC record available at https://lccn.loc.gov/2023016990

British Library Cataloguing-in-Publication Data
A catalogue record for this book is available from the British Library.

For any available supplementary material, please visit
https://www.worldscientific.com/worldscibooks/10.1142/13350#t=suppl

Desk Editor: Vanessa Quek ZhiQin

Typeset by Stallion Press
Email: enquiries@stallionpress.com

https://doi.org/10.1142/9789811274213_fmatter

About the Editors

Professor Yuefei Wang: Director of Tea Research Institute of Zhejiang University, China.

Professor Youying Tu: Tea Research Institute of Zhejiang University, China.

Professor Yushun Gong: Department of tea, College of horticulture, Hunan Agricultural University, China.

Associate Professor Jihong Zhou: Tea Research Institute of Zhejiang University, China.

Dr. Ying Chen: Department of tea, College of horticulture, Hunan Agricultural University.

Dr. Guoliang Jie: General Manager of Huangshan Maofeng Tea Group Co., Ltd., China.

Caisheng Lan: National first-class tea appraiser, general manager of Yibaozhai Tea Industry Co., Ltd., China.

Jianping Sun: National first-class tea appraiser, chairman of Yibaozhai Tea Industry Co., Ltd., China.

Professor Baolu Zhao: State Key Laboratory of brain and cognition, Institute of Biophysics, Chinese Academy of Sciences, China.

Preface

In recent years, many research articles about tea polyphenols have been published in world academic journals. However, so far, there is no scientific, comprehensive, and objective monograph on tea polyphenols, oxidative stress, and health. The author, Professor Baolu Zhao, formerly chairman of the Asian Free Radical Society and director of the Chinese Free Radical Biology and Medicine Committee of China, was selected as the world's top scientists with global lifetime impact in 2022. The author has been engaged in the research of free radicals and tea polyphenols for decades and has published more than 300 research papers and 6 monograph books, of which more than 30 have directly studied tea polyphenols. In addition, Professor Yuefei Wang, Professor Youying Tu, and Associate Professor Jihong Zhou of Zhejiang University of China, Professor Yushun Gong of Hunan Agricultural University China, Dr Ying Chen of Hunan Agricultural University China, Dr Jie Guoliang the general manager of Huangshan Maofeng Tea Group Co., Ltd. China, national first-class tea evaluator Caisheng Lan, and Jianping Sun the chairman and general manager of Wuyishan Yibaozhai Tea Co., Ltd. China, were invited to participate in the writing of this book. They have studied tea polyphenols for many years and have published many books and papers in scientific journals. On the basis of summarizing these research results, this book refers to the literature on tea polyphenols in recent years and attempts to comprehensively introduce the research results of tea polyphenols, oxidative stress, and health. Therefore, this is not only a monograph on the research results of tea polyphenols, but also a popular science book. This book focuses on the properties and structural characteristics of tea polyphenols and systematically expounds the biological functions of tea polyphenols, especially the free radicals scavenging and antioxidant effects of tea polyphenols, the regulation of oxidative stress, the protection of nerve cells, anticancer effects, strengthening body immunity, preventing cardiovascular and cerebrovascular diseases, anti-inflammatory effects, reducing blood lipid, blood glucose and anti-arteriosclerosis, myocardial protection, especially Alzheimer's disease and Parkinson's disease, anti-aging, improving memory, anti-radiation effect of tea polyphenols, etc. In addition, the sources and safety of tea polyphenols and the precautions for drinking tea and using tea polyphenols are also included.

This book can be read and referred by the majority of scientific researchers in free radical, biology, chemistry, and medicine and teachers and students from colleges and universities of relevant majors. It can also be used as a reference for technicians engaged in the research and development of free radicals and antioxidants, tea processing and sales, and tea lovers and drinkers. At the same time, this book is popular science. Therefore, the publication of this book will serve as a good teacher and friend for the health of the people.

Most of the contents of this book come from published research papers by authors. The authors have sorted and summarized these research papers. The book has also collected hundreds of literature articles, and the list of research articles are placed at the end of the relevant chapters as references.

The author would like to express his sincere gratitude to all his colleagues who participated in writing this book! The author would like to thank to Professor Liping Du and her students at Tianjin University of Science and Technology for providing some useful and meaningful materials for the first chapter of this book. Additionally, the author would like to thank to Associate Professor Chunai Zhang for some valuable suggestions for this book. Finally, the author would like to thank to World Scientific Publishing Co. Pte. Ltd. for editing and publishing this book, so that this book can be met with a wide audience.

Contents

Chapter 1

Origin, History and Species of Tea

Caisheng Lan, Jianping Sun

Yibaozhai Tea Group Co., Ltd., Fujian Province, China

Baolu Zhao

Institute of Biophysics, Chinese Academy of Sciences, Beijing, China

1.1 Introduction

Besides water, tea is the most popular drinking in the world. In 2019, the total global tea production was 6,497,443 metric tons, of which the five major producing countries were China (42.9% of the total production), India (21.4%), Kenya (7.1%), Sri Lanka (4.6%) and Vietnam (4.1%) [FAOSTAT, 2021]. At present, black tea (fermented tea) and green tea (non-fermented tea) are the most popular in the world. Black tea accounts for more than 90% of the total tea sales in western countries. The world's top black tea includes Qimen black tea in China, Darjeeling black tea and Assam black tea in India, and UVA black tea in Sri Lanka. The most popular green tea is undoubtedly all kinds of green tea in China. The top 10 famous teas in the world are all made in China. West Lake Longjing tea is the most famous green tea [Pan *et al.*, 2022]. Tea is not only the most consumed beverage in the world, but also the most beneficial beverage for human health. A large number of studies have shown that drinking tea can increase the antioxidant level in the body and prevent various diseases caused by various oxidative stress [Zhao, 2003, 2007]. When people appreciate the delicacy of tea and enjoy the health of drinking tea, people cannot help thinking about the source and consider the origin of tea and where is the hometown of tea. This chapter will discuss and prove that China is the hometown of tea through a variety of research data. It is the Chinese ancestors who first discovered the tea tree and the efficacy of tea, domesticated and cultivated various kinds of tea trees, and developed various kinds of tea varieties. Then, tea gradually spread from China to all countries in the world.

The origin of tea, tea culture, the main varieties of tea, tea polyphenols, and the story of Wuyishan Dahongpao tea will discussed in this chapter.

1.2   China is the hometown of tea

China is the hometown of tea, tea drinking, and tea culture. Tea trees originated in southwest China 60 million or 70 million years ago. Written records show that Chinese ancestors had begun drinking tea over 3 million years ago. China has the oldest wild and planted tea trees in the world, fossil of a tea leaf from 35,400,000 years ago, and abundant tea-related literatures and art works. Moreover, tea may be the first Chinese herbal medicine used by Chinese people in ancient times. Tea drinking has many benefits to our physical health via its antioxidant, anti-inflammatory, immuno-regulatory, anticancer, cardiovascular-protective, anti-diabetic, and anti-obesity activities. Following from ancient works, the origin and breeding of tea domestication, archaeological discoveries, and linguistics, expansion history proved that tea originated in China.

1.2.1   *Ancient books and document works proved that tea originated in China*

There is a long record of tea drinking in Chinese history. It is impossible to find out exactly when it was, but there is a saying about the general era. Moreover, evidence can also be found that the habit of drinking tea in many parts of the world is indeed passed down from China. Therefore, many people believe that tea drinking is the first creation by Chinese people. The tea drinking habits and tea planting habits in other parts of the world are directly or indirectly passed down from China.

Yu Lu (about 733–804) wrote in his book of tea, "Tea is a drink, which originated from Shen Nong." [Lu, 1974]. Though there is a general consensus about the center of origin of the tea plant, the evolutionary origin and expansion history of the species remain shrouded in controversy, with studies often reporting conflicting findings. Based on Chinese legends and ancient writings about tea, a legend from 2,737 B.C. credits the origin of tea drinking to Shen Nong, a Chinese emperor. Moreover, the earliest literary reference to tea, an ancient Chinese dictionary, is dated 350 B.C., and was followed by the first monograph on tea in 780 A.D., later translated to English [Lu, 1974]. Subsequent works supported these earlier sources; for instance, the various tea types were subsequently dispersed southwards to the Assam-Myanmar-Yunnan area and southeastwards to eastern China [Lu, 1974]. Based on geological and biogeographical evidence, it was demonstrated that the tea plant originated from a narrow region between Wenshan and Honghe in Yunnan province, China (between 22°40′-24°10′N and 103°10′–105°20′E) [Yu, 1986].

It is said that Shen Nong tasted all kinds of herbs and was poisoned. While he was boiling water under the tea tree, leaves from the tree fell and landed in the water. When Shen Nong drank the water in the pot, it tasted bitter and dried up. After drinking, the poison was removed. So, Shen Nong found that the leaves of tea tree could detoxify when drunk. According to the textual research of historians, Shen Nong is actually the praise of later generations for the invention of medicine and farming by ancient ancestors for the benefit of mankind. A hypothetical figure based on the speculation of the primitive social scene. Shen Nong era is a long historical stage from matriarchal clan society to patriarchal society. When there was no agriculture, people mainly depended on collection and hunting for a living. From the original pottery unearthed in China, it is speculated that China's agriculture originated more than 10,000 years ago, so China has used tea for at least 10,000 years. As recorded in "Shen Nong Materia Medica", it also has a history of more than 4,000 years. Tea was first made from wild leaves and used raw or boiled as food for medicine. Later, it gradually developed into antidote, sacrifice, and beverage [Lu, 1974]. Some people think that when Shen Nong boiled water in a cauldron in the wild, just a few leaves floated into the cauldron. The boiled water was slightly yellow. It produced fluid to quench thirst and refresh the mind. It was found that it was a kind of medicine based on Shen Nong's past experience of tasting herbs. This is the most common saying about the origin of tea in China.

After the Qin and Han Dynasties, the wind of tea spread gradually and there were tea drinks from time to time, indicating that the Qin Dynasty had used tea as a drink. Later, due to the increase of consumption, it gradually developed into artificial cultivation. In the Tang Dynasty, Lu systematically compiled the world's first tea monograph, "Tea Classic". The history, planting, processing, production, and tea drinking customs of tea were described in the book, which greatly promoted the development of tea production [Lu, 1974].

1.2.2 *The tea breeding and domestication proved that tea originated in China*

Domesticated over 3,000 years ago, tea is one of the earliest tree crop species in China [Yamanishi, 1995]. Tea breeding has a long history in China and can be traced back over 1,000 years [Wu, 1987]. The tea plant is currently grown in over 52 countries, and China and India are the two largest global tea producers [FAOSTAT, 2015]. To date, over 5,100 accessions of tea germplasm have been selected and conserved in China and India [Chen *et al.*, 2007; Das *et al.*, 2012; Yao and Chen, 2012] and all genetic stocks for most tea growing countries were directly or indirectly introduced from either China or India [Meegahakumbura *et al.*, 2016]. Dated phylogenies of Theaceae indicate that Tribe Theeae, Genus Camellia, and Species diverged from their ancestors ~50 Mya (million years ago), ~15 Mya, and ~625 Kya, respectively [Yu *et al.*, 2017].

From the evolutionary type of tea tree, tea tree always tends to evolve in the long history of its phylogeny. Therefore, all areas where primitive tea trees are concentrated should belong to the origin of tea trees. The wild tea trees in the three provinces of Southwest China and their adjacent areas have the morphological and biochemical characteristics of primitive tea trees, which also proves that southwest China is the center of tea origin. It seems plausible that southeastwards to eastern China (adapted to the relatively colder higher latitudes) was selected from the wild Assam-Myanmar-Yunnan area (adapted to a sub-tropical climate) in order to enhance southeastwards to eastern China survival in the harsher northern temperate climate. One of the key traits that was likely selected in southeastwards to eastern China is the small leaf size, which is thought to confer a selective advantage to plants at higher latitudes [Wright *et al.*, 2017]. Assam-Myanmar-Yunnan area spread northwards within China, and later to Korea and Japan, cold tolerance became a necessary trait for survival, thus persistent selection for this attribute eventually produced southeastwards to eastern China. According to the classification system of Ming (2000), the cultivated tea plant is currently treated as two varieties, i.e., C. sinensis var. sinensis (China type tea) and C. sinensis var. assamica (Masters) Chang (Assam type tea). Apart from C. sinensis, there are 11 more species of Camellia sect. Thea that occurred in China and several species were used as beverage "tea" in Yunnan, Southwest China [Ming, 2000; Ming and Bartholomew, 2007]. In a recent study based on nuclear microsatellite (nSSR) data, two domestication events of tea in China and one in India have been detected [Meegahakumbura *et al.*, 2016].

From the natural distribution of tea plants, there are 23 genera and more than 380 species of Camelliaceae found, while there are 15 genera and more than 260 species in China, and most of them are distributed in Yunnan, Guizhou, and Sichuan. More than 100 species of Camellia have been found, including more than 60 species in Yunnan Guizhou Plateau, of which tea species account for the most important position. From the perspective of Botany, the origin centers of many genera are concentrated in a certain area, which indicates that this area is the origin center of this flora. The high concentration of Theaceae and Camellia in Southwest China shows that southwest China is the birthplace of Camellia, which should be the birthplace of tea.

A paper studied the domestication origin and breeding history of the tea plant (*Camellia sinensis*) in China and India Based on Nuclear Microsatellites and cpDNA Sequence Data. To determine the origin and historical timeline of tea domestication in these two countries, we used a combination of 402 samples and three cpDNA regions (101 samples) to genotype domesticated tea plants and its wild relative. Based on a combination of demographic modeling, NewHybrids and Neighbour joining tree analyses, three independent domestication centers were found. In addition, two origins of Chinese Assam type tea were detected: Southern and Western Yunnan of China. Results from demographic modeling suggested that China type tea and Assam type tea first diverged 22,000 years ago during the last glacial maximum and subsequently split into the Chinese Assam type tea and Indian Assam type tea lineages 2,770 years ago, corresponding well with the early record of tea usage in Yunnan, China [Meegahakumbura *et al.*, 2000].

The haplotypes of Chinese Assam type tea from Southern Yunnan of China formed a distinct clade with haplotypes of C. taliensis, whereas the haplotypes of Chinese Assam type tea from Western Yunnan grouped together with haplotypes of Indian Assam type tea. We speculate that the Southern Yunnan Chinese Assam type tea possibly had a different maternal parent than the Western Yunnan Chinese Assam type tea. This pattern may be indicative of two possible independent origins for Assam type tea in Yunnan, China. Although the cpDNA haplotypes of Chinese Assam type tea in Western Yunnan were genetically close to Indian Assam type tea, with an exception of the hybrid TF 5 with haplotype H19, no haplotype was shared between each other. This result agreed with previous results and confirmed the existence of independent domestication events of China type and Indian Assam type teas [Meegahakumbura *et al.*, 2016]. These results indicated that Chinese Assam type tea in Western Yunnan and Indian Assam type tea in Assam of India may have arisen from a single ancestral population from an area where Southwest China, Indo-Burma, and Tibet meet [Kingdon-Ward, 1950]. From this origin of domestication, consecutive independent domestication events likely occurred in Western Yunnan of China and Assam of India. In addition, the results indicated that China type tea is a distinct genetic lineage, which suggests a separate domestication event for this tea type [Meegahakumbura *et al.*, 2016]. The divergence time calculated between Chinese Assam type tea and Indian Assam type tea (2,770 years ago) is consistent with early historical records of tea usage in Yunnan [Zhao & Yin, 2008] and might be indicative of an early tea domestication event in China, much earlier than the documented usage of tea in India [Das *et al.*, 2012]. This suggests that the largely "pure" genetic composition of China type tea might be attributable to the limited gene flow between China type tea and its wild relatives; the wild species is not in the close vicinity of major growing areas of China type tea in China. The cpDNA results revealed that Chinese Assam type tea possibly has two origins in China (Southern and Western Yunnan) and possesses high haplotype diversity.

However, some people can find evidence that the habit of drinking tea is not only invented by the Chinese, but also invented in other parts of the world, such as India and Africa, but it still needs to be confirmed. Of course, there are records of wild big tea trees in China, which are concentrated in the southwest, and the records also include individual areas in Gansu and Hunan province of China. Tea tree is an ancient dicotyledonous plant, which is closely related to people's life.

1.2.3 *Archaeological discoveries proved that tea originated in China*

Recently, a bowl containing charred suspected tea remains was unearthed from the early stage of Warring States period tomb in Zoucheng City, Shandong Province, China. A paper studied the analysis and identification of the charred suspected tea remains unearthed from Warring State Period Tomb. To identify the remains is significant in understanding the origin of tea and tea drinking culture. Scientific investigations of the remains were carried out by using calcium phytoliths analysis, Fourier transform infrared spectroscopy, Gas Chromatograph Mass Spectrometer, and Thermally-assisted hydrolysis-methylation Pyrolysis Gas Chromatography Mass Spectrometry techniques. The infrared spectra of the archaeological remains was found similar to modern tea residue reference sample. The archaeological remains in the bowl are tea residue after boiling or brewing by the ancient [Jiang *et al.*, 2021]. It indicated that the Chinese drank tea from bowls for more than 3,000 years. Recent excavations of the "Han Yangling Mausoleum" in Xi'an, Shaanxi province, China, revealed that tea drinking was popular among emperors in the Han dynasty more than 2,100 years ago [Lu *et al.*, 2016] providing strong evidence that China type tea originated in China and has a long history of utilization and cultivation in the region. Consistently, tea plant remains from Chang'an (Xi'an, China) indicated that cultivation of tea begun at least 2,100 years ago [Lu *et al.*, 2016]. The famous Tea-Horse road, winding through southwest China (Sichuan, Yunnan, and Tibet), was an important route for trading tea and other products, and is one of China's major sites of cultural heritage today [Sigley, 2013].

The history of tea plantations in India is relatively short, less than 200 years ago [Das *et al.*, 2012]. It is known that China type tea germplasm was introduced to India in 1836 [Sharma *et al.*, 2010]. In particular, the Chinese Assam type tea from Southwestern Yunnan province; China was recently found to represent a distinct gene pool [Meegahakumbura *et al.*, 2016] and both landraces and cultivars of Chinese Assam type tea possessed a high proportion of rare alleles and private haplotypes.

1.2.4 *Linguistics proved that tea originated in China*

Linguistics, for instance, could be useful in understanding the expansion history of the tea crop and the tea drinking culture. Origin of the word tea and its extension to designate different hot infused drinks. When it was imported from China to Europe during the 17th century, its consumption spread rapidly among town people. Used by the Chinese traders operating in the Fou-Kien area, the vulgar word "te" set up in Great Britain, the Netherlands, Germany, and France [Trepardoux & Delaveau, 1999]. This is possible because there are only two main phonetic forms of the names that refer to tea across the world ("Cha" -derived and "Te" -derived forms), with only a few exceptions. The Cantonese "Cha" form, which was/is the most popular name for tea in China, might have spread along the Silk Road to Persia, and onwards to East Africa (possibly by Arab traders), and Eastern Europe. It is also likely that the "Cha" form spread along the Tea Horse Road to India via Tibet, and along the southern Silk Road to Laos and Thailand. On the other hand, the Min Nan (Fujian Province) "Te" form, then spoken in Fujian Province, China, was likely spread by the seafaring Europeans through the Dutch and British trading companies to Southeast Asia, Sri Lanka, Southern Africa, and Eastern Europe. The British East India Company could also be responsible for the introduction of the "Cha" form to East Africa from India, as the company's trade route was often along the East African coast.

Tea has been transformed from medicine into habitual drink, and the strict meaning of "tea" has emerged. Its typical symbol is the emergence of the sound of "Cha". Guo Pu's note in *Erya Shimu*, "The tree is as small as a gardenia, with winter leaves, which can be boiled as a soup. Tea is picked in the morning and is drunk in the evening. It is a bitter tea in the people of Sichuan province." It can be seen that the word "tea" in the Han Dynasty has a pronunciation that specifically refers to the beverage "tea". Tea "was embarked on the road of" independent development. However, the emergence of the word "tea" is accompanied by the development of tea affairs and the increasing frequency of commercial activities. Until the middle Tang Dynasty, it is also in line with the law of the change of characters after the emergence of new symbols in people's social life.

1.2.5 *Expansion history of tea from China to all over the world*

It might be difficult to distinguish between the spread of the tea plant and that of the tea products/drinking culture. These are the biological (artificial) dispersal, trading of tea products, as well as introduction of the tea-drinking culture around the world. The spread of tea within the present-day China was mainly facilitated by the Yunnan-Tibet Tea Horse Road, which was established in the 6th century. Subsequently, spread from China to the rest of the world began in the early 8th century. Within Asia, tea was first introduced to Japan and onwards to Indonesia [Mondal, 2009], with China being the likely ultimate source. Tea was then introduced to Europe in 1768 [Booth, 1830] and later to Sri Lanka in 1839, from India in both cases. From Europe, the crop spread to Africa at the end of the 19th century, though certain anonymous records indicate an earlier date of introduction [Kamunya *et al.*, 2012].

The evidence from the above ancient works, linguistics, tea breeding and domestication, archaeological discoveries, and the expansion history of tea around the world, showed that China is the hometown of tea and the birthplace of tea.

1.3 Tea culture

Chinese tea culture has a long history, which proves that China is the birthplace of tea from another angle. Moreover, the study of tea culture has entered all levels of society, including temple system. The characteristics of tea culture have historical, contemporary, regional, international, and modern.

1.3.1 *Tea culture of past dynasties*

According to the records of the "Book of Songs" and other relevant documents, in the early stage of history, "tea" generally refers to all kinds of bitter wild plant food/raw materials. Only after discovering other values of tea did it have an independent name "tea". In the historical era of the integration of food and medicine, it is not difficult for people to find the medicinal functions of tea vegetables, such as quenching thirst, exciting nerves, help digestion, and treat malaria and constipation. However, there must be some special factors, that is, some specific needs in people's actual life. Bashu region (Sichuan province) is a place of "miasma", which causes frequently diseases. If the people lack tea, they will get sick. Bashu people often eat spicy food, which has been used for thousands of years and still exists today. It is this kind of regional natural conditions and people's dietary customs that make Bashu people first take "fried tea" to eliminate miasma and detoxify heat. After taking tea for a long time, the purpose of medicine gradually disappeared, and tea became a daily drink. When Qin people came to Bashu, they probably saw this tea drinking custom as a daily drink [Kong, 2000 BC].

Chinese Han people pay attention to the word "taste" when drinking tea. When guests come, the etiquette of making and serving tea is essential. When a guest comes to visit, you can ask for advice and choose the best tea set that suits the guest's taste. When offering tea to guests, it is also necessary to mix tea appropriately. When the host accompanies the guests to drink tea, he should pay attention to the residual amount of tea in the guests' cups and pots. Generally, he makes tea with a teacup. If half of the tea has been drunk, he should add boiled water as he drinks, so that the concentration of tea is basically consistent and the water temperature is appropriate. When drinking tea, it can also be appropriately supplemented with tea, candy, and dishes to adjust the taste and dessert.

Tea culture is very important in the life of the Chinese Han nationality. As another example, in the Han Dynasty, tea became a special tonic for Buddhist "meditation". Tea drinking in the Wei and Jin Dynasties has been popular in the north and south of Chaina. In the Sui Dynasty, tea was widely consumed by the whole people. In the Tang Dynasty, the tea industry flourished, and tea became "people can't have no tea for a day". Teahouses, tea banquets, and tea parties appeared, advocating guests to offer tea. In the Song Dynasty, tea fighting, tribute tea, and gifting tea were popular [Kong, 2000 BC].

1.3.2 *Temple tea culture*

Drinking tea has also become an important part of the temple system. After thousands of years of honing, the humanistic factors in the connotation of tea are increasing, and the natural attributes of tea are gradually contained in the humanistic factors. The method of Buddhist practice determines the inseparable relationship between monks and tea, and precepts, calmness, and wisdom are the main methods of Buddhist practice. Therefore, it needs a kind of nutritional supplement that not only conforms to the Buddhist precepts, but also can eliminate the fatigue caused by sitting and sleepy for a long time and make up for "not eating at noon" and "abstaining from meat and vegetarianism". As it happens, tea has become an ideal drink for monks because of its refreshing and beneficial thinking, the pharmacological function of generating saliva and relieving thirst, and its rich nutrients. Over time, as monks drank and practiced tea, it gradually became their unique Temple tea culture.

There is a "tea hall" in the temple, which is a place for Zen monks to debate Buddhism, entertain donors, and taste fragrant tea. The "tea drum" in the temple hall is the drum that calls the monks to drink tea, and the "tea head" is in charge of boiling water, boiling tea, and offering tea to guests. The "tea monk" in front of the temple is dedicated to giving tea. "Temple tea" refers to the tea in Buddhist temples, which is usually used for Buddhism, hospitality, and self-worship. According to the regulations, tea soup is offered every day in front of the Buddha, ancestors, and spirits, which is called "drinking tea". Drinking tea according to the number of years of abstinence is called "abstinence wax tea". All monks are invited to drink tea, which is called "Pu tea". The tea obtained by begging for food is called "Huacha". When monks meditate, tea is also essential. At each stage, after burning one incense, the supervisor of the temple should "make tea" and "walk four or five turns of tea" so as to refresh the mind and eliminate the fatigue caused by long-term meditation. Within the Buddhist monasteries in Tibet, the offering of tea-ceremonies was a requirement for new monks, as well as a mark for certain key milestones for the serving monks [Snellgrove *et al.*, 2003].

Broadly speaking, tea culture is divided into two aspects: 1) the natural science of tea and 2) the humanities of tea. It refers to the sum of the material wealth and spiritual wealth related to tea created in the historical practice of human society. In a narrow sense, it focuses on the humanities of tea, which mainly refers to the function of tea to spirit and society. Since the natural science of tea has formed an independent system, the often talked about tea culture focuses on the humanities.

1.3.3 *The characteristics of Chinese tea culture*

Chinese tea culture has distinct characteristics, with its historical, contemporary, regional, international, and modern characteristics and more monographs [Yu, 1999].

1. Historical characteristics
The history and development of Chinese tea is not only the process of forming a simple food culture, but also reflects the spiritual characteristics of a nation with a history of 5,000 years. The formation and development of tea culture has a very long history. During the period of time when King of Wu kingdom attacked Zhou kingdom, tea has been used as a tribute. In the later period of the primitive commune, tea became an exchange of goods. In the Warring States period, tea had a certain scale. Tea is recorded in the collection of the "Book of Songs" before Qin Dynasty.

In the Han Dynasty, tea became a special tonic for Buddhist "meditation". Tea drinking in the Wei and Jin Dynasties had been popular in the north and south of China. In the Sui Dynasty, tea was widely consumed by the whole people. In the Tang Dynasty, the tea industry flourished, and tea became "people have tea everyday". Teahouses, tea banquets, and tea parties appeared, advocating guests to offer tea. In the Song Dynasty, tea fighting, tribute tea and gift tea were popular. In the Qing Dynasty, opera and art entered teahouses and tea foreign trade developed. Tea culture was born with the emergence of commodity economy and the formation of urban culture. Tea culture in history pays attention to cultural ideology, focusing on elegance, poetry, calligraphy and painting, tea drinking, singing, and dancing. In the formation and development of tea culture, it integrates the philosophical color of Confucianism, Taoism, and Buddhism, and evolves into the etiquette and customs of various nationalities. It has become an integral part of excellent traditional culture and a unique cultural model [Yu, 1999].

2. Epochal character

The development of material civilization and spiritual civilization has injected new connotation and vitality into tea culture. The connotation and manifestation of tea culture are constantly expanding, extending, innovating, and developing. In the new era, tea culture is integrated into the essence of modern science and technology, modern news media, and market economy, which makes the value function of tea culture more significant and further enhances its role in modern society. The value of tea is the core of tea culture. The consciousness is further established, and international exchanges are becoming more and more frequent. In the new era, the mode of transmission of tea culture shows the trend of large-scale, modernization, socialization, and internationalization. Its connotation expands rapidly and its influence grows, which has attracted the attention of the world [Yu, 1999].

3. Regional character

Famous tea, famous mountains, famous waters, celebrities, and scenic spots breed distinctive regional tea culture. China has a vast area, a wide variety of tea varieties, different tea drinking customs, and differences in history, culture, life, and economy, forming a tea culture with local characteristics. In large cities with economic and cultural centers, with its unique advantages and rich connotation, it also forms a unique urban tea culture. Since 1994, Shanghai has held four consecutive international tea culture festivals, showing the characteristics and charm of urban tea culture [Yu Y, 1999]. Since the 18th session in 2010, Henan Xinyang tea culture festival has been officially renamed "China Xinyang International Tea Culture Festival". The first China Wuyi Mountain tea culture and Art Festival was held in November 2003, and it was held at the same time as the 6th Fujian Wuyi Mountain rock tea Festival. Every year, various tea culture festivals are held in Huangshan. The West Lake has a long history of tea production. The West Lake Longjing began in the Song Dynasty, carried forward to the Ming Dynasty, and flourished in the Qing Dynasty. It combines the four wonders of "green color, fragrant, sweet taste and beautiful shape". It is world famous throughout the country. It integrates famous mountains, famous temples, famous lakes, and famous springs. It is famous both at home and abroad. It is a huge intangible asset of Hangzhou's urban brand. It is held at the end of March and the beginning of April in Longjing tea township of West Lake every year [Yu, 1999].

4. International character
The ancient Chinese traditional tea culture is combined with the history, culture, economy, and Humanities of various countries in the world. Chinese tea culture is the cradle of tea culture of various countries in the world. It has evolved into British tea culture, Japanese tea culture, Korean tea culture, Russian tea culture, and Moroccan tea culture. In Britain, drinking tea has become a part of life. It is a kind of etiquette for Britons to express their gentleman style. It is also an essential procedure in the Queen's life and a necessary instrument in major social activities. Japanese tea ceremony originated from China. Japanese tea ceremony has strong Japanese national customs and forms a unique tea ceremony system, genre, and etiquette [Yu, 1999].

5. Modern development
After the founding of new China, the annual output of Chinese tea increased from 7,500 tons in 1949 to more than 600,000 tons in 1998. The

substantial increase in the material wealth of tea has provided a solid foundation for the development of Chinese tea culture. In 1982, the first social group with the purpose of promoting tea culture, "Tea man's home", was established in Hangzhou, the "Luyu (Shandong and Henan provinces) Tea Culture Research Association" was established in Hubei provinces in 1983, the "Chinese tea man fellowship" was established in Beijing in 1990, and the "China International Tea Culture Research Association" was established in Huzhou in 1993. In 1991, the China Tea Museum was officially opened in West Lake Township, Hangzhou. In 1998, the China International Peace tea culture exchange hall was completed. With the rise of tea culture, there are more and more tea houses in various places. The international tea culture seminar has been held for the fifth time. All provinces, cities, and main tea-producing counties have hosted "tea Festivals", such as the Rock Tea Festival in Wuyi City, Fujian Province, the Pu'er Tea Festival in Yunnan province, and the tea festivals in Xinchang, Taishun, Hubei province and Yingshan and Xinyang, Henan province. All take tea as the carrier to promote the all-round development of economy and trade [Yu, 1999].

6. Monograph of tea

There are many monographs about tea in Chinese history. From many monographs about tea, we can also prove that China is worthy of being the birthplace and hometown of tea. The most famous is the "Tea Classic" [Lu, 1974]. It is the earliest, most complete, and most comprehensive monograph on tea in China and even the world. It is known as the "Encyclopedia of tea", written by Lu Yu, the founder of Chinese tea art in the Tang Dynasty. This book is a comprehensive treatise on the history, origin, current situation, production technology, tea drinking skills, and tea ceremony principles of tea production. It is an epoch-making monograph on tea science. It is not only an incisive agricultural work, but also a book on tea culture. It upgraded ordinary tea to a wonderful cultural and artistic ability. It is an important work on tea in ancient China, which promoted the development of Chinese tea culture. It is also an incisive work on agronomy and a book on tea culture. Upgrading ordinary tea affairs to a wonderful culture and art has promoted the development of Chinese tea culture.

At the age of 21, Lu Yu decided to write the book of tea. Therefore, he began to travel and investigate tea. He was hungry for food and thirsty for tea. He went through Yiyang and Xiangyang to Nanzhang and Wushan of Sichuan province. Everywhere he went, he discussed tea with local villagers, made various specimens of various kinds of tea, wrote down the tea anecdotes he learned on the way, and made a lot of "tea notes". After more than 10 years of on-the-spot investigation in 32 states, Lu Yu finally lived in seclusion in Tiaoxi (now Huzhou, Zhejiang Province) and began to study and write about tea. It took five years to write the first draft of the "Tea Classic". In the following five years, it was supplemented and revised before it was officially finalized. At this time, Lu Yu was 47 years old. It took a total of 26 years to finally complete the world's first outstanding and masterpiece monograph on tea, the "Tea Classic".

On the basis of personal investigation and practice, Lu Yu carefully summarized and studied the production experience of tea predecessors at that time and completed the founding work "Tea Classic". Therefore, people honored him as the tea god and tea immortal. The book of tea is divided into 3 volumes and 10 sections, with about 7,000 words. Volume 1: The source of tea. First part talks about the origin, shape, function, name, and quality of tea. Second part talks about the utensils for picking and making tea, such as tea picking basket, tea steaming stove, tea baking shed, etc. The third part discusses the types and harvesting methods of tea. Volume 2: The four utensils describe the utensils for boiling and drinking tea, that is, 24 kinds of tea drinking utensils, such as air stove, tea kettle, paper bag, wood mill, tea bowl, etc. Volume 3 covers five topics related to tea: the methods of brewing tea and the quality of water in different regions, the customs and traditions surrounding tea drinking, and historical and contemporary anecdotes, origins, and benefits of tea (Fig. 1-1).

The book classifies the distribution of tea districts in the Tang Dynasty into eight regions: Shannan (south of Jingzhou), southern Zhejiang province, western Zhejiang province, Jiannan, eastern Zhejiang province, Central Guizhou province, Jiangxi province, and Lingnan (Guangdong province). The advantages and disadvantages of tea produced in various places are discussed. According to the analysis of tea picking and tea making tools, some tools can be omitted according to the current environment.

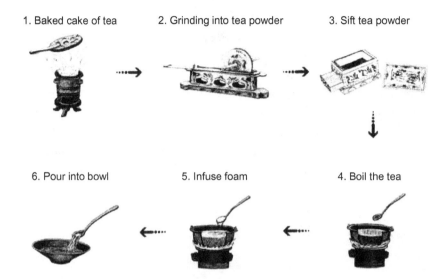

1. Baked cake of tea   2. Grinding into tea powder   3. Sift tea powder

6. Pour into bowl   5. Infuse foam   4. Boil the tea

Figure 1-1. How to boil tea in Tang Dynasty [Lu, 1974].

The book also has 10 paintings painted on silk to teach people how to learn and practice the content of the "Tea Classic". It appeared in all corners of the country and was preserved after witnessing. The book of tea systematically summarized the experience of tea collection and drinking at that time, comprehensively discussed the problems related to the origin, production, and drinking of tea, disseminated the scientific knowledge of tea industry, promoted the development of tea production, and pioneered the Chinese tea art. Besides, the book of tea is the most complete tea book in ancient China. All kinds of contents related to tea are described. Later, tea books were all based on his book. The book of tea is a set of rules summarized by Lu Yu on the basis of observing the growth law of tea and the processing of tea by tea farmers in major tea areas, further analyzing the quality of tea and learning the good methods of folk tea cooking. In addition, Lu Yu also pays attention to the production of civilian tea sets and tea utensils and makes his own unique set of tea sets. Lu Yu spent his whole life in studying tea affairs, and his footsteps spread all over the major tea areas in the country of China.

Since Lu Yu wrote the book of tea, tea monographs have been published one after another, which further promoted the development of

Chinese tea affairs [Dong, 2003]. For example, Tang Dynasty: Zhang Youxin's *Fried Tea* and Su Lin's *Sixteen Soups*. Song Dynasty: Cai Xiang's *Tea Record* (1049–1053), Zhao Ji's *Daguan tea theory* (1107), Xiongfan's Xuanhe *Beiyuan tribute tea record* and Shen'an's *Picture praise of tea sets*. Ming Dynasty: Zhu Quan's *Tea manual*, Gu Yuanqing's *Tea manual*, Tu Long's *Tea theory*, Xu Cishu's *Tea Shu*, Feng Shike's *Tea record*, Wen Long's *Tea note*, Gao Qi's *Dongshan tea Department*, Zhou's *Yang Xian tea pot Department*, Qian Chunnian's *Tea manual*, and Gu Yuanqing's *Tea spectrum*. Qing Dynasty: Liu Yuanchang's *History of tea*, Yu Huai's *Supplement to the history of tea*, and Fangxiang's *Money for tea exchange* [Dong, 2003].

In recent years, China has published many books on tea, such as *The Chinese tea classic* written by Zongmao Chen [Chen, 1992], *The Chinese tea affairs* written by Zheng Guojian [Zheng, 2016], Zhu Zizhen's and Shen Dongmei's *Integration of ancient Chinese tea books* [Zhu & Shen, 2010], *Wuyi tea classics* by Xiao Tianxi [Science Publishing House, 2008], tea road course by Yu Yue — *A brief history of the evolution of Chinese tea culture* [Yu, 1999], *Chinese tea history* by Guo Mengliang [Guo, 2003], *General history of tea industry* by Chen Chuan [Chen, 1970], *Tea history* by Dong Shangsheng [Dong, 2003] Wang Runxian's *et al. The formation and evolution of Chinese tea varieties and processing methods* [Wang, 2011], Guo Yuanchao's *Origin, propagation and evolution of tea plants* [Guo, 1996].

1.4 Varieties of tea

China's tea varieties are very complex and diverse. In recent decades, the combination of tea science and botany research has made a more detailed and in-depth analysis and demonstration of the origin of tea trees from different angles such as tree species, geological change, and climate change, which further proves that southwest China is the origin of tea trees.

From the natural distribution of tea plants, there are 23 genera and more than 380 species of Camelliaceae found, while there are 15 generas and more than 260 species in China, and most of them are distributed in Yunnan, Guizhou, and Sichuan provinces. More than 100 species of Camellia have been found, including more than 60 species in Yunnan and

Guizhou province Plateau, of which tea species account for the most important position [Chen, 1992; Pan, 1995; Peng, 1998].

Tea is divided into green tea, black tea, dark tea, oolong tea, white tea, yellow tea Pu'er tea. Tea can also be classified according to its fermentation degree, including full fermentation, semi fermentation, and non-fermentation. Fermentation can be divided into hot fermentation and cold fermentation. Green tea and white tea are not fermented, black tea and dark tea are fully fermented, oolong tea is semi fermented, black tea and dark tea are hot fermented, and Pu'er tea is cold fermented [Chen, 1992; Pan, 1995; Peng, 1998].

In the 1980s, there were 211 famous tea varieties: green tea, 15 black tea, 10 yellow tea, 10 flour tea, 5 black tea, and 17 green tea. There are 264 kinds of tea in total (other famous teas are still increasing.

In different periods, there are many statements about China's top 10 famous teas:

In 1915, the Panama world expo listed Biluochun, Xinyang Maojian, Xihu Longjing, Junshan silver needle, Huangshan Maofeng, Wuyi rock tea, Qimen black tea, Duyun Maojian, Lu'an Guapian, and Anxi Tieguanyin as the top 10 famous tea in China.

In 1959, Dongting Biluochun, Nanjing Yuhua tea, Huangshan Maofeng, Lushan Yunwu Tea, Lu'an Guapian, Junshan silver needle, Xinyang Maojian, Wuyi rock tea, Anxi Tieguanyin, and Qimen black tea were listed as China's top 10 famous tea [1].

In 2001, the associated press and the "New York Daily" listed Huangshan Maofeng, Dongting Biluochun, Mengding manna, Xinyang Maojian, Xihu Longjing, Duyun Maojian, Lushan Yunwu, Anhui Guapian, Anxi Tieguanyin, and Suzhou jasmine as the top 10 famous teas in China.

In 2002, "Hong Kong Wen Wei Po" listed West Lake Longjing, Jiangsu Biluochun, Huangshan Maofeng, Hunan Junshan silver needle, Xinyang Maojian, Qimen black tea, Anhui Guapian, Duyun Maojian, Wuyi rock tea, and Fujian Tieguanyin as the top 10 famous tea in China.

The book introduces China's top 10 famous teas:

Yu's *Handbook of Chinese tea* introduces the ten famous teas in China: Longjing tea, Biluochun, Huangshan Maofeng, Junshan silver needle,

Qimen black tea, Lu'an Guapian, Xinyang Maojian, Duyun Maojian, Wuyi rock tea, and Tieguanyin [Yu, 1998].

Chen's *Tea talk* introduces the top 10 famous teas in China: Shifeng Longjing tea, Dongting Biluochun, Lu'an Guapian, Junshan silver needle, Huangshan Maofeng, Xinyang Maojian, Taiping monkey Kui, Lushan cloud, Mengding manna, and Guzhu purple bamboo shoots [Chen, 1982].

Zhao's *Junshan silver needle* introduces the top 10 famous teas in China: West Lake Longjing tea, Dongting Biluochun, Huangshan Maofeng, Junshan silver needle, Qimen black tea, Lu'an Guapian, Xinyang Maojian, Duyun Maojian, Wuyi rock tea, and Anxi Tieguanyin [Zhao, 2000].

Li's *Top ten famous teas in China* introduces that the top 10 famous teas in China are: West Lake Longjing, Lushan Yunwu, Huangshan Maofeng, Dongting Biluochun, Lu'an Guapian, Junshan silver needle, Xinyang Maojian, Wuyi rock tea, Anxi Tieguanyin, and Qimen black tea [Li, 2015].

1.4.1 *Green Tea*

There are many varieties of green tea in China. The main green tea includes fried green tea, baked green tea, sun dried green tea, and steamed green tea. Each variety includes many brands. The following varieties are just a few common brands.

Fried green tea is divided into eyebrow tea, pearl tea, tender fried green tea, Dafang, Biluochun, Yuhua tea, manna, pine needle, etc.

Roasted green tea is divided into ordinary roasted green tea and delicate roasted green tea, etc.

Sun green tea is divided into Sichuan green tea, Yunnan green tea and Shaanxi green tea, etc.

Steamed green tea is divided into fried tea, jade dew, etc.

1.4.2 *Black tea*

There are many varieties of black tea in China, mainly divided into small black tea, Gongfu black tea, broken black tea, etc.

Varieties of black tea include Zhengshan varieties tea, tobacco varieties tea, etc.

Gongfu black tea includes Chuanhong (Golden manna tea, red manna tea, etc.), Qihong, Dianhong tea, Minhong tea (Golden Junmei tea, etc.).

Broken black tea includes leaf tea, broken tea, slice tea and end tea etc.

1.4.3 *Oolong Tea*

There are many varieties of Oolong tea in China, mainly divided into:

Oolong tea in Northern Fujian Province: Wuyi Rock Tea, Dahongpao tea, Narcissus tea, Cinnamon tea, Bantianyao tea, Qilan, Baxian, and Jianou Jianyang oolong, etc.

Oolong in Southern Fujian Province: Tieguanyin tea, Qilan tea tea, Narcissus, huangjingui tea, etc.

Guangdong Oolong tea: Fenghuang danzong tea, Fenghuang Narcissus tea, Lingtou danzong tea, etc.

Taiwan Oolong tea (Frozen top oolong tea, Baozhong, etc.) Alishan Gaoshan tea, etc.

1.4.4 *White tea*

There are many varieties of white tea in China, mainly divided into white bud tea and white leaf tea, where white bud tea mainly refers to silver needle, etc. and white leaf tea mainly refers to white peony, hongmei, etc.

1.4.5 *Yellow tea*

There are many varieties of yellow tea in China, mainly divided into yellow bud tea, yellow tea, yellow tea:

Yellow bud tea including Mengding yellow bud tea, Junshan silver needle tea, etc.

Huang Xiaocha tea including Beigang Maojian tea, Heshan Maojian tea, Wenzhou huangtang tea, etc.

Huangda tea including Huoshan huangda tea, Guangdong dayeqing tea, etc.

1.4.6 *Dark tea*

There are many varieties of dark tea in China, mainly divided into:

Hunan black tea (Anhua black tea, etc).
Hubei old green tea (Puqi old green tea, etc).
Sichuan side tea (South Road side tea and West Road side tea, etc).
Yunnan Guangxi black tea (Liubao tea, etc).
Shaanxi black tea (Jing Wei Fu tea, etc.)

1.4.7 *Pu'er tea*

Pu'er tea is found in Xishuangbanna, Lincang Yuannan province, and other areas of Yunnan Province. Pu'er tea has unique quality characteristics. According to its processing technology and quality characteristics, Pu'er tea can be divided into two types: raw Pu'er tea and cooked Pu'er tea.

1.5 Tea polyphenols

The main antioxidant active components of tea are tea polyphenols. Green tea and white tea mainly contain green tea polyphenols, while black tea and dark tea mainly contain poly tea polyphenols (theaflavins), and oolong tea and yellow tea contain two components, green tea polyphenols and theaflavins [Yang *et al.*, 2003]. China also leads the world in the research, development, production, and application of tea polyphenols. Every year, Chinese researchers publish a large number of research papers about tea polyphenols at home and abroad. China has established a number of tea polyphenol production enterprises and factories, produced a large number of tea polyphenols every year, researched and developed various tea polyphenol products in the laboratory, and achieved gratifying results in many industries such as drugs, health products, and feeding materials every year, which also proves that China is worthy of the birthplace and hometown of tea in China [Zhao, 1999; Yang *et al.*, 2003; Zhao, 2008].

The second and third chapters of this book will discuss these two kinds of tea polyphenols in detail respectively, so we will not discuss them in detail here.

1.6 The story about Wuyi-Mountain Dahongpao tea

As Lu said in the book *Tea classic*, with the development of society, tea has gradually shifted from Southwest China to the more developed eastern region where the climate and geography are more suitable for tea [Lu, 1974]. Among them, Fujian and Zhejiang provinces are rising stars. I have been visited to many places rich in tea, such as Longjing in Hangzhou West Lake, Xinyang in Henan, Wuyi Mountain in Fujian provinces, etc. Only Wuyi Mountain has left a deep impression on me. Wuyi Mountain has good mountains, water, soil, and climate. It is also a National Park with World Natural and Cultural Heritage (China National Park and National Forest Park, world natural heritage reserve). It can also reflect a clearer concept that China is the hometown of tea through the development of Dahongpao tea in Wuyi Mountain, Fujian Province.

Wuyi Mountain in Fujian Province is the birthplace of tea and black tea in the world, rich in Dahongpao tea and other Oolong tea. I visited Wuyi Mountain twice. The first time was in 2001, I participated in the National Fault Archaeology Acceptance Meeting held there. The second time was in 2008, Mr. Sun J, the national first-class tea evaluator, chairman of Yibaozhai, and Mr. Lan C, the general manager, invited me to visit Wuyi Mountain. There, we went to see the earliest rock tea trees growing on the cliff again. We also went to the tea garden to watch how tea farmers to pick tea and visited the whole process of tea processing — how to ferment

Figure 1-2.   Author visited Wuyi Mountain and drifted with national first-class tea critic, Mr. Sun Jianping and Mr. Lan Caisheng in 2018.

Figure 1-3.   Author talked about tea and Zen with Huang Zhenguo, one of the founders of commodity Dahongpao tea, Buddhist Zedao, national first-class tea critic, Mr. Sun J, Mr. Lan C in Dahongpao tea ancestral court in 2018 at Wuyi-Mountain city.

Figure 1-4.   Author inspected Dahongpao tea base with national first-class tea critic, Mr. Sun J, and Mr. Lan C in 2018 at Wuyishan-Mountain.

the tea in the workshop of the tea factory. We also went to tea farmers' homes to taste new tea. In Wuyi Mountain, I was influenced in this aspect and learned a lot of knowledge about tea art, tea ceremony, and tea culture. We drank tea and talked with tea masters every day. Specially trained ladies boil water and perform the brewing process of tea for us. They also accompanied me to the imperial teahouse where Emperor Qianlong (the Qing Dynasty) had tea to taste at the palace tea. Sitting in the imperial teahouse, sipping tea and looking out from the window, seemed to be in the landscape of Wuyi Mountain. They also accompanied me to Tianxin Yongle Zen Temple, where I drank Kaiguang tea with a monk who had knowledge on tea ceremony and talked with him about tea ceremony and tea science (Figs. 1-2, 1-3, 1-4). Finally, they also decided to provide tea for the International Free Radical Conference held in 2018 at Beijing.

1.6.1 *Wuyi Mountain in Fujian Province is the birthplace of black tea*

The tea people in Wuyi-Mountain have offered the production methods of Oolong tea and black tea among the six tea categories to the world, thus opening a new era of tea production. The tea people in Wuyi-Mountain have contributed a lot to the colorful tea drinks all over the world. Black tea plays a very important role in the international tea market. The earliest exported black tea is Zhengshan small black tea produced in Wuyi -Mountain City, Fujian Province. In 1650, the Dutch East India Company imported Chinese black tea to Europe for the first time. In 1669, the British government stipulated that tea should be monopolized by the British East India Company. Since then, the British government purchased Wuyi-Mountain tea from Xiamen to replace green tea as the main tea in Europe. In Britain, tea was called "Cha" in the early days, but since Xiamen exported tea, it is called "tea" according to Xiamen pronunciation, and it is also called "black tea" because Wuyi Rock Tea (belonging to oolong tea) is black and thick. Monk Wuyi tea has a long history. It has more than 2,000 years since it was handed down [Xiao, 2008].

Wuyi tea evolved from different stages of history. Before the Yuan Dynasty, it was mainly round cake tea and steamed green and sun green loose tea. Fried green tea appeared in the Ming Dynasty. Wuyi Rock Tea (special name of Oolong tea) and black tea were developed in the early Qing Dynasty. Wuyi Mountain, the birthplace of Oolong tea and black tea, has four of the six major tea categories in China. Therefore, Wuyi tea plays a very important role in the history of Chinese tea and even the history of world tea [Xiao, 2008].

As early as the Shang and Zhou dynasties, Wuyi tea was presented to King Wu of the Zhou, when it was allied with the monarch of the "Pumin nationality". Since the Han Dynasty, Wuyi Mountain has been regarded as a part of sacrifice by the central government. Emperor Wu of the Han Dynasty named Wuyi Mountain a famous mountain. In the process of offering sacrifices to Wuyi Mountain, the abbot of the temple entertained guests with tea, worshipped tea as a sacred object after the officials saw it, offered it to the emperor in Jin Dynasty, and then paid tribute with tea. In the Han Dynasty, fresh leaves were pounded into cake-shaped tea balls with wooden sticks, dried for storage. When drinking, the tea balls were crushed and eaten, which was more classified as medicine. Therefore, the production technology of tea was not mature at that time. After receiving the tribute of Wuyi-Mountain tea, Emperor Wu of the Han Dynasty highly appreciated it. Soon, he abandoned the old tea and welcomed the new tea, and replaced it with Wuyi tea, performing a legend. Su Dongpo, a great poet of the Song Dynasty, wrote a prose which personified Wuyi-Mountain tea as a belle [Su, 2005].

In 784, Emperor Xuanzong of the Tang Dynasty proclaimed Wuyi Mountain as a "famous mountain and great river". Taoism also listed Wuyi Mountain as the 16th elevation of the "thirty-six cave heaven", which was transformed into a true cave heaven. Buddhist temples also flourished, and Wuyi tea was regarded as a treasure. In the poem "Hui wax tea" (the earliest tea poem in Fujian Province, more than 1,100 years old) written by Xu, a scholar and Secretary of the provincial government in the first year of Qianning of the Tang Dynasty (894). The whole poem not only spoke highly of Wuyi tea, but also explained that Wuyi tea at that time had changed from paste to wax and was printed with decorative patterns such as flying magpies [Xiao, 2008].

The way of drinking tea in the Song Dynasty became more and more exquisite, changing from boiling to tasting, reaching its peak. At that time, the main purpose of brewing famous tea was not to quench thirst, but to appreciate it. Emperor Song Huizong's tea book *Daguan tea theory* said that the tea of Wuyi and other places are extremely rich in the essence of selection, the workmanship of production, and the taste and the beauty of cooking points. Famous mountains, famous waters, famous tea, plus the praise and exaggeration of celebrities, Wuyi rock tea is even more valuable. After the Southern Song Dynasty, more and more people came to Wuyi Mountain to visit and give lectures. At that time, it was popular to enjoy tea, fight tea, and divide tea in the mountains. Moreover, it also has a close relationship with Confucianism and Taoism. Zhu Xi, a neo-Confucianism scholar, took an apprentice and gave lectures about tea in Wuyi Mountain. He woke up his heart with tea, solved his difficulties with tea, made friends with tea, and had a special preference for Wuyi tea [Xiao, 2008].

In the Yuan Dynasty, Wuyi-Mountain tea officially became a tribute. From the Yuan Dynasty to the 16th year (1279), Wuyi tea supervised the production of "stone milk" tea and offered several kilograms to the Imperial Palace, which was deeply appreciated by the emperor. So, in the Yuan Dynasty, Wuyi tea officially became tribute tea. In the fifth year of Yuan Dynasty (1301), Gao Jiu, the manager of Shaowu road in Fujian Provinc, was ordered to supervise the production of tribute tea in Wuyi Mountain and built a royal tea garden to make tribute tea. Zhang Duan, the magistrate of Chong'an County and Sun Yu, supervised the production. There were 250 workers, making 5,000 dragon and Phoenix cake teas for tribute, while Wuyi tea has been paying tribute for 255 years. The legendary ceremony of "shouting at the mountain" began here. According to the cloud, after officials and field workers worship and shout, the tea sprouts and the well water rises significantly, which shows that "the emperor's grace is vast" and the mountain name is becoming more and more inflamed. Although it was harsh to local governments and tea farmers, objectively speaking, it also expanded the influence of Wuyi tea [Xiao, 2008].

The tea making technology of Wuyi tea in the Ming Dynasty was greatly improved. With the founding emperor Zhu Yuanzhang's edict of "Reform the loose tea", Wuyi tea gradually changed from steamed green regiment cake tea to sun green and steamed green loose tea. Loose tea retains the original color, aroma, shape, and taste of tea, which improves the interest of tea drinking and promotes the development of tea art. At the same time, due to the simplification of tea making process, the cost of tea is greatly reduced, which also lays the foundation for the production of rock tea. In the late Ming Dynasty, it developed to the production of fried green tea. The emergence of roasted green tea was a great leap in tea making skills. Therefore, Wuyi tea gained a great reputation at that time. In the book *Tea carding*, Xu said that "today's tribute tea is the best tea before the rain in Wuyi tea" [Xu, 1597].

Oolong tea came into being at the beginning of the Qing Dynasty. Wuyi Rock Tea (Oolong tea) originated in the late Ming Dynasty. With the development of time, its production technology gradually changed in the exploration of many tea makers and formed the prototype of rock tea in the late Ming Dynasty and early Qing Dynasty. Ruan Minxi, a monk in Wuyi Mountain wrote *Wuyi tea song* about the history, planting, management, picking, making, frying, baking, and tasting of Wuyi tea. After tea is picked, it is evenly spread in bamboo baskets and put on shelves in the wind. It is called sun green tea. After its green color gradually recovers, it is then fried and baked by fire. Longjing is fried but not baked, so its color is pure green. Wuyi tea is fried and baked alone. When it is cooked, it is half green and half red. Green is fried and red is baked. Tea is picked and spread, accelerated, and the aroma is fried. It is not allowed to be out of date. Both fried and baked are not allowed to be pick out and made into one color. In this regard, Mr. Wu, the contemporary tea saint, also commented that "up to now, the preliminary preparation method of Wuyi rock tea, which belongs to oolong tea, is inseparable from the above basic characteristics" [Wu, 1992]. Mr. Zhang, a tea expert and oolong tea leader, also said, "Oolong tea is a semi fermented tea after Wuyi tea. The origin of Oolong tea is spread from Wuyi to Jianou, Anxi and Taiwan." Due to the extremely fine production technology and the excellent internal quality of Wuyi tea, Wuyi rock tea has neither the bitterness of green tea nor the intensity of black tea, but it also takes the fragrance of green tea and the mellow of black tea, so it has been highly praised by tea makers in the world [Zhang, 1978].

Wuyi tea market town is full of fragrance, with boats and rafts parked on the shore, red wine and green lights, and pedestrians rubbing their shoulders. Tea operators are from Zhangquan, Chaoshan, and Guangzhou. They buy tea from Northern Fujian Province, then sell the tea along the coast, or transports it abroad for the needs of overseas Chinese. Therefore, rock tea has the elegant name of "overseas Chinese selling tea". Those who are good at business first order from the factory, and then buy the factory to produce and make their own system. They integrated production and marketing and make great profits. According to the reports, after Quanzhou businessman Zhang bought Huiyuan East tea factory, he was complacent and boasted that "If you get this tea factory, you will be succeed in the world!" Its abundant profits can be seen. Boss Zhang's rock tea is sold by himself, never wholesale, and a notice is posted that "there is only one, there is no others". It is still popular to invite customers with sincerity. At that time, the price of tea was equal to that of silver. Large tea transactions were converted into gold and silver, which can be said to be precious [Xiao, 2008].

The hot tea market has attracted tea merchants from the surrounding areas. "Tea doesn't taste good if it is not in Wuyi" also came out. The mountain people made the best of the time and the right place to prosper, which added a bright spot to the history of Wuyi tea. In 1610, Dutch merchant ships first came to China and brought back a small amount of Wuyi Mountain black tea. Since the early 17th century, the Dutch have trafficked Wuyi black tea to Europe and been recognized by the upper class to be a symbol of noble status: "Wuyi tea is as red as agate, and its quality is better than that of India and Ceylon. Anyone who treats guests with Wuyi tea will stand up and pay tribute. Wuyi black tea was exported to the Netherlands, which also led to the export of Wuyi tea in the late Ming and early Qing Dynasties [Xiao, 2008].

During the period of the Republic of China, Wuyi-Mountain became a tea research base. After the outbreak of the war of resistance against Japan, due to the trampling of most tea bases along the coast, a group of knowledgeable people in the tea industry chose Wuyi Mountain, which produces famous tea and is beyond the reach of the Japanese aggressors, as the development research base. Some famous tea masters and manufacturers crossed mountains and came to Wuyi Mountain. After 12 years of painstaking operation of the "demonstration tea farm", they prepared to set up the "Tea Research Institute" and devoted themselves to the Chinese tea business. They conducted experiments in the Wuyi Mountain tea garden and achieved many results, such as the successful trial production of the "September 18" tea kneading machine and the book *Production, manufacturing, transportation and marketing of Wuyi tea* written by Lin Fuquan during this period, which raised the technology of Wuyi rock tea into theory and contributed to the development of the tea industry [Lin, 1943].

After the founding of new China, especially the national reform and opening up, Wuyi rock tea has experienced a rapid rise and development stage from recovery and development to now. Today, Wuyi tea covers an area about 10,000 hectare, with a total output of nearly 10 million kg, more than 13 times higher than that in 1978 and 13,000 kg more than that in 1948. In 2002, Wuyi rock tea was listed as the national "product species protected by geographical indications". In 2003, Wuyi Mountain was awarded the title of "Hometown of Chinese tea Culture and Art" by the Ministry of Culture, China. In June 2006, the handicraft technology of Wuyi Rock Tea (Dahongpao tea) was recognized as the first batch of "National Intangible Cultural Heritage" by the Ministry of Culture. Wuyi-Mountain Dahongpao tea and Zhengshan small black tea have been approved as certification trademarks by the State Trademark Office, further enhancing the reputation and status of Wuyi tea. Tea industry has become a pillar industry in Wuyi-Mountain city [Xiao, 2008].

1.6.2 *Wuyi Mountain, the hometown of tea*

Wuyi-Mountain has beautiful peaks, green streams, and flowing springs. It has a mild climate, abundant rainfall, and myriad meteorology. It is especially suitable for the growth of tea trees. For thousands of years, Wuyi-Mountain tea people have used valleys, gullies, rock depressions, and some gentle slopes to build stone ladders and open gardens to grow tea. This kind of potted tea garden is mostly located in the mountain forest. The trees between the tea gardens form a natural and good biological chain. The tea garden is surrounded by cliffs and steep cliff. The bamboo trees on the cliffs are lush, and the gap spring water is jetting. The air is filled with strong flower fragrance all year round, forming a unique small environment of rock tea. A naturalist of Fujian nationality in the late Qing Dynasty said in his book, "All tea produced in other counties is slightly cold in nature; those produced in the rocks in Wuyi-Mountain are only warm in nature." The reason why Wuyi tea is different, is precisely due to the influence of the geographical environment of Wuyi Mountain; "The rocks are straight, and there is abundant wind, sun, rain and dew; the sweetness of the water spring is better than that of other mountains; the grass is fragrant and strong." It means that in a landscape like Wuyi

Mountain, even grass has strong fragrance, not to mention tea [Xiao, 2008]!

1. Unique mountain

On 1 December 1999, Wuyi-Mountain was listed as a World Cultural and Natural Heritage by the UNESCO World Heritage Committee. Wuyi-Mountain has clear water and green mountain. Within the geographical protection scope of 2,798 square kilometers of Wuyi-tea origin, it has beautiful mountains and rivers and excellent ecological environment. In the scenic area alone, there are 36 peaks, 72 caves, 99 rocks, steep peaks, and deep gullies surrounded by mountains, just like the immortal landscape on earth and Penglai yaochi. Mrs. Balko, former chairman of the Executive Committee of the World Tourism Organization, happily wrote an inscription after visiting Wuyi-Mountain, "The unpolluted Wuyi-Mountain scenic spot is a model of world environmental protection." Wuyi Mountain has a unique, rare, and wonderful natural landscape. It is a rare natural beauty zone. It is representative of the harmony and unity between human beings and natural environment [Xiao, 2008].

The unique natural conditions of Wuyi-Mountain are created by the uncanny workmanship of nature. Wuyi-Mountain is a key area for global biodiversity conservation. It is distributed with the most complete, typical, and largest primary forest ecosystem in the middle subtropical zone at the same latitude in the world; the high mountains are perfectly combined with water and the humanities and nature are organically integrated. It enjoys a high reputation for its beautiful scenery, long history and culture, and numerous cultural relics and historic sites, such as beautiful water, strange peaks, deep valleys, and dangerous gullies. It is listed as a "Man and Nature Biosphere Reserve" by UNESCO. In addition, Wuyi-mountain, known as the "Roof of East China" in the national key nature reserve, is known as the "Paradise of Birds", "Kingdom of snakes", "World of insects", and "Subtropical zoo", which can be described as the window of life in the world [Xiao, 2008].

2. Soil conglomerate thick soil

Travel notes of Jiang Shunan said, "Wuyi-mountain produces tea, which is famous all over the world. The soil is mixed with gravel. Because it is located in a deep valley, there is little sunlight and more rain and dew, so the tea is good, like Dahongpao tea, it is the best." Wuyi-Mountain tea garden is mostly in the rock gullies and secluded streams, surrounded by mountains. The tea garden soil is composed of volcanic conglomerate, red sandstone, and shale. The soil is loose, with good water permeability and ventilation, thick surface rot, high content of organic matter and rich in nutrients. Lu said in his book of tea that for the soil of tea in mountain, "There are differentiated stones at the top, gravelly soil in the middle and loess at the bottom". The soil of Wuyi-mountain tea garden is rotten stones or gravelly soil, as described in Xu's book *Tea examination* (Ming Dynasty), "The soil in Wuyi-Mountain is suitable for tea trees" [Xiao, 2008].

   The rock walls around the tea tree are covered with miscellaneous trees and weeds, forming an environment that not only has sunlight, but also avoids the direct irradiation of ultraviolet rays, which is professionally called "diffuse light". As there are many rocks and thick soil layers, there are many wild four-season orchids and calamus near the rock walls and streams. The bushes are full of osmanthus and azalea, and the air is always filled with fresh flower fragrance all year round.

3. The fresh air changes all the time

Zhu Xi once described "The cloud of Wuyi Mountain flying around and floating, it will change at any time, for a while it will be seen and for a while invisible". Wuyi-Mountain is shrouded with morning clouds and evening mist all day long. The fog covers the green mountains and the smoke cage water. The clouds are flying like mountains and the fog is floating like mountains. Wuyi-Mountain tea area has a mild climate, warm in winter and cool in summer. Due to the high terrain and towering peaks in the northwest, it blocks the invasion of the cold current in the north, with subtropical climate characteristics, and the annual average temperature is between 18–18.5°C. Wuyi-Mountain has abundant rainfall, with an annual rainfall of about 2,000 mm. Jiuqu River, Chongyang River, and Huangbai River are crisscrossed with peaks and hills, forming a unique regional climate. Wuyi-Mountain tea garden is mainly distributed in the hills and low mountains below 500 meters. The mountains are filled with clouds and fog all the year round, the air is humid, and the annual average relative humidity is about 80%. In the mountain streams and rock gullies, the sunshine is short and foggy, forming a large amount of scattered light, there is no frost damage and wind damage. As early as the Ming Dynasty, Xu said, "Wuyi-Mountain tea is the best before the rainy season in spring" in his *Tea* book [Xiao, 2008].

4. The water is clear and sweet

The stream flows in a zigzag way and the mountain light is soaked with clear ripples. Wuyi-Mountain has 36 peaks and 99 rocks. The peaks and rocks are crisscrossed, and the streams are crisscross. The nine winding streams run through them. They meandered for 15 miles, three big turns, and nine small turns. There is a legend that more than 3,000 years ago, Peng Zu lived for longevity because he ate Ganoderma lucidum and drank the spring water from Wuyi-Mountain. "The dew absorbed the fragrance from flowers, and the water absorbs the spring in the stone." Around Wuyi-Mountain tea tree, the water flows from peak to peak and the mountain spring is purified by plants and trees. Along the rock cracks of the cliff, it reaches the tea tree and gradually infiltrates into the root of the tea tree. In this way, there are clear water, red mountain, secluded springs in high mountains, fog, and rain. In this way, Wuyi-Mountain tea receives the moisture of air, enjoys the benefits of nature, and contributes to people's unique "rock and flower fragrance". Tea drinks water from rain and fog and people drink the tea. Wuyi-Mountain tea magically connects man and nature. Water is the spirit of Wuyi-Mountain. Tea is an excellent gift from nature to Wuyi-Mountain.

1.6.3 *The king of Wuyi tea-Dahongpao tea*

Wuyi Dahongpao (means large red robe) tea is a wonderful flower in China's famous tea and is known as the "No. 1 scholar in tea". It is also known as the "king of rock tea" and can be called a national treasure [Xiao, 2008].

1. History of Dahongpao Tea

The mother tree of Dahongpao-tea was discovered and harvested in the late Ming and early Qing Dynasties, with a history of 380 years. It has been famous for hundreds of years and has many legends. The lush moss growing on the cliffs of Wuyi, Fujian Province, makes the withered moss grow on the cliffs of Wuyi Mountain, which is rich in the nutrients of tea trees. In 1995, after more than 10 years of repeated experiments on Dahongpao tea, Wuyi-Mountain Tea Research Institute succeeded in asexual breeding, avoided the worry of dating, maintained the characteristics of Dahongpao tea, and passed the appraisal by the Provincial Science and Technology Commission. In 2002, Wuyi-Mountain rock tea was listed as the Nnational Protection Pproducts of Ggeographical Indications, which standardized a series of production, and product standards. In June 2006, the handicrafts of Wuyi-Mountain Rock Tea (Dahongpao tea) were recognized as the first batch of "National Intangible Cultural Heritage" by the Ministry of Culture. Wuyi-Mountain Dahongpao and Zhengshan small black tea have been approved as Certification Trademarks by the State Trademark Office, further improving the reputation and status of Wuyi-Mountain tea. In 2010, the "Wuyi-Mountain Dahongpao tea" declared by Wuyi-Mountain Municipal Government was newly recognized as a well-known trademark in China by the State Administration for Industry and Commerce [Xiao, 2008].

2. Characteristics of Dahongpao tea

Dahongpao tea mother tree grows on the cliffs of Wuyi-Mountain. On both sides of the cliffs, the sunshine is short and the temperature is appropriate. There are trickling springs to nourish the tea tree all year round. The organic matter formed by the decay of dead leaves, moss and other plants, and fertile soil, replenishes nutrients for the tea tree, making Dahongpao tea gifted, unique, and superior quality.

(1) Dahongpao tea has "three colors", "toad back", and "three red and seven green":

"Three sections of color" refers to the characteristics of dry tea; the head of dry tea is dark brown and the tail is light red, therefore, "three section color" is a typical feature of Wuyi rock tea. "Toad's back" is one of the characteristics of the traditional baking fire attack of tea. Generally, the rock tea with toad's back is a traditional rock tea. After a long time of baking, local heat expands and bulges small bubbles on the surface of the tea; it is easy to observe on the bottom of fried rice cake leaves. "Three red and seven green" refers to the fermentation degree of tea; when observing the leaf bottom of rock tea, it can be found that the periphery of the leaf is red, the middle is cyan, and the three sides are divided into seven green leaves, also known as "green leaves with red edges".

The picking standard of Dahongpao tea tree is that the growth and maturity of new shoots, buds, and leaves (three or four leaves on the open surface) should have no water on the leaf surface, no damage, and be fresh and uniform. Fresh leaves should not be too tender. If they are too tender, they will become tea with low aroma and bitter taste; also not to be too old, as too old will have a weak taste and poor aroma; moreover, picking in rainy days and with dew should be avoided as far as possible.

The production technology of Dahongpao tea combines the technology of green tea and black tea. It is the tea with the most processes, the highest technical requirements and the most complex; its production process is withering — making green — killing green — rolling — baking in water — drying — Mao tea.

(2) Dahongpao tea is a treasure:

Dahongpao tea produced in Wuyi rock, Fujian Province, belongs to oolong tea in Northern Fujian. It is a kind of semi fermented tea; it is also the best quality of Wuyi rock tea. It is a special famous tea in China. Dahongpao tea has a long history. It was used as a tribute for the royal family as early as the Ming Dynasty; because it grows on the cliff, it is difficult to pick and the quantity is rare, so it is called "treasure".

In history, Wuyi-Mountain tea was the tribute tea of emperors of man dynasties. It has won the national Oolong Tea Gold Award for many consecutive years and the gold award of the first China International Tea Expo. The State Council and the Ministry of Culture officially announced the first batch of national intangible cultural heritage list, among which the production technology of Dahongpao tea is listed. Dahongpao tea making technology is the only tea making technology among manual skills. Dahongpao tea is the king of tea and is known as "National Treasure".

Wuyi-Mountain has superior soil, climate, humidity, and other natural conditions, forming a unique climate. Coupled with the production technology rated as "Intangible Cultural Heritage", the growing Dahongpao tea has superior quality, unique characteristics, fragrant aroma, sweet aftertaste, and obvious "Rock Charm". Tea gardens are managed according to national green organic standards.

The brand status of Dahongpao tea is determined by historical accumulation, cultural heritage, excellent quality, and other factors. In addition, with the introduction of Wuyi-Mountain tourism routes over the years, most people who come to Wuyi-Mountain have seen the Dahongpao tea tree, heard the magical legend of Dahongpao tea, and tasted the famous brand Dahongpao tea. The fame of Wuyi-Mountain Dahongpao tea has spread all over the world.

Wuyi-Mountain rock tea has been regarded as the best beverage for health preservation and health care since ancient times. It is said that Shennong tea once used it to detoxify 72 poisons and was named for curing difficult and miscellaneous diseases such as county magistrate and scholar queen. Now through scientific research and determination, Wuyi-Mountain rock tea is rich in trace elements such as potassium, zinc, and selenium. It can be separated and obtained the mixtures of oxalic acid, tannic acid, tea polyphenols, and flavonoids, which is of great benefit to human health.

At the 7th China Wuyi-Mountain Dahongpao tea Cultural Festival and the best auction of "Dahongpao tea" held in Wuyi-Mountain, a world heritage site, 20 grams of Dahongpao tea were sold at an amazing price of 208,000 yuan.

3. Legend of Dahongpao tea

Since Dahongpao tea has a long history and good quality, there are many stories and legends about Dahongpao tea in Wuyi-mountain.

(1) Monkey picking tea

Dahongpao tea trees grow on steep precipices and cliffs that no one can climb. Every year, when picking tea, Temple monks use fruit as bait and train monkeys to pick it. Therefore, some people call it "monkey picking tea".

(2) Dahongpao tea trees are planted by immortals.

Every New Year's Day, monks burn incense and worship devoutly to Dahongpao tea. They soak a cup of tea for the Buddha. Tea can take care of themselves. Anyone who steals it would immediately have abdominal pain. They can only heal after watering the tea trees. So, people think the tea trees are planted by immortals.

(3) The Dahongpao tea tree has been granted the imperial title and was given its name by the emperor.

The local magistrate personally visits Dahongpao tea trees every spring, takes off his red robe and covers it on the tea trees, and then pays homage. Amid the smoke, everyone shouted in unison, "Tea sprouts! Tea sprouts! When the red robe was taken off, the tea tree sprouted as expected! The tea buds are as red as dye".

(4) Imperial tribute tea:

A queen of a Dynasty was ill and had not been cured for a long time. The crown prince followed his mother's order to find the secret recipe of fairy grass. On the way, he met an old man who fell under the tree and was nearly killed by a tiger. The crown prince saved him bravely. They talked with each other. In order to repay the grace of saving his life, the old man accompanied the crown prince to Wuyi Mountain, picked tea leaves, wrapped them in cloth, and gave to the crown prince. The crown prince went down the mountain quickly and urged his horse to go to the capital day and night to boil the tea and drink it to his mother. After drinking for several days, the queen improved day by day and recovered from her illness. The emperor was overjoyed and issued two decrees: one was to give a Dahongpao to keep the tea trees warm in the cold winter every year, and the other was to make the old man a general to protect the trees. He inherited his position from generation to generation and collected and paid tribute every year. Since then, Wuyi-Mountain has called these three tea trees Dahongpao.

(5) Tribute tea treasures:

Another story is about a scholar who went to Beijing to take an exam. When he passed Wuyi-Mountain, he fell ill on the road. He happened to meet the old abbot of Tianxin Temple in Wuyi-Mountain, who went down the mountain for alms, so he carried him back to the temple. Seeing that the scholar was pale, thin, and bloated, he brewed the tea harvested from

Jiulongke in Wuyi-Mountain to the scholar with boiling water. After drinking several bowls of tea, he felt that his abdominal distention decreased and recovered in a few days. The scholar said goodbye to the abbot and said, "You saved me. If I succeed, I will come back here to thank you." Soon, the scholar succeeded and became the top scholar. With the favor of the emperor, he went straight to the Tianxin temple in Wuyi-Mountain to meet the abbot and said, "I have come to thank you for your great kindness." The abbot said, "This is not a miraculous elixir, but the tea from Jiulongke." The champion believed that divine tea could cure diseases and wanted to bring some back to Beijing to pay tribute to the emperor. At this time, it was the spring tea mining season. The old abbot led the monks to pick tea, made tea, and packed the tea in tin cans. The champion brought it back to the capital. After that, the scholar sent people to renovate the Tianxin temple. However, when the champion returned to the court, the queen fell ill and all the doctors failed to find a cure. The champion took out the pot of tea and presented it to the queen. After drinking, the queen gradually recovered. The emperor was very happy and gave a red robe. He ordered the champion to go to Jiulongke in person and drape the bred robe on the tea trees to show his gratitude. At the same time, he sent someone to take care of it. He harvested it every year and paid all the tribute. It was not allowed to hide it privately. Since then, these three Dahongpao tea trees have become tribute tea. The dynasties have changed, but the guards of Dahongpao tea trees have never stopped.

1.7 Conclusion

Through the above discussion on ancient works, the origin and reproduction of tea, domestication, archaeological discoveries, linguistics, history of development, tea culture, tea products, research and production of tea polyphenols, and even the story of Dahongpao tea in Wuyi Mountain, we can clearly draw a conclusion that China is the hometown of tea and the tea of the world comes out of China. The first is the cultivation, processing, and production of tea in Yuannan province of southwest China, which spreads through and other places in Southwest China to India, and Sri Lanka, while the trade of tea reaches western countries such as Britain by sea, which gradually spreads all over the world. China has a long history of tea cultivation and cultivation, a rich tea culture and a rich variety of tea, and many tea monographs, especially the earliest tea monograph of Lu Yu, which is well-known in the world. Tea is rich in natural antioxidants and tea polyphenols, which are beneficial to human health, can inhibit oxidative stress damage and prevent many diseases. Now tea has become the healthiest and most popular drink in the world.

Acknowledgement

Thanks for providing some data and literature in this chapter from Professor Liping Du and her students, Tian Xie and Hanlu Li (Tianjin University of science and technology, Tianjin, China).

References

Booth WB. (1830) History and description of the species of camellia and Thea. *Hortic Soc*, **7**, 519–562.

Chen C. (1970) General history of tea industry. China agriculture publishing house.

Chen L, Zhou ZX, Yang YJ. (2007) Genetic improvement and breeding of the tea plant (*Camellia sinensis*) in China: from individual selection to hybridization and molecular breeding. *Euphytica*, **154**, 239–248.

Chen X. (1982) Tea talk. Guangxi People's publishing house.

Chen Z. (1992) The Chinese tea classic. Shanghai Culture publish, Shanghai.

Chen Z-M. (1992) *Chinese tea classic*. Shanghai Culture Publishing House.

Das SC, Das S, Hazarika M. (2012) Breeding of tea plant (*Camellia sinensis*) in India. In *Global Tea Breeding: Achievements Challengers and Prospective*, Chen L, Apostolides Z, Chen ZM (eds.), Hangzhou: Zhejiang University Press and Berlin Heidelberg: Springer-Verlag, 69–124.

Dong S. (2003) Tea history. Zhejiang University Publishing House.

FAOSTAT. (2015) *FAO Database*. Food and Agriculture Organization, United Nations. Available online at: http://faostat3.fao.org/download/Q/QC/E (accessed 10 May 2015).

FAOSTAT. (2021) *Crops and livestock products*. Available online at: http://www.fao.org/faostat/en/#data/QC (accessed 8 April 2021).

Guo M. (2003) Chinese tea history. Shanxi ancient books publishing house.

Guo Y. (1996) Origin, propagation and evolution of tea plants. Tea science and technology, Beijing.

Jiang J, Lu G, Wang Q, Wei S. (2021) The analysis and identification of charred suspected tea remains unearthed from Warring State Period Tomb. *Sci Rep*, **11**(1), 16557.

Kamunya SM, Wachira FN, Pathak RS, Muoki RC, Sharma RK. (2012) Tea improvement in Kenya. In *Global Tea Breeding Achievements, Challenges and Perspectives*, Chen L, Apostolides Z, Chen ZM (eds.), Hangzhou: Zhejiang University Press and Berlin Heidelberg: Springer-Verlag, 178–226.

Kingdon-Ward F. (1950) Does wild tea exist? *Nature*, **165**, 297–299.

Kong Q. (2000 BC) Book of songs.

Li J. (2015) Top ten famous teas in China. Shanghai Science and Technology Press.

Lin FQ. (1943) Production, manufacturing, transportation and marketing of Wuyi tea.

Liu J, Liu BY, Li DZ, Gao LM. (2018) Domestication origin and breeding history of the tea plant (*Camellia sinensis*) in China and India based on nuclear microsatellites and cpDNA sequence data. *Front Plant Sci*, **8**, 2270.

Lu HY, Zhang JP, Yang YM, Yang XY, Xu BQ, Yang WZ, *et al.* (2016) Earliest tea as evidence for one branch of the silk road across the Tibetan Plateau. *Sci Rep*, **6**, 18955.

Lu Y. (1974) *The Classic of Tea*. Boston, MA: Little, Brown & Co.

Meegahakumbura MK, Wambulwa MC, Thapa KK, Li MM, Möller M, Xu JC, *et al.* (2016) Indications for three independent domestication events for tea plant (Camellia sinensis (L.) O. Kuntze) and new insights into the origin of tea germplasm in China and India revealed by nuclear microsatellites. *PLoS ONE*, **11**, e0155369.

Meegahakumbura MK, Wambulwa MC, Li MM, Thapa KK, Sun YS, Möller M, Xu JC, Yang JB, Ming TL. (2018) Monograph of the genus Camellia. Kunming: Yunnan Science and Technology Press.

Ming TL. (2000) Monograph of the genus Camellia. Kunming: Yunnan Science and Technology Press.

Ming TL, Bartholomew B. (2007) Theaceae. In *Flora of China*.

Mondal T. (2009) Tea breeding. In *Breeding Plantation Tree Crops: Tropical Species*, Jain SM, Priyadarshan MP (eds.), New York, NY: Springer Science+Business Media, 545–587.

Pan G. (1995) All tea. China Agricultural Press.

Pan SY, Nie Q, Tai HC, *et al.* (2022) Tea and tea drinking: China's outstanding contributions to the mankind. *Chin Med*, **17**(1), 27.

Peng C. (1998) Tea culture (2) — Types and components of tea. Health. Wuyi tea classic. Science Press.

Sharma RK, Negi MS, Sharma S, Bhardwaj P, Kumar R, Bhattachrya E, *et al.* (2010) AFLP-based genetic diversity assessment of commercially important tea germplasm in India. *Biochem Genet*, **48**, 549–564.

Sigley G. (2013) The ancient tea horse road and the politics of cultural heritage in southwest China: regional identity in the context of a rising China. In *Cultural Heritage Politics in China*, Blumenfield T, Silverman H (eds.), New York, NY: Springer, 235–246.

Snellgrove D, Richardson H. (2003) *A Cultural History of Tibet.* Bangkok: Orchid Press.

Su D. (2005) 1037–1101 Biography of Ye Jia. In *Selected Works of Su Shi's prose*, Author Xu berong, Baihua Literature and Art Publishing House.

Trepardoux F, Delaveau P. (1999) Origin of the word tea, and its extension to designate different hot infused drinks. *Rev Hist Pharm (Paris)*, **47**(322), 247–253.

Wang R, *et al.* (2011) The formation and evolution of Chinese tea varieties and processing methods. Agricultural Archaeology, Beijng.

Wright IJ, Dong N, Maire V, Prentice IC, Westoby M, Díaz S, *et al.* (2017) Global climatic drivers of leaf size. *Science*, **357**, 917–921.

Wu JN. (1992) A study on the origin of tea plants. *Journal of the Chinese agricultural society*, **37**.

Wu JN. (1987) *Review on 'Cha Ching'.* Beijing: Agriculture Press.

Xiao TX. (2008) Wuyi tea classic. Science Press.

Xu C. (1579) Tea carding.

Yamanishi T. (1995) Special issue on tea. *Food Rev Int*, **11**, 371–546.

Yang X-Q, Wang Y-F, Chen L-J. (2003) Tea polyphenol chemistry. Shanghai Science and Technology Press.

Yao MZ, Chen L. (2012) Tea germplasm and breeding in China. In *Global Tea Breeding: Achievements Challengers and Prospective*, Chen L, Apostolides Z, Chen ZM (eds.), Hangzhou: Zhejiang University Press and Berlin Heidelberg: Springer-Verlag, 13–68.

Yu FL. (1986) Discussion on the originating place and the originating center of tea plants. *J Tea Sci*, **6**, 1–8.

Yu G. (1998) Handbook of Chinese tea people. China Forestry Publishing House.

Yu XQ, Gao LM, Soltis DE, Soltis PS, Yang JB, Fang L, *et al.* (2017) Insights into the historical assembly of East Asian subtropical evergreen broadleaved forests revealed by the temporal history of the tea family. *New Phytol*, **215**, 1235–1248.

Yu Y. (1999) The course of Tea Road — a brief history of the evolution of Chinese tea culture. Guangming Daily Press.

Zhang TF. (1978) A textual research on the history of Fujian tea. *Tea Science Bulletin*, 15–18.

Zhao B-L. (1999) Oxygen free radicals and natural antioxidants. Beijing, Scientific Press.

Zhao B-L. (2003) Free radicals and natural antioxidants and health. Hung Kang: Chinese Scientific and Culture Press.

Zhao FR, Yin QY. (2008) The Khmer Meng nationalities in China earliest domesticated cultivated tea. *J Simao Teacher's College*, **24**, 28–34.

Zhao Z-T. (2000) Junshan silver needle. Hunan Science and Technology Press.

Zheng G. (2016) The Chinese affairs. Light Industry Publishing House, Beijing.

Zhu Z and Shen D. (2010) Integration of ancient Chinese tea books. Shanghai Culture Publishing House, Shanghai.

Chapter 2

Green Tea Polyphenols

Jihong Zhou, Yuefei Wang, Lin Chen, Yueling Zhao,
Shasha Guo, Shiqi Zhao

Tea Research Institute, Zhejiang University, Hangzhou, China

2.1 Introduction

Tea polyphenols, as the main secondary metabolites in the tea plant, are the important factors that determine the sensory and nutritional quality of plants and the content of which accounts for about 18–36% of the dry weight of tea leaves. The biological synthesis of tea polyphenols in the tea plant is mainly synthesized through the phenylpropanoid pathway and flavonoids synthetic pathway. According to the different content of polyphenols in tea, tea polyphenols are divided into four types: catechins, flavonoids and flavonols, anthocyanins, and phenolic acids condensed phenolic acids. Tea polyphenols are also regarded as the index of the basic tea classification system, in which six types of tea were classified based on the different processing procedures, especially the degree of oxidation of polyphenols. Green tea polyphenol is a well-known natural antioxidant and antibacterial agent. The antioxidant effect, as well as the beneficial effects on cardiovascular disease, cancer prevention, anti-cancer, etc., remains tea polyphenols broad space in the application of human health. Nowadays, tea polyphenols have been widely used in the fields of daily cosmetics, food industry, livestock production, environmental treatment, and so on.

In this chapter, green tea polyphenols are well introduced from the view of chemical construction, the synthesis and metabolism in tea plants, the applications in processing and production of green tea polyphenols like extraction and preparation, and the metabolism and bioavailability of tea polyphenols *in vivo*. Furthermore, the advice on the way of green tea polyphenols dietary intake based on function and security is mentioned in this chapter as well.

2.2 Types and properties of green tea polyphenols

The polyphenols, as the main secondary metabolites in plants, are the compounds belonging to tannins and important factors that determine the sensory and nutritional quality of plants. The structures of these compounds include benzene rings with one or more hydroxyl groups, ranging from simple phenolic molecules to complex high-molecular polymers. Haslam classified plant polyphenols into two basic types — "Condensed proanthocyanidins" and "esters of gallic acid and their metabolites", which are those that have been widely called condensed tannins and hydrolyzable tannins [Haslam E, 1982].

The fresh tea leaves are rich in polyphenols (Fig. 2-1), the content of which accounts for about 18–36% of the dry weight of tea leaves. Most of the polyphenols in tea are condensed tannins, which are also called water-soluble tannins because most of them are dissolved in water. According to the different content of polyphenols in tea, tea polyphenols are considered as a general term of 1) catechins, 2) flavonoids and flavonols, 3) anthocyanins, and 4) phenolic acids condensed phenolic acids. Except for phenolic acids and condensed phenolic acids, others have the basic skeleton of C6-C3-C6 configuration (a derivative of benzopyran or chromane), which

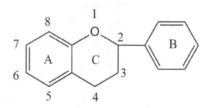

Figure 2-1.   The basic skeleton structure of tea polyphenols.

essentially consists of two aromatic rings A and B, connected by a 3-carbon bridge, usually in the form of a heterocyclic ring C.

2.2.1 *Catechins*

The most important part of tea polyphenols is the type of catechins, which accounts for approximately 70–80% of the total polyphenols. Catechins are characterized by di- or tri-hydroxyl group substitution of the B ring and the meta-5,7-dihydroxy substitution of the A ring. There are eight major catechins isolated and identified in green tea, including: epicatechin (EC), epigallocatechin (EGC), epicatechin gallate (ECG), epigallocatechin gallate (EGCG), catechin (C), gallocatechin (GC), catechin gallate (CG), and gallocatechin gallate (GCG). The structure of the eight catechins is shown in Figure 2-2. EGCG and ECG have galloyl group attached to the carbon at the 3rd position of the C ring, which is called ester catechins or complex catechins, while EGC and EC have hydroxyl group attached to the carbon at the 3rd position of the C ring, which is called non-ester catechins or simple catechins. EGCG, EC, EGC, and ECG are called epi-catechins, while GCG, C, GC, and CG are called

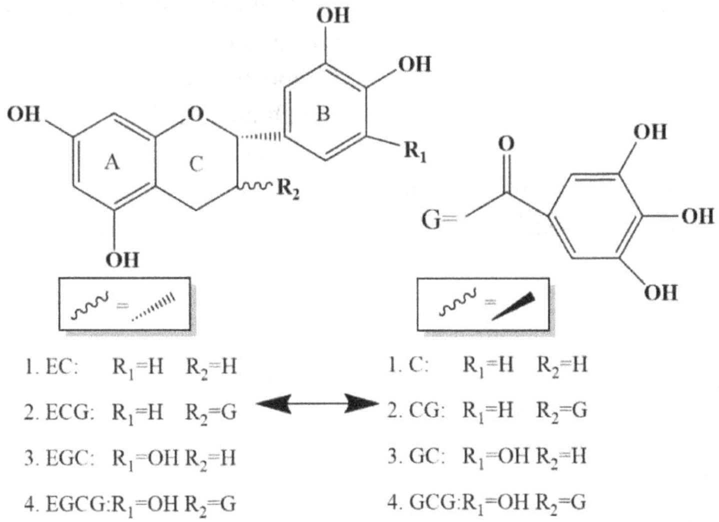

1. EC: R_1=H R_2=H
2. ECG: R_1=H R_2=G
3. EGC: R_1=OH R_2=H
4. EGCG:R_1=OH R_2=G

1. C: R_1=H R_2=H
2. CG: R_1=H R_2=G
3. GC: R_1=OH R_2=H
4. GCG:R_1=OH R_2=G

Figure 2-2.   The structure of the eight major catechins.

non-epi-catechins, and they are the corresponding epimer. Among the major catechins, EGCG is the most abundant substance, accounting for 50–75% of the total catechins content. The content of green tea polyphenols varies greatly among different teas (green tea, black tea, oolong tea, white tea, yellow tea, and dark tea).

2.2.2 *Flavonoid and Flavonols*

Flavones are a class of yellow pigments widely found in nature, and their typical structural characteristics are the existence of a carbonyl group attached at position four of the C ring in the C6-C3-C6 skeleton (Fig. 2-3). The position three of the C ring is easily hydroxylated to form flavonols. The major flavonols in tea including kaempferol, quercetin, and myricetin. The structure of major flavonols is shown in Figure 2-3. Flavonols in tea are mostly combined with sugars to form flavonoid glycosides. Flavonoid glycosides accounts for approximately 3–4% of the dry weight of tea leaves and consist of mono-, di-, and tri-glycosides that are based on kaempferol, quercetin, and myricetin and are conjugated with glucose, galactose, rhamnose, arabinose, and rutinose. There are various flavonoid glycosides in tea leaves, among which rutin (0.05–0.15%), quercitrin (0.2–0.5%), and kaempferin glycosides (0.16–0.35%) are in high content.

Flavonoid glycosides are considered to be an important component in the taste of tea. Flavonoid glycosides exhibit bitterness and astringency, which could produce a velvety and mouth-coating feeling. Since the astringency threshold of flavonoid glycosides (e.g., approximately 0.0015–19.8 μmol/L) is much lower than those of catechins (e.g., 190–930 μmol/L) and theaflavin (TFs) (e.g., 13–16 μmol/L), the contribution

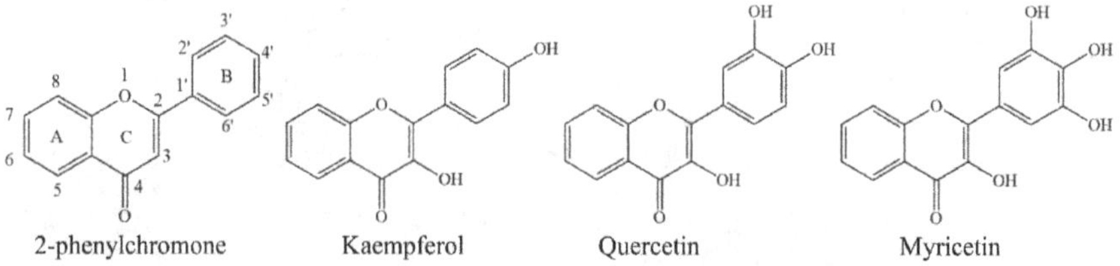

2-phenylchromone   Kaempferol   Quercetin   Myricetin

Figure 2-3.   The structure of 2-phenylchromoneand major flavonols.

of flavonoid glycosides to astringency in tea was considered to be higher than those of catechins and TFs. Studies have also reported that flavonoid glycosides could enhance the bitterness and astringency intensities of EGCG and caffeine.

2.2.3 *Anthocyanidins*

Anthocyanins are a kind of natural water-soluble pigment widely existing in plants, and their structures are highly conjugated and belong to chromogen derivatives, having a variety of isomers. Anthocyanins in plants are often condensed with glucose, galactose, and rhamnose to form glycosides due to the hydroxyl group at the C3 position. Generally, anthocyanins in tea leaves account for about 0.01% of the dry weight, while in some purple-colored cultivars such as "Zijuan" and "Ziyan", the content can reach 0.5–1.0%. The major anthocyanidins in tea include delphinidin, cyanidin, pelargonidin, and its glycoside. The structure of major anthocyanidins is shown in Figure 2-4.

2.2.4 *Phenolic acids and condensed phenolic acids*

Phenolic acids, a class of aromatic compounds with carboxyl and hydroxyl groups in the molecule, are widespread plant secondary metabolites, virtually derived from benzoic and cinnamic acids and have been studied in tea since the middle of the 19th century. Phenolic acids commonly occur as free acids and their esters, glycosides, and bound complexes. Tea contains a variety of phenolic acids and condensed phenolic acid substances, including: theogallin, gallic acid, caffeic acid, chlorogenic acid, ellagic

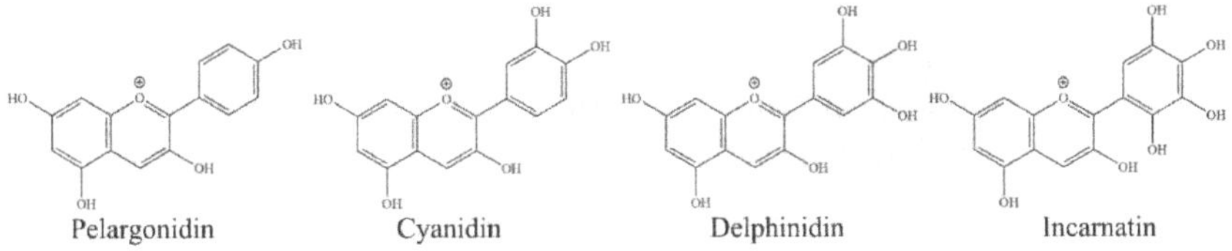

Figure 2-4. The structure of major anthocyanidins.

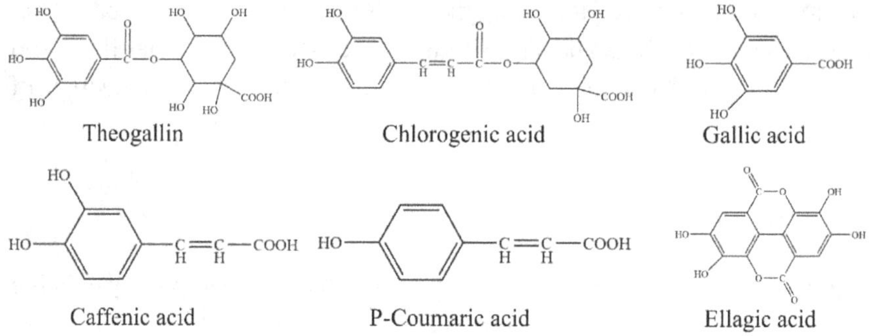

Theogallin   Chlorogenic acid   Gallic acid

Caffenic acid   P-Coumaric acid   Ellagic acid

Figure 2-5.   The structure of major phenolic acids and condensed phenolic acids.

acid, etc. Phenolic acids and condensed phenolic acids account for approximately 5% of the dry weight of tea leaves. In general, gallic acid has the highest content (1–2%), followed by gallic acid (0.5–1.4%), chlorogenic acid (0.3%), and caffeic acid and ellagic acid content are lower. The structure of major phenolic acids and condensed phenolic acids are shown in Figure 2-5.

2.3 Synthesis and metabolism of green tea polyphenols

The biological synthesis of tea polyphenols in the tea plant is mainly synthesized through the phenylpropanoid pathway and flavonoids synthetic pathway (Fig. 2-6).

Many secondary metabolites in plants come from the branched metabolic pathway of phenylpropanoid. This pathway mainly includes the following reactions. Phenylalanine ammonia-lyase (PAL) catalyzes the non-oxidative deamination of phenylalanine to form cinnamic acid, the first step of the phenylpropanoid pathway [Watts *et al.*, 2006]. A cytochrome P450 monooxygenase, cinnamate 4-hydroxylase (C4H), is the second key enzyme in the phenylpropanoid pathway. C4H catalyzes the hydroxylation of cinnamic acid to form 4-coumarate [Li *et al.*, 2015]. 4-coumaroyl-CoA ligase (4CL) catalyzes 4-coumarate with coenzyme A to form 4-Coumaroyl-CoA [Gross and Zenk, 1974]. 4CLs belong to the larger family of acyl-activating enzymes (AAEs). 4CL is mainly divided into two categories:

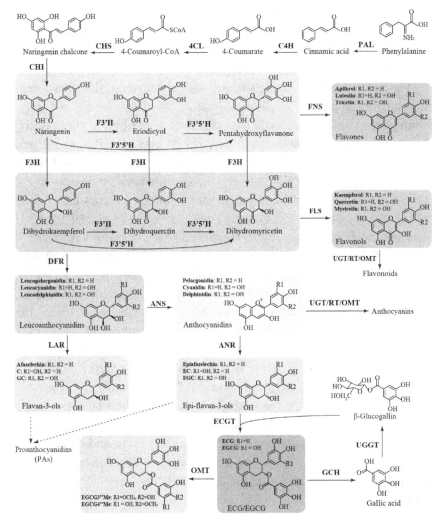

Figure 2-6. Synthesis and metabolism of green tea polyphenols.

I and II. Group I may be related to the synthesis of lignin, and Group II is mainly related to the synthesis of flavonoids [Cukovica *et al.*, 2001].

A molecule of 4-coumaroyl-CoA and three molecules of malonyl CoA are condensed under the action of chalcone synthase (CHS) to form naringenin chalcone [Heller and Hahlbrock, 1980]. This is the first key step of the biosynthesis of flavonoids and isoflavones. Chalcone isomerase (CHI) catalyzes the closure of the chalcone C ring to convert naringenin [Koes *et al.*, 1994]. Naringenin is the direct precursor for the synthesis of all flavonoids, so CHI is one of the key enzymes controlling the content of catechins in the tea plant.

Flavanone (naringenin, eriodictyol, and pentahydroxyflavanone) generate dihydroflavonols (dihydrokaempferol, dihydroquercetin, and dihydromyricetin) under the catalysis of flavanone 3-hydroxylase (F3H) [Singh *et al.*, 2008]. Flavone synthase (FNS) catalyzes flavanones into flavones [Lam *et al.*, 2014]. Flavonoid 3-hydroxylase (F3′H) and flavonoid 3′,5′-hydroxylase (F3′,5′H) are monooxygenases belonging to the cytochrome P450 enzyme family dependent monooxygenases that require NADPH as a co-factor [Forkmann, 1991; de Vetten *et al.*, 1999]. F3′H catalyzes the hydroxylation of naringenin and dihydrokaempferol at the 3′ position of the B ring to produce eriodictyol and dihydroquercetin, respectively. F3′,5′H catalyzes the hydroxylation of the 5′ or both 3′ and 5′ of the B ring of compounds, such as naringenin, eriodictyol, dihydrokaempferol, and dihydroquercetin. Dihydroflavonols generate flavonols under the action of flavonol synthase (FLS), and then generate flavonoids under the catalysis of transferase, such as UDP-Glucose flavonoid 3-O-glucosyl transferase (UGT), rhamnosyl transferase (RT), and O-Methyltransferase (OMT) [Britsch *et al.*, 1981; Zhao *et al.*, 2020a].

Dihydroflavonols are reduced to the corresponding leucocyanidins under the action of dihydroflavonol reductase (DFR), which is a key enzyme in the synthesis of procyanidins, anthocyanins, and catechins in the flavonoid pathway [Stafford and Lester, 1985; Punyasiri *et al.*, 2004]. Leuacoanthocyanidin reductase (LAR) converts leucoanthocyanidins to flavan-3-ols (such as C and GC), while anthocyanin synthase (ANS) catalyzes leucoanthocyanidins into anthocyanidins, which are then reduced to epi-flavan-3-ols (such as EC and EGC) by anthocyanidin reductase (ANR) [Stafford and Lester, 1985; Stracke *et al.*, 2009; Xie *et al.*, 2003]. Epicatechin:1-O-galloyl-b-D-glucose O-galloyltransferase (ECGT) is involved in the galloylation of EC/EGC to form ECG/EGCG [Liu *et al.*, 2012]. The latest study found that co-expression of two members of serine carboxypeptidase-like acyltransferases (SCPL-ATs) (SCPL4 and SCPL5) is likely responsible for the galloylation [Yao *et al.*, 2022]. Gallic acid can generate β-glucogallin under the catalysis of UDP-glucose:galloyl-1-O-b-D-glucosyltransferase (UGGT) to participate in the biosynthesis of gallated catechins [Liu *et al.*, 2012]. However, the gallated catechins could be hydrolyzed to non-gallated catechins and gallic acid with the gallated catechins hydrolase (GCH, such as tannase) action [Dai *et al.*, 2020]. Using S-adenosyl-L-methionine (SAM) as a methyl donor, O-methyltransferase (OMT) can catalyze EGCG into O-methylated EGCGs [Zhang *et al.*, 2015; Kirita *et al.*, 2010].

2.4 Processing of green tea polyphenols

Based on the different processing procedures (Fig. 2-7), especially the degree of oxidation of polyphenols, teas are classified as green tea, black tea, oolong tea, white tea, yellow tea, and dark tea (National Standard GB/T 30766-2014). Green tea is non-fermented tea and the critical process is fixation, which can make the enzymes inactive and thus prevent the oxidation of tea polyphenols. White tea is slightly-fermented tea and is subjected to the simplest processing procedure, including only a prolonged withering and drying process without any steps of enzyme deactivation or fermentation. Black tea is fully-fermented tea, generally made through four major steps: withering, rolling, fermentation, and drying, where polyphenols undergo dramatic oxidation and transformation during fermentation. Yellow tea is non-fermented tea and has a typical yellow color of tea leaves and infusions due to its unique "overbrewing" manufacturing process. Dark tea is a post-fermented tea and polyphenols are destroyed by microorganisms and Maillard reactions during the microbial

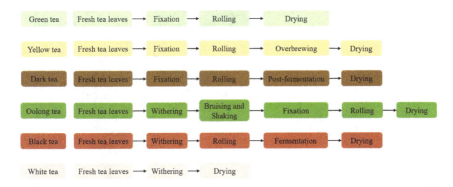

Figure 2-7. Classification of tea according to the degree of polyphenol oxidation.

post-fermentation. Oolong tea is semi-fermented tea and some polyphenols are partially oxidized and transformed during the critical bruising and withering processing.

Through the fermentation, catechins in fresh tea leaves are oxidized and polymerized to higher molecular weight polyphenols under the catalysis of endogenous polyphenol oxidase (PPO) and peroxidase (POD) (Fig. 2-8).

In the 1950s, Roberts E.A.H experimentally confirmed that during the processing of black tea, tea polyphenols, mainly catechins, are catalyzed by endogenous polyphenol oxidase (PPO) and peroxidase (POD) to form oxidation products theaflavins (TFs) and thearubigins (TRs). Moreover, recent studies have more clearly explored the transformation pathways of tea polyphenols in black tea processing [Roberts, 1958].

Oxidation of green tea leaves into black tea comprises different levels, which are described by the "oxidative cascade hypothesis" [Kuhnert *et al.*, 2013]. Furthermore, the different oxidation levels are oligomerization, rearrangement, and hydroxylation. All reactions combined are considered to result in the extensive molecular diversity of TRs in black tea. The polyphenols and enzymes, which are present in different compartments in

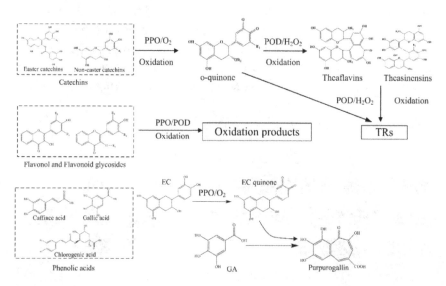

Figure 2-8. Transformation of green tea polyphenols during black tea processing.

leaves, are mixed during the rolling process, which initiates a series of enzyme-catalyzed oxidative reactions during fermentation. Catechins are firstly oxidized by PPO to form o-quinones, and o-quinones are unstable. Then, catechins and o-quinones could polymerize to form various oxidation products, such as TFs and theasinensins (TSs). Additionally, catechins and TFs could also be further oxidized by PPO and peroxidase (POD) to form TRs. PPO is an oxygen oxidoreductase enzyme that can oxidize substrates such as flavanols using oxygen, while POD catalyzes substrate oxidation in the presence of hydrogen peroxide and cannot use oxygen from the air. Thus, PPO activity is mostly associated with the formation of TFs, whereas POD activity has been associated with subsequent oxidation of TFs into TRs.

Expect for catechins, the flavonoid glycoside structure contains phenolic hydroxyl groups and can be oxidized by the oxidase catalyst in tea. PPO is the most active enzyme catalyzing the oxidation of flavonoid glycoside in tea and the coexistence of POD significantly intensified the PPO-induced oxidation of flavonoid glycoside. Also, aglycone was the critical moiety that affected the oxidation reaction and the presence of sugar moiety also positively contributed to the reaction [Guo *et al.*, 2021]. Phenolic acids are unstable and easily oxidized, while their sensitivities to oxidative enzymes differ due to their different structures. The enzymes in tea can directly oxidize caffeic acid while gallic acid can be coupled and oxidized, which is driven by epigallocatechin o-quinone or epicatechin o-quinone that have relatively high redox potentials in tea leaves.

2.5 Extraction of green tea polyphenols

The extraction of green tea polyphenols depends on various factors including solubility, pH, extraction time, and temperature [Pasrija & Anandharamakrishnan, 2015]. However, the selection of extraction technique is influenced by the niche of the compound of interest and the extent of purity required.

Solvent-based extraction gives a higher yield and different solvents have been used for the extraction of polyphenols from green tea including water, methanol, ethanol, acetonitrile, CO_2, and acetone.

2.5.1 *Hot water extraction technique*

Hot water extraction is a conventional technique that is being used for the extraction of green tea polyphenols. In this technique, dried leaves are simply placed in a boiling water bath and the temperature is between 80–100°C. Temperature and extraction time could influence the total polyphenol content and antioxidant activity. Many studies have optimized the conditions such as temperature and time for hot water extraction of green tea polyphenols.

2.5.2 *Organic solvent extraction technique*

In the organic solvent extraction technique, two aspects check on the process, namely, equilibrium state and mass transfer rate. Factors affecting the extraction efficiency of green tea polyphenols include the nature of the solvent, pH of extraction medium, temperature, number of extraction steps, the volume of solvent, and particle size and shape. The pH influences the solubility of the compounds and an increase in temperature increases the efficiency. The applications of solvent extraction have been foreseen in a wide area of industry, with better reproducibility and efficiency. Polyphenols may occur in combination with sugars, proteins, and sometimes get polymerized with each other. The solvent extraction technique is found to be more efficient and provides the required conditions in terms of temperature and pH to extract the polyphenols present in derived forms.

2.5.3 *Supercritical fluid extraction technique (SCF)*

The use of carbon dioxide as a solvent for extraction purposes has reduced the cost of using chemicals to a great extent as well as the threat of thermal degradation. The main advantage of SCF was the possibility of carrying out extractions at a temperature near to ambient temperature, thus preventing the substance of interest from thermal denaturation, while the technique was restricted to a very small range of products due to high capital investment and unawareness of the method. With advances in equipment design, the application of SCF is increasing.

Solvent extraction is the basic technique used for the extraction of green tea polyphenols. However, the use of organic solvent is not considered safe for consumption. In contrast, after SCF extraction of green tea, liquid CO_2 can be converted to gas by releasing the pressure and giving solvent-free polyphenols. Moreover, solvent extraction requires a long time with a huge amount of solvent for extraction and post-process purification, which limits this technique. The limitation of a long extraction process can be overcome by some new technologies — microwave-assisted and ultrasonication extraction techniques.

2.5.4 *Microwave-assisted water extraction (MWE)*

Microwave-assisted water extraction is a technique with higher efficiency requiring less time. Nkhili *et al.* optimized two parameters, temperature, and extraction time of MWE to estimate the higher extraction of polyphenols and confirmed polyphenols' chemical composition by HPLC-MS analysis. Process optimization and modifications for MWE in recent times have allowed wide exploitation of the technique for the extraction of various compounds from different sources. All these studies have concluded MWE as a better technique than all other conventional techniques with better recovery of bioactive, less degradation, and higher product integrity in terms of purity [Nkhili *et al.*, 2009].

2.5.5 *Ultrasonic-assisted extraction technique*

Ultrasonic-assisted extraction technique possesses enough potential than conventional techniques for the extraction of plants' secondary metabolites. Ultrasonication overcomes the disadvantages of high cost, post-process concentration, low recovery, and others. The use of the ultrasonication technique for the extraction of polyphenols increases the efficiency of the process and circumvents the degradation of polyphenols. It increases the mass transfer kinetics and quasi-equilibrium can be achieved by using this technique.

Different techniques for the extraction of green tea polyphenols and antioxidants have emerged with certain merits and demerits. The technique used for extraction, duration of extraction, the solvent used, solvent to solid ratio, and intensity of waves can influence the polyphenol extraction in terms of content, efficiency, composition, and purity of polyphenols. Microwave-assisted and ultrasonication extraction techniques seem to be more promising and have shown a greater potential and better efficiency for the extraction of polyphenols as compared to other techniques. Extracted polyphenols are very sensitive to the environment and prone to epimerization at a higher temperature and exposure to light. Hence, it is necessary to protect them from degradation. Techniques such as spray drying, freeze drying, emulsification, coacervation, inclusion, complexation, and nanoprecipitation are mainly used for encapsulation purpose, which serves the need to avoid the degradation of polyphenols [Ezhilarasi *et al.*, 2013]. However, there is a need to reduce or avoid the use of organic solvents for extraction and exploit other techniques with better efficiency.

2.6 Function of green tea polyphenols

It has been found that green tea polyphenols have many functions. Most of the later chapters of this book will introduce and discuss in detail. Here, only some basic functions of green tea polyphenols are briefly introduced.

2.6.1 *Antioxidant effect*

The antioxidant effect of polyphenols and their oxidation products mostly refers to their free radical scavenging effect. Free radicals can be produced by the automatic oxidation of some proteins, lipids, and low molecular compounds, and the oxidation-reduction of peroxides and some metal ions. Free radicals in organisms are in the balance between the biogenic system and the biological protection system. Once the balance is destroyed, it will endanger the body and cause disease. It needs exogenous antioxidants to remove free radicals and protect the normal operation of the body.

Tea polyphenols are a class of compounds that contain multiple phenolic hydroxyl groups, which are easier to oxidize and provide protons, and have the permeability of phenolic antioxidants, which can provide H^+

combined with free radicals, which can be reduced to inert compounds or more stable free radicals, thereby directly scavenging free radicals and avoiding oxidative damage. Additionally, tea polyphenols can also act on the related enzyme that produces free radicals, complexing metal ions that can scavenge free radicals indirectly, thereby playing a dual role in prevention and chain scission. The antioxidant and free radical scavenging effects of tea polyphenols have received much attention at home and abroad, and extensive research has been conducted on their effects and mechanisms [Ying *et al.*, 2018a].

Tea catechins have anti-lipid peroxidation and thus are beneficial in delaying body aging. When the organism has increased free radicals due to a variety of causes or the content of vitamin E and coenzyme Q is insufficient, the phospholipids in the biological membrane undergo peroxidation, and the cells destroy, leading to the senescence of cells. Studies have shown that tea catechin can inhibit lipid peroxidation, and its effect is more obvious than vitamin E [Wiseman *et al.*, 1997].

Polyphenols have abundant reactive hydroxyl groups, which can combine with biomacromolecules such as proteins through hydrogen bonding and thus affect many physiological processes. It can affect the enzymatic activity of many enzymes. For instance, it enhances the activities of antioxidant enzymes (e.g., glutathione peroxidase, catalase, quinone reductase, etc.) and phase II enzymes (e.g., glutathione sulfotransferase) and inhibits the activities of various enzymes such as ornithine decarboxylase, cyclooxygenase, squalene epoxidase xanthine oxidase, collagenase, interstitial multiple proteinases, protein kinase C, aromatase, NADPH cytochrome c reductase, telomere terminal transferase, etc. Thereby, exerting a significant inhibitory effect on many pathological processes; in the soybean lipid system antioxidant experiment, it was found that the polarity part of catechin and lecithin liposome membrane had an extremely strong affinity, playing a protective role [Huang and Frankel, 1997].

Polyphenols can complex metal ions that induce oxidation. The latest *in vitro* studies showed that the antioxidant capacity of EGCG was greatly improved after complexation with Cu^{2+}, which is more than twice the original antioxidant capacity. In addition, the antioxidant capacity of Mn^{2+}, Cr^{3+}, Mg^{2+}, and Al^{3+} to EGCG is also improved [Kumamoto *et al.*, 2001].

2.6.2 *Effects on cardiovascular disease*

Cardiovascular disease is predicted by numerous scientists to be a human health disease of the highest incidence and great harm in the 21st century. Among them, atherosclerosis is a hot spot for scientists to pay more attention to, and its occurrence is closely related to plasma lipids. Low-density lipoprotein (LDL) is atherogenic, whereas high-density lipoprotein (HDL) acts antagonistically. Apolipoprotein deficiency and abnormalities can affect blood lipid transport and metabolism, oxidative modification of LDL can damage vascular endothelial cells, and atherosclerosis occurs by deposition of cholesterol into the vessel wall. Epidemiological investigations have found a close relationship between natural polyphenol content and the incidence of coronary heart disease. Tea is rich in polyphenols, which can act on cardiovascular diseases through many mechanisms, such as regulating blood lipid metabolism, anticoagulant and inhibiting platelet aggregation, inhibiting arterial smooth muscle cell proliferation, and affecting hemorheological properties.

1. Regulation of lipid metabolism

Tea polyphenols reduce blood triglycerides (TG), cholesterol (CH), and low-density lipoprotein cholesterol (LDL-C) and increase high-density lipoprotein cholesterol (HDL-C). Experiments showed that a high dose of black tea extract reduced lipid levels in plasma, promoted the excretion of total lipids and cholesterol, and thus reduced cholesterol *in vivo*. In addition, tea polyphenols can prevent the oxidation of unsaturated fatty acids in food, and unsaturated fatty acids can promote the flow and transformation of cholesterol, promote the transformation of cholesterol into cholic acid, reduce the content of serum cholesterol and the deposition of cholesterol on the intima of blood vessels, so as to play an anti-atherosclerotic role by inhibiting the oxidation of unsaturated fatty acids.

2. Anticoagulant and profibrinolytic

The generation of atherosclerosis and thrombus is closely related to arachidonic acid metabolism. 12-lipoxygenase and cyclooxygenase are present in platelets. Transport of 12-lipoxygenase through tunica media smooth muscle can cause atherosclerosis and produce substances that lead to allergic diseases. Cyclooxygenases can metabolize arachidonic acid to form coagulation factors, which in turn promote platelet coagulation for the appearance of a thrombus. Natural polyphenols such as tea polyphenols can inhibit 12-lipoxygenase and cyclooxygenase enzymes, alter arachidonic acid metabolism, increase prostacyclin, and reduce thromboxane synthesis while playing an inhibitory role in platelet aggregation, anticoagulation, and fibrinolysis.

3. Prevention of cardio-cerebrovascular disease

Hypertension is the most common cardiovascular disease in China and one of the risk factors for cerebral stroke and coronary heart disease. It has been reported that tea polyphenols reduce blood pressure in rabbits and increase perfusate outflow in the hind limbs of rats. It is indicated that tea polyphenols can reduce peripheral vascular resistance and directly dilate blood vessels. Tea polyphenols also act as antihypertensive agents by promoting the formation of endothelium-dependent relaxing factors, relaxing vascular smooth muscle, enhancing vascular wall, and modulating vascular wall permeability; tea polyphenols can inhibit the activity of angiotensin I-converting enzyme, resulting in some preventive effects on hypertension.

2.6.3 *Anti-allergy and regulating immune function*

Immunity is the biological response process of the immune system to antigenic substances. It has the function of "identifying" and "eliminating" antigenic foreign bodies and maintaining the physiological balance of the body. From the perspective of clinical medicine, when the body comes into contact with pathogens as antigenic substances again, the response of defense and resistance to pathogens is called the immune response. On the other hand, when the body comes into contact with antigenic substances again, the body may react too strongly and cause damage to the body itself. This kind of specific immune reaction that is unfavorable to the body is called allergy.

Tea has a strong anti-allergy ability, so it has attracted the attention of many researchers at home and abroad. Seventy-two kinds of foods such as tea and their components have the anti-allergy ability. Hyaluronidase is a lysosomal enzyme that can decompose mucopolysaccharides, which is related to the inflammation of the permeable area of blood vessels. Various drugs that affect histamine release from mast cells can inhibit the activity of hyaluronidase. It was found that green tea, oolong tea, and black tea could significantly inhibit enzyme activity. The inhibitory effect is green tea < oolong tea < black tea. With the oxidative fermentation of tea polyphenols, the anti-allergy ability of tea showed a downward trend.

Tea polyphenols have the ability to alleviate the excessive allergic reaction of the body, but they can promote the overall immune function of the body. After taking tea polyphenols, the content of plasma immunoglobulin (Ig), especially IgM and IgA, increased in cancer patients receiving chemotherapy and radiotherapy. The effect of tea polyphenols on nonspecific immune function was determined by an animal macrophage phagocytosis test. The results showed that tea polyphenols greatly improved the nonspecific immune function of the body. At the same time, the effect of tea polyphenols on cellular immune function was studied by mouse peripheral lymphocyte transformation test. The results showed that when the dose of tea polyphenols reached 33 g/L, the lymphocyte transformation rate increased significantly ($p < 0.01$), that is to say, tea polyphenols effectively increased the cellular immune function of the body.

2.6.4 *Cancer prevention and anticancer*

Epidemiological investigation shows that tea has an anti-tumor effect (including experimental animal models and *in vitro* tests), and tea polyphenols are the main component. Tea polyphenols have anti-mutagenic activity *in vitro* and can inhibit carcinogenic-induced tumors of the skin, lung, stomach, esophagus, duodenum, and colon in rodents.

1. Inhibitory effect on liver cancer

The four catechin monomers, tea pigment, and oolong tea extract can significantly reduce the number and area of glutamyl transpeptidase in liver precancerous lesions, and show that tea pigment has the same chemopreventive effect on liver cancer as catechin monomers. Green tea polyphenols significantly inhibited the production of nitric oxide in mouse hepatocytes induced by 12-oxygen-tetrasunoyphorbol-13-acetic acid (TPA), and the inhibition rate was as high as 90% [Srivastava *et al.*, 2000]. This was postulated to be related to the inhibitory effect of green tea polyphenols on the activity of nitric oxide synthase, independent of the antioxidant capacity of the polyphenols.

2. Inhibitory effect on skin tumor and cancer
It was found that topical EGCG, EGC, and EGC inhibited TPA-induced skin inflammation in mice. Research shows that excessive exposure to solar radiation, especially ultraviolet B (UVB), can easily lead to skin cancer. Oral and topical tea polyphenols can resist the carcinogenic effect of UVB, among which EGCG has the strongest effect. The application of green tea extract (mainly tea polyphenols) can eliminate the skin photochemical damage caused by ultraviolet A (UVA) and bone fat supplement and reduce skin hyperplasia and horny hyperplasia.

The anti-tumor mechanism of tea polyphenols may involve inhibiting the activity, edema, and proliferation of skin ODC, cyclooxygenase, and lipoxygenase induced by tumor promoters. By studying the effect of topical EGCG on TPA-induced mouse skin oncogene expression, it was shown that topical EGCG could inhibit the transcription levels of ODC, PKC, and c-myc induced by TPA [Surh *et al.*, 2001]. In addition, tea polyphenols can inhibit the growth of skin cancer. *In vitro* studies showed that EGCG, ECG, and EGC significantly inhibited the growth of oral squamous cell carcinoma cells.

3. Inhibitory effect on colorectal tumors
Japanese scholars have studied the effect of tea polyphenols on colorectal carcinogenesis induced by 1,2-dimethylhydrazine (DMH) in male rats [Kan *et al.*, 1996]. It shows that tea polyphenols can significantly reduce the incidence of tumors and the number of Clostridium perfringens (which can produce carcinogenic harmful substances) in cecal contents. Tea polyphenols can also stimulate the transcription of phase II detoxification enzyme through Antioxidant Response Element (ARE) in the plasmid, so as to achieve the effect of anti-colorectal tumor [Mukhtar and Ahmad,

1999; Mukhtar and Ahmad, 2000]. Research also found that (−)-Epigallocatechin-3-gallate and EZH2 inhibitor GSK343 have similar inhibitory effects and mechanisms of action on colorectal cancer cells [Ying *et al.*, 2018b].

4. Inhibitory effects on prostate tumors

Both results of epidemiological investigation and animal experiments showed that drinking green tea, which contains a large number of tea polyphenols, effectively suppressed prostate tumorigenesis. The inhibitory effect of green tea polyphenols on ornithine decarboxylase (ODC) in prostate tumor cells, activated by testosterone induction, was investigated by *in vitro* (LNCaP cells) and *in vivo* experiments, and feeding 0.2% green tea polyphenols effectively inhibited 20–54% of ODC enzyme activity. EGCG inhibited cell growth and androgen-regulated expression of PSA and hK2 genes, while also significantly repressing the androgen inducibility of PSA enhancers. EGCG may also repress the expression of this gene by repressing the Sp1 site on the androgen receptor enhancer gene [Chen *et al.*, 2004].

5. Inhibitory effects on breast tumors

High consumption of green tea increased progesterone and estrogen receptor expression and decreased the number of axillary lymph node metastases in premenstrual stage I and II mastoma patients. Another study found that EGCG interferes with the activity of tNOX, a NOx enzyme implicated in cancer tumors, while it has no effect on normal NOx activity. The experiments showed that EGCG limits the growth of breast cancer cells but seems to have no effect on normal healthy breast cells [Ying *et al.*, 2018a; Wei *et al.*, 2018].

6. Inhibitory effect on gastric cancer

In vitro experiments on gastric tumor cell lines showed that polyphenols and their oxidation products inhibited the growth of gastric cancer cells by inhibiting TNF gene expression and TNF release, and the inhibitory ability was ECG > EGCG > EGC > TF > EC. Furthermore, Zhao *et al.* found that Epigallocatechin gallate reverses gastric cancer by regulating the long noncoding RNA LINC00511/miR-29b/KDM2A axis [Zhao *et al.*, 2020b].

7. Inhibitory effect on lung cancer

The research on rodent lung cancer, induced by tobacco-specific nitrosamine NNK, shows that green tea and black tea can inhibit DNA oxidative damage induced by NNK and inhibit the occurrence of lung cancer [Chung, 1999].

2.6.5 *Antibacterial, antiviral, and bactericidal effects*

As early as the Shennong period, tea was used for sterilization and anti-inflammatory. Tea polyphenols have a broad antibacterial spectrum, strong antibacterial ability, and excellent selectivity. It has a certain inhibitory ability on almost all animal and plants pathogenic bacteria in nature [Wang *et al.*, 2011]. It can change the physiology of bacteria through biochemical and physicochemical mechanisms, interfere with the metabolism of bacteria, maintain the balance of normal flora, and promote the proliferation of some beneficial bacteria. In addition, tea polyphenols will not cause bacteria to develop drug resistance. The concentration of tea polyphenols required for bacteriostasis is low.

2.6.6 *Anti-inflammatory, detoxifying, and anti-allergic effects*

Since tea polyphenols can precipitate caffeine and heavy metal salts (such as Pb^+, Hg^{2+}, Cr^{6+}, etc.), drinking tea can disinfect drinking water. In addition, tea polyphenols can also be used as an antidote for alkaloids and heavy metal salt poisoning, which can alleviate the toxic effect of these heavy metal ions. EGCG and ECG had significant inhibitory effects on oxazoline-induced type IV allergy in female ICR rats, and the effect of methyl derivatives of EGCG was higher than that of EGCG. In addition, oral administration of EGCG and tea polyphenols can also prevent the occurrence of type IV allergy.

Tea polyphenols can significantly inhibit the cell mutation and chromosome damage induced by cigarette agglutinate, and its inhibitory effect is better than that of vitamin C, vitamin E, and β-Carotene. In addition, modern medical research has proved that the antioxidant effect of tea polyphenols can prevent damage to the liver after ethanol free radical and ethanol oxidation to acetaldehyde, so as to reduce menstrual alcoholism.

2.6.7 *Anti-radiation effect*

Tea polyphenols can prevent the effects of various radiation, such as solar radiation, radioactive elements, and X-ray. Scholars from the former Soviet Union irradiated mice with strontium-90 and then regularly fed concentrated catechin. The results showed that the mice in the experimental group still survived, while the mice in the control group died of radiation disease. It has been confirmed that gallate in catechins has the strongest anti-radiation effect. In addition, tea polyphenols can also prevent the damage of ultraviolet A and B to the skin [Chen *et al.*, 2004; Katiyar *et al.*, 2001]. Tea polyphenols have the function of scavenging singlet oxygen and free radicals, so as to prevent the lesion of plastids caused by UVA [Buer *et al.*, 2010].

2.7 Application of green tea polyphenols

At present, tea polyphenols have been widely used, including in industry, agriculture, medicine. Tea polyphenols are also widely used in daily cosmetics and even in wastewater treatment. As the research and development of tea polyphenols will be increasingly standardized in more aspects, it will create a wealth for national economy and make greater contributions to human health.

2.7.1 *Application in the food industry*

With the rapid advancements of modern society, human demand for diet has prompted the urgent development of healthy foods. The antioxidant activity of GTPs shows great potential for food application. The primary consideration in adding GTPs is the improvement in flavor, nutritional value, quality, and safety. The addition of GTPs into dairy products exhibited the possibility of eliminating the odor of dairy products, preventing the oxidation of oil and fat, and improving the flavor. The study confirmed that a 0.3% addition of GTPs increased the foaming property and foam stability of egg juice for cake bakeries, which can attribute to the combination between the hydroxyl of GTPs and the protein of egg juice.

Apart from being a flavor additive, GTPs added to candies, biscuits, cakes, mooncakes, instant noodles, milk powder, and other products show more function of inhibiting the growth of bacteria and prolonging shelf life. Numerous studies have demonstrated that GTPs show excellent safe, natural, antioxidant, and anti-bacterial properties for widely application on fruit, vegetable, meat, and aquatic products preservation. The enzymatic reaction brings browning for fruits and vegetables, hence requires a radical scavenger. GTPs can significantly prolong the shelf life of fresh fruits and vegetables by using them as food film or packaging [Ramsaha *et al.*, 2015]. The use of GTP positively affected the preservation of ascorbic acid and carotenoid content of the samples [Martín-Diana *et al.*, 2008]. Lan *et al.* observed that TPs strengthened the antimicrobial property of polyvinyl alcohol-TP films by delaying the loss of titratable acids and soluble solids in wrapping strawberries and extending the shelf life of these fruits [Lan *et al.*, 2019]. The application of a coating containing green tea extract and chitosan has been shown to effectively inhibit lipid oxidation and fungal growth during the storage of walnut kernels [Sabaghi *et al.*, 2015]. Zhang and Jiang also prepared chitosan film added with TPs to reduce water content and capture reactive oxygen [Zhang & Jiang, 2020].

Meat and aquatic products are rich in proteins and fatty acids, which could cause quality deterioration resulting in lipid oxidation. Jia *et al.* [2018] found out that TP improved the quality and prolonged the shelf-life of silver carp fillets by modulating the microbiota composition. Researchers explored the oxidation resistance of TPs by maintaining the appearance of pork under oxidizing conditions [Jia *et al.*, 2018]. Their results illustrated that a low concentration of TPs inhibits the formation of methemoglobin and maintains the appearance color of pork by chelating with Fe^{2+}. Green tea extracts added to chitosan active film were used in pork sausage packages, resulting in enhancing the antioxidant activity, inhibiting microbial growth, maintaining the quality, and prolonging the shelf life of pork sausages [Siripatrawan & Noipha, 2012]. Moreover, Xi *et al.* investigated the beneficial effects of tea extracts containing phenolic compounds on reducing the growth of bacteria and extending the shelf life of oysters [Xi *et al.*, 2012]. TPs could effectively inhibit the microorganisms and pathogen's growth and improve the safety of chilled mutton [He *et al.*, 2016]. Besides, a novel bioactive film has been developed by adding TPs to pectin and cellulose and figured out the potential utilization on food preservation due to its anti-bacteria property against *Escherichia coli*, *Candida albicans*, and *Staphylococcus aureus* [Ye *et al.*, 2019].

2.7.2 *Application for human health*

Green tea polyphenols are found as a good antioxidant and antiinflammation medicine in both scientific research and medical practice. Cardiovascular and cerebrovascular diseases (CVD) including hyperlipidemia, atherosclerosis, and hypertension showed the highest cause of death every year in the world. In the rat model of cardiac hypertrophy, EGCG has been found to prevent the shortening of telomere and inhibit cardiac myocyte apoptosis [Sheng *et al.*, 2013]. Oral EGCG also increases the activity of the antioxidant enzyme in rats to avoid the occurrence of myocardial infarction [Devika & Prince, 2008]. An investigation considered that people with regular consumption of green tea ≥ 500 mL showed a lower risk for CVD compared to those with consumption of less than 100 mL [Kuriyama, 2008]. Recently, a similar conclusion was obtained by Shanghai Health Study, that middle-aged and elderly adults having the habit of drinking green tea face less risk of CVD [Zhao *et al.*, 2017]. Further exploration of the anti-cardiovascular activity of GTP revealed that the reactive oxygen species (ROS) generation in vascular endothelial cells causing CVD can be inhibited by EGCG through the prevention of NADPH oxidase production [Ahn *et al.*, 2010]. Up to now, GTPs registered as healthcare products have been frequently used in China. The basic reason is that GTPs have better lipid-lowering functions by adjusting the levels of triglyceride (TG), lipid peroxide (LPO), low-density lipoprotein cholesterol (LDL-C), high-density lipoprotein cholesterol (HDL-C), and other indicators in the blood.

GTPs can also alleviate the symptoms of Parkinson's disease (PD), owing to the phenolic hydroxyl and ring structure to inhibit the oxidation of dopamine [Zhou *et al.*, 2019]. PD is the second largest neurodegenerative disease with the highest incidence rate. Many studies have shown that GTPs can prevent PD through anti-oxidation, iron chelation, and regulation of the intracellular signal pathways. Preliminary clinical research indicated that GTPs can effectively alleviate the early clinical symptoms of PD patients, showing a good clinical application prospect. Besides, GTPs have protective effects on cerebral ischemia-reperfusion injury, brain injury, and traumatic brain injury after cardiopulmonary resuscitation.

Cancer is a huge threat to human health. The accumulation of malondialdehyde and DNA mutations increase the risk of breast cancer. EGCG was reported to inhibit disease development by regulating the activity of matrix metalloproteinases in breast cancer cells [Deb *et al.*, 2015]. EGCG significantly lowered the number of aberrant crypt foci and alleviate colon cancer in a rat model. GTPs have also been shown to prevent respiratory cancer, digestive cancer, lung cancer, liver cancer, and urinary cancers. The anti-cancer mechanism of GTPs can be summarized in the following five ways: antioxidation, regulation of key enzymes in carcinogenesis, blocking information transmission, anti-angiogenesis, and promoting apoptosis.

GTPs have good anti-inflammatory and anti-infection activities. Di Paola *et al.* treated the mouse with 25 mg/mL tea polyphenol extract and found that tea polyphenol can treat lung injury caused by acute inflammation. A liver inflammation study showed that GTPs can significantly reduce the levels of serum alanine aminotransferase and IL-6, the infiltration of polymorphonuclear leukocytes, the expression of cell adhesion molecules, and the phosphorylation of IκBa, thus preventing liver injury in rats [Zhou *et al.*, 2018]. GTPs have a broad spectrum against bacteria, which are favorable as natural bacteriostatic. The effects of EGCG on the growth and virulence of *Chlamydomonas nucleoli* were investigated by Ben Lagha *et al.* and found to be used in related disease treatment. The combination of GTPs with antibiotic medicines was also worth noticing. Periodontitis is the sixth most prevalent disease in the world. GTPs can effectively inhibit oral bacteria, reduce the growth of cariogenic bacteria, and alleviate periodontitis in the oral cavity [Ben Lagha *et al.*, 2017]. Hannig *et al.* observed a clear decrease in initial bacterial adhesion influenced by green tea onto the intraoral pellicle [Hannig *et al.*, 2009]. It is considered that flavanols like EGCG and ECG have strong interaction with oral cells, and EGCG could change the structure of the pellicle [Rehage *et al.*, 2017]. Song *et al.* found that EGCG can effectively inhibit pneumolysin activity and potentially be an anti-virulence agent for *Streptococcus pneumoniae* infection [Song *et al.*, 2017]. Zimmermann *et al.* confirmed these observations and demonstrated that EGCG showed no effect on the negative charge of *in situ* pellicles [Zimmermann *et al.*, 2019].

2.7.3 *Other applications*

In addition to its wide application in industry, agriculture, and medicine, tea polyphenols are also widely used in environmental treatment, animal husbandry, and daily cosmetics.

1. GTPs in environmental treatment

Research on GTPs combined with ozone as a disinfectant for water treatment showed that the total number of bacteria was controlled at 80 CFU/mL for 48 hours, which proved that the combined disinfection process of GTP and ozone had a good application prospect. Besides, ultraviolet (UV) rays were also applied to combine with GTP in water treatment and exhibited better performance than UV-chlorine disinfection. Adsorption is a common treatment for wastewater management, while GTPs are favorable for heavy metal adsorption. GTPs were extracted from waste tea and added to copper sulfate solution, resulting in a decrease of copper ions. The adsorption capacity was 38.50 mg/g. Furthermore, GTPs are natural green reducing compounds that can be used as accelerators in the Fenton reaction to degrade antibiotics. The Fe^{3+}/Fe^{2+} cycle was promoted by GTPs due to the destruction of the chelate structure [Ouyang *et al.*, 2019]. Zhang *et al.* prepared polyvinylidene fluoride membrane with the modification of GTPs and found excellent anti-oxidation performance and potential ability in removing dye from wastewater. In addition to wastewater management, GTPs also present availability in soil remediation. The antibacterial property of GTPs can effectively solve the problem of soil fertility loss caused by soil microorganisms. For instance, the utilization of urea in soil usually resulted in air pollution due to the hydrolysis of urease. Studies have shown that GTPs can inhibit the growth and metabolism of urease fungi by affecting the cell membrane. Researchers also found that GTPs could inhibit soil nitrification by preventing the growth of ammonia-oxidizing bacteria and nitrite-oxidizing bacteria with a rate of about 31.2–92.6% [Zhang *et al.*, 2020].

2. GTPs in livestock production

Intensive livestock production leads to the improvement of market weight, but the decline of meat quality and the utilization of antibiotics at the same time. Tea polyphenols, as a safe and efficient natural antioxidant, can improve animal growth performance, relieve stress, improve meat quality, and reduce the use of antibiotics. The addition of GTPs can not only improve growth performance, antioxidant capacity, and immune function, but also inhibit bacterial reproduction, delay pork spoilage, and improve pork quality after slaughter. Piglets often suffer from anorexia and lethargy under oxidative status, which leads to growth inhibition. The intake of GTPs can increase the ratio of CD^{4+}/CD^{8+} and the level of IL-4, which indicate that the oxidative stress caused by immune damage has been relieved and restored [Deng *et al.*, 2010]. Suzuki *et al.* reported that feeding growing pigs with a diet supplemented with 3% tea can not only increase the content of vitamin E in meat, but also significantly increase the content of linoleic acid and linolenic acid in pork [Suzuki *et al.*, 2002].

In addition to the improvement of the growth performance of meat and decrease of the stress symptoms, tea polyphenols also have a good regulating effect on the growth and laying performance of poultry. Bioactive substances including ovotransferrin, lysozyme, and biotin-binding protein in eggs, which have important physiological functions, are easy to be decomposed with the extent of storage and lead to low-quality of eggs. The effects of GTPs on egg quality have been found to increase the content of linoleic acid, docosahexaenoic acid, and other functional fatty acids in eggs, and reduce the content of cholesterol and thiobarbituric acid in egg yolk, with the best addition amount of 2%.

3. GTPs in daily cosmetics

Nowadays, GTPs have been widely used in the fields of daily cosmetics including toiletries, skin care products, textiles, and so on. Daily cosmetic products refer to daily chemicals, including shampoo, shower gel, skin-care products, washing powder, hand sanitizer, etc. In recent years, product quality, safety, and environmental friendliness have become the main demand in the transformation and upgrading of the daily chemical industry. Tea polyphenol has obvious functions of anti-radiation, anti-aging, scavenging excess free radicals, inhibiting bacteria, resisting viruses, eliminating the peculiar smell, and showing huge potential as additives in daily cosmetic products.

At present, the daily cosmetic products made from tea polyphenols include facial cleanser, toner, lotion, cream, bath lotion, shampoo, toothpaste, chewing gum, deodorant, etc. The efficacy of tea polyphenols is the main factor for people to choose daily cosmetic products containing tea polyphenols. The application of tea polyphenols can significantly improve the economic benefits for tea enterprises, promote the transformation and upgrading of tea deep processing, and reduce the waste of tea resources.

2.8 Metabolism and bioavailability of tea polyphenols *in vivo*

Bioavailability is very important for a drug and nutrient, and bioavailability determines the actual biological effect of the drug and nutrient. This is because if its bioavailability is very low, no matter how much you eat, it has no practical value. Bioavailability depends on metabolic processes. Therefore, in this section we discuss the metabolism and bioavailability of tea polyphenols *in vivo*.

2.8.1 *Absorption and efflux of tea polyphenols*

Some compounds have large volume (high molecular weight), large surface area (formation of large-area hydrated membrane) and strong polarity, which are usually difficult to be absorbed by oral administration, and the bioavailability of tea polyphenols also follows this rule. Both human and animal experimental studies showed that the bioavailability of EC and (+)-catechin (both containing five phenolic hydroxyl groups and relative molecular weight of 290) was significantly higher than that of EGCG (containing eight phenolic hydroxyl groups and relative molecular weight of 458) [Yang *et al.*, 2016]. When two or three cups of green tea are given orally at the same dose, the peak level of catechins in plasma (including bound catechins) is generally 0.2–0.3 μmol/l. The peak plasma concentrations of oral pharmacologically high-dose EGCG in humans and mice were 7.5 μmol/l and 2–9 μmol/l, respectively [Lambert *et al.*, 2006; Chow *et al.*, 2005].

Most catechins are absorbed by passive diffusion in the intestine, and monocarboxylic acid transporter (MCT) is involved in the absorption of catechins [Vaidyanathan *et al.*, 2003]. Active efflux limits the bioavailability of many phenols, including catechins. Multidrug resistance-associated protein (MRP) is an ATP-dependent efflux transporter expressed in many tissues, especially in many human tumor tissues. MRP1 exists in almost all tissues and is located on the basolateral side of the cell. The physiological function of this protein is to transport compounds inside the cell to the blood or cell gap. On the contrary, MRP2 protein is located on the surface of the intestine and liver and regulates the transport of compounds to the lumen and bile respectively [Leslie *et al.*, 2005]. Studies on human colon cancer cells have shown that EGCG and its methylated or glucuronized metabolites are the substrates of MRP2 [Lambert *et al.*, 2006; Chen *et al.*, 2021]. These studies also showed that co-culture with selective inhibitors of MRP could increase the accumulation of EGCG and its metabolites in cells. Since MRP2 is located at the apical membrane of epithelial cells, regardless of whether EGCG is methylated or glucuronidated, MRP2 can transport EGCG from intestinal epithelial cells back to the lumen of the small intestine through active transport, thus limiting the bioavailability of EGCG. The rest of EGCG is absorbed into the human portal vein circulation, then into the human liver, and then discharged by MRP2 located on the microtubule membrane of hepatocytes. In contrast, MRP1, located on the basal membrane of intestinal cells, hepatocytes and other tissues, pumps EGCG inside the cells into the blood or intestine. It is speculated that MRP1 is beneficial to improve the bioavailability of EGCG *in vivo*. Nevertheless, the bioavailability of MRP1 and MRP2 to EGCG *in vivo* is likely to depend on the distribution of each efflux pump protein in the tissue. It is reported that in human jejunum, the transcription level of MRP2 is even 10 times higher than that of MRP1. Therefore, the efflux of EGCG through MRP2 may play a major role in the intestine, which leads to the decline of EGCG bioavailability.

In the liver, MRP2 pumps EGCG and its metabolites into bile. Therefore, EGCG, ECG, and their metabolites are mainly excreted in feces, but almost not in urine. In fact, this phenomenon has been confirmed in humans and rats. However, after mice were given tea or EGCG, low levels of EGCG metabolites (bound form) were detected in urine. In addition, higher concentrations of EGC, EC, and their metabolites were detected in urine after drinking tea. EGC and EGCG seem to exist only in tea, while EC can be found in many vegetables and fruits.

2.8.2 *Pharmacokinetics of tea polyphenols*

The peak concentration of catechin in human and animal plasma is generally submicromolar level. Some pharmacokinetic studies of tea catechins in rats and mice have been reported. After intragastric administration of 200mg/kg caffeine-free green tea to rats, the half-lives of EGCG, EGC, and EC in plasma were 165 minutes, 66 minutes, and 67 minutes respectively, and the absolute plasma bioavailability were 0.1%, 14%, and 31% respectively [Lambert *et al.*, 2009]. Correspondingly, the absolute bioavailability of serum EGCG was much higher after intragastric administration of 75 mg/kg EGCG, in which more than 50% of EGCG was glucosidic acid conjugate. After gavage, the concentrations of EGCG in the small intestine and colon were 20.6 μg/g and 3.6 μg/g respectively, and the concentration in other tissues is less than 45.8 ng/g. Adding 0.6% green tea polyphenol extract to the drinking water of rats increased the plasma concentration after 14 days, and the concentrations of EGC and EC were higher than EGCG. Plasma concentrations of these catechins decreased over the next 14 days. When the same polyphenol extract was applied to mice, the concentration level of catechin reached the peak on day 4 and decreased to less than 20% of the peak on days 8–10. These results showed that prolonging tea drinking time enhanced the biotrans-formation and elimination of catechin.

We first reported the levels of catechin in the blood and urine of volunteers after taking different doses of green tea, and also reported the results of the pharmacokinetics of tea polyphenols in volunteers. For example, when people take 20 mg/kg of green tea orally, the time for catechin to reach the peak plasma concentration is 1.4–1.6 hours, and the maximum plasma concentrations of EGC, EC, and EGCG are 223 ng/ml, 124 ng/ml, and 77.9 ng/ml respectively. The elimination half-lives of EGCG, EGC and EC were 3.4 hours, 1.7 hours, and 2 hours respectively. EC and EGC in serum mainly exist in the form of a bound state, while 77% of EGCG exists in the form of a free state. EGC and EC in serum mainly exist in the form of glucosidic acid and sulfate, and only a small part exists in the form of a free state. Methylated EGCG and methylated EGC were also present in the serum. After drinking tea, the maximum plasma concentration of 4',4"-di-O-methyl EGCG in serum is 20% of that of EGCG, but the 24-hour cumulative excretion of 4',4"-di-O-methyl EGCG in urine is more than 10 times higher than that of EGCG.

EGCG and ECG mainly exist in the form of a free state in serum, while EGC and EC mainly exist in the form of glucosidic acid or sulfate. After the next night, when Polyphenon E was ingested on an empty stomach, the concentrations of EGCG and ECG in serum were 3–5 times higher than those taken during eating. The study also showed that the levels of EGCG and EGC in plasma increased significantly with the increase of intake dose. These observed different metabolic kinetic characteristics may be related to the food interaction in the satiety state, the saturation state before system metabolism, and the consumption of phase II enzymes/cofactors in the overnight fasting state. The results also showed that the concentration of EGCG in the blood increased after a single intake of 800 mg of green tea polyphenols per day for four weeks, but this phenomenon did not occur in the experimental treatment of 400 mg twice a day. Similarly, the area under the curve of EGC and EC did not change significantly. The reasons for these contradictions are not clear [Chow *et al.*, 2011].

2.9 The security of green tea polyphenols

Green tea polyphenol is a well-known natural antioxidant and antibacterial agent, which is expected to be widely used in food and healthcare products. Therefore, the efficacy, toxicology, and security of green tea polyphenols should be thoroughly discussed. The assessment of potential toxicity is vital for green tea polyphenols, which has already been carried out in the early 1990s. Green tea polyphenols applied in the acute toxicity test showed that the oral LD_{50} for rats was 2469 ± 326 mg/kg, which represented the hypotoxicity of green tea polyphenols. However, some studies believed that a high concentration of green tea extracts brings harmful effects, which are not only because of caffeine, but also tea polyphenols. A high concentration of green tea extract has been reported to exert acute toxicity in rat liver cells, while EGCG seems to be the key factor causing this damage [Schmidt *et al.*, 2005]. Two rats were observed with gastric erosions treated with green tea catechin at 2000 mg/kg/day, thus, the no-observed-adverse-effect level (NOAEL) in rats should be no higher than this level of daily intake [Chengelis *et al.*, 2008]. Additionally, it has been observed that high oral doses of the green tea polyphenol EGCG might have hepatotoxic effects in mice, potentially involving oxidative stress in the liver.

In consideration of the popularity and convenience of green tea and its derived products, an objective evaluation of the toxicity of green tea polyphenols has attracted close attention. A report from US Pharmacopeia in 2008 concluded that the safety information for green tea, arising from diverse sources, provides a signal for the possibility of liver damage caused by products that contain concentrated green tea extracts and assigned green tea as safety class 2 [Sarma *et al.*, 2008]. Joint research by Japan and Britain also considered that green tea catechin presents no significant genotoxic concern under the anticipated conditions of use after performing *in vivo* experiment on mice and rats administered with a high dose of the green tea catechin up to 2000 mg/kg, and *in vitro* bacterial reverse mutation assay (Ames test) in cultured Chinese hamster lung cells [Ogura *et al.*, 2008]. Similarly, oral administration of standardized green tea catechin at a dose of 2000 mg/kg/day for 28 days was safe for rats under clinical conditions [Chengelis *et al.*, 2008]. Their results indicated that standardized green tea polyphenols (containing 800 mg EGCG, or decaffeinated

Polyphenon E containing 800 mg EGCG, 148 mg EGC, 124 mg EC, and other green tea polyphenols) for humans for 28 days are also safe.

A series of studies hold the view that green tea extracts or polyphenols have no or very low toxicity even at a very high dose. Safety studies on EGCG showed that oral administration of 2000 mg EGCG/kg to mice did not induce micronuclei formation in bone marrow cells, and administering 1200 mg EGCG/kg/day in their diet for 10 days produced plasma EGCG concentrations comparable to those reported in human studies [Isbrucker *et al.*, 2006a]. Also, a NOAEL of 500 mg EGCG preparation/kg/day for rats was established, which was equivalent to 30 g administrated EGCG for 60 kg human body weight [Isbrucker *et al.*, 2006b]. A higher concentration of administered green tea extract (2500 mg/kg/d) for 28 days showed nonobvious damage to body weight, hematology, urinalysis, or biochemistry in mice [Hsu *et al.*, 2011].

Furthermore, Saleh *et al.* proved the safety of green tea polyphenols over moderate consumption, but a high dose of EGCG still caused concern due to its mild liver injury [Saleh *et al.*, 2013]. Nevertheless, green tea polyphenols, mainly catechins, easily bind to certain drugs and lead to the inhibitory of adsorption and bioactivities, although this has not been observed broadly in human cases [Yang & Pan, 2012].

Based on these toxicological research, the current established NOAEL value of green tea polyphenols is higher than the daily intake for humans by traditional consumption of green tea. DSI EC (Dietary Supplement Information Expert Committee) of US Pharmacopeia had given the advice that dietary supplement products derived from green tea extracts could carry the following labeling statement, "Take with food. Discontinue use and consult a healthcare practitioner if you have a liver disorder or develop symptoms of liver trouble such as abdominal pain, dark urine, or jaundice." It should be safe to control the dietary intake of green tea polyphenols under the reported reference lower the limit of safe dose and avoid fasting taking.

2.10 Conclusion

Originating from China, tea has gained the world's taste in the course of 5,000 years. Green tea has been discovered to have health-beneficial effects by ancient Chinese and is now considered one of the most promising dietary agents for the prevention and treatment of many diseases. For being studied extensively worldwide, numerous studies in a variety of experiments showed that green tea polyphenols (GTPs) were the staff of green tea's healthy effects.

As shown in this chapter, GTPs have been the focus of research owing to their multiple protective effects in anti-cancer and other diseases such as diabetes, and neurological and cardiovascular diseases. With more and more research studying on GTPs, the chemical structure, the metabolism in tea plants, and the function *in vivo* were clarified to instruct the employment of GTPs in both processing, human health, and the food industry, etc.

References

Ahn HY, Kim CH, Ha TS. (2010) Epigallocatechin-3-gallate regulates NADPH oxidase expression in human umbilical vein endothelial cells. *Korean J Physiol Pharmacol*, **14**(5), 325–329.

Ben Lagha A, Haas B, Grenier D. (2017) Tea polyphenols inhibit the growth and virulence properties of Fusobacterium nucleatum. *Sci Rep*, **7**(1), 1–10.

Britsch L, Heller W, Grisebach H. (1981) Conversion of flavanone to flavone, dihydroflavonol and flavonol with an enzyme system from cell cultures of parsley. *Zeitschrift für Naturforschung C*, **36**(9–10), 742–750.

Buer CS, Imin N, Djordjevic MA. (2010) Flavonoids: New roles for old molecules. *J Integr Plant Biol*, **52**(1), 98–111.

Chen D, Daniel KG, Kuhn DJ, *et al.* (2004) Green tea and tea polyphenols in cancer prevention. *Front Biosci*, **9**(4), 2618–2631.

Chen L, Cao H, Huang Q, *et al.* (2021) Absorption, metabolism and bioavailability of flavonoids: a review. *Crit Rev Food Sci Nutr*, 1–13.

Chengelis CP, Kirkpatrick JB, Regan KS, *et al.* (2008) 28-Day oral (gavage) toxicity studies of green tea catechins prepared for beverages in rats. *Food Chem Toxicol*, **46**(3), 978–989.

Chow HHS, Hakim IA. (2011) Pharmacokinetic and chemoprevention studies on tea in humans. *Pharmacol Resi*, **64**(2), 105–112.

Chow HHS, Hakim IA, Vining DR, *et al.* (2005) Effects of dosing condition on the oral bioavailability of green tea catechins after single-dose administration of Polyphenon E in healthy individuals. *Clin Cancer Res*, **11**(12), 4627–4633.

Chung FL. (1999) The prevention of lung cancer induced by a tobacco-specific carcinogen in rodents by green and black tea. *Proceedings of the Society for Experimental Biology and Medicine*, **220**(4), 244–248.

Cukovica D, Ehlting J, Ziffle JAV, *et al.* (2001) Structure and evolution of 4-coumarate: Coenzyme A ligase (4CL) gene families. *Biol Chem*, **382**(4), 645–654.

Dai X, Liu Y, Zhuang J, *et al.* (2020) Discovery and characterization of tannase genes in plants: roles in hydrolysis of tannins. *New Phytol*, **226**(4), 1104–1116.

de Vetten N, ter Horst J, van Schaik HP, *et al.* (1999) A cytochrome b_5 is required for full activity of flavonoid 3′,5′-hydroxylase, a cytochrome P450 involved in the formation of blue flower colors. *Proceedings of the National Academy of Sciences*, **96**(2), 778–783.

Deb G, Thakur VS, Limaye AM, *et al.* (2015) Epigenetic induction of tissue inhibitor of matrix metalloproteinase-3 by green tea polyphenols in breast cancer cells. *Mol Carcinog*, **54**(6), 485–499.

Deng Q, Xu J, Yu B, *et al.* (2010) Effect of dietary tea polyphenols on growth performance and cell-mediated immune response of post-weaning piglets under oxidative stress. *Arch of Anim Nutr*, **64**(1), 12–21.

Devika PT, Prince PSM. (2008) Protective effect of (−)-epigallocatechin-gallate (EGCG) on lipid peroxide metabolism in isoproterenol induced myocardial infarction in male Wistar rats: a histopathological study. *Biomed Pharmacother*, **62**(10), 701–708.

Ezhilarasi PN, Karthik P, Chhanwal N, *et al.* (2013) Nanoencapsulation techniques for food bioactive components: a review. *Food Bioproc Tech*, **6**(3), 628–647.

Forkmann G. (1991) Flavonoids as flower pigments: the formation of the natural spectrum and its extension by genetic engineering. *Plant Breeding*, **106**(1), 1–26.

Gross GG, Zenk MH. (1974) Isolation and properties of hydroxycinnamate: CoA ligase from lignifying tissue of Forsthia. *European J Biochem*, **42**(2), 453–459.

Guo XY, Lv YQ, Ye Y, *et al.* (2021) Polyphenol oxidase dominates the conversions of flavonol glycosides in tea leaves. *Food Chem*, **339**, 128088.

Hannig C, Spitzmüller B, Hannig M. (2009) Characterisation of lysozyme activity in the in situ pellicle using a fluorimetric assay. *Clin Oral Investig*, **13**(1), 15–21.

Haslam E. (1982) The Flavonoids. Chapman and Hall, London.

He L, Zou L, Yang Q, *et al.* (2016) Antimicrobial activities of nisin, tea polyphenols, and chitosan and their combinations in chilled mutton. *J Food Sci*, **81**(6), M1466–M1471.

Heller W, Hahlbrock K. (1980) Highly purified "flavanone synthase" from parsley catalyzes the formation of naringenin chalcone. *Arch Biochem Biophys*, **200**(2), 617–619.

Hsu YW, Tsai CF, Chen WK, *et al.* (2011) A subacute toxicity evaluation of green tea (*Camellia sinensis*) extract in mice. *Food Chem Toxicol*, **49**(10), 2624–2630.

Huang SW, Frankel EN. (1997) Antioxidant activity of tea catechins in different lipid systems. *J Agric Food Chem*, **45**(8), 3033–3038.

Isbrucker RA, Bausch J, Edwards, JA, Wolz E. (2006a) Safety studies on epigallocatechin gallate (EGCG) preparations. Part 1: Genotoxicity. *Food Chem Toxicol*, **44**, 626–635.

Isbrucker RA, Edwards JA, Wolz E, Davidovich A, Bausch J. (2006b). Safety studies on epigallocatechin gallate (EGCG) preparations. Part 2: Dermal, acute and short-term toxicity studies. *Food Chem Toxicol*, **44**, 636–650.

Jia S, Huang Z, Lei Y, *et al.* (2018) Application of Illumina-MiSeq high throughput sequencing and culture-dependent techniques for the identification of microbiota of silver carp (Hypophthalmichthys molitrix) treated by tea polyphenols. *Food Microbiol*, **76**, 52–61.

Kan H, Onda M, Tanaka N, *et al.* (1996) Effect of green tea polyphenol fraction on 1, 2-dimethylhydrazine (DMH)-induced colorectal carcinogenesis in the rat. *Nihon Ika Daigaku Zasshi*, **63**(2), 106–116.

Katiyar SK, Bergamo BM, Vyalil PK, *et al.* (2001) Green tea polyphenols: DNA photodamage and photoimmunology. *J Photochem Photobiol B Biol*, **65**(2–3), 109–114.

Kirita M, Honma D, Tanaka Y, *et al.* (2010) Cloning of a novel O-methyltransferase from Camellia sinensis and synthesis of O-methylated EGCG and evaluation of their bioactivity. *J Agric Food Chem*, **58**(12), 7196–7201.

Koes RE, Quattrocchio F, Mol JNM. (1994) The flavonoid biosynthetic pathway in plants: function and evolution. *BioEssays*, **16**(2), 123–132.

Kuhnert N, Dairpoosh F, Yassin G, *et al.* (2013) What is under the hump? Mass spectrometry based analysis of complex mixtures in processed food–lessons from the characterisation of black tea thearubigins, coffee melanoidines and caramel. *Food Funct*, **4**(8), 1130–1147.

Kumamoto M, Sonda T, Nagayama K, *et al.* (2001) Effects of pH and metal ions on antioxidative activities of catechins. *Biosci, Biotechnol Biochem*, **65**(1), 126–132.

Kuriyama S. (2008) The relation between green tea consumption and cardiovascular disease as evidenced by epidemiological studies. *J Nutr*, **138**(8), 1548S–1553S.

Lam PY, Zhu FY, Chan WL, *et al.* (2014) Cytochrome P450 93G1 is a flavone synthase II that channels flavanones to the biosynthesis of tricin O-linked conjugates in rice. *Plant Physiol*, **165**(3), 1315–1327.

Lambert JD, Yang CS, Gelderblom WCA, *et al.* (2009) Tea and its constituents. In *Chemoprevention of Cancer and DNA Damage by Dietary Factors*, 595–633.

Lambert JD, Lee MJ, Diamond L, *et al.* (2006) Dose-dependent levels of epigallocatechin-3-gallate in human colon cancer cells and mouse plasma and tissues. *Drug Metab Dispos*, **34**(1), 8–11.

Lan W, Zhang R, Ahmed S, *et al.* (2019) Effects of various antimicrobial polyvinyl alcohol/tea polyphenol composite films on the shelf life of packaged strawberries. *Lebensmittel-Wissenschaft & Technologie*, **113**, 108297.

Leslie EM, Deeley RG, Cole SPC. (2005) Multidrug resistance proteins: role of P-glycoprotein, MRP1, MRP2, and BCRP (ABCG2) in tissue defense. *Toxicol Appl Pharmacol*, **204**(3), 216–237.

Li Y, Kim JI, Pysh L, *et al.* (2015) Four isoforms of Arabidopsis 4-coumarate: CoA ligase have overlapping yet distinct roles in phenylpropanoid metabolism. *Plant Physiol*, **169**(4), 2409–2421.

Liu Y, Gao L, Liu L, *et al.* (2012) Purification and characterization of a novel galloyltransferase involved in catechin galloylation in the tea plant (Camellia sinensis). *J Biol Chem*, **287**(53), 44406–44417.

Martín-Diana AB, Rico D, Barry-Ryan C. (2008) Green tea extract as a natural antioxidant to extend the shelf-life of fresh-cut lettuce. *Innov Food Sci Emerg Technol*, **9**(4), 593–603.

Mukhtar H, Ahmad N. (1999) Green tea in chemoprevention of cancer. *Toxicol Sci*, **52**(Suppl 1), 111–117.

Mukhtar H, Ahmad N. (2000) Tea polyphenols: prevention of cancer and optimizing health. *Am J Clin Nutr*, **71**(6), 1698S–1702S.

Nkhili E, Tomao V, El Hajji H, *et al.* (2009) Microwave-assisted water extraction of green tea polyphenols. *Phytochem Anal*, **20**(5), 408–415.

Ogura R, Ikeda N, Yuki K, *et al.* (2008) Genotoxicity studies on green tea catechin. *Food Chem Toxicol*, **46**(6), 2190–2200.

Ouyang Q, Kou F, Zhang N, *et al.* (2019) Tea polyphenols promote Fenton-like reaction: pH self-driving chelation and reduction mechanism. *Chem Eng J*, **366**, 514–522.

Pasrija D, Anandharamakrishnan C. (2015) Techniques for extraction of green tea polyphenols: a review. *Food Bioproc Tech*, **8**(5), 935–950.

Punyasiri PAN, Abeysinghe ISB, Kumar V, *et al.* (2004) Flavonoid biosynthesis in the tea plant Camellia sinensis: properties of enzymes of the prominent epicatechin and catechin pathways. *Arch Biochem Biophys*, **431**(1), 22–30.

Ramsaha S, Aumjaud BE, Neergheen-Bhujun VS, *et al.* (2015) Polyphenolic rich traditional plants and teas improve lipid stability in food test systems. *J Food Sci Technol*, **52**(2), 773–782.

Rehage M, Delius J, Hofmann T, *et al.* (2017) Oral astringent stimuli alter the enamel pellicle's ultrastructure as revealed by electron microscopy. *J Dent*, **63**, 21–29.

Roberts EAH. (1958) The chemistry of tea manufacture. *J Sci Food Agric*, **9**(7), 381–390.

Sabaghi M, Maghsoudlou Y, Khomeiri M, *et al.* (2015) Active edible coating from chitosan incorporating green tea extract as an antioxidant and antifungal on fresh walnut kernel. *Postharvest Biol Technol*, **110**, 224–228.

Saleh IG, Ali Z, Abe N, *et al.* (2013) Effect of green tea and its polyphenols on mouse liver. *Fitoterapia*, **90**, 151–159.

Sarma DN, Barrett ML, Chavez ML, *et al.* (2008) Safety of green tea extracts. *Drug Saf*, **31**(6), 469–484.

Schmidt M, Schmitz HJ, Baumgart A, *et al.* (2005) Toxicity of green tea extracts and their constituents in rat hepatocytes in primary culture. *Food Chem Toxicol*, **43**(2), 307–314.

Sheng R, Gu ZL, Xie ML. (2013) Epigallocatechin gallate, the major component of polyphenols in green tea, inhibits telomere attrition mediated cardiomyocyte apoptosis in cardiac hypertrophy. *Int J Cardiol*, **162**(3), 199–209.

Singh K, Rani A, Kumar S, *et al.* (2008) An early gene of the flavonoid pathway, flavanone 3-hydroxylase, exhibits a positive relationship with the concentration of catechins in tea (Camellia sinensis). *Tree Physiol*, **28**(9), 1349–1356.

Siripatrawan U, Noipha S. (2012) Active film from chitosan incorporating green tea extract for shelf life extension of pork sausages. *Food Hydrocoll*, **27**(1), 102–108.

Song M, Teng Z, Li M, *et al.* (2017) Epigallocatechin gallate inhibits Streptococcus pneumoniae virulence by simultaneously targeting pneumolysin and sortase A. *J Cell Mol Med*, **21**(10), 2586–2598.

Srivastava RC, Husain MM, Hasan SK, *et al.* (2000) Green tea polyphenols and tannic acid act as potent inhibitors of phorbol ester-induced nitric oxide generation in rat hepatocytes independent of their antioxidant properties. *Cancer Lett*, **153**(1–2), 1–5.

Stafford HA, Lester HH. (1985) Flavan-3-ol biosynthesis: The conversion of (+)-dihydromyricetin to its flavan-3, 4-diol (leucodelphinidin) and to (+)-gallocatechin by reductases extracted from tissue cultures of Ginkgo biloba and Pseudotsuga menziesii. *Plant Physiol*, **78**(4), 791–794.

Stracke R, De Vos RCH, Bartelniewoehner L, *et al.* (2009) Metabolomic and genetic analyses of flavonol synthesis in Arabidopsis thaliana support the in vivo involvement of leucoanthocyanidin dioxygenase. *Planta*, **229**(2), 427–445.

Surh YJ, Chun KS, Cha HH, *et al.* (2001) Molecular mechanisms underlying chemopreventive activities of anti-inflammatory phytochemicals: Down-regulation of COX-2 and iNOS through suppression of NF-κB activation. *Mutat Res-Fund Mol M*, **480**, 243–268.

Suzuki K, Kadowaki H, Hino M, *et al.* (2002) The influence of green tea in pig feed on meat production and quality. *Japanese Journal of Swine Science (Japan)*, **39**, 59–65.

Vaidyanathan JB, Walle T. (2003) Cellular uptake and efflux of the tea flavonoid (-) epicatechin-3-gallate in the human intestinal cell line Caco-2. *J Pharmacol Exp Ther*, **307**(2), 745–752.

Wang YF, Shao SH, Xu P, *et al.* (2011) Catechin-enriched green tea extract as a safe and effective agent for antimicrobial and anti-inflammatory treatment. *Afr J Pharm Pharmacol*, **5**(12), 1452–1461.

Watts KT, Mijts BN, Lee PC, *et al.* (2006) Discovery of a substrate selectivity switch in tyrosine ammonia-lyase, a member of the aromatic amino acid lyase family. *Chem Biol*, **13**(12), 1317–1326.

Wei R, Mao L, Xu P, *et al.* (2018) Suppressing glucose metabolism with epigallocatechin-3-gallate (EGCG) reduces breast cancer cell growth in preclinical models. *Food Funct*, **9**(11), 5682–5696.

Wiseman SA, Balentine DA, Frei B. (1997) Antioxidants in tea. *Crit Rev Food Sci Nutr*, **37**(8), 705–718.

Xi D, Liu C, Su YC. (2012) Effects of green tea extract on reducing Vibrio parahaemolyticus and increasing shelf life of oyster meats. *Food Control*, **25**(1), 368–373.

Xie DY, Sharma SB, Paiva NL, *et al.* (2003) Role of anthocyanidin reductase, encoded by BANYULS in plant flavonoid biosynthesis. *Science*, **299**(5605), 396–399.

Yang CS, Zhang J, Zhang L, *et al.* (2016) Mechanisms of body weight reduction and metabolic syndrome alleviation by tea. *Mol Nutr Food Res*, **60**(1), 160–174.

Yang CS, Pan E. (2012) The effects of green tea polyphenols on drug metabolism. *Expert Opin Drug Metab Toxicol*, **8**(6), 677–689.

Yao S, Liu Y, Zhuang J, *et al.* (2022) Insights into acylation mechanisms: co-expression of serine carboxypeptidase-like acyltransferases and their non-catalytic companion paralogs. *Plant J*, **2022**.

Ye S, Zhu Z, Wen Y, *et al.* (2019) Facile and green preparation of pectin/cellulose composite films with enhanced antibacterial and antioxidant behaviors. *Polymers*, **11**(1), 57.

Ying L, Kong D, Gao Y, *et al.* (2018a) In vitro antioxidant activity of phenolic-enriched extracts from Zhangping Narcissus tea cake and their inhibition on growth and metastatic capacity of 4T1 murine breast cancer cells. *J Zhejiang Uni-Sci B*, **19**(3), 199–210.

Ying L, Yan F, Williams BRG, *et al.* (2018b) (-)-Epigallocatechin-3-gallate and EZH 2 inhibitor GSK 343 have similar inhibitory effects and mechanisms of action on colorectal cancer cells. *Clin Exp Pharmacol Physiol*, **45**(1), 58–67.

Zhang R, Liu Y, Li Y, *et al.* (2020) Polyvinylidene fluoride membrane modified by tea polyphenol for dye removal. *J Mater Sci*, **55**(1), 389–403.

Zhang W, Jiang W. (2020) Antioxidant and antibacterial chitosan film with tea polyphenols-mediated green synthesis silver nanoparticle via a novel one-pot method. *Int J Biol Macromol*, **155**, 1252–1261.

Zhang Y, Lv H, Ma C, *et al.* (2015) Cloning of a caffeoyl-coenzyme A O-methyltransferase from *Camellia sinensis* and analysis of its catalytic activity. *J Zhejiang Uni-Sci* , **16**(2), 103–112.

Zhao Y, Chen X, Jiang J, *et al.* (2020b) Epigallocatechin gallate reverses gastric cancer by regulating the long noncoding RNA LINC00511/miR-29b/KDM2A axis. *Biochim Biophys Acta Mol Basis Dis*, **1866**(10), 165856.

Zhao J, Li P, Xia T, *et al.* (2020a) Exploring plant metabolic genomics: chemical diversity, metabolic complexity in the biosynthesis and transport of specialized metabolites with the tea plant as a model. *Crit Rev Biotechnol*, **40**(5), 667–688.

Zhao LG, Li HL, Sun JW, *et al.* (2017) Green tea consumption and cause-specific mortality: Results from two prospective cohort studies in China. *J Epidemiol*, **27**(1), 36–41.

Zhou J, Mao L, Xu P, *et al.* (2018) Effects of (−)-epigallocatechin gallate (EGCG) on energy expenditure and microglia-mediated hypothalamic inflammation in mice fed a high-fat diet. *Nutrients*, **10**(11), 1681.

Zhou ZD, Xie SP, Saw WT, *et al.* (2019) The therapeutic implications of tea polyphenols against dopamine (DA) neuron degeneration in Parkinson's disease (PD). *Cells*, **8**(8), 911.

Zimmermann R, Delius J, Friedrichs J, *et al.* (2019) Impact of oral astringent stimuli on surface charge and morphology of the protein-rich pellicle at the tooth–saliva interphase. *Colloids Surf B*, **174**, 451–458.

https://doi.org/10.1142/9789811274213_0003

Chapter 3

Polymerized Tea Polyphenols (Thearubigin and Theaflavin)

Youying Tu

Tea Research Institute of Zhejiang University, Huangzhou, Zhejiang province, China

3.1 Introduction

The growing interest in the potential health benefits of tea, together with its popularity as a beverage, have prompted a large number of investigations on the chemical constituents of tea and their biological activities. The considerable works were focused on green tea whereas reports on black tea are relatively less. The black tea soup has a bright red color, a rich aroma, and a strong mellow taste. Black tea is one of the most popular drinks not only in China, but also in western countries, especially in the United Kingdom. It is reported that it has many health benefits. Black tea can help gastrointestinal digestion, promote appetite, diuresis, eliminate edema, and strengthen the heart function. Recently, more studies on black tea and its polyphenols are being carried out, such as studying the formation pathways, structures and functions of theaflavins (TFs), and theaflavin monomers. The production, separation, and extraction process of theaflavins, the antioxidant properties and other biological functions of theaflavins. In this chapter, we will discuss these issues. Since there are few studies on thearubigin, this chapter mainly discusses thearubigin. A large part of the content of this chapter comes from the research work of our laboratory, and some from the literature.

Polyphenols in tea are important active substances in tea. They are the general name of various phenolic compounds in tea and the main component of fresh tea leaves. Catechins are the main component, accounting for 60–80% of the total polyphenols. They are closely related to the color, taste, and aroma of tea. The water-soluble products of polymerized tea polyphenols are theaflavins, thearubigins, and theabrownine. Theafuscin is the general name of a class of products with very complex structure formed by the oxidation and polymerization of catechins. During the processing of Pu'er tea, 80% of theaflavins and thearubigins are oxidized and polymerized to form theabrownine, which doubles its content, thus significantly reducing the astringency and bitter taste of the tea soup. In addition, the high content of sugar and soluble water extracts forms the material basis for the mellow taste of Pu'er tea and the red brown color of the soup.

From the tea color, theaflavin is an important component of "bright" color, but it is greatly reduced due to oxidative polymerization during tea processing. Thearubigin is an important component of black tea soup color "red", the main substance of taste intensity, and is related to the concentration of tea soup; theabrownine element is the main reason for the dark color of the soup. When the content of theabrownine element reaches 6–8%, the color of the soup can be red brown and bright. When the content of theabrownine is lower than 5%, it means that the fermentation is insufficient, and the soup color is red and orange bright.

Thearubigins is a kind of orange-brown pigment in black tea, which is the product of tea fermentation. It is an important component of the color "red" of Pu'er tea soup, the main substance of taste intensity, and is related to the concentration of tea soup. Theabrownine is a strong antioxidant, which can help the aging body resist biological oxidation. In biochemistry, thearubigin is a kind of heterogeneous red or brown-red phenolic substances with great molecular differences, but it is difficult to extract. Thearubigin, an orange-brown compound, accounts for 6–8% of the weight of dry tea. Theabrownine plays a certain role in the taste, color, and luster of tea soup. Thearubigin accounts for about 35% of the total color, and also plays an important role in the browning of finished tea.

Theaflavins increase the level of leal conjugated bile acids (Bas), thereby inhibiting the intestinal FXR-FGF15 signal pathway, leading to increased liver production and fecal excretion of Bas, decreased liver cholesterol, and reduced fat production [Huang *et al.*, 2009].

3.2 Theaflavins and thearubigins

Theaflavin was first discovered by Roberts in 1957 during the fermentation of black tea. It is a kind of yellow substance that can be dissolved in ethyl acetate and is formed from the oxidation and condensation of tea polyphenols. Theaflavin not only plays an important role in the color and taste of black tea, but also plays important roles such as anti-oxidation, anti-cancer, anti-inflammatory, and anti cardiovascular and cerebrovascular diseases. It is the soul of black tea and has a variety of health functions. Theaflavins exist in all fermented and semi fermented teas, such as black tea, oolong tea, yellow tea, etc. Theaflavins are called "soft gold".

3.2.1 *Theaflavins definition and its functions*

Theaflavins are orange or orange-red in color and possess a benzo-tropolone skeleton which is formed from the co-oxidation of selected pairs of catechins, one with a vic-trihydroxyphenyl moiety and the other with an ortho-dihydroxyphenyl structure [Geissman, 1962; Runeckles *et al.*, 1972]. In addition to the four major theaflavins such as theaflavin (TF1), theaflavin 3-gallate (TF2A), theaflavin 3′-gallate (TF2B) and theaflavin 3,3′-digallate (TFDG), some of other stereoisomers of theaflavins and a number of theaflavin derivatives, including theaflavic acids and theaflavates from black tea, have also been reported [Collier *et al.*, 1973; Wan *et al.*, 1999; Lewis *et al.*, 1998]. It is known that theaflavins, which account for 2–6% of the dry weight of solids in brewed black tea [Takino *et al.*, 1964], contribute greatly to the quality of tea in terms of color [Roberts, 1962], "mouthfeel" [Millin *et al.*, 1969], and extent of tea cream formation [Powell, 1993]. Recently, theaflavins have attracted considerable interest because of their potential benefits for human health, including antimutagenicity [Apostolides *et al.*, 1997], suppression of cytochrome P450 1A1 in cell culture [Feng *et al.*, 2002], anticlastogenic effects in bone marrow cells of mice [Gupta *et al.*, 2001], suppression of extracellular signals and cell pro-liferation [Liang *et al.*, 1999], and anti-inflammatory and cancer chemical preventive action [Pan *et al.*, 2000]. Theaflavins also scavenge H_2O_2 [Lin *et al.*, 2000] and have been shown to inhibit lipid oxidation [Shiraki *et al.*, 1994; Yoshino *et al.*, 1994], LDL oxidation [Clark *et al.*, 1998], DNA oxidative damage, and xanthine oxidase activity [Lin *et al.*, 2000].

To date, more than 28 theaflavin (TF) derivatives have been isolated. TF complexes were generally believed to be the major antioxidant constituents of black tea, inhibiting free radical generation [Miller *et al.*, 1996], inhibiting pro-oxidative enzyme activities [Lin *et al.*, 1999; Lin *et al.*, 2000; Yang *et al.*, 2008], and chelating transition metal ions to prevent lipid peroxidation *in vitro* and *in vivo* [Rice-Evans *et al.*, 1997]. TF3 has also been shown to possess a higher antioxidative activity than catechins, including (–)-epigallocatechin gallate (EGCG) in HL-60 cells [Lin *et al.*, 2000; Yang *et al.*, 2008]. However, the differences among the four main TF derivatives in scavenging reactive oxygen species (ROS) *in vitro* are still unknown, partly because of their low abundance and technical limitations in their separation process. The mixture of TF2A and TF2B was usually used in previous studies, due to the difficulty of separation of TF2A from TF2B. In the present work, we used the optimized semi-preparative HPLC separation system, and efficiently obtained four purified individual TFs. The scavenging abilities of these four TF derivatives for ROS (superoxide radical, singlet oxygen, hydrogen peroxide, and the hydroxyl radical) were systematically investigated *in vitro*, and their abilities in preventing DNA damage from the hydroxyl radical were also studied.

3.2.2 *Structures of theaflavins and theaflavin derivatives*

The structures and chemical molecular modes of the main theaflavins and theaflavin derivatives are showed as Figures 3-1 and 3-2.

3.2.3 *Properties of theaflavins*

PPO catalyzes the oxidation of catechins into quinones by consuming molecular oxygen, the quinones from the oxidation of B-ring ehydroxylated catechins condense with quinones arising from the B-ring trihydroxylated catechins to give different theaflavins and thearubigins. The major tea catechins include (–)epicatechin (EC), (–)epicatechin gallate

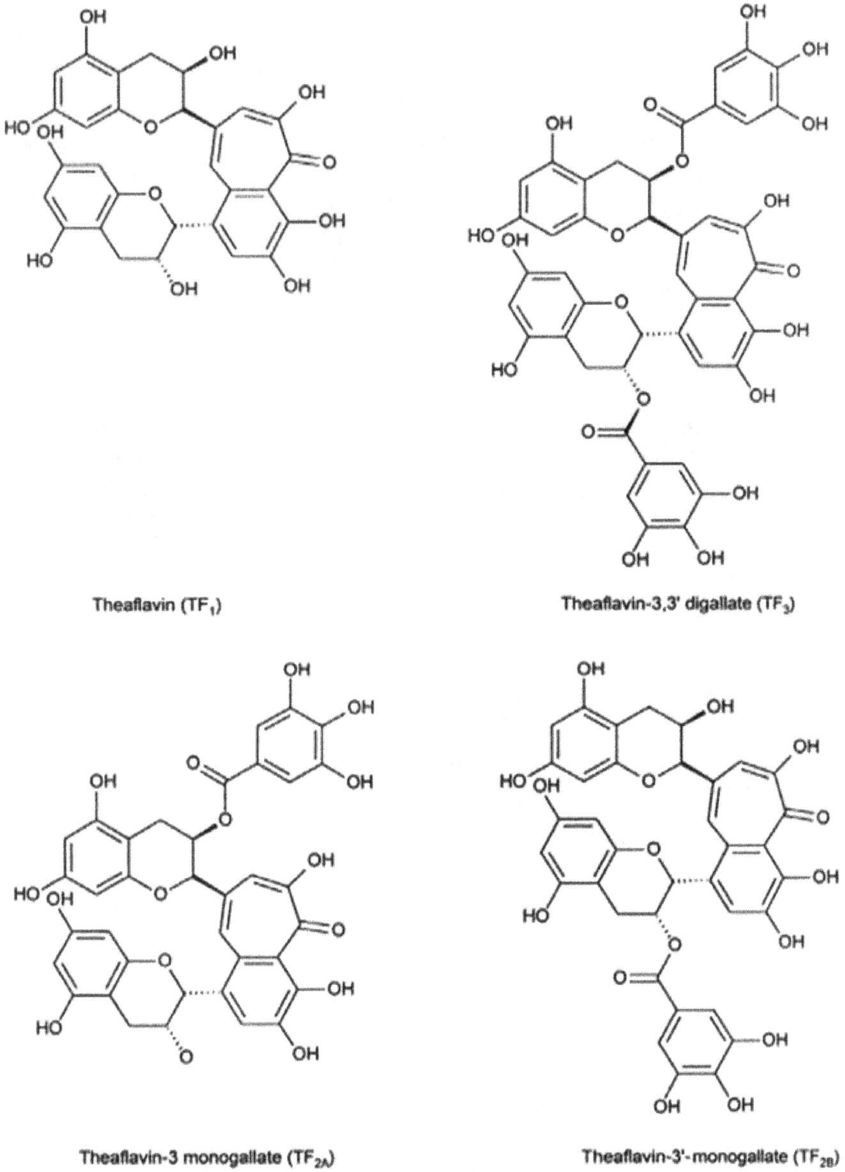

Theaflavin (TF₁)

Theaflavin-3,3′ digallate (TF₃)

Theaflavin-3 monogallate (TF₂ₐ)

Theaflavin-3′-monogallate (TF₂ᵦ)

Figure 3-1.   Structures of individual theaflavins and theaflavin derivatives.

(a) Theaflagallines/epitheaflagallines; (b) theaflavic/epitheaflavic acids; (c) theaflavin ($R^1 = R^2 = H$), theaflavin 3-gallate (R^1 = gallate, R^2 = H), theaflavin 3′-gallate (R^1 = H, R^2 = gallate), theaflavin 3,3′-digallate ($R^1 = R^2$ = gallate); (d) benzoditropolone; (e) theacitrin A.

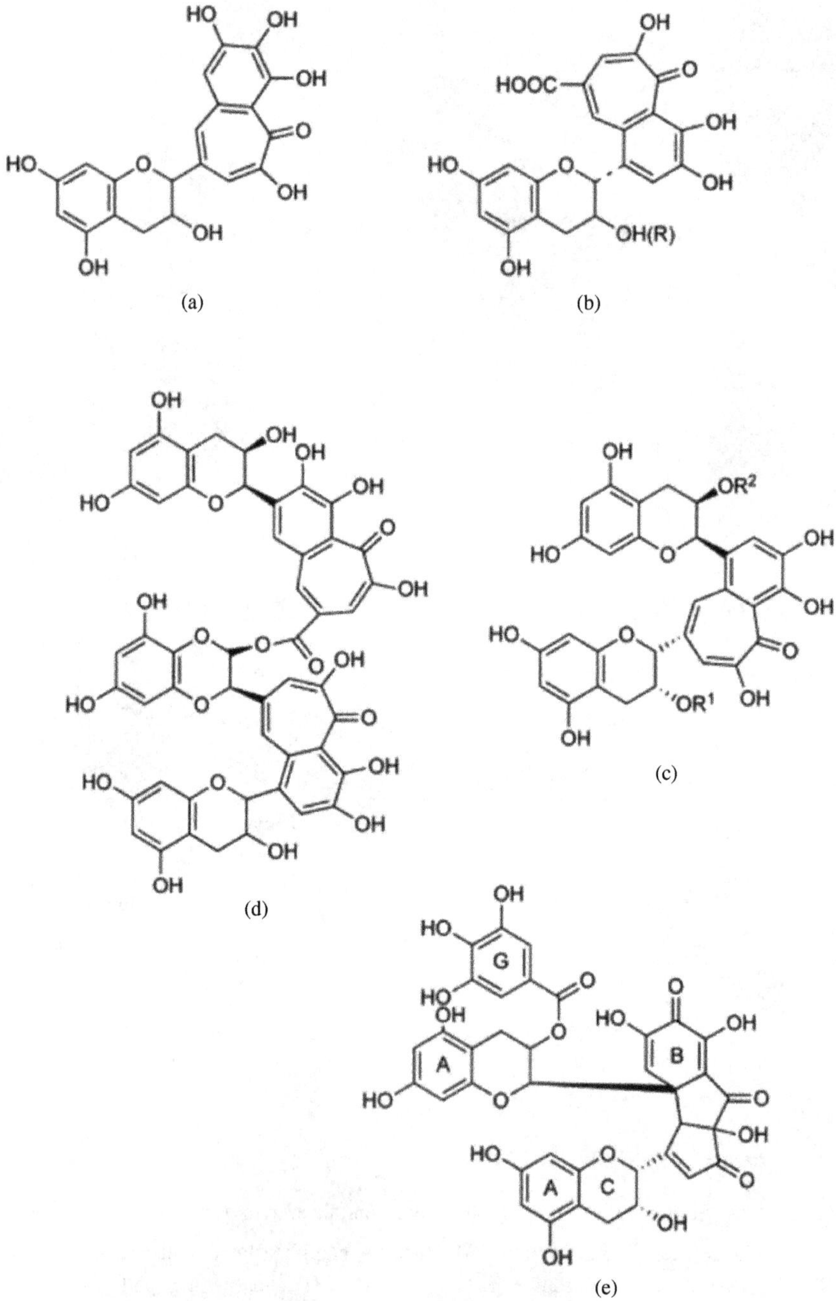

(a)

(b)

(c)

(d)

(e)

Figure 3-1.   (*Continued*)

(ECG), (+)catechin (C), (−)epigallocatechin (EGC), and (−)epigallocate-
chin gallate (EGCG). That is why the properties such as the maximum
absorption wavelength and the taste in black tea of TFs and TFs derivatives are obviously different. Some parameters of TFs and TFs derivatives
are showed in Table 3-1.

In our previous studies, we concerned with the extract method [Tu,
1994], functions, and properties of green tea polyphenols [Chen *et al.*,

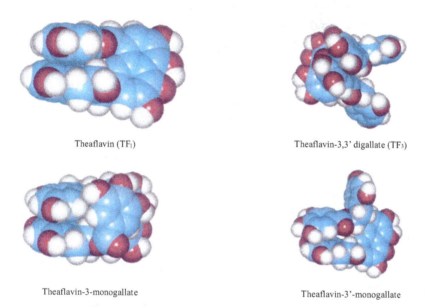

Theaflavin (TF₁)

Theaflavin-3,3' digallate (TF₃)

Theaflavin-3-monogallate

Theaflavin-3'-monogallate

Figure 3-2. Chemical molecular mode of TF1, TF2A, TF2B, TFDG [Clark *et al.*, 1998].

Table 3-1. Parameters of theaflavins derivatives [Yang *et al.*, 2003].

Component	Maximum absorption wavelength (nm)			Molecular weight	TF in black tea (%)	TF formula
TF	461	378	270	565	0.22	$C_{29}H_{24}O_{12}$
TF-3-G	455	376	272	717	0.77	$C3_6H_{28}O_{16}$
TF-3'-G	452	376	278	717	0.43	$C_{36}H_{28}O_{16}$
TF-3,3'-DG	455	378	278	869	0.63	$C_{43}H_{32}O_{20}$
TFA		398	278	593		
ETFA		400	280	593		
ETFA-3-G		398	279	745		

2002], and the role of polyphenol oxidase (PPO) in the black tea process-ing [Tu *et al.*, 1992], which provides a good basis for us to study the isola-tion of theaflavins and theaflavin monomers.

3.3 Biosynthesis of theaflavins and thearubigins

A large number of studies have been carried out on the composition of the products formed by the condensation of theaflavins and black tea fermen-tation, and the production and purification processes of theaflavins and thearubigins have been found. Tea polyphenols are the main substances condensed during black tea fermentation and the pigments formed during tea fermentation; various methods and technologies for separating theafla-vins and theaflavins from black tea have also been established. Enzymes play this important role in the process of tea fermentation, especially polyphenol oxidase, so it will be introduced and discussed as a focus. In addition, the pathway and mechanism of catechins forming TFs and TRs, and the preparation and separation of theaflavins from black tea by high-speed countercurrent chromatography will also be discussed.

3.3.1 *Phenol oxidase category*

According to the currently accepted enzyme nomenclature classes, hydroxylating phenol oxidase is called tyrosinase or phenolase [Yoshida *et al.*, 1999]. Phenol oxidase as monophenol monooxygenase (EC 1.14.18.1), o-diphenols oxidizing phenol oxidase as catechol oxidase, catecholase, diphenol oxidase, o-diphenolase, or polyphenol oxidase (EC 1.10. 3.1) [Rompel *et al.*, 1999], p-oxidizing phenol oxidase as laccase (EC 1.10.3.2), monophenol monooxygenase catalyzes hydroxylation of monophenols.

PPO exists in some bacteria and fungi, most plants, some arthropods, and all mammals. In all cases, PPO is associated with dark pigmentation in the organism and seems to have a protective function. Due to theafla-vins' low abundance and challenging purification procedure, previous research on theaflavins have focused on using mixtures. Alternatively, chemical synthesis could allow preparation of large quantities of pure compounds for biological assays. It is well known that PPO and peroxidase (POD) are key enzymes in pigment generation during the process of making black tea [Dix *et al.*, 1981]. Many studies have been carried out on the PPO-catalyzed formation of black tea oxidation products [Robertson and Bendall, 1983; Robertson, 1983a; Robertson, 1983b; Opie *et al.*, 1990; Opie *et al.*, 1993; Opie *et al.*, 1995]. Model oxidation systems have also been used to compare the oxidation products obtained from tea PPO with that obtained from horseradish POD.

PPO, a Cu-containing enzyme [Tsushida and Takeo, 1981; Mayer, 1987], is located in the chloroplast in tea (*Camelia sinensis*) leaves. It catalyzes the oxidation especially catechins into o-quinones and water, o-quinones are further polymerized into theaflavins and thearubigins. PPO is characterized by having the ability to act as a catalyst in two different reactions: monophenol hydroxylation into o-diphenols (hydroxylase activity) and dehydrogenation of o-dioxysubstituted polyphenols (catechol oxidase activity).

1. Molecular weight and specification of PPO

Phenol oxidase, found in the microbial and fungal kingdom, can be successfully used in biotransformation of phenols [Burton *et al.*, 1998]. Catechol oxidase of M. sterilia IBR 35 219/2 with molecular weight (MW) 240,000 vigorously catalyzed the oxidation of (–) EGC, (–) EGCG, and natural mixture of tea catechins. Pruidze *et al.* [2003] reported that it had the same substrate specificity with tea catechol oxidase with MW 250,000. This result gave a possibility to use the M. sterilia IBR 35 219/2 catechol oxidase in enzymatic processes while producing black tea.

The MW and specification of PPO in Kolkhida tea plant and sterilia IBR 35219/2 are studied in Table 3-2. Mushroom PPO is generally thought to contain four subunits with a total MW of 128 kDa. It is well known that the subunit of enzyme has the same function with the whole enzyme according to their MW and substrate specifications. The MW 118 kDa and MW 250 kDa protein are nearly twice and four times to compare with MW 58 kDa, respectively. That implied that there are four PPO subunits in Kolkhida tea plant, the lowest MW of the subunit of PPO protein is 58 kDa and the subunit of PPO displays the same substrate specifications with MW 118 kDa and MW 250 kDa.

Table 3-2.   Specification and optimum pH of PPO in Kolkhida tea plant and *sterilia IBR 35219/2*.

Origin	Kolkhida tea plant					sterilia IBR 35219/2			
MW ($\times 10^3$ kDa)	28	41	58	118	250	41	58	93	240
Substrate	–	–	–	–	–	–	–	–	–
Hydroxyase activity	–	–	–	–	–	–	–	–	–
Oxidation of o-diphenols	–	–	+	+	–	–	–	–	–
(-)EGC	–	–	+	+	+	–	–	–	+
(-)EGCG	–	–	+	+	+	–	–	–	+
Optimum pH			6.2–6.5					4.6–5.4	

2. PPO gene

The PPO gene has been cloned, the nucleotide acid sequence of PPO cDNA and deduced amino acid sequence are showed in the Figure 3-3 (primer regions are indicated with shadow) [Zhao *et al.*, 2001].

There are two conserved amino acid sequence regions in the PPO sequence of tea. The two regions seem to correspond to the active site of the enzyme and show good correlation with the accepted enzymatic mechanism, and most of histidines are present in these regions. Site-directed mutagenesis of histidine residues 62–189 is important in Cu binding.

3.3.2 *Pathways and mechanisms of TFs and TRs formation from catechins*

The catechins vary not only in chemical structures, but also in reduction potentials [Bajaj *et al.*, 1987]. These properties of catechins are either according to the number of hydroxyl groups on the B-ring or gallated or non-gallated [Sanderson *et al.*, 1976].

Seven equations (1)–(7) in which catechins and GA were oxidized into the correspondent theaflavins and theaflavins derivatives. C, ECG, EGC, EGCG, and GA are the main catechins in production of TFs and TFs derivatives (Fig. 3-4).

```
TGGCTCTTCTTTCCGTTCCATAGATTCTATCTCTACTTCTTCGAAAAGATTTTGGGAATGCTGCTCGAT      69
W  L  F  F  P  F  H  R  F  Y  L  Y  F  F  E  K  I  L  G  M  L  L  D         23
GATCCAGCGTTTGCAATTCCTTTTTGGAATTGGGATTCTCCGGCGGGCATGAAAATACCCGCCATGTAT     138
D  P  A  F  A  I  P  F  W  N  W  D  S  P  A  G  M  K  I  P  A  M  Y         46
GCGGACATAAATTCGCCACTCTATAACCGTCTCCGTGACGCCAAACACCAGCCACCGACATTAATTGAC     207
A  D  I  N  S  P  L  Y  N  R  L  R  D  A  K  H  Q  P  P  T  L  I  D         69
CTTGACTACAATTTGACTGACCCGAAAAATGTCGATGAGGAGAAGCAGAAGTTGAGAAATCTAACTATA     276
L  D  Y  N  L  T  D  P  K  N  V  D  E  E  K  Q  K  L  R  N  L  T  I         92
ATGTACCGACAAGTGGTGGCGGGTGGGAAAACGCCTCGGCTTTTCCTTGGAAGCTCGTACCGTGCGGGA     345
M  Y  R  Q  V  V  A  G  G  K  T  P  R  L  F  L  G  S  S  Y  R  A  G        115
GATGACCCGGACCCAGGTGCCGGGTCCCTGGAGAACATCCCGCATGGTCCGGTTCACATATGGTGCGGA     414
D  D  P  D  P  G  A  G  S  L  E  N  I  P  H  G  P  V  H  I  W  C  G        138
GACCGCACCCAGCCGAATCTAGAAGACATGGGGAACTTCTACTCTGGGGGACGAGATCCGATCTTCTAC     483
D  R  T  Q  P  N  L  E  D  M  G  N  F  Y  S  G  G  R  D  P  I  F  Y        161
GGTCATCACGCGAACGTCGATCGGATCTCGACGGTGTGGAAGACATTAGAAGGAAAACGAAACGATTTC     552
G  H  H  A  N  V  D  R  I  W  T  V  W  K  T  L  E  G  K  R  N  D  F        184
AAGGATCCGGATTGGTTGAATTCAGAGTTCACCTTTTACGACAAAAATGCTCAGGTTGTGACTGTAAAA     621
K  D  P  D  W  L  N  S  E  F  T  F  Y  D  K  N  A  Q  V  V  T  V  K        207
GTAAAAGAGAGTTTGGATCATCGAAAACTCGGCTACGTCTACCAAGACGTGGAAATTCCATGGCTAAAC     690
V  K  E  S  L  D  H  R  K  L  G  Y  V  Y  Q  D  V  E  I  P  W  L  N        230
ACTCGACCCAGTCCTCGTATTTCGAATTTTTTTCGAAAAATAAAGAACAAGGCCGGGAGAGCAATGGCG     759
T  R  P  S  P  R  I  S  N  F  F  R  K  I  K  N  K  A  G  R  A  M  A        253
ACAGAGACACTGGATTCTGCTGCCATTGTATTCCCAAGAAAGCTTGATGAGGTGGTGAAGGTGGTGGTG     828
T  E  T  L  D  S  A  A  I  V  F  P  R  K  L  D  E  V  V  K  V  V  V        276
AAGCGGCCGACGAAGTCGAGGAGGGGAGAGAGAGGAAGAAGAGGAGGAGGTGGTGGTAGTGGAGGGGATA     897
K  R  P  T  K  S  R  R  E  R  E  E  E  E  E  V  V  V  V  E  G  I           299
GAGGTGGAGAGAGATGTGTCTGTGAAGTTTGATGTGTTTATTAACGACGAAGACGAGGCGGCGAGTGGG     966
E  V  E  R  D  V  S  V  K  F  D  V  F  I  N  D  E  D  E  A  A  S  G        322
CCGGAGAACCACAGAGTTTGCCGGCAGCTTCGTCAGCGTC                                1006
P  E  N  H  R  V  C  R  Q  L  R  Q  R                                     335
```

Figure 3-3. Nucleotide acid sequence of PPO.

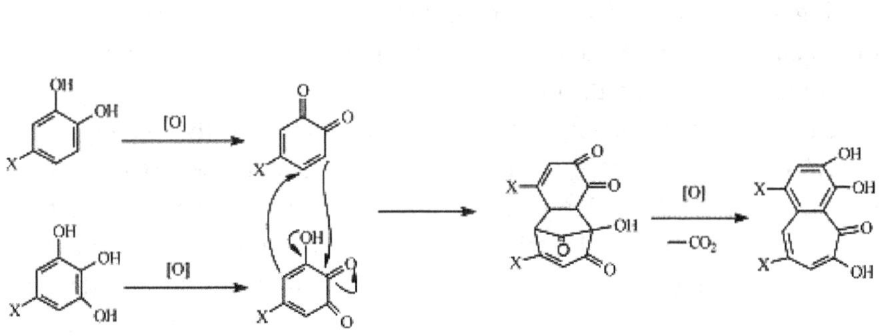

Figure 3-4. Possible mechanism of benzo para ketone structure formation.

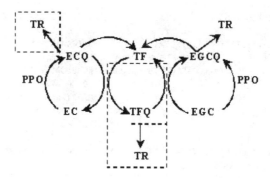

Figure 3-5. Pathway of TFs and TRs formation from catechins oxidation reactions [Goodsall *et al.*, 1996].

(A) Theaflavin
(1) (–)-EC + (–)-EGC → TF
(2) (–)-EC + (–)-EGCG → TF-3-G
(3) (–)-ECG + (–)-EGC → TF-3′-G
(4) (–)-ECG + (–)-EGCG → TF-3, 3′-DG
(B) Theaflavins derivatives
(5) (–)-EC + GA → (–)-ETFA
(6) (+)-C + GA → TFA
(7) (–)-ECG + GA → (–)-ETFA-3′-G

Figures 3-4 and 3-5 demonstrate that during black tea processing, the PPO catalyzes the oxidation of catechins into quinones by consuming molecular oxygen, and the quinones from the oxidation of B-ring hydroxylated catechins condense with quinones arising from the B-ring trihydroxylated catechins to give different theaflavins [Brown *et al.*, 1966; Brown *et al.*, 1969; Takino *et al.*, 1964]. Due to differences in reduction potential, quinones will also take part in the oxidation-reduction equilibration reactions during fermentation, causing different catechins to deplete at different rates [Bajaj *et al.*, 1987]. The B-ring trihydroxylated catechins (EGC and EGCG), deplete at much faster rates than the B-ring hydroxylated catechins (EC, ECG, C). The simple, non-gallated tea catechins EC, EGC, and C are not as astringent as the gallated catechins ECG and EGCG. Thus, the catechin composition changes as fermentation proceeds, the taste and colour characteristics of black tea also change [Owuer *et al.*, 1986a; 1986b; 1986c].

3.3.3 *Preparative separation of theaflavin monomers from black tea using high-speed counter-current chromatography*

High-speed counter-current chromatography (HSCCC) is an all-liquids chromatographic technique operated under gentle conditions and allows non-destructive isolation even of unstable natural compounds. Due to the absence of any solid stationary phase, adsorption losses are minimized, hence, a 100% sample recovery is guaranteed. HSCCC has been success-fully applied to the isolation of carotenoids from *Gardenia jasminoides* [Degenhardt *et al.*, 2001], flavonol glycosides from black tea, endive, and shallots [Degenhardt *et al.*, 2000a; 2001; 2004], catechins and proantho-cyanidins from black tea [Degenhardt *et al.*, 2000a], isoflavones from soy flour [Degenhardt and Winterhalter, 2001] and lignans from flaxseed [Degenhardt *et al.*, 2002]. It was also possible to isolate anthocyanins from various sources by using HSCCC, in amounting up to several hun-dred milligrams of pure compounds [Degenhardt *et al.*, 2000b; 2000c; 2001].

1. HSCCC separation procedure
Solvents are delivered by an HPLC pump and the sample is injected via sample loop. Separation takes place in a multilayer coil and is monitored by UV-Vis detection. Fractions are collected with a fraction collector. Three theaflavin monomers were separated from the oxidation product of polyphenols by IPPO using HSCCC. Separation columns made of polyte-trafluoroethylene (PTFE) tubing with a 260 ml capacity were used. Separation was performed with a two-phase solvent system composed of hexane-ethyl acetate-methanol-water (3:1:1:4, v/v/v/v) and (3:1:3:4, v/v/v/v) by eluting the lower aqueous phase at 1.2 ml/min at a revolution speed of 880 rpm. A total of 57 mg of 97.7% theaflavin, 26 mg of 61.2% theaflavin-3′-gallate, 64 mg of 84.5% theaflavin-3-3′-gallate, and 33 mg of 92.3% TFDG and TF2B complex were obtained from 250 mg of the crude TFs sample. HSCCC has been successfully used to separate the

complex compounds that have the similar chemical properties and as an all-liquid chromatographic technique operated under gentle conditions thus allowing non-destructive isolation of labile natural compounds. In the current study, HSCCC was used to separate TF monomer.

The HSCCC experiments were performed at a revolution speed of 880 rpm with a two-phase solvent system composed of hexane-ethyl acetate-methanol-water (3:1:1:4, v/v/v/v), where the upper organic phase was used as the stationary phase and the lower aqueous phase as the mobile phase. The upper organic phase were delivered by Amersham Pharmacia AKTA prime (including a pump, UV detector, fraction collector, and a recorder REC 112). The sample solution was prepared by dissolving the TFs complex in a 1:1 mixture of each phase and loaded into the column by loop injection and eluted with the aqueous phase at a flow-rate of 1.2 ml/min. The effluent was monitored with a UV-Vis detector at 254 nm and fractions were collected at every 5-minute interval.

2. HPLC analysis of catechins and theaflavins
A 0.4 mg tea polyphenol extract and 0.4 mg TFs complex were dissolved separately in 1 ml methanol each. The concentration of theaflavin and catechins was detected using a Shimadzu LC-2010A (Shimadzu, Kyoto, Japan) equipped with a 4.6×250 mm C18 column, the column was maintained at 40°C. The mobile phase was acetic acid/acetonitrile/water (0.5:30:69.5, v/v/v). Flow rate was 1 ml/ min, a 10 μl injection volume, and monitored at 280 nm. The TFs standard sample was used as an authentic standard and to identify peaks of TF monomer and calculate the concentration of TFs complex. Standard samples GA, C, EC, ECG, EGCG, and EGC were used to identify peaks and calculate the concentration of catechins in tea polyphenol extract and in TFs complex.

The HPLC chromatogram of TFs obtained from catechins oxidization is showed in Figure 3-6. The peak area of the obtained TFs sample was 67.5% as that of the standard TFs.

3. Isolation of the theaflavin monomers by HSCCC
The separation results of TFs complex by HSCCC method are showed in Figure 3-7. After some thearubigins were removed by using $NaHCO_3$, HSCCC separation profile appeared six peaks, suggesting that the

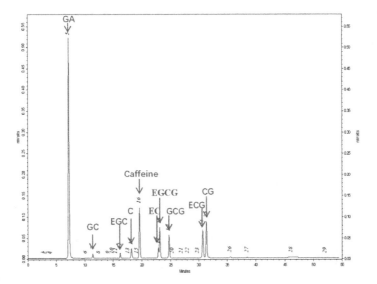

Figure 3-6.   HPLC chromatogram of catechins standard sample.

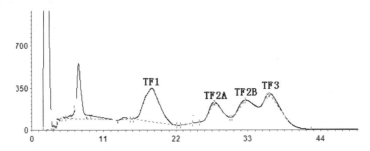

Figure 3-7.   HSCCC separation profile of TFs complex.

pre-treatment of TFs complex with $NaHCO_3$ was very necessary to improve the isolation efficiency of TF monomers.

4. HPLC analysis of the crude TF monomers

Six fractions were collected according to the HSCCC separation profile of TFs complex as shown in Figure 3-8, and each fraction was detected with HPLC and the component was analyzed.

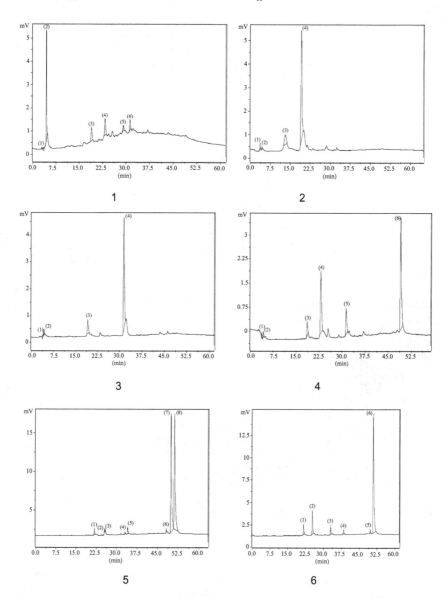

Figure 3-8. Responding HPLC chromatograms of six fractions of theaflavins elution from HSCCC.

The compounds in the six fractions were identified by comparing them with the HPLC chromatograms of the standard catechins and theaflavins. The compound in each fraction in the corresponding HPLC chromatograms were identified as the following: 1: GA (No.2), 2: ECG (No.4), 3: caffeine (No.4), 4: TF2B (No.6), 5: TFDG (No.7) and TF2B (No.8), and 6: TFDG (No.6).

Each fraction obtained from HSCCC was dried and weighed. A totle of 57 mg of 97.7% theaflavin, 26 mg of 61.2% theaflavin-3′-gallate, 64 mg of 84.5% theaflavin-3-3′-gallate, and 33 mg of 92.3% TFDG and TF2B complex were obtained from 250 mg of the crude TFs sample. Due to the similarity of the chemical properties of TFDG and TF2B, TF monomers could not be completely separated under one solvent system to obtain TF monomer, therefore, the solvent system was regulated, and the HSCCC separation profile of TFs under the regulated solvent system is presented in Figure 3-9. The HSCCC separation curve of TFs under the solvent system is shown in Figure 3-10. The corresponding HPLC chromatogram showed one peak, according to the retention time of TF monomer, the peak was identified as TF monomer, thus, 57 mg of 97.7% theaflavin was obtained. The result indicated that increasing the polarity of the solvent system TF monomer could be separated from TFs.

The solvent system was hexane-ethyl-acetate-methanol-water (3:1:3:4, v/v/v/v) and experimental conditions were: rotation speed: 880 rpm; mobile phase: lower aqueous phase; flow-rate: 1.2 ml/min; sample size: 250 mg; retention of the stationary phase: 51%.

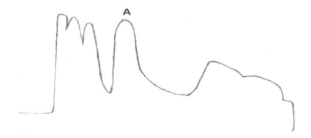

Figure 3-9. HSCCC separation profile of TF1.

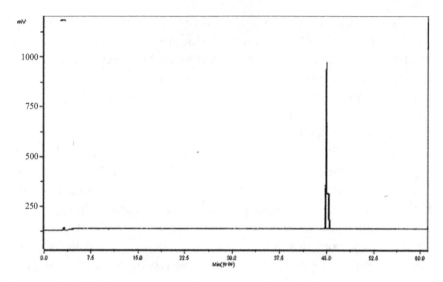

Figure 3-10.   HPLC chromatogram of fraction A (indicated in Fig. 3-9).

5. Selection of the optimum parameters of HSCCC separation system

Based on the above experiment results, various operating parameters, including injection volume and elution speed, were evaluated to optimize the HSCCC separation system [Xu *et al.*, 2010]. It was clear that the separation conditions of profile A and C were better than that of C and D profiles, even the elution speed was different between A and C (Fig. 3-11), which meant injection volume is more important than elution speed for separating TF monomer in this study. The optimum conditions for separating TF monomer in here was an injection volume of 15 ml, an elution speed of 1.2 ml/min at the same rotation speed of 880 rpm.

6. Result analysis

Gallic acid (GA) was separated successfully from the n-butanol extract of the fruit of *Cornus officinalis* Sieb. et Zucc by using HSCCC in two steps under two solvent systems composed of ethyl-acetate-ethanol n-butanol-water (5:1.8:6, v/v/v) and ethyl acetate-ethanol-water (5:0.5:6, v/v/v). From 1 g of n-butanol extract, 60 mg of gallic acid at a purity of 97% was produced [Tian *et al.*, 2000]. In the present study we used one step to remove GA by HSCCC, but we pretreated the TFs mixture by several

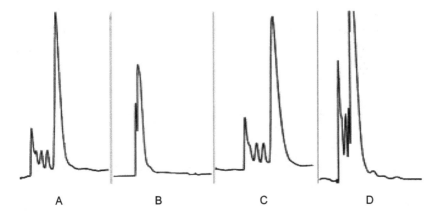

Figure 3-11. HSCCC separation profile of TFs in different separation conditions solvent system: hexane-ethyl-acetate-methanol-water (3:1:3:4, v/v/v/v) and at a rotation speed of 880 rpm.

A: injection volume: 15 ml, flow-rate: 1.2 ml/min
B: injection volume: 20 ml, flow-rate: 1.2 ml/min
C: injection volume: 15 ml, flow-rate: 1.0 ml/min
D: injection volume: 20 ml, flow-rate: 1.0 ml/min

steps and made the TFs partial purification, which made easier to remove GA from TFs mixture. From 250 mg of crude sample, the column yielded 57 mg of 97.7% theaflavin, 26 mg of 61.2% TF2B, 64 mg of 84.5% TFDG, and 33 mg of 92.3% TFDG and TF2B complexes. To get better separation of individual TF monomer, pretreatment with $NaHCO_3$ to remove thearubigin was helpful and of significance in practice.

Separated theaflavins of black tea by HSCCC, TFDG, TF2B, and TF2A were separated using one solvent system (hexane-ethyl-acetate-methanol-water (1:3:1:6, v/v/v/v)), but they had not showed the HPLC chromatogram of HSCCC separation fractions in the paper. Unfortunately, we have not gotten the same results to compare the HPLC chromatogram of TF monomers with the description of separation compounds by Du *et al.* [2001] although we got the same curve of HSCCC separation by using the same method. The similar principle and method was reported by Schwarz *et al.* [2003] for separating anthocyanins under two solvent systems. Anthocyanins are polar polymeric pigments, the purification of this kind of polar pigments is usually carried out by changing more or less polar solvent system. In the present study, we modified the solvent system more polar for separation TF1. However, TF2A had not obtained in the separation system, which suggests that solvent composition or some other conditions should be investigated further.

3.4 Antioxidant properties of theaflavins and thearubigins

Everyone knows that green tea polyphenols have strong antioxidant properties. In Chapter 2, we discussed the antioxidant properties of green tea polyphenols. So, does green tea polyphenols have antioxidant properties after fermentation and polymerization? We studied the antioxidant properties of four major theaflavin derivatives. The results showed that theaflavin could effectively scavenge superoxide anion radicals, hydroxyl radicals, and singlet oxygen, and inhibit the oxidation of hydrogen peroxide.

3.4.1 *Effects of TF derivatives on superoxide radical*

Superoxide is the precursor of many excited and toxic oxygen species. Usually, superoxide anions are transformed to hydrogen peroxide by superoxide dismutase (SOD) and SOD-like antioxidants; otherwise, oxidative damage would occur. In this work, a pyrogallol auto-oxidation system, used for superoxide anion generation, and chemiluminescence signals determination were applied for testing the inhibition effects of four main TF derivatives on superoxide anions (Table 3-3).

The effects of the test samples on the superoxide radical were analyzed, and the dose-response equations and IC_{50} values for TF derivatives (TF1, TF2A, TF2B, and TF3) and EGCG are shown in Table 3-1. The degree of superoxide radical scavenging increased as the sample concentrations increased. Their dose-dependent effects on superoxide anions quenching were detected at the range of 5.00–50.00 μmol/L. Among all test samples, EGCG had the highest IC_{50} value of 45.80 μmol/L, which presented the weakest inhibition of superoxide radical. All TF derivatives had lower IC_{50} values than EGCG. Among the TF

Table 3-3. Inhibition capacities of four main TF derivatives on the chemiluminescence signal caused by the superoxide radical.

Sample	Regression equation	R^2	Linear range (µmol/L)	IC_{50} (µmol/L)
EGCG	$y = 0.2059\text{Ln}(x) - 0.2875$	0.9674**	5.00–50.00	45.80
TF_1	$y = 0.2403\text{Ln}(x) - 0.1424$	0.9936**	5.00–50.00	14.50
TF_2A	$y = 0.2412\text{Ln}(x) - 0.2421$	0.9858**	5.00–50.00	21.70
TF_2B	$y = 0.2555\text{Ln}(x) - 0.2472$	0.9956**	5.00–50.00	18.60
TF_3	$y = 0.2551\text{Ln}(x) - 0.3382$	0.9925**	5.00–50.00	26.70

[a]y, inhibition rate (%); x, corresponding concentrations of different samples.
** $p < 0.01$.

derivatives, TF1 was the most efficient superoxide inhibitor with an IC_{50} value of 14.50 µmol/L, only 31.66% that of EGCG. The IC_{50} values of TF derivatives with monogallate and digallate were 40.61% to 47.38% that of EGCG (Table 3-3). The monogallate TFs were more efficient than TF3 in superoxide radical scavenging, with the scavenging capacity of TF2B> TF2A> TF3.

Polyphenols was used as a control. The half maximal inhibitory concentration (IC_{50}) was determined from the plotted graph of scavenging activity versus the concentration of the test samples, which was used to measure the efficiency of different compounds in inhibiting ROS.

3.4.2 *Effects of TF derivatives on singlet oxygen*

Singlet oxygen ($^1O_2$) is related to oxidation of LDL cholesterol and result-anticardiovascular effects [Wang *et al.*, 2002]. The generation of $^1O_2$ from lipid hydroperoxide involves a cyclic mechanism from a linear tetraoxide intermediate [Miyamoto *et al.*, 2003]. It is unstable and deleterious, and usually occurs from the degradation of biological systems. Three independent experiments were performed to determine the effect of the test samples on $^1O_2$ over a wide concentration range. All samples could scavenge $^1O_2$ at low concentrations, and their inhibition abilities were detected to increase with the increasing concentration from 0.50–10.00 µmol/L (Table 3-2). The IC_{50} value of EGCG was 0.87 µmol/L, it was almost similar to that of TF2A. TF2B had the lowest IC_{50} ($^1O_2$) value of 0.55 µmol/L, which was 63.22% that of EGCG. The scavenging ability of $^1O_2$ decreased in the order of TF2B > TF1 > TF3 > TF2A > EGCG. TF2B had the strongest scavenging activity of $^1O_2$ among all TF derivatives.

3.4.3 *Effects of TF derivatives on hydrogen peroxide*

The hydroxyl radical is one of the most harmful radicals in nature. The hydroxyl free radical can cause damage by cell oxidation, particularly erythrocytes (or red blood cells), DNA, lipids, and proteins. The scavenging activities of TF derivatives on hydrogen peroxide and the hydroxyl radical were determined with the chemiluminescence assay *in vitro* (Table 3-4). The effects of four main TF derivatives and EGCG on hydrogen peroxide and the hydroxyl radical increased in a dose-dependent manner. All TF derivatives efficiently inhibited hydrogen peroxide at low concentrations from 0.50–10.00 µmol/L, while they were effective on scavenging the hydroxyl radical at the lowest concentration of 10.00 µmol/L. Most TF derivatives had higher scavenging capacities on H_2O_2 and •OH than that of EGCG. TF2B was found to be equally effective to TF3 with the IC_{50} value of 0.39 µmol/L for quenching H_2O_2. The capacities of scavenging H_2O_2 were TF3 = TF2B > TF2A > TF1 > EGCG (Table 3-4). Meanwhile, TF3 was found to be the most effective •OH scavenger, followed by TF2B, TF2A, EGCG, and TF1.

Table 3-4.   Inhibition capacities of four main TF derivatives on hydrogen peroxide.[a]

Sample	Regression equation	R^2	Linear range (µmol/L)	IC_{50} (µmol/L)
EGCG	$y = 0.1598Ln(x) + 0.5653$	0.9632**	0.50–10.00	0.66
TF_1	$y = 0.1550Ln(x) + 0.6101$	0.9531**	0.50–10.00	0.49
TF_2A	$y = 0.1521Ln(x) + 0.6227$	0.9725**	0.50–10.00	0.45
TF_2B	$y = 0.1566Ln(x) + 0.6484$	0.9539**	0.50–10.00	0.39
TF_3	$y = 0.1562Ln(x) + 0.6452$	0.9500**	0.50–10.00	0.39

[a]y, inhibition rate (%); x, corresponding concentrations of different samples.
** $p < 0.01$.

3.4.4 *Inhibition of hydroxyl radicals and DPPH*

Percent hydroxyl radical scavenging activities of oxidized phenolic compounds were dose-dependent. Quenching ability of oxidized phenolic compounds on the CL signal, which indicated their potentials to scavenge hydroxyl radicals produced in the Fe (II)–H_2O_2–luminol system, was ranked by IC_{50}. The lower the IC_{50} value, the higher was the activity of oxidized phenolic compounds. From regression analysis of scavenging rate (%) and the natural logarithm of oxidized phenolic compounds concentration, a good linear relationship was observed, and the regression equations and correlation coefficients are listed in Table 3-5. With the regression equations derived, it was easy to calculate the IC_{50} values of each compound. Comparing the IC_{50} values of each sample (EGCG as a positive control), it appears that TF shows the greatest quenching of hydroxyl radicals, followed by TB and TR. TF acts as a competitive inhibitor and is the most potent inhibitor of hydroxyl radicals produced in the Fe(II)–H_2O_2–luminol system. Similar to the above experiment on hydroxyl radicals, TF, TR, and TB also showed a good linear relationship between scavenging rate (%) and the natural logarithm of oxidised phenolic concentrations in the DPPHassay. The order of DPPH scavenging ability was TF > TB > TR (Table 3-6), which was similar to the inhibition of hydroxyl radicals (Table 3-5). The main sites of catechin antioxidant

Table 3-5. Quenching ability of oxidized phenolic compunds to quench the chemiluminescence signal caused by hydroxyl radicals.

Sample	Regression equation	R^2	Linear range (µg/ml)	IC_{50} (µg/ml)
EGCG[b]	$y = 51.1x + 16.6$	0.988**	0.5–5.0	1.9
TF	$y = 22.3x + 13.1$	0.964**	0.5–25	5.2
TR	$y = 20.9x - 8.9$	0.903*	1.0–50	16.8
TB	$y = 21.5x - 2.5$	0.948**	1.0–50	11.4

[a]y, scavenging rate (%); x, natural logarithm values of corresponding concentrations of oxidized phenolic compounds.
[b]EGCG was used as a control.
*$p < 0.05$.
**$p < 0.01$.

Table 3-6.   DPPH-scavenging activity of oxidized phenolic compounds.

Sample	Regression equation	R^2	Linear range (µg/ml)	IC_{50} (µg/ml)
EGCG[b]	$y = 34.8x - 10.5$	0.998**	2.5–20	5.7
TF	$y = 32.2x - 17.5$	0.967**	2.5–20	8.1
TR	$y = 19.9x - 21.3$	0.895*	2.5–40	36.1
TB	$y = 26.8x - 27.4$	0.964**	2.5–40	18.0

[a]y, scavenging rate (%); x, natural logarithm values of corresponding concentrations of oxidized phenolic compounds.
[b]EGCG was used as a control.
*$p < 0.05$.
**$p < 0.01$.

action appear to be the o-dihydroxy B-ring, or vic-trihydroxy B-ring, or gallate group through the one electron transfer or H-atom abstraction mechanism [Jovanovic *et al.*, 1997]. Theaflavins posses a benzotropolone benzotropolone skeleton that is formed from co-oxidation of appropriate pairs of catechins, one with a vic-trihydroxy moiety and the other with an o-dihydroxy structure. Therefore, the chemical structures of theaflavins are bulkier and have two A-rings of flavanols linked by a fused seven-member ring. These structural features may provide more interaction sites for radicals, which could partly explain that TF have the most potent antioxidant activity among these oxidised phenolic compounds. Despite the fact that TR is the most abundant of the polyphenolic oxidation products in black tea, the papers concerned with the chemistry of TR at present are limited. Therefore, it is not easy to explain from the chemical structure of TR that TR showed no potent antioxidant activity in this study. To date, there are no definitive data on TB structures. We speculated that some polyphenols in TB make the most contribution to free radical scavenging activity of TB.

3.4.5 *Result analysis of TF antioxidant properties*

ROS accelerates membrane damage, DNA base oxidation, DNA strand break, and chromosome aberration. They also play a causative role in aging and several degenerative diseases, such as cancer [Hussain *et al.*, 2003], atherosclerosis [Cooke *et al.*, 2003], and cataract [Finkel and Holbrook, 2000]. The supplement of antioxidants, such as vitamin E and vitamin C from foods and beverages, has been an attractive therapeutic strategy for reducing the risk of these diseases.

Black tea is widely preferred in India and the Western countries, not only as a popular beverage, but also as an antioxidative agent. It is generally believed that polyphenols such as TFs and thearubigins, as well as catechins, as major constituents of black tea, are mainly responsible for antioxidant actions. Since the abundant constituent of catechins, EGCG has been proven to be the most effective antioxidant among green tea polyphenols. Its superiority is attributed to the numbers of hydroxyl groups in its chemical structure.

TFs, the characteristic compositions in black tea, are formed via the co-oxidation of pairs of epimerized catechins, one with a vic-trihydroxy-phenyl moiety, and the other with an ortho-dihydroxyphenyl structure. Different from epimerized catechins, four main TF derivatives reserved two A-rings, two C-rings from their precursors, and possess a characteristic element of the fused seven-member benzotropolone ring. It has been suggested that the existence of resonance formed in the benzotropolone moiety might be responsible for electron donation [Jovanovic *et al.*, 1997]. Jhoo *et al.* [2005] suggested that the benzotropolone moiety of TFs might play an important role in affording antioxidant protection for the preferred oxidation site in the oxidant models of DPPH and hydrogen peroxide. Accordingly, we suggest that the higher number of phenyl hydroxyl groups in TF derivatives, which could interact with ROS, might increase their antioxidant capacities, and their benzotropolone moiety might have played an important role in scavenging superoxide radical, singlet oxygen, hydrogen peroxide, and the hydroxyl radical in our *in vitro* models.

It is noteworthy that four major TF derivatives were obviously detected to have different antioxidant capacities when facing with different ROS. TF3 containing two gallate groups has been reported to be a prior inhibitor on ABTS radical [Miller *et al.*, 1996], DPPH radical [Yang *et al.*, 2008], and Cu^{2+}-mediated LDL oxidation [Leung *et al.*, 2001]. Consistent with previous works, we found that TF3 could efficiently scavenge hydrogen peroxide and the hydroxyl radical with the lowest IC_{50}

values (Tables 3-4 and 3-5). Similarly, TF2A and TF2B, which have one gallate group, had stronger activities than TF1, which has no gallate group. The gallate ester seems to be positive in reacting with hydrogen peroxide and the hydroxyl radical.

Meanwhile, the difference in antioxidant activities between TF2A and TF2B was observed in the present work. TF2A and TF2B are a pair of monogallate TFs with the same basic skeleton and one gallate ester in different positions. They were thought to have equal bioactivities for the same structure. Actually, TF2B showed the highest activity in suppressing chemiluminescence signal of singlet oxygen. Moreover, it had more effective antioxidant capacities in inhibiting superoxide radical, hydrogen peroxide, and the hydroxyl radical than TF2A. Therefore, the monogallate ester of 3'-position in TFs seems to play an important role in scavenging ROS, such as $^1O_2$, H_2O_2, and •OH. Although the mechanism is still unknown, these results suggest that TF2B might possess a higher biological activity in other aspects.

Interestingly, compared with other three TF monomers, TF1 was not so efficient in scavenging the hydroxyl radical and free radicals arising from hydrogen peroxide. However, it showed the most effective inhibition on the superoxide radical among four major TF derivatives and EGCG. This result was highly consistent with their reaction rates order with superoxide anions; the reaction rates of TF derivatives with the superoxide radical were found to be almost an order of magnitude higher than that of gallocatechins, and TF1 had the highest reaction rates of 1.0×10^6 M-1 s-1 [Jovanovic *et al.*, 1997]. The benzotropolone skeleton of TF derivatives was thought to be important in scavenging the superoxide radical with the capacities of charge separation and one-electron abstraction. The present work also provided the possibility that TFs gallate derivatives inhibits free radical induced DNA oxidative damage, which is responsible for various diseases. The preventive effects of TF derivatives on •OH-induced DNA damage were in accord with their capacities of inhibiting •OH. It is suggested that the positive role of TF derivatives in scavenging free radicals might contribute to their effects on DNA oxidative damage. Green tea polyphenols have been shown to directly interact with DNA radicals to repair DNA by a mechanism of electron transfer [Anderson *et al.*, 2001].

The precise mechanism of TF gallate derivatives in preventing DNA oxidative damage is still unclear and worthy of further study.

In summary, the activity differences of four major TF derivatives provide some new insights into the antioxidant potency of these derivatives. More detailed study of their antioxidant effects on ROS and DNA oxidative damage is required.

3.5 The inhibiting effect of theaflavins on free radicals to protect cells from oxidative damage

The above results indicate that theaflavins can not only scavenge free radicals and have antioxidant effects, but also protect cells from oxidative damage. The typical pigments theaflavins (TF) in black tea is mentioned above. Thearubigins (TR) is the most abundant of black tea polyphenols, at approximately 12–18% of solid extracts of black tea liquors. It embraces a number of indeterminate structures, from the "monomeric" to the "polymeric", derived by enzymatic oxidation of the flavan-3-ols. It is rust-brown in colour and contributes the reddish colour and richness in taste, totally termed "body" to black tea [Roberts, 1962]. Recently, TR has attracted considerable interest because of its beneficial health properties, including antiproliferative effect on A432 cells and mouse NIH3T3 fibroblast cells [Liang *et al.*, 1999], inhibition of intestinal carcinogenesis, antimutagenic effects *in vitro* [Gupta *et al.*, 2001], antileukemic effects in U-937 cell and leukemic cells [Das *et al.*, 2002], anti-inflammatory, antineurotoxin effects in mice, and inhibition of lipid peroxidation under biological conditions [Yoshino *et al.*, 1994]. However, little progress has been made toward an understanding of the chemical nature of the TR. Unlike the well-characterised TF, theabrownins (TB) is very poorly characterized. TB is brown and very soluble in water. At best, TB may be taken as a term to embrace amylose, protein, nucleic acid, and polyphenols, derived from the oxidation of TF and TR as a result of the excessive withering and anoxic fermentation during black tea processing. TB endows tea liquor and leaf with a dark brown colour, which has a negative effect on tea quality.

3.5.1 *Effect on H₂O₂-induced or normal HPF-1 cell viability*

The cell viability was expressed as MTT conversion rate. The effects of oxidized phenolic compounds, at different concentrations, on H_2O_2-induced loss of HPF-1 cell viability, are depicted in Figure 3-12. Treatment with 600 μM H_2O_2 for 24 hours, decreased the viability of HPF-1 cells by about 26–32% relative to the negative control. Upon pretreatment with the TF, TR, and TB at different concentrations (0.72 μg/ml, 1.43 μg/ml, 2.87 μg/ml, 5.73 μg/ml, and 11.5 μg/ml), the cell viability was almost dose-dependently ameliorated. Oxidized phenolic compounds had a significantly protective effect on H_2O_2-damaged HPF-1 cell ($p < 0.05$), magnitude of the increased viability was about 10% compared with the positive control group. It was equivalent to recovery by almost 30%, of the cell loss. Compared with EGCG, TF and TB showed an inferior protective effect at the concentration of 2.87~11.5 μg/ml. However, at 0.72~1.43 μg/ml, TF and TB was more effective than was EGCG. TR acts as an inferior protector among these oxidized phenolic compounds in this system. The

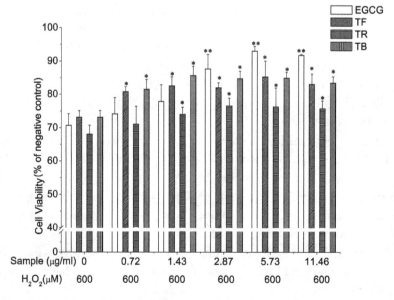

Figure 3-12. Effect of oxidized phenolic compounds on H_2O_2-induced decrease of HPF-1 cell viability.

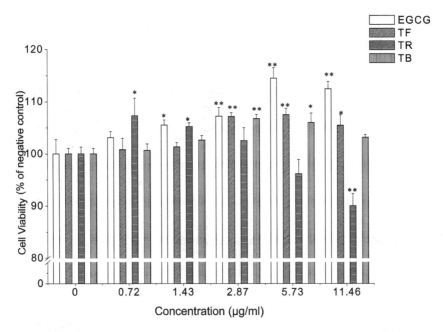

Figure 3-13.   Effect of oxidized phenolic compounds on normal HPF-1 cell viability.

cell viability result is shown in Figure 3-13, when the cells were treated with oxidized phenolic compounds alone. TF and TB caused almost dose-dependent proliferation, whereas the TR result was dose-dependently reduced. TR showed a negative effect on normal HPF-1 cells at 11.5 µg/ml ($p < 0.01$). HPF-1 cells, human normal diploid fibroblasts, exhibit finite proliferative potential *in vitro*, the so-called Hayflick limit. They undergo a limited number of population doublings before entering a state of permanent growth arrest, referred to as "replicative senescence," "cellular senescence", or "cellular aging", in which they remain alive and metabolically active but are completely refractory to mitogenic stimuli. HPF-1 offers a typical model for studying the process of aging *in vitro*. Various oxidative stresses have been used to study the onset of cellular senescence. The early onset of cellular senescence, induced by oxidative stresses, is termed as stress-induced premature senescence (SIPS) and H_2O_2 has been the most commonly used inducer of SIPS, which shares features of replicative senescence: similar morphology, senescence-associated ß-galactosidase activity, and cell cycle regulation.

3.5.2 *Effect on the accumulation of ROS in H_2O_2-induced or normal HPF-1 cell*

ROS is the main cause of oxidative stress, which results in decreasing cell viability. The level of DCF fluorescence is an indicator of ROS production. As shown in Figure 3-14, 600 μM H_2O_2 was able to produce these toxic species in the HPF-1 cells; exposure to H_2O_2 for one hour increased the DCF fluorescence intensity by about 60–70% over the negative control. The increase in DCF fluorescence intensity was partly eliminated when the cells were co-treated with different concentrations of EGCG, TF, TR, and TB. The decrease in fluorescence intensity was almost dose-dependent. At the concentration of 0.96–3.84 μg/ml, TF and TB showed stronger ability to eliminate the fluorescence intensity than compared to

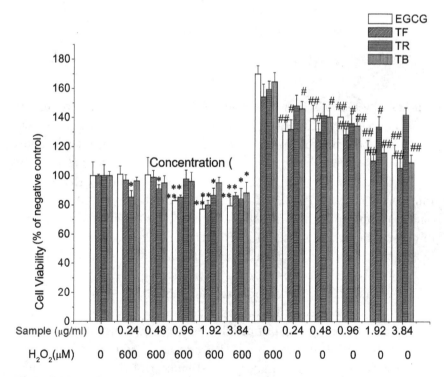

Figure 3-14.   Effect on the accumulation of ROS in H_2O_2-induced or normal HPF-1 cell by oxidized phenolic compounds.

EGCG. In addition, with no exposure to H_2O_2, the DCF fluorescence intensity in the HPF-1 cells, after the cells were treated with oxidized phenolic compounds at the concentration of 0.96–3.84 µg/ml for one hour, also showed some decrease. These results implicate the involvement of ROS in H_2O_2-induced phenolic compounds in HPF-1 cells. It is well known the generation of ROS is associated with the initiation and promotion of carcinogenesis. Excessive amounts of ROS increase oxidative stress in the body. In recent years, many studies have suggested that green tea polyphenols (catechins) are good scavengers of ROS *in vitro*. TF (monomer), as an oxidized phenolic product, proved more effective than EGCG in the suppression of intracellular reactive oxygen species in HL-60 cells [Lin *et al.*, 2000]. Some studies have indicated that TR protects against oxidative DNA damage *in vivo*, by scavenging ROS. TR was usually regarded as "tea pigments or black tea extract" in these reports. Bioavailability of thearubigins as oxidative phenol compounds has not been described. Unlike the well-examined and published antioxidant properties of TF, bioactivity of TB has hardly been discussed. TB has attracted little interest so far due to the complexity of its ingredients and its negative effect on black tea quality.

TF, TR, and TB, as the major oxidized phenolic products during the fermentation stage of black tea processing, have the most effect on the quality and make the most contribution to the antioxidant activity of black tea. In this study, TB's beaviour is outstanding. There is a need for more detailed studies to enhance the harmony between quality and bioactivity in the black tea process.

3.5.3 *Protective effects of TF derivatives from hydroxyl radical induced DNA damage*

Evaluation of the antioxidant effects of four main theaflavin derivatives through chemiluminescence and DNA damage. Four purified TF derivatives were prepared at different concentrations for further analysis. EGCG is the most abundant and effective compound in green tea.

A hydroxyl radical-induced DNA breaking system was applied to test the effects of TF derivatives on DNA oxidative damage *in vitro*. When the super-coiled plasmid DNA is attacked by a hydroxyl radical generated from the Fenton reaction, it will be transformed into three forms: super-coiled (SC), open circular (OC), and linear (Linear). The open circular DNA and linear DNA represent the damaged DNA. The extent of undamaged DNA is represented by the percentage of the SC form in DNA bands, used to demonstrate the antioxidant effects of the test samples by comparing the percentage of the SC DNA in the test samples and in the control (DNA treated with •OH). The conversion of the super-coiled form of DNA to open-circular (OC) and linear DNA has already been used as an index of DNA damage [Lewis *et al.*, 1988].

The effects of different TF derivatives on •OH-induced DNA damage were compared in parallel, and EGCG was also used for comparison (Fig. 3-15). For the different speeds of different DNA forms in the electrophoresis gel, the upper band is an open-circular DNA resulted from

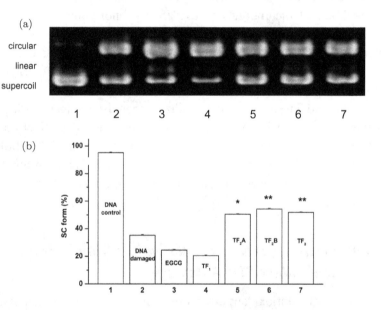

Figure 3-15. Agarose gel electrophoretic analysis of •OH-induced plasmid DNA damage withdifferent samples. (a) Lane 1: plasmid DNA; Lane 2: DNA damage control (DNA treated with $FeSO_4$ and H_2O_2), 0.25 µg of pUC19 DNA was incubated with 2 mmol/L $FeSO_4$ and 1 mol/L H_2O_2 at 37°C for one hour; Lanes 3–7: DNA treated with $FeSO_4$ and H_2O_2 in the presence of EGCG, TF1, TF2A, TF2B, and TF3 at a concentration of 0.025 mmol/L. (b) Comparison of percentages of super-coiled form in the test samples to DNA damage control. Experiments were performed three times and the values are represented as mean ± SD. * $p < 0.05$ and ** $p < 0.01$ when compared with EGCG.

•OH-induced DNA damage and the lower is super-coiled DNA. Lane 1 was pUC19 plasmid DNA without •OH treatment (SC form 95.23%). This DNA control was broken into OC and Linear forms by the •OH generated with 35.31% SC form being kept (Lane 2). However, there was no significant damage on DNA when being treated by H_2O_2 or Fe^{2+} [Tian and Hua, 2005]. Except for TF1, there were significantly protective effects by TF2A, TF2B, and TF3 compared with EGCG ($p < 0.05$). The protective effects of TFs (TF2A, TF2B, and TF3) on DNA could be inferred from the higher recovery (50.60%, 54.43%, and 52.07%) of the SC form (Fig. 3-15b). This result was also consistent with chemiluminescence assay result of their scavenging abilities on the hydroxyl radical.

3.6 The inhibiting effect of theaflavins on ATP synthase and the respiratory chain without increasing superoxide production

Several plant polyphenols, including tea catechins, have been shown to be inhibitors of the mitochondrial ATP synthase. In the lab of Walker, the bovine F1 sector of the ATP synthase was co-crystallized with three different polyphenols: resveratrol, piceatannol, and quercetin. All three were bound inside the enzyme alongside the rotary shaft of the gamma subunit, and not at the nucleotide binding sites. The ATP synthase from *E. coli*, known to be a valuable model for the mitochondrial enzyme, was also found to be inhibited by many similar compounds. Treatment with piceatannol, morin, silymarin, baicalein, silibinin, rimantadin, amantidin, and epicatechin resulted in complete inhibition, while resveratrol, quercetin, quercetrin, quercetin-3-β-D-glucoside, hesperidin, chrysin, kaempferol, diosmin, apigenin, genistein, and rutin were partially inhibitory, in the range of 40~80%.

3.6.1 *Inhibition of the respiratory chain by theaflavins does not contribute to superoxide production*

Polyphenols from black tea, the theaflavins, can inhibit the ATP synthase from *E. coli*, and are more potent than EGCG [Matsushita *et al.*, 1987]. In a similar fashion, these compounds are also inhibitory towards Complex I

Table 3-7.   Inhibition of ATPase anctivity by theaflavins.

Inhibitor	Membrane-bound ATPase activity[a]		F_1-ATPase activity[b]	
	IC_{50} (μM)[c]	Maximum inhibition $(\%)$[c]	IC_{50} (μM)[c]	Maximum inhibition $(\%)$[c]
TF1	60	85	4.0	85
TF2A	20	95	3.0	90
TF2B	15	95	1.5	95
TF3	10	90	0.7	90
EGCG	30	95	4.5	90

[a]Membrane-bound ATPase activity was about 0.7 μmoles ATP/min/mg protein.
[b]F_1-ATPase activity was about 1.8 μmoles ATP/min/mg.
[c]IC_{50} concentrations and maximum inhibition % were determined from two replicates of five to six concentrations of inhibitor. The data are displayed in Supplementary Figure 3-1.

(NDH-1) of the electron transport chain. No evidence of increased production of superoxide was found in the presence of these inhibitors. Four
theaflavin monomers have been isolated from black tea, and their effects
on oxidative phosphorylation and superoxide production in a model system (*E. coli*) have been examined. The esterified theaflavins were all
potent inhibitors of the membrane-bound ATP synthase, inhibiting at least
90% of the activity, with IC_{50} values in the range of 10~20 μM (Table 3-7).
ATP-driven proton translocation was inhibited in a similar fashion, as was
the purified F1-ATPase, indicating that the primary site of inhibition was
in the F1 sector (Fig. 3-16). Computer modeling studies supported this
interpretation. All four theaflavins were also inhibitory towards the electron transport chain, whether through Complex I (NDH-1) or the alternative NADH dehydrogenase (NDH-2) [Ohnishi *et al.*, 2010]. Inhibition of
NDH-1 by TF3 appeared to be competitive with respect to NADH, and
this was supported by computer modeling studies. Rates of superoxide
production during NADH oxidation by each dehydrogenase were measured. Superoxide production was completely eliminated in the presence of
about 15 μM TF3, suggesting that inhibition of the respiratory chain by
theaflavins does not contribute to superoxide production.

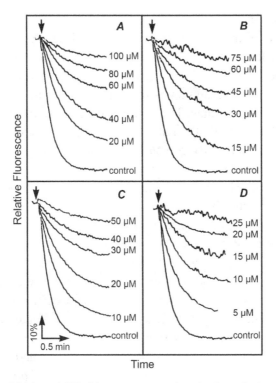

Figure 3-16.   Inhibition of ATP-driven proton translocation by polyphenols.

Proton translocation by membrane vesicles is indicated by the quenching of the fluorescence of ACMA upon the addition of ATP. Membranes without the addition of polyphenols are indicated by control. In each case, the fluorescence quenching could be eliminated by the addition of FCCP (carbonyl cyanide p-(trifluoromethoxy) phenylhydrazone), a protonophore (results not shown). The final concentrations of inhibitors are indicated in the panels.

3.6.2 *Inhibition of the ATP synthase*

The four major theaflavin compounds: TF1, TF2A, TF2B, and TF3 were tested as inhibitors of the membrane-bound F1Fo-ATPase, along with

EGCG. Several catechins, including EGCG, have been previously shown to be inhibitors of the rat liver mitochondrial ATPase, but the theaflavins are somewhat larger than EGCG, and therefore might not be capable of binding to the same sites. Membrane vesicles were prepared from wild-type cells, and ATP hydrolysis was measured in the presence of increasing amounts of the polyphenols. The results showed the theaflavins with gallate esters (TF2A, TF2B, TF3) all inhibited to an extent of 90~95% with IC_{50} values in the range 10–20 μM. TF1 and EGCG were somewhat less potent, with IC_{50} values of about 60 μM and 30 μM, respectively. Complete inhibition data are shown in Supplementary Figure 3-15. Previous work has indicated that the sites of inhibition by polyphenols are in the F1 sector of the ATP synthase. Since the theaflavins are twice as large as many other polyphenols, it was possible that they might bind to other regions of the enzyme. For that reason, isolated F1-ATPase was examined as well.

This can be detected by monitoring the fluorescence quenching of an acridine dye, ACMA. In this assay, the rate and extent of fluorescence quenching is indicative of the generation of a proton gradient across the membrane. Each of the four theaflavins was tested as an inhibitor of ATP-driven proton translocation, using membrane vesicles, and the results are presented in Figure 3-15. In each case, the inhibition of proton transloca-tion seemed to follow that of the inhibition of ATP hydrolysis, and so there was no evidence for additional inhibitory binding sites in the membrane sector of the enzyme. The binding site for several polyphenols in the F1-ATPase has been well established by the co-crystallization of the bovine enzyme with resveratrol, piceatannol, and quercetin, each binding to the same position inside the enzyme near the membrane-distal end. Considering the size of the theaflavin polyphenols, especially the digallate ester (TF3), it seemed unlikely that they could bind to the same site as the smaller flavonol quercetin. Computer modeling and docking studies were carried out with each of the theaflavins, along with EGCG, to see where they might bind in the F1-ATPase. Using the bovine F1-ATPase structure (Protein Data Bank file 1efr), stripped of ligands as the template, querce-tin was docked to a position very lose to the actual binding site discovered by co-crystallization studies (Fig. 3-17A). In contrast, TF3 was docked to a totally different site, closer to the membrane, and overlapping with the actual binding site of efrapeptin (Fig. 3-17B).

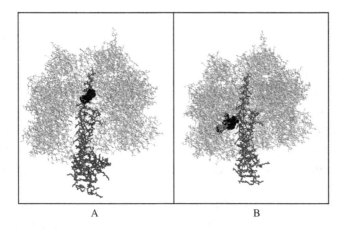

A   B

Figure 3-17. Comparison of actual and predicted inhibitor binding sites in bovine. F1 is shown in gray wireframe, the gamma subunit is in thick wireframe, and the inhibitors are shown in black space-filling. (A) The model shown is from Protein Data Bank file 2jj2. Bound nucleotides are also visible. (B) The results from a docking study of TF3 and the F1-ATPase from Protein Data Bank file 1efr are shown.

F1-ATPase is shown in gray wire frame (Fig. 3-17), the gamma subunit is in thick wire frame, and the inhibitors are shown in black space filling. A) The model shown is from Protein Data Bank file 2jj2. B) The results from a docking study of TF3 and the F1-ATPase from Protein Data Bank file 1efr are shown. In both cases, the inhibitors are bound to the gamma subunit, but the TF3 is docked much lower and below the nucleotide binding sites. The TF3 site in panel B is accessible from the bottom of the structure, as shown, rather than from the top, as in panel A.

3.6.3 *Inhibition of electron transport*

Since the polyphenol resveratrol has previously been shown to inhibit the mitochondrial respiratory chain, the theaflavins were also tested as inhibitors of the electron transport chain in membrane vesicles. *E. coli* has two distinct NADH dehydrogenases in its plasma membrane. Both use the oxidation of NADH to reduce ubiquinone, which is re-oxidized by one of two quinol oxidases. NDH-1 is homologous to the mitochondrial Complex I. It is a multi-subunit, proton translocating enzyme, and is similar to the

ATP synthase in that it has two large sectors, one embedded in the membrane and another peripheral to it. The peripheral sector binds NADH, while the membrane sector is involved with ion translocation. In contrast, NDH-2 is a single polypeptide that is monotopically bound to the membrane and does not translocate ions. In *E. coli*, they can be distinguished by the use of strains in which one enzyme is knocked out genetically, or by the use of deamino-NADH, which only NDH-1 is able to use. Each NADH dehydrogenase was tested with the four theaflavins and EGCG, and the results of the inhibition are presented in Table 3-3. For NDH-1, the IC_{50} values for the five polyphenols range from about 10~50 μM and the inhibition reaches 90~95%. In contrast, the theaflavins inhibit NDH-2 with lower IC_{50} values in the range of 1~4 μM, while the extent of inhibition is 75~95%. However, the IC_{50} for EGCG is about the same with respect to both NADH dehydrogenases. Clearly, the inhibitors have different targets in these two respiratory chains.

To support the finding that the NDH-1 was competitively inhibited by TF3, computer modeling studies were carried out using the peripheral sector of the Thermus thermophilus Complex I (Protein Data Bank file 3i9v). Results showed that the highest affinity binding occurred on the surface of the enzyme at a site that partially overlapped with the NADH binding site. In addition, the activity of NDH-1 was assayed by the use of an artificial electron acceptor, hexamine ruthenium (HAR), which is thought to draw off electrons before they reach the membrane sector. In this case, using TF3 as the inhibitor, the results showed uncompetitive inhibition, where the Dixon plots have parallel lines and the Cornish-Bowden lines intersect in the upper left quadrant. The Km was estimated to be 0.16 μM. This appeared to be in conflict with the earlier finding that TF3 inhibits competitively at the NADH site.

3.6.4 *Modulation of the rate of superoxide production*

Inhibitors of electron transport typically cause an increase in superoxide production during the consumption of NADH. Since TF3 was found to be a rather potent inhibitor of both NDH-1 and NDH-2, rates of superoxide production were measured in membrane vesicles with increasing levels of TF3. The results, presented in Figure 3-18, showed that NADH oxidation by

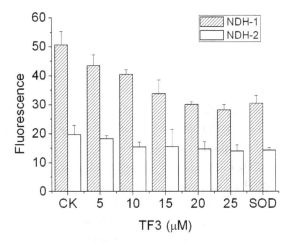

Figure 3-18. The effect of increasing levels of TF3 on superoxide production by NDH-1 and NDH-2.

NDH-1 caused higher rates of superoxide production than by NDH-2. Addition of superoxide dismutase reduced the fluorescence, and so, indicated the amount that was due to superoxide. When TF3 was added at levels in the range of the IC_{50} value, superoxide did not increase, but rather decreased. This shows that while TF3 is an inhibitor of the electron transport chain in *E. coli* membranes, it does not cause an increase in superoxide production.

Superoxide was measured by the increase in fluorescence of the Mitosox reagent, in the absence of NADH (CK) and in the presence of superoxide dismutase (SOD), which reflects the background fluorescence. To differentiate between NDH-1 and NDH-2, strain MWC215 was used for NDH-1 (shaded bars) and strain BA14 was used for NDH-2 (open bars). The results shown are the mean of three measurements, with the error bars indicating the standard deviation.

3.6.5 *Result analysis*

Flavonoid polyphenols are dietary compounds from plants that are known to interact directly with enzymes. The ATP synthase was previously found to be inhibited by several simple polyphenols, including resveratrol, quercetin, and EGCG. In this report, theaflavins, a class of polyphenols found exclusively in black tea, were also shown to be inhibitors of the ATP synthase. The four theaflavins tested varied in size according to the presence of 0, 1, or 2 gallate esters. The apparent affinities, as indicated by IC_{50} values, had a four-fold range of values and correlated well with overall size. The importance of the gallate group is also suggested by the inhibitory properties of EGCG. It is smaller than TF1 but contains both a gallate ester and a galloyl group and has an IC_{50} value that is similar to that of TF1. Furthermore, in another inhibition study of the *E. coli* ATP synthase, epicatechin, lacking both the galloyl and gallate groups, was found to have a substantially larger IC_{50} of 4 mM. Similar to the results here, rat liver mitochondrial ATP synthase was found to be inhibited by EGCG with an IC_{50} of 17 μM. The extents of inhibition also correlated with the size of the polyphenols. That is probably related to the fact that the ATP synthase is a rotary motor and many of its inhibitors appear to be mechanical inhibitors. For example, the polyphenols that have been co-crystallized with the enzyme, for example, quercetin or resveratrol are found to bind between the shaft of the rotary gamma subunit and the stationary alpha and beta subunits, which bind the nucleotides. Since the theaflavins are somewhat larger than quercetin and much larger than resveratrol, it could be expected that they would not bind to the same site even though they are related chemically. Indeed, the computer modeling studies suggested a different binding site, one which overlaps with a previously identified site for the antibiotic efrapeptin, from crystallographic studies. This site is accessible from the opposite end of the enzyme, and is a region that undergoes large conformational changes during rotary catalysis. It is possible that even with full occupancy, some of the smaller polyphenols fail to inhibit completely because they do not totally block rotation but only hinder it. The lower IC_{50} values obtained relative to the membrane-bound enzyme might reflect complexities in the activity of the enzyme. The rate of ATP hydrolysis by the membrane-bound enzyme is quite sensitive to pH and other conditions, which cause it to be released from the membrane sector. The results shown do not preclude additional binding sites in the membrane sector. Other inhibitors, such as the antibiotic oligomycin, are known to act there. However, the assays of proton translocation showed no evidence for additional inhibitor sites in the membrane sector. While it cannot be ruled out that lower affinity sites exist, the polarity of the theaflavins, especially those with gallate esters, might be too great to allow binding in the lipid phase. Two different enzymes allow NADH to initiate electron transport in *E. coli*, and both were found to be inhibited by all five polyphenols tested. In the case of NDH-1, the IC_{50} values followed the same pattern as with the ATP synthase: affinity correlated with size, except that TF1 had a larger IC_{50} than did EGCG. This again indicates the likely importance of the gallate groups in binding. The computer docking of TF3 suggested important interactions of the benzotropolone group and one of the gallates (3′) with the enzyme, supporting that conclusion and the relatively low IC_{50} for TF2B. Overall, the polyphenols had relatively high extents of inhibition (90–95%), which is consistent with a competitive mode of inhibition, as was found for TF3, in which substrate binding is precluded. The inhibition of NDH-2 followed the pattern of larger theaflavins having lower IC_{50} values, but in this case, EGCG had the highest IC_{50} value of all by a factor of 5~10 mM. This suggests importance of the benzotropolone group, in conjunction with at least one gallate. Overall, the extents of inhibition were somewhat lower than for NDH-1, with values in the range of 65–95%. Since the mode of inhibition by TF3 was determined to be uncompetitive, it suggests that for some of the inhibitors, the enzyme-inhibitor-NADH complexes retained some activity. It was not expected that TF3 would be a competitive inhibitor with respect to NADH, since it does not resemble NADH structurally. Moreover, it also does not closely resemble ubiquinone, so there was no basis to predict that it would bind in the membrane sector. The finding of its competitive mode of inhibition of NADH oxidase activity was in conflict with the finding that it was uncompetitive with respect to NADH using the HAR reductase assay. The former result requires TF3 binding in the absence of NADH, while the later requires binding in the presence of NADH. An important consideration is that the mechanism of the HAR reaction is unknown. A recent paper by Birrell *et al.* [2009] suggests that HAR might draw off electrons through the protein from one of the Fe-S centers that is close enough to the surface. Given such uncertainties about this artificial electron acceptor, we favor the interpretation that the inhibition is competitive with respect to NADH. Supporting that view are the findings by computer modeling whereby a potential binding site exists that partially overlaps with the entrance to the NADH binding site. In this way, the binding of TF3 could block the binding of NADH without significant structural similarity to it.

Many inhibitors of the electron transport chain are significant contributors to superoxide production in the mitochondrion, because they tend to keep particular electron carriers in a reduced state. For example, some Complex I inhibitors are known to increase the rate of superoxide production. Superoxide has been suggested to be formed at various sites in Complex I, including at theaflavin, at an Fe-S center, or at a ubisemiquinone site. The mode of the inhibition by TF3 is consistent with its lack of stimulation of superoxide production. A competitive inhibitor will tend to prevent reduction of the electron carriers, rather than to keep them in a reduced state. In addition, theaflavins are known to be superoxide scavengers. In this study, TF3 caused a decrease in superoxide production while it inhibited the electron transport chain in *E. coli* membranes at levels in the range of the IC_{50} value. The bioavailability of catechin compounds in the human body is thought to be low, not with standing the possibility of higher bioavailability for some catechins such as EGCG, or in some tissues such as mouth and esophagus. Thus, given the micromolar IC_{50} values measured here and previous studies on the health benefits of theaflavins as antioxidants *in vivo*, we hypothesized that theaflavins had dual effects on cells in a dose-dependent way. At low doses, theaflavins play an important role as antioxidants, and the physiological consequences of the inhibition of ATP synthase or Complex I by theaflavins might be limited. At high dosage, theaflavins might have some effect on the electron transport chain. These results suggest the possibility of theaflavins used as drugs targeting Complex I or ATP synthase of some pathogens such as *Mycobacterium tuberculosis*, or in some diseases such as obesity. In addition, the results presented here suggest that there are modes of inhibition of these enzymes that might be common to a variety of polyphenols, or related compounds, found in the human diet.

3.7 The inhibitory effect of theaflavins and thearubigins on the growth of cancer cells *in vitro*

Black tea, one of the most popular beverages in the UK, has been reported to have a number of health benefits. In the Ames test, black tea effectively decreased the mutagenic activity of structurally diverse chemical carcinogens [Bu-Abbas *et al.*, 1996]. It is proposed that black tea exercises its anti-initiation activity by modulating the xenobiotic-metabolising enzyme systems in such a way as to limit the availability of genotoxic metabolites. Theaflavins, catechin dimmers, have been shown to antagonise the carcinogenicity of nitrosamines in mice [Yang *et al.*, 2002; Shukla & Taneja, 2002] and to possess antimutagenic activity, and inhibit lipid oxidation as well [Apostolides *et al.*, 1997]. Indeed, exposure of rats to black tea resulted in up-regulation of CYP1A2 and a selective increase in the Phase II enzyme systems glutathione S-transferases and UDPGA-glucuronosyl transferases [Sohn *et al.*, 1994; Bu-Abbas *et al.*, 1998; Bu-Abbas *et al.*, 1999]. As a result of these effects, black tea alters the metabolism of carcinogens such as 2-amino-3-methylimidazo-(4,5-f) quinoline (IQ), leading to reduced excretion of mutagens and promutagens in rats treated with a single oral dose of this carcinogen [McArdle *et al.*, 1999].

The biological effects of tea and tea constituents have been studied by many investigators and reviewed in previous publications. The inhibitions of tea and tea components against chemically induced carcinogenesis have been demonstrated in animal models, including cancers of the skin, lung, esophagus, stomach, liver, small intestine, pancreas, colon, bladder, prostate, and mammary glands [Gupta *et al.*, 2004]. Most studies are involved in green tea, whereas limited investigation has been carried out with black tea. Theaflavins (TFs), pigment complex in black tea, is transformed from tea catechins (a kind of colorless component in tea) through oxidation catalyzed by enzyme and chemical condensation. TFs play an important role in particular color and taste of black tea infusion. It is mainly consisted of the following four components: TF1, TF2A, TF2B, and TFDG. It has been proved that TFs possess evident effects such as anti-oxidation, anti-cancer, scavenging or inhibiting free radicals, treatment of blood vessel, and heart diseases [Li, 2001; Xie & Xu, 2000; Gong *et al.*, 2000; Han & Gong, 1999; Wan & Li, 2001; Hasaniya *et al.*, 1997; Shiraki *et al.*, 1994]. TF-1 and TF2B have been reported to be capable of inducing apoptosis in human lymphoid leukemia cells and stomach tumor cells [Hibasami *et al.*, 1998], differential effects of theaflavin monogallates on cell growth, apoptosis, and Cox-2 gene expression in cancerous versus normal cells [Lu *et al.*, 2000]. Recently, studies on the mechanisms of the action of TF2A, TFDG, TF2B, and catechins to NF-kB macrophages, mice liver cancer cells and Human Fibrosarcoma HT 1080 Cells were reported [Hibasami *et al.*, 1998; Lu *et al.*, 2000]. TF1, TF2A, TF2B, and TFDG showed strong inhibition on cell growth and activator protein 1 activity. TFDG inhibited the phosphorylation of p44/42 (extracellular signal-regulated kinase 1 and 2) and c-jun without affecting the levels of phosphorylated-c-jun-NH2-terminal kinase. TFDG also inhibited the phosphorylation of p38 [Chung *et al.*, 1999]. TF2B (10–50 μM) inhibited the growth of SV40 transformed WI38 human cells (WI38VA) and Caco-2 colon cancer cells but had little effect on the growth of their normal counterparts [Lu *et al.*, 1997]. Liang *et al.* [1999] had provided evidence for a mechanism of antitumor proliferation by TFDG and EGCG. The inhibition of EGF receptor (or PDGF receptor) kinase activity by TFDG and EGCG may be mediated through blocking of EGF (or PDGF) by binding to its receptor. The IC_{50} of TFDG was as low as 15 μM and 18 μM for the growth of NIH3T3 and A431 cells, respectively. The IC_{50} values for EGCG were 26 μM and 28 μM for the growth of NIH3T3 and A431 cells, respectively. Other important studies on theaflavins were concerned with protaste cancer. Nomura *et al.* [2000] found that theaflavins inhibited a tumor promoter 12-O-tetradecanoylphorbol-13-acetate-induced NF-kB activity in a concentration-dependent manner in the JB6 mouse, the results suggested that theaflavins is also an important reason in accounting for the anti-tumor promotion effects. TFDG also has been found to have antiviral activity against influenza A and B [Green, 1994; Nakayama *et al.*, 1990; Nakayama *et al.*, 1993]. However, the effects of theaflavins on the human liver cancer cells (BEL-7402), gastric cancer cells (MKN-28), acute promyelocytic leukemia (LH-60), and human lung cancer cell (A549) were poorly known. In this chapter, the inhibition effects of TFs and theaflavins monomers on the four kinds of cancer cells were studied. Interaction of EGCG and AA was reported. To measure the growth of breast (MDA-MB-231), colon (HCT116), and skin (melanoma, A2058) when ascorbic acid, proline, and lysine combined with EGCG were added in the medium, the lower concentration of EGCG was effective, the proliferation was reduced to 74% and colon cancer cells HCT116 to 69% compared to the unsupplemented medium [Netke *et al.*, 2003]. It implies that there is good synergism between AA, praline, lysine, and EGCG against human cancer cell. EGCG has demonstrated chemopreventive and chemothera-peutic actions in cellular and animal models of cancer and selectively induces apoptosis in human carcinoma cell lines [Du *et al.*, 2001; Borska *et al.*, 2003; Lambert & Brand, 2004]. Therefore, EGCG was designed as a control in this test. However, rare study on synergism between organic acid and TFs was published; in the present study, the synergisms between ascorbic acid and TFs, TF2A+TF2B, or EGCG against lung cancer cell A549 were considered, and the apoptosis of cancer cell was observed by transmission electron microscope.

We studied the inhibition effects of TFs, TFDG, TF2B, and CAFFEINE on the growth of human liver cancer cells (BEL-7402), gastric cancer cells (MKN-28), and acute promyelocytic leukemia (LH-60). The results showed that TF2B significantly inhibited the growth of the three kinds of cancer cells, whereas TFs, TFDG, and CAFFEINE had little activity on LH-60. TF2B, TFDG, and CAFFEINE had stronger inhibition effects on BEL-7402 and MKN-28 than TFs. The relationship coefficients between concentration and inhibition rate against MKN-28 and BEL-7402 were 0.87 and 0.98 for TF2B, 0.96 and 0.98 for CAFFEINE, respectively. The IC_{50} values were 0.18 mM, 0.11 mM, and 0.16 mM on BEL-7402, and 1.11 mM, 0.22 mM, and 0.25 mM on MKN-28 for TFS, TF2B, and TFDG, respectively. Five concentrations such as 0.0174 µg/ml, 0.174 µg/ml, 1.74 µg/ml, 17.4 µg/ml and 174 µg/ml of TF1, TFs, TFDG, TF2A with TF2B complex (TF2A+TF2B) and EGCG were selected for measuring the inhibition on the growth of human lung cancer cell line A549 growth. The results showed that the effects of TFs, TFDG, TF2A+TF2B, and EGCG on the growth of A549 were significant when the concentration of the compounds was above 17.44 µg/ml except for that of TF1. EGCG and TFDG appeared to have the highest inhibition towards A549 among the five compounds. Synergism inhibition experiments of TF1, TF2A+TF2B, TFs, TFDG, and EGCG combined with ascorbic acid (AA) against the growth of A549 were designed and detected. There were the strong synergism inhibitions between TFs or TF2A+TF2B and ascorbic acid against the A549 growth, EGCG as well [Li *et al.*, 2010]. Especially, the inhibition of combination of TF2A+TF2B with 0.1 mM ascorbic acid (AA) was increased by 29.5% than that of the control. The images induced the apoptosis of lung cancer cell A549 by the combination of 0.29 mg/ml

TFs with 0.1 mM AA and TFs alone were observed with electron microscopy. The ultrastructural changes of A549 control were found, cells exhibited morphologically normal and apoptosis body, but no chromatin margination in the control cancer cell was found. Chromatin marginations were significantly observed in both treatments, and especially, blocks appeared in the treatment of the combination of TFs with ascorbic acid.

3.7.1 *Inhibition of TFs and TF monomers against cancer cells*

The cell population was determined using 3-(4, 5-dimethylthiazoly1-2)-2, 5-diphenytrazotium bromide (MTT) assay according to the method described by Xu and Zhen [2003]. Cells of BEL-7402, MKN-28, and LH-60, as described above at a density of 5 million cells per ml, were separately seeded into 96-well plates in 0.1 ml per well. TFs, TFDG, TF2B, and CAFFEINE obtained by using HSCCC were added into the 96-well plates with 20 μl MTT separately per well. The final five concentrations of TFs, TF1, TFDG, TF2B, and CAFFEINE were set as 3.9 μg/ml, 16 μg/ml, 63 μg/ml, 250 μg/ml, and 1000 μg/ml for each compound. The A549, as described above at a density of 5 million cells per ml, were respectively seeded into 96-well plates in 0.2 ml per well. TFs, TF2A+TF2B, TFDG, TF1, and TF2B obtained by using HSCCC were added into the 96-well plates with 20 μl MTT separately per well. The final concentrations of TFs, TF1, TFDG, TF2B, TF2A+TF2B, and EGCG were set as 3.9 μg/ml, 16 μg/ml, 63 μg/ml, 250 μg/ml, and 1000 μg/ml, respectively. After incubation 72 hours at 37°C in an incubator, the incubator was equilibrated with 95% air plus 5% CO_2. Five replicates were set for each treatment. Effects on HL-60, BEL-7402, Ca A549, and MKN-28 were measured using Sulfur Rodamine B method. The inhibiting activities of theaflavins were expressed as the mean effective concentrations in mg/ml that inhibited 50% of cell growth amounts (IC_{50}). The IC_{50} was calculated by applying a linear logarithms equation using power of regression. The significance of the relationship coefficients between concentration and inhibition rate against cancer cell lines was tested using t-test under the SAS system for windows developed by SAS Institute, Cary, NC, USA.

1. Effect of theaflavins and monomers on human cancer cell lines

The results showed that the inhibition activities of TFs and individual monomers against BEL-7402 and MKN-28 cancer cells complied with the first order dynamics equation. The inhibition rates of TFs and TF monomers on four cancer cell lines are presented in Figures 3-19 and 3-20. The relationship coefficients were greater than 0.86 and were significant for all treatments. Particularly, the inhibition effects of CAFFEINE and TF2B against BEL-7402 were the most significant with a relationship coefficient of 0.98.

Figure 3-19 shows that at the different compound concentrations of 3.9 μg/ml, 16 μg/ml, 63 μg/ml, 250 μg/ml, and 1000 μg/ml, the inhibition rates on BEL-7402 were determined, as described in the materials and methods. Data were presented as the means of six independent experiments performed in five replicates.

Figure 3-20 shows that at the different compound concentrations of 3.9 μg/ml, 16 μg/ml, 63 μg/ml, 250 μg/ml, and 1000 μg/ml, the percentages of inhibition on MKN-28 were determined, as described in the materials and methods. Data were presented as the means of six independent experiments performed in five replicates.

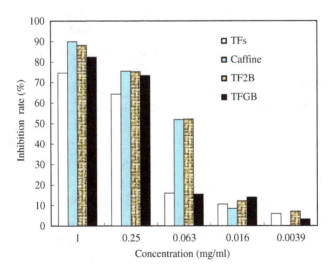

Figure 3-19. Inhibition rates against human liver cancer cells by TFs, CAFFEINE, TF2B, and TFDG.

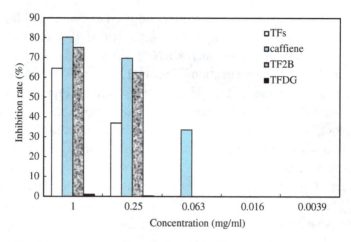

Figure 3-20.   Inhibition rates against gastric cells by TFs, CAFFEINE, TF2B, and TFDG.

The highest inhibition rates of TFDG, CAFFEINE, TF2B, and TFs were 82.4%, 90.0%, 88.2%, and 74.7% against BEL-7402, and 76.5%, 80.2%, 75.0%, 64.5% against MKN-28, respectively, which indicated that both TFs and theaflavin monors were more effective in inhibiting BEL-7402 among the tested cancer cells. CAFFEINE exhibited the strongest inhibitory effect among the tested compounds against BEL-7402 and MKN-28, whereas TFs had the lowest. The inhibition rate of TF2B against the growth of LH-60 was about 100% at the concentration of 1.0 mg/ml, but only 26.5% at 0.25 mg/ml. TFs, TFDG, and CAFFEINE had less than 10% inhibition rate on LH-60 even at the concentration of 1.0 mg/ml, which implied that the inhibition of theaflavins had some selection on different cancer cells. Meanwhile, the results demonstrated that TF2B had more multifaceted bioactivity functions whereas TFs, TFDG, and CAFFEINE had less. According to the pre-test result, A549 was much more sensitive than BEL-7402, MKN-28, and LH-60 to TF complex or TF monomers, even at lower concentration of TF complex or TF monomers which was efficient in inhibiting the growth of A549. Therefore, we selected five concentrations: 0.0174 µg/ml, 0.174 µg/ml, 1.74 µg/ml, 17.4 µg/ml, and 174 µg/ml of TF and EGCG for an inhibiting experiment on A549 growth. The results showed that the effects of TFs, TFDG, TF2B, TF2A+TF2B, and EGCG on the growth of A549 were significant when

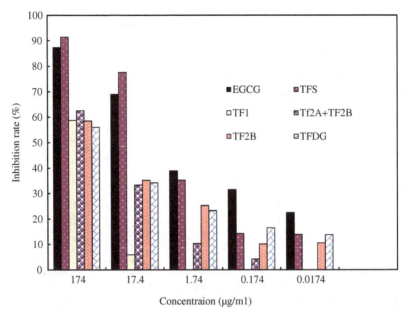

Figure 3-21. Effect of TFs and TF monomers on A549.

concentrations of compounds were above 17.44 µg/ml except for that of TF1. EGCG and TFDG appeared to have the highest inhibition activities towards A549 among the six tested compounds even at very low concentration of 0.0174 µg/ml.

Figure 3-21 showed that at the different compound concentrations of 1.74×10^{-2} µg/ml, 1.74×10^{-1} µg/ml, 1.74 µg/ml, 17.4 µg/ml, 174 µg/ml, the percentages of inhibition on A549 were determined, as described in materials and methods. Data were presented as the means of six independent experiments performed in five replicates.

2. IC$_{50}$ of theaflavins against BEL-7402 and MKN-28

For convenience of comparison with other literatures, the unit of IC$_{50}$ values of TFs, TF2B, and TFDG was converted into mM. The IC$_{50}$ values derived from the corresponding linear equations are listed in Table 3-8. The IC$_{50}$ values of TF2B, TFDG, CAFFEINE, and TFs on BEL-7402 were lower than those on MKN-28, suggesting that these components

Table 3-8. IC_{50} of theaflavins on human liver cancer cell and gastric cancer cell.

Compound	TF_S (mM)	CAFFEINE (mg/ml)	TF2B (mM)	TFDG (mM)
BEL-7402	0.18 (r = 0.93**)	0.08 (r = 0.98**)	0.11 (r = 0.98**)	0.16 (r = 0.93**)
MKN-28	1.11 (r = 0.89*)	0.14 (r = 0.96**)	0.22 (r = 0.87*)	0.25 (r = 0.90**)

*The relationship coefficients are significant at $p < 0.05$ level.
**The relationship coefficients are highly significant at $p < 0.01$ level.

Table 3-9. IC_{50} of theaflavins on human lung cancer cell A549.

Compound	TF_S (μM)	TF2B (μM)	TFDG (μM)	EGCG (μM)	TF2A+TF2B (μM)	TF1 (μM)
IC_{50}	4.07	207	308	4.95	202	368
R	r = 0.96**	r = 0.94*	r = 0.94**	r = 0.97**	r = 0.94**	r = 0.91*

*The relationship coefficients are significantly different at $p < 0.05$ level.
**The relationship coefficients are significantly different at $p < 0.01$ level.

have better inhibiting activities against BEL-7402 than against MKN-28. The IC_{50} values (mg/ml) of TF2B and CAFFEINE on BEL-7402 were the lowest among all the treatments, implying that TF2B and CAFFEINE were the most effective components among the tested TFs and monomers. On the contrary, TFs showed one of the lowest IC_{50} 4.07 μM against A549 in the present study (Table 3-9), the IC_{50} 4.95 μM of EGCG was very close to the IC_{50} of TFs, which meant that TFs and EGCG had a good inhibition on A549. However, the IC_{50} of TFDG, TF2B, and TF2B complex were obviously higher than those of TFs and EGCG. To compare the data, we found that TFs and EGCG showed the lowest IC_{50} against A549, TF2B, and TFDG, and showed the lower IC_{50} against BEL-7402 and MKN-28.

All the values of determination coefficient showed a good correlation between the treatment concentration of compound and inhibition growth on A549 in Table 3-9.

The synergism inhibitions of TFs or TF2A+TF2B and EGCG with ascorbic acid against A549 were measured. The results are showed in Table 3-10. It was obvious that in a range of AA concentration from

Table 3-10. Synergism inhibitions of TFs, TF2A+TF2B, and EGCG with AA towards A549.

Treatment AA(mM)	Inhibitions of TFs (%)		Inhibitions of TF2A+TF2B (%)		Inhibitions of EGCG (%)	
	A	B	A	B	A	B
0.50	67.36	15.00	64.26	9.51	45.44	13.12
0.25	76.48	24.12	78.51	23.76	51.52	19.2
0.10	74.77	22.41	84.22	29.47	59.13	26.81
0.05	70.78	18.42	59.70	4.95	47.90	15.58
0.00	52.36		54.75		32.32	

0.1–0.25 mM, the inhibition rates of the three group treatments against A549 growth were increased by about 20%.

"A" expressed the inhibitions rate of the combinations of TFs (8.14 μM), TF2A+TF2B (404 μM) and EGCG (9.15 μM) and AA towards A549; "B" expressed the increased the inhibitions (the synergism inhibition rate of compounds with AA towards A549 reduced inhibition rate of correspondent compound towards A549).

The total inhibition rates either TFs or TF2A+TF2B at the same ascorbic acid concentration was at least 20% higher than those of EGCG, which suggested that the synergism inhibitions between TFs, TF2A+TF2B, and AA against A549 were much higher than those of EGCG.

3.7.2 *Microscope imaging of A549 cell lines apoptosis*

As apoptosis could be a major cause for growth inhibition, the effects of the TFs on the apoptosis of lung cancer cell A549 were measured.

The change of apoptosis induced by combination of 0.1 mM AA and 0.29 mg/ml and TFs was observed with electron microscopy. Under electron microscopy, early apoptosis, advanced apoptosis, late apoptosis including cell shrinkage, nuclear chromatin condensation, formation of membrane blebs, and apoptotic bodies of A549 cell were frequently observed in TFs treatment and the combination of TFs and ascorbic acid treatment (TFA). The microscope images of A549 cell apoptosis are presented in Figures 3-22 and 3-23.

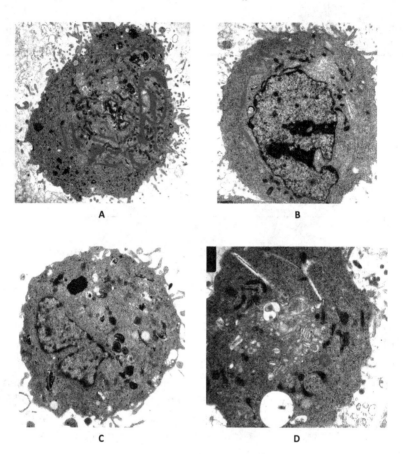

Figure 3-22.   Electron micrographs of control A549. Control A549: PBS-treated control cell. (A) Normal nucleoli and normal morphology; (B) Chromatin condensation into dense blocks; (C) Formation of apoptotic bodies; (D) Cell disruption. Scale bar 500 μm in figure A and 2 μm in figures B, C, and D.

3.7.3 *Result analysis*

We discerned the ultrastructural changes of A549 control cells exhibited morphologically normal — apoptosis body appeared but no chromatin margination. On contrary, chromatin imaginations were significantly observed in Figures 3-23 and 3-24, and especially blocks only appeared in Figure 3-24. In this study, all the tested theaflavins showed inhibition activities against BEL-7402 and MKN-28, and there was a good negative

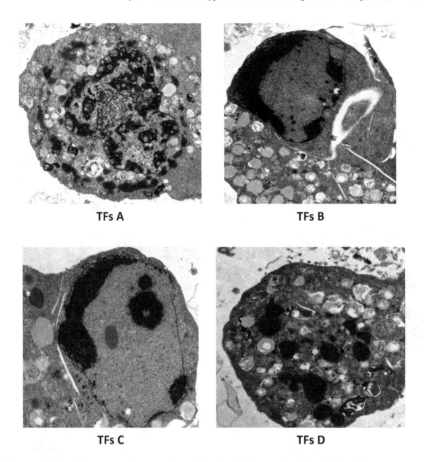

Figure 3-23. Electron micrographs of Ca A549 induced with TFs. (TFsA) Normal nucleoli and normal morphology; (TFsB) Chromatin margination; (TFsC) Typical apoptotic morphology including and condensation into dense granules; (TFsD) Formation of apoptotic bodies. Scale bar of figures TFs A to TFs D are 2 μm.

relationship between the concentration of the tested compounds and growth of cancer cells. After a comparison of the inhibition concentration of TFs with catechins against the cancer cells, we found that the effective concentration of TFs was lower than catechins. The theaflavins showed a strong inhibitory effect on AP-1 activity with an estimated IC_{50} value of 5 μM for TF2A, TF2B, and TFDG. TF1 was less effective than its gallate derivatives at 20 μM [Gupta *et al.*, 2004]. However, the inhibition

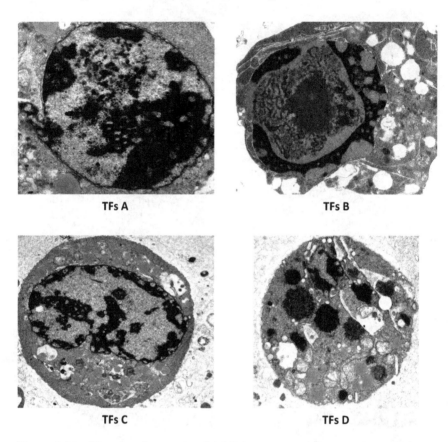

Figure 3-24. Electron micrographs of A549 induced by the combination of TFs and ascorbic acid. (TFA A) Exhibiting intact plasma membranes, clear cytoplasm, and normal nucleoli; (TFA B) Chromatin margination; (TFA C) Formation of apoptotic bodies including condensation into dense granules and blocks; (TFA D) Formation of apoptotic bodies. Scale bar of figures TFs A to TFs D are 1 μm.

concentration of the three most potent green tea components EGCG, GC, and EGC against four different human cancer cell lines such as MCF-7 breast carcinoma, HT-29 colon carcinoma, A-427 lung carcinoma, and UACC-375 melanoma were 10 mM, 100 mM, and 1000 mM [Gupta *et al.*, 2004]. Lung *et al.* [2002] reported that EGCG suppressed the proliferation of the EoL-1 cells in a dose-dependent manner, with an estimated IC_{50} value of 31.5 mM. Yang *et al.* [1998] reported that EGCG and EGC displayed strong growth inhibitory effects against lung tumor cell lines H661 and H1299, with an estimated IC_{50} values of 22 mM, which are much higher than the TFs concentration range, mentioned above, and the TFs concentration in this test. Another study also proved that the result IC_{50} of TF-2 for the growth inhibition of WI38 and WI38VA cells were 300 μM and 3 μM TF-1 and TF2B, respectively. The tested concentration of theaflavins in this study was also lower than the effective concentration of 1 mg/ml in inhibiting mice liver cancer cells of AH109A and antibavterium reported by [Maeda *et al.*, 1999]. Although their IC_{50} values are higher than 20.0 μg/ml, it is still worth exploring and designing new natural anticancer medicine based on TFs structures.

A comparison of the inhibition concentration of nine compounds showed in Tables 3-9 and 3-10 implied that the order of inhibition ability against Ca A549 is the complex of TFs with AA > the complex of TF2A+TF2B with AA > the complex of EGCG with AA > TFs > EGCG > TF2A+TF2B > TF2B > TFDG > TF1. The present experiment showed that the IC_{50} values (mg/ml) of TF2B and CAFFEINE on BEL-7402 were the lowest among all the tested compounds, we proposed that CAFFEINE might be the isomer of theaflavin-gallate according to the study reported by Lewis *et al.* [1998] and the structures of some other tea polyphenol oxides reported in recent years. CAFFEINE will be purified further, and its structure will be identified in our near future work. Netke *et al.* [2003] reported that a specific combination of ascorbic acid 100 μM, proline (P) 140 μM, and lysine (L) 400 μM had a significant antiproliferative and antimetastatic effect against breast (MDA-MB-231), colon (HCT116), and skin (melanoma, A2058). Their results showed that the invasion was reduced to zero using 20 μg/ml EGCG with AA, P, and L in breast cancer cells and melanoma cells. Ascorbic acid is very cytotoxic to malignant cell lines [Koh *et al.*, 1998; Roomi *et al.*, 1998] and exerts antimetastatic action [Liu *et al.*, 2000; Netke *et al.*, 2003]. Our studies using the combination of ascorbic acid with TFs, TF2A+TF2B, and EGCG indicated that this combination exerted a potent inhibition on A549 cancer cell line in cell culture studies. It was a very interesting result showing that there were strong synergism inhibitions between TFs or TF2A+TF2B and ascorbic acid against the A549 growth. When TFs, TF2A+TF2B, and EGCG were used alone at 8.14 μM/l, 404 μM/l, and 9.45 μM/l, the invasion of A549 was reduced to 52.36%, 54.75%, and 32.32%, respectively. The inhibition rates of A549 increased up to 74.77%, 84.22%, and 59.13% when 100 mM/l ascorbic acid were used together with TFs, TF2A+TF2B, and EGCG in the inhibition experiments against the A549 growth. Especially, the inhibition of the combination of TF2A+TF2B with 0.1 mM ascorbic acid was increased by 29.5%. The inhibition of the combination of TFs, TF2A+TF2B with ascorbic acid from 0.1~0.25 mM against A549, were higher than of that of EGCG.

It is easy observed that at the same concentration of AA, the synergistic inhibition effects between 0.1 mM ascorbic acid and 8.14 μM TFs or 9.45 μM EGCG on A549 cancer cells are stronger than on breast cancer cells. Thus, it suggests that A549 cancer cells might be more sensitive than breast cancer cells to combination of AA and TFs or EGCG. TF2A, but not TF1 or TF2B, induced apoptosis in transformed WI38VA cells but not in normal WI38 cells, suggesting that apoptosis was responsible, at least in part, for the differential growth-inhibitory effect of TF2A [Lu *et al.*, 2000]. Our study confirmed that the treatment of the combination of TFs with AA was more efficient to induce apoptosis of A549 than the treatment of TFs alone *in vitro*, even the later treatment showed significant induction apoptosis of A549.

TFs is a synthetic form of two B rings. Since there are three OH groups for this structure, it is likely that catechins have scavenging action for radicals and good interaction with ascorbic acid, proline, and lysine. The gallic acid moiety is important for TFs to express antioxidative activity and antimutagenicity. Thus, TFs may be expected to play an important role in chemoprevention, for which lipid peroxides or active oxygen are important. The present studies indicated that theaflavins and some monomers were efficient to inhibit the growths of BEL-7402, MKN-28, A549, and LH-60 cancer cell line even at much lower concentration. There were strong synergism inhibitions between TFs or TF2A+TF2B and AA against the A549 growth, EGCG as well. The images of the apoptosis of lung cancer cell A549 induced by TFs and TFs combined with 0.1 mM ascorbic acid were observed with electron microscopy.

3.8 Effects of theaflavins and thearubigins on obesity and lipid metabolism

Obesity is a disarray of energy balance and primarily considered as a disorder of lipid metabolism [Birari *et al.*, 2011]. It is one of the major public health problems in the world because of its association with an increased risk of various chronic diseases, including cardiovascular diseases, type 2 diabetes, hypertension, dyslipidemia, and cancers [Chang *et al.*, 2011; Huang *et al.*, 2009]. Only two drugs, orlistat and sibutramine, are currently approved for long-term obesity reatment. However, both have undesirable side effects, including increased blood pressure, dry mouth, constipation, headache, and insomnia [Yun, 2010]. At present, the potential of natural products for the treatment of obesity has been explored and might be an excellent alternative strategy for the development of safe and effective antiobesity drugs [Birari *et al.*, 2010]. So far, black tea has been shown to reduce body fat and serum cholesterol levels in rats fed a high-cholesterol diet and improve blood glucose level in streptozotocin-diabetic rats [Chen *et al.*, 2009; Gomes *et al.*, 1995; Matsumoto *et al.*, 1998]. However, functional components accounting for the antiobesity effect of black tea are not completely understood. Therefore, this study was designed to evaluate the antiobesity and lipid lowering effects of theaflavins on high-fat diet induced obese rats. Furthermore, the effects of theaflavins on insulin sensitivity, leptin level, and the activities of alanine transaminase (ALT), hepatic lipase (HL), and superoxide dismutase (SOD) were investigated.

3.8.1 *Effects of BTE, TFs, and TF1 on growth parameters of rats in different groups*

Rats were randomly divided into five groups as follows: 1) normal diet (ND), 2) high-fat diet (HFD), 3) HFD + BTE, 4) HFD + TFs, 5) HFD + TF1, each group consisting of eight animals. Effects of samples on serum lipid profiles total cholesterol (TC), triacylglycerol (TG) and low-density lipoprotein cholesterol (LDL-C), and high-density lipoprotein cholesterol (HDL-C) levels in serum and AI value are shown in Table 3-5. Feeding HFD caused the elevations of serum TC and TG concentrations compared with that of the ND group. TC levels in the HFD + BTE and HFD + TF1 groups were slightly decreased ($p > 0.05$), and that in the HFD + TFs group was dramatically decreased by 26.5% ($p < 0.05$). Consuming BTE, TFs, and TF1 significantly reduced TG content to 56.9%, 50.8%, and 52.3% of the value seen in the HFD control group, respectively ($p < 0.05$). No significant difference in the HDL-C content between three treatment groups and the HFD controls was observed in this study ($p > 0.05$). Administration of BTE,

TFs, and TF1 remarkably decreased the serum LDL-C level by 69.6%, 71.7%, and 43.5% respectively, and reduced AI value by 78.0%, 68.3%, and 41.5%, respectively, compared with that of the HFD controls ($p < 0.05$).

3.8.2 *Effects of TF on blood glucose level and insulin sensitivity*

The effects of BTE, TFs, and TF1 on fasting serum glucose, fasting serum insulin, and ISI of rats in different groups are shown in Table 3-6. At the end of the experiment, all the groups had similar blood glucose concentration ($p > 0.05$). A reduction of insulin sensitivity induced by feeding with HFD was characterized by an evident decrease in ISI of the HFD controls compared with that of the ND group ($p < 0.05$). Consuming BTE, TFs, and TF1 reduced the serum insulin concentration to 44.2%, 83.2%, and 74.7% of the value detected in the HFD group respectively ($p < 0.05$), and significantly increased the ISI value compared with that of the HFD group ($p < 0.05$). The effects of BTE, TFs, and TF1 on leptin level in liver are measured. Leptin concentration of the HFD group was a little higher than that of the ND group, but the difference was not significant ($p > 0.05$). Consuming BTE, TFs, and TF1 reduce the leptin level by 18.4%, 13.0%, and 20.6%, respectively ($p > 0.05$), in comparison with that of the HFD controls.

The effects of BTE, TFs, and TF1 on HL activity showed that HL activity of the HFD controls was higher than that of the ND group, but the difference was not significant.

3.8.3 *Effect of TF on serum ALT activity*

The effects of BTE, TFs, and TF1 on serum ALT activity are obtained. Although no significant difference in serum ALT activity was observed among animal groups ($p > 0.05$), the mean value of ALT activity was increased after consuming HFD, and was reduced to 67.0%, 77.5%, and 78.6% of the value detected in the HFD group by BTE, TFs, and TF1, respectively.

3.8.4 *Effect of samples on serum SOD activity*

The effects of BTE, TFs, and TF1 on serum SOD activity are shown in Figure 3-21. Feeding HFD slightly decreased serum SOD activity compared with that of the ND group ($p > 0.05$). Consuming BTE, TFs, and TF1 increased SOD activity by 8.5%, 6.0%, and 13.8%, respectively [Chen *et al.*, 2009].

3.8.5 *Result analysis*

Caffeine is considered to be another component that accounts for antiobesity effect of tea,and may act independently or synergistically with catechins to stimulate thermogenesis and increase fat oxidation [Han *et al.*, 1999; Huang *et al.*, 2009; Rains *et al.*, 2011]. Black tea is the most oxidized and contains the lowest catechin content among varieties of teas. Although approximately 80% of the tea produced and consumed worldwide is black tea [Luczaj & Skrzydlewska, 2005], studies on the components that account for its antiobesity effect are limited. Black tea polyphenols extract (BTPE) has been shown to prevent diet-induced obesity by inhibiting intestinal lipid absorption, and the major active component in the BTPE is the polymerized polyphenol fraction [Uchiyama *et al.*, 2011]. Among the black tea polymerized polyphenols, theaflavins have been well characterized in chemical structure and are generally considered to be the more effective components [Li *et al.*, 2012]. To provide further assurance of antiobesity effect for theaflavins, the high-fat diet induced obese rat model was employed in this study, and the effects of highly purified theaflavins and black tea extract on body and tissue weights, food intake, serum lipid profiles, insulin sensitivity, leptin level, and activities of HL, ALT, and SOD were investigated. The present study showed the body weights of rats were slightly reduced by BTE and TFs ($p > 0.05$) and was significantly decreased by TF1 ($p < 0.05$) relative to the HFD control group (Table 3-11). All samples slightly reduced liver weight ($p > 0.05$) and remarkably decreased the adiposity index ($p < 0.05$) (Table 3-11). Previous studies showed that BTE had antiobesity effect by suppressing body weight gain and adipose tissue formation in rat model [Chen *et al.*,

Table 3-11. Effect of BTE, TFS, and TF1 on growth parameters of rats in different groups.

	ND	HFD	HFD + BTE	HFD + TFs	HFD + TF1
Body weight (g)					
Initial	87.40 ± 5.31^A	87.16 ± 5.98^A	87.01 ± 6.35^A	87.46 ± 5.56^A	87.56 ± 5.27^A
Final	368.71 ± 12.87^B	402.86 ± 16.69^A	$385.50 \pm 23.24^{B,A}$	401.50 ± 17.14^A	364.00 ± 17.52^B
Total food intake (g/30 days)	531.80 ± 4.81^B	589.00 ± 8.20^A	520.50 ± 7.42^B	524.88 ± 23.51^B	536.33 ± 43.95^B
Liver weight					
Absolute (g)	10.98 ± 0.75^A	12.02 ± 0.66^A	11.51 ± 0.74^A	11.75 ± 0.64^A	10.98 ± 0.72^A
Relative (%)	3.09 ± 0.13^A	3.14 ± 0.15^A	3.04 ± 0.06^A	2.97 ± 0.12^A	2.99 ± 0.07^A
Adipose tissues					
Epididymal (g)	5.51 ± 0.57^B	7.68 ± 0.98^A	5.40 ± 1.48^B	$5.87 \pm 1.24^{B,A}$	$6.34 \pm 1.15^{B,A}$
Perirenal (g)	3.40 ± 0.64^B	6.38 ± 1.48^A	$4.84 \pm 0.66^{B,A}$	$5.31 \pm 1.11^{B,A}$	$5.29 \pm 1.13^{B,A}$
Adiposity index (%)	2.64 ± 0.35^B	3.75 ± 0.42^A	2.59 ± 0.47^B	2.83 ± 0.45^B	2.82 ± 0.27^B

Values are expressed as mean x SD ($n = 8$).

Values in a row followed by different letters are significantly different ($p < 0.05$).

Relative liver weight: the ratio of liver to terminal body weight.

2009; Ramadan *et al.*, 2009], which was in agreement with our results. Furthermore, our study firstly demonstrated theaflavins reduce body weight by eliminating fat *in vivo*. BTE, TFs, and TF1 were observed to have appetite suppressing effect on the HFD induced obese rats in this work, the results indicated the weight and lipid lowering effects of the samples may be at least partly attributed to the reduction of energy intake. It was reported that Keemun black tea extract could reduce food intake of obese rats via oral administration [Du *et al.*, 2005]. However, other reports indicated that black tea decreased body weight and adipose tissue mass without altering the food intake [Chen *et al.*, 2009; Ramadan *et al.*, 2009]. So far, it is known that caffeine may modify the food intake in rats through influencing corticotropin-releasing factor and the sympatho-adrenal system [Racotta *et al.*, 1994], but appetite suppressing effect of catechins remains controversial [Rains *et al.*, 2011]. Leptin, a truly pleiotropic hormone, exerts its influence on food intake, energy expenditure, body weight, and neuroendocrine function through actions on neuronal targets in the hypothalamus [Lee *et al.*, 2002]. In rodent studies, obesity is associated with a state of leptin resistance characterized by a failure of elevated circulating leptin to counteract obesity. The sustained high concentration of leptin results in leptin desensitization. The pathway of leptin control in obese people might be flawed at some point, so the body does not adequately receive the satiety feeling after eating [Ricci & Bevilacqua, 2012]. BTE, TFs, and TF1 all slightly decrease the concentration of leptin in liver in comparison to that of the HFD controls ($p > 0.05$), which is similar to the previous report that black tea significantly suppressed the increase of the leptin level in HFD-fed mice [Nishiumi *et al.*, 2010]. These results revealed that theaflavins and black tea might reduce the body weight and food intake via alleviating leptin resistance in HFD induced obese rats. Obesity is an independent risk factor that increases cardiovascular disease (CVD) risk and overall mortality that warrants medical attention. The relationship between obesity and CVD is not fully understood but involves a number of mechanisms including dyslipidemia and insulin resistance [Charakida & Nicholas, 2012]. High levels of TC, TG, and LDLC and low level of HDL-C in serum are strongly associated with increased risk of coronary heart disease [Schneider *et al.*, 2011]. The AI, defined as the ratio of LDL-C and HDLC, is believed to be an important risk factor of

Table 3-12. Effect of BTE, TFS, and TF1 on serum lipid profiles of rats in different groups.

	ND	HFD	HFD + BTE	HFD + TFs	HFD + TF1
TC (mmol/L)	1.42 ± 0.09^A	1.70 ± 0.10^A	1.67 ± 0.43^A	1.25 ± 0.17^B	$1.44 \pm 0.12^{B,A}$
TG (mmol/L)	0.41 ± 0.1^B	0.65 ± 0.07^A	0.37 ± 0.1^b	0.33 ± 0.09^B	0.34 ± 0.08^B
HDL-C (mmol/L)	1.06 ± 0.10^B	$1.11 \pm 0.17^{B,A}$	1.46 ± 0.36^A	1.05 ± 0.15^B	$1.11 \pm 0.17^{B,A}$
LDL-C (mmol/L)	0.28 ± 0.04^B	0.46 ± 0.08^A	0.14 ± 0.04^C	0.13 ± 0.01^C	0.26 ± 0.07^B
A1	0.26 ± 0.06^B	0.41 ± 0.14^A	0.09 ± 0.01^C	$0.13 \pm 0.02^{B,C}$	0.24 ± 0.10^B

A1: atherogenic index.
Values are expressed as mean x SD ($n = 8$).
Values in a row followed by different letters are significantly different ($p < 0.05$).

atherosclerosis [Khaled *et al.*, 2012]. The present study showed the HDL-C content was not influenced by the samples (Table 3-12), which was consistent with the previous report that none of the green, oolong, and black tea extracts effectively altered plasma HDL-cholesterol concentrations in hyperlipidemia rats fed high-sucrose diet [Yang *et al.*, 2001]. BTE, TFs, and TF1 could reduce the serum levels of TC, TG, and LDL-C, and decreased the AI values in rats fed HFD (Table 3-12). These results were in agreement with the previous study on the hypolipidemic effect of black tea in rats [Kuo *et al.*, 2005], and suggested that theaflavins and black tea had the potential to reduce the incidence of CVD by improving dyslipidemia in obese patients. The roles of theaflavins and black tea in lipid metabolism have been studied. Theaflavins could decrease intestinal cholesterol absorption via inhibition of dietary mixed micelle formation [Vermeer *et al.*, 2008], and suppressing fatty acid synthase (FAS) to affect lipogenesis [Yeh *et al.*, 2003]. Black tea extract and theaflavins were able to inhibit the activity of the pancreatic lipase *in vitro*, and the inhibitory activity was stronger than that of green tea extract and EGCG [Birari & Bhutani, 2007; Lin & Lin-Shiau, 2009; Uchiyama *et al.*, 2011]. In addition, black tea could inhibit the expression of lipoprotein lipase in rats fed HFD [Chen *et al.*, 2009]. HL is a glycoprotein belonging to the lipase super family that includes pancreatic lipase and lipoprotein lipase. It hydrolyses phospholipids and triacylglycerols of plasma lipoproteins and plays an important role in the metabolic processing of HDL-C and LDL-C [Tan *et al.*, 2001]. Higher level of HL activity is associated with the increased risk of coronary artery disease (CAD), insulin resistance, and type 2 diabetes mellitus [Baynes *et al.*, 1991; Brunzell *et al.*, 2012; Tan *et al.*, 1999]. The present study firstly reported the inhibition of HL activity by theaflavins and black tea extract *in vivo*. All above reports indicated theaflavins and black tea might induce hypolipidemic effect by suppressing lipogenesis, lipid absorption and digestion.

Numerous epidemiologic studies indicate that obesity is associated with an increased risk of liver disease. Obesity predisposes hepatocytes to lipid peroxidation and oxidative stress, thus increasing the possibility of hepatic injury, fibrosis, cirrhosis, and even hepatocellular carcinoma as a result of hepatic inflammation [Marchesini *et al.*, 2008]. Serum ALT activity is a commonly used surrogate marker for the evaluation of hepatocellular damage. It was reported that obesity and diabetes were important risk factors for elevated ALT activity [Wang *et al.*, 2010]. This study showed BTE, TFs, and TF1 reduced serum ALT level of rats fed HFD, indicating these samples could attenuate hepatocellular damage induced by obesity. This result was consistent with the previous report that BTE protected against hepatocellular damage in rats with HFD-induced non-alcoholic steatohepatitis, and this action was associated with reversing the changes in the pro-oxidant and antioxidant status of the liver [Karmakar *et al.*, 2011]. Reactive oxygen species (ROS) play critical roles in the pathogenesis of various diseases including the metabolic syndrome, CVD, cancer, and neurodegenerative diseases [Higdon & Frei, 2003]. The oxidative stress in adipocytes induced by fat overload might be the origin of the metabolic syndrome caused by obesity. ROS production is stimulated by fatty acids via NADPH oxidase activation. Oxidative stress in adipocytes caused dysregulate production of adipocytokines, impaired insulin signals, and decreased insulin-stimulating glucose uptake [Furukawa *et al.*, 2004]. Oxidative stress is an early event in the evolution of hyperlipidemia [Yang *et al.*, 2008]. The oxidative modification of LDL-associated lipids is directly involved in the initiation of the atherosclerotic process [Albertini *et al.*, 2002]. The potential health benefits associated with tea polyphenols have been partially attributed to their antioxidative property [Khan & Mukhtar, 2007]. It was reported that epigallocatechin gallate

(EGCG), the most abundant catechin in green tea, could ameliorate adipose insulin resistance by improving oxidative stress [Yan *et al.*, 2012] and improve serum lipid profile and cardiac tissue antioxidant parameters in Wistar rats fed an atherogenic diet [Ramesh *et al.*, 2008]. Theaflavins have been found to quench ROS such as hydroxyl radicals, 2,2-diphenyl-1-picrylhydrazyl (DPPH) and superoxide *in vitro*, and their scavenging ability is higher than that of EGCG [Jovanovic *et al.*, 1997; Sarkar & Bhaduri, 2001; Yang *et al.*, 2007]. Theaflavins also show strong antioxidant activity against LDL oxidation in mouse macrophage cells and suppression of intracellular ROS in HPF-1 cells [Yang *et al.*, 2007; Yoshida *et al.*, 1999]. The chemical structures contributing to effective antioxidant activity of polyphenols include the vicinal dihydroxy or trihydroxy structure. Theaflavins have two A-rings of flavanols linked by a fused seven-member ring. These structural features may provide more interaction sites with radicals. In addition, the benzotropolone moiety of theaflavin plays an important role in affording antioxidant protection for the preferred oxidation site and might be responsible for electron donation because of the existence of resonance forms [Khan and Mukhtar, 2007; Yang *et al.*, 2007]. Except for the direct free radical-scavenging ability, theaflavins are able to inhibit the formation of ROS through the competitive inhibition of xanthine oxidase or NADPH oxidase [Lin *et al.*, 2002]. The SOD can remove superoxide anion radicals and is an important antioxidant defense in nearly all cells exposed to oxygen. The previous study showed that SOD activity in the serum was increased in the rats fed with tea leaves, and the order was black tea > oolong tea > green tea > puerh tea [Kuo *et al.*, 2005]. In this study, consuming BTE, TFs, and TF1 increased SOD activity in the serum of rats fed with HFD. These results suggested that theaflavins and black tea may at least partly serve as an antioxidant to attenuate hepatocellular damage, improve insulin resistance and hyperlipidemia, and then prevent type 2 diabetes and CVD in obese patients.

3.9 Metabolism and bioavailability of theaflavin and thearubigin

Normally, polyphenols are presented in dietary in terms of ester, glycosides, or polymeric forms, which will not be absorbed directly. These

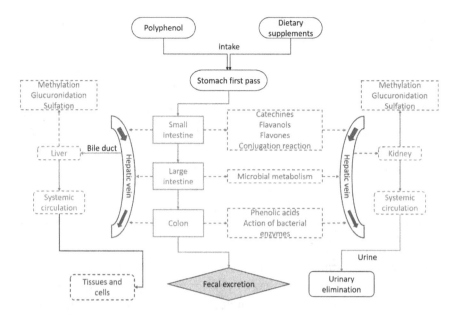

Figure 3-25. Schematic diagram of metabolites of tea polyphenols *in vivo.*

compounds should be hydrolyzed by enzymatic reaction before absorption. The large portion of polyphenolic compounds consist of several hydroxyl groups, which are enzymatic catalyzed by methylation, glucuronidation, or sulfatation. As shown in Figure 3-25, the structure of polyphenolic compounds are modified by enzymatic reaction and reached to liver via portal vein by active, passive, or facilitated transportation. However, only 5–10% of the total polyphenolic compounds may be taken in the small intestine, and these compounds may undergo further extensive metabolism. The rest of polyphenols may be accumulated in the large intestinal and excreted in the feces. The consumption of polyphenolic compounds may be significantly different based on the nature of foods.

In order to explore the absorption regularity of TF1 and EGCG in the organism, in our previous study, we research used the Caco-2 cell monolayer model *in vitro* to simulate the absorption of TF1 and EGCG in the small intestine (Table 3-13). The results illustrated that the absorptions of the TF1 and EGCG in Caco-2 model showed that the apparent permeability coefficient raised with the increasing of two compound concentrations in the range of 10–100 μM. The efflux rates of the TF1 and EGCG

Table 3-13. Papp of TF_1 and EGCG at different concentrations across Caco-2 cell monolayers.

(μM)	TF₁			EGCG		
	Papp (1×10⁻⁷) (cm/s)			Papp (1×10⁻⁷) (cm/s)		
	AP → BL	BL → AP	R_{Papp}	AP → BL	BL → AP	R_{Papp}
10	ND[a]	2.60 ± 0.21	—	0.91 ± 0.15	2.05 ± 0.20	2.26
20	0.70 ± 0.15	3.06 ± 0.47	4.39	0.96 ± 0.13	2.91 ± 0.39	3.03
30	0.91 ± 0.10	3.24 ± 0.20	3.56	1.01 ± 0.17	3.91 ± 0.40	3.88
50	1.16 ± 0.07	5.26 ± 0.48	4.51	1.12 ± 0.10	5.45 ± 0.66	4.87
100	2.07 ± 0.09	10.01 ± 0.93	4.82	1.85 ± 0.15	11.80 ± 1.20	6.38

showed the same rules as absorption. However, the increasing range of efflux rate was higher than that of absorption rate. The values of Papp about TF1 and EGCG in the cell model were lower than 1×10^{-6} cm/s, which indicated that both belonged to the kind of drugs that were difficult to absorb. However, to compare the absorption rate, TF1 was higher than EGCG in this model. Both efflux transports showed the passive process because the efflux rates of TF1 and EGCG were higher than in cell model. Since the efflux regulation of TF1 was in accordance with EGCG by temporal variation (Fig. 3-26), it suggested that both were the substrate of the same efflux protein.

Furthermore, the bioavailability of polyphenols is not only affected by their ability to cross a membrane but also by the maintenance of structural integrity in humans. A large proportion of polyphenol are metabolized in the small intestine and then modified by the liver and other organs. Some others (small intestine unabsorbed) pass through to the large intestine, where they undergo further modification by colonic microflora. All *in vitro* studies using single compound or polyphenol-rich dietary have to comprehensively survey bioavailability of them. For example, tea catechins in human plasma always showed their forms of methylated, sulfated, and glucuronidated conjugates. Moreover, according to "the drug discovery rule of 5", the molecular weight of compound for dietary polyphenol should be no more than 500 Da, less than five OH bond donors, and less than 10 H-bond acceptors to be absorbed into the

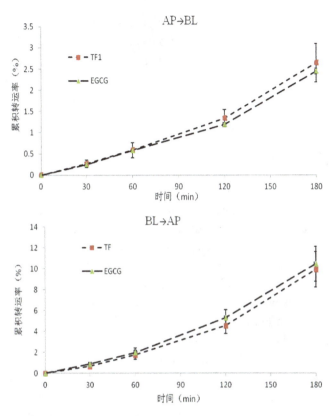

Figure 3-26. Effect of temporal variation on the transport of TF$_1$ and EGCG L-theanine in Caco-2 monolayers.

human body. To date, most investigation on biological activity of dietary polyphenols is based on cultured cells as tissue models. Many previous studies reported that EGCG, GC, and CGC in tea are the active components based on its strong activity *in vitro*. If one measures the concentration of EGCG in serum, the actually absorbed EGCG is rather low, absorption, in turn, depends on the chemical structure of a given property. Another question depends on the concentration levels of sample used *in vitro*, which are usually several folds above the plasma concentration. Such analogical inaccuracy could be also found in some *in vivo* works to evaluate bioavailability, in which given doses exceed the average recommended dietary intake.

The *in vivo* bioactive effects of polyphenols depend on their respective intakes, absorption, metabolism, and bioavailability, which could be quite different. Although polyphenols are abundant in food, some of them are very poorly utilized and others are not even absorbed at all for human circulation, thus their *in vivo* effects are restricted. Intakes of many flavonoids are rather low, and plasma concentrations rarely exceed 1 µmol/L such as flavonols, flavones, and flavanols (limited absorption to rapid elimination). Flavanones and isoflavones are the flavonoids with the best bioavailability profiles, and plasma concentrations may reach 5 µmol/L. However, the distribution of these substances is restricted to citrus fruit and soya. Finally, some phenolic acids are found in our diet, however, esterified after ingestion, decreasing the absorption. In general, polyphenols are eliminated very fast from plasma, which indicates that consumption of natural product on a daily basis is necessary to maintain high concentrations of metabolites in the blood. Recent studies have greatly increased our knowledge of the plasma concentrations and urinary excretion of polyphenol metabolites in humans. However, values for these variables seem not to be well correlated with concentrations measured in tissues. Collected data, especially those obtained from *in vivo* studies, revealed that some phenolic metabolites may accumulate in certain gut organ or tissue. The metabolites also may present differently in target tissue or plasma, and the characteristic of the polyphenol metabolites therefore should be further evaluated. More available data are required to dig intracellular metabolism and the accumulation of polyphenol metabolites in specific organs or tissues. Additionally, it should be noticed that the differences may exist between animals and humans in some metabolic processes, especially the conjugation process. *In vivo* bioavailability of polyphenols combines several factors, as mentioned above, intestinal absorption, the characteristics of circulating metabolites, hepatic enzyme metabolism, plasma kinetics, excretion of glucuronides, metabolism by microorganism, cellular uptake, and intracellular metabolism. The challenge to evaluate health effects of phenolic compounds is by combining all the information and relating all the variables. These assignments are difficult because of the relative weight of each factor, however, they depend on the polyphenol considered. Some phenolic compounds may be less efficiently absorbed than others but nevertheless reach equivalent plasma concentrations because of lower secretion and relatively lower elimination. Therefore, better understanding of bioavailability is indispensable for investigating the health effects of polyphenols, whatever the approach used. The fact that most phenolic aglycones are not important metabolites in blood because of their comprehensive conjugation in the digestive tract, has thus far been largely ignored. In addition, many *in vitro* studies dealing with the mechanisms of biological action of polyphenols continue to concentrate on aglycones or glycosides rather than their corresponding metabolites, often at concentrations that cannot realistic be attained in the body. Therefore, it is more essential to confirm the effects observed with aglycones through studies using physiologic concentrations of the metabolites actually found in the body. Additionally, the activities of microbial metabolites must be examined in further studies to determine active structures, available concentrations, and potential modulation of the capacity of the microflora to produce such metabolites. Clinical research plays an important role in evaluation of health benefit effect of phenolic by providing reliable information on the prevention of related diseases. Better knowledge of some factors influenced the bioavailability, such as the kinetics of absorption, accumulation, and elimination, will take the challenge out of the design of such studies. Moreover, more exact data on the nature of the circulating metabolites and on metabolism by the microflora can now be used for interpretations. Research on polyphenol bioavailability must finally allow us to correlate polyphenol intakes with one or several accurate measures of bioavailability (such as concentrations of key bioactive metabolites in plasma and tissues) and with potential health effects in epidemiologic studies. Knowledge of these correlations must be attained despite the difficulties linked to the high diversity of polyphenols, their different bioavailability, and the high interindividual variability observed in some metabolic processes, especially those in which the microflora is involved.

3.10 The security of theaflavins and thearubigins

Theaflavins and thearubigins are natural compounds with less toxicity. According to 158~316 times of the daily dose (10~20 mg/kg BW) of adults, the acute oral LD50 of theaflavin in mice is more than 100 times of the adult dose. However, too high dose can also cause acute toxicity and chronic toxicity.

3.10.1 *Acute toxicity of theaflavin and thearubigin*

The acute toxicity of theaflavins have been assessed in mice and rats. After 30 minutes of theaflavin (40% purity) administration by gavage at 2150 mg/kg BW, 4640 mg/kg BW, and 10000 mg/kg BW, the male and female mice both appeared several unnormal symptoms including decreased mobility, dyskinesia, and nervousness. Several deaths of mice were observed within 2~24 hours after intragastric administration. At the dose of 1000 mg/kg BW, all the mice had normal appearance and activity, without no deaths. The female mouse acute oral LD50 of theaflavins extracts was 2710.0 mg/kg BW, which was 136~271 times of adult daily dose (10~20 mg/kg BW). The male mouse acute oral LD50 of theaflavin extracts was 3160.0 mg/kg BW, which was 158~316 times of adult daily dose (10~20 mg/kg BW). In summary, the male and female mouse acute oral LD50 of theaflavins was more than 100 times of adult dosage, suggesting that these compounds were less toxic in accordance with the acute toxicity evaluation standards of health food [Zhang *et al.*, 2014].

Xu *et al.* [1996] studied the oral acute toxicity of theaflavins in rat (2200~11300 mg/kg BW) and mice (2330~23156 mg/kg BW). The symptoms of poisoning in rats and mice were basically the same. After exposure to the drugs, central nervous system excitement symptoms appear rapidly, which are manifested as increased activity, deepening, and speeding up breathing. After 30 minutes, the excitement of animals was inhibited, with decreased motion and drowsiness. Most of them had morphine-like symptoms, such as embracing forelimbs, muscular rigidity, and tubular tail. The dead rats were dissected and diffuse hemorrhagic foci in both lungs were found. The LD50 of rats was 5230 mg/kg, with the 95% confidence limit of 4366~6265 mg/kg, while the LD50 of mice was 6081 mg/kg, with the 95% confidence limit of 4147~8924 mg/kg. According to the tea polyphenol, LD50 of rats was 2496 mg/kg [Yang *et al.*, 1992], which meanings the toxicity of theaflavins is lower than tea polyphenol's.

3.10.2 *Cumulative toxicity of theaflavin and thearubigin*

Forty Kunming mice, half male, and half female, were randomly divided into two groups. The experimental group was treated by gavage with regular increasing dose. The control group was given distilled water of equal volume. The poisoning symptoms and deaths of animals were observed and recorded every day.

When the cumulative exposure dose reached 2.6 folds of LD_{50}, some rats died. When the cumulative exposure dose reached 5.3 folds of LD_{50}, the total death number was five. When the cumulative exposure dose was more than 5.3 folds of LD_{50}, the transient activity increased, the respiration deepened, the central nervous system excitatory symptoms such as muscle stiffness and tail erectness and the sustained activity are significantly reduced, and the central nervous system inhibitory symptoms such as lethargy and diet decrease. Pathological examination showed that diffuse hemorrhagic foci occurred in both lungs [Xu *et al.*, 1996].

3.10.3 *Mutagenicity of theaflavin and thearubigin*

For micronucleus test, 25 Kunming mice were randomly divided into five groups. Three doses of theaflavins (375 mg/kg, 750 mg/kg, and 1500 mg/kg) were administered orally for two consecutive days. A 40 mg/kg of cyclophosphamide was administered intraperitoneally in the positive control group and normal saline was administered orally in the negative control group. The animals were killed six hours after the last administration, the sternal bone marrow was taken for routine sectioning, and 1000 PCE micronucleus rates were counted. The results showed that the PCE micronucleus rates at three doses of theaflavins were 0.20%, 0.28%, and 0.24%, respectively. No significant difference were observed among three groups, indicating that the micronucleus test of theaflavins were negative.

For Ames test, four different histidine-deficient (his−) Salmonella typhimurium strains TA97, TA98, TA100, and TA102 were used in the reverse mutation test both with and without S9 mixture. Five theaflavin concentrations of 312.5 μg/plate, 625 μg/plate, 1250 μg/plate, 2500 μg/plate, and 5000 μg/plate were designed in this test. All theaflavin treatment showed no evidence of mutagenic activity in this bacterial system with MR value less than 2.

For the chromosome aberration test of mouse primary spermatocyte, 25 male Kunming mice were randomly divided into five groups. Theaflavins was administered orally for five consecutive days at three doses (375 mg/kg BW, 750 mg/kg BW, 1500 mg/kg BW). The positive control group received intraperitoneal injection of cyclophosphamide (50mg/kg) for five consecutive days. The negative control group was given normal saline by gavage. On the 13th day after administration, the animals were killed, and their testes were taken for sectioning according to the conventional method. A total of 100 metaphase primary spermatocytes were analyzed for each animal. The Poisson test showed that there were significant differences between the 1500 mg/kg-dose group and the control group in terms of sex chromosome and autosomal univalent ($p < 0.05$), while there was no significant difference in the rate of chromosome aberration between the theaflavin treated group and the control group ($p > 0.05$).

3.11 Conclusion

According to many published reports, theaflavins and thearubigin are bioactive compounds as similar as tea polyphenols. To compare with other natual products, theaflavins and thearubigin not only have many biological functions but also good safety. At low doses, theaflavins play an important role as antioxidants, and the physiological consequences of the inhibition of ATP synthase or Complex I by theaflavins might be limited. At high dosage, theaflavins might have some effect on the electron transport chain. Unfortunately, the bioavailability of theaflavins and thearubigin are still low now. For future utilization of TF and TR, scientists have to improve the bioavailability of theaflavins through nanotechnology or other ways.

References

Albertini R, Moratti R, De Luca G. (2002) Oxidation of low-density lipoprotein in atherosclerosis from basic biochemistry to clinical studies. *Curr Mo Med*, **2**, 579–592.

Anderson RF, Fisher LJ, Hara Y, *et al.* (2001) Green tea catechins partially protect DNA from •OH radical-induced strand breaks and base damage through fast chemical repair of DNA radicals. *Carcinogenesis*, **22**(8), 1189–1193.

Apostolides Z, Balentine DA, Harbowy ME, *et al.* (1997) Inhibition of PhIP mutagenicity by catechins, and by theaflavins and gallate esters. *Mutat Res*, **389**(2-3), 167–172.

Bajaj KL, Anan T, Tsushida T, *et al.* (1987) Effects of (–) epicatechin on oxidation of theaflavins by polyphenol oxidase from tea leaves. *Agri Biol Chem*, **51**(7), 1767–1772.

Balentine DA, Wiseman SA, Bouwens LCM. (1997) The chemistry of tea flavonoids. *Crit Rev Food Sci*, **37**(8), 693–704.

Baynes C, Henderson AD, Anyaoku V, *et al.* (1991) The role of insulin insensitivity and hepatic lipase in the dyslipidaemia of type 2 diabetes. *Diabetic Med*, **8**, 560–566.

Berrisford JM, Sazanov LA. (2009) Structural basis for the mechanism of respiratory complex I. *J Biol Chem*, **284**, 29773–29783.

Birari R, Javia V, Bhutani KK. (2010) Antiobesity and lipid lowering effects of Murraya koenigii (L.) Spreng leaves extracts and mahanimbine on high fat diet induced obese rats. *Fitoterapia*, **81**, 1129–1133.

Birari RB, Bhutani KK. (2007) Pancreatic lipase inhibitions from natural sources: Unexplored potential. *Drug Discov Today*, **12**, 879–889.

Birari RB, Gupta S, Mohan CG, *et al.* (2011) Antiobesity and lipid lowering effects of Glycyrrhiza chalcones: Experimental and computational studies. *Phytomedicine*, **18**, 795–801.

Birrell JA, Yakovlev G, Hirst J. (2009) Reactions of the flavin mononucleotide in complex I: A combined mechanism describes NADH oxidation coupled to the reduction of APAD(+), ferricyanide, or molecular oxygen. *Biochemistry*, **48**, 12005–12013.

Borska S, Gebarowska E, Wysocka T, *et al.* (2003) Induction of apoptosis by EGCG in selected tumour cell lines in vitro. *Folia Histochem Cyto*, **41**, 229–232.

Brown AG, Eyton WB, Hohnes A, *et al.* (1969) The identification of thearubigins as polymeric proanthocyanidins. *Phytochemistry*, **8**, 2333–2340.

Brown AG, Falsaw CP, Haslam E, *et al.* (1966) The constitution of theaflavins. *Tetrahedron Lett*, **11**, 1193–1204.

Brunzell JD, Zambon A, Deeb SS. (2012) The effect of hepatic lipase on coronary artery disease in humans is influenced by the underlying lipoprotein phenotype. *BBA-Mol Cell Biol L*, **1821**, 365–372.

Bu-Abbas A, Clifford MN, Walker R, *et al.* (1998) Contribution of caffeine and flavanols in the induction of hepatic phase II activities by green tea. *Food Chem Toxicol*, **36**, 617–621.

Bu-Abbas A, Clifford MN, Walker R, *et al.* (1999) Modulation of hepatic cytochrome P450 activity and carcinogen bioactivation by tea: contribution of caffeine and flavanols. *Environ Toxicol Phar*, **7**, 41–47.

Bu-Abbas A, Nunez X, Clifford MN, *et al.* (1996) A comparison of the antimutagenic potential of green, black and decaffeinated teas: contribution of flavanols to the antimutagenic effect. *Mutagenesis*, **11**, 597–603.

Burton SG, Bosho A, Edwards W, *et al.* (1998) Biotransformation of phenols using immobilized polyphenol oxidase. *J Mol Catal B Enzym*, **5**, 411–416.

Calhoun MW, Gennis RB. (1993) Demonstration of separate genetic loci encoding distinct membrane-bound respiratory NADH dehydrogenases in Escherichia coli. *J Bacteriol*, **175**, 3013–3019.

Calhoun MW, Oden KL, Gennis RB, *et al.* (1993) Energetic efficiency of Escherichia coli: effects of mutations in components of the aerobic respiratory chain. *J Bacteriol*, **175**, 3020–3025.

Chang HP, Wang ML, Chan MH, *et al.* (2011) Antiobesity activities of indole-3-carbinol in high-fatdiet-induced obese mice. *Nutrition*, **27**, 463–470.

Charakida M, Nicholas F. (2012) Drug treatment of obesity in cardiovascular diseaser. *Am J Cardiovasc Drug*, **12**, 93–104.

Chen LY, Tu YY, Chen X, *et al.* (2002) A study on the stability of catechins under acidic conditions. *J Tea*, **28**(2), 86–88.

Chen N, Bezzina R, Hinch E, *et al.* (2009) Green tea, black tea, and epigallocatechin modify body composition, improve glucose tolerance, and differentially alter metabolic gene expression in rats fed a high-fat diet. *Nutr Res*, **29**, 784–793.

Chung JY, Huang CS, Meng XF, *et al.* (1999) Inhibition of activator protein 1activity and cell growth by purified green tea and black tea polyphenols in H-ras-transformed Cells: structure-activity relationship and mechanism involved. *Cancer Res*, **59**(18), 4610–4617.

Clark KJ, Grant PG, Sarr AB. (1998) An in vitro study of theaflavins extracted from black tea to neutralize bovine rotavirus and bovine coronavirus infections. *Vet Microbiol*, **63**, 147–157.

Collier PD, Bryce T, Mallows R, Thomas PE. (1973) Tetrohedron. **29**, 125–142.

Cooke MS, Evans MD, Dizdaroglu M, *et al.* (2003) Oxidative DNA damage: mechanism, mutation, and disease. *FASEB J*, **17**(10), 1195–1214.

Degenhardt A, Engelhardt UH, Lakenbrink C, *et al.* (2000a) Preparative separation of polyphenols from tea by high-speed countercurrent chromatography. *J Agr Food Chem*, **48**, 3425–3430.

Degenhardt A, Habben S, Winterhalter P. (2002) Isolation of the lignan secoisolariciresinol diglucoside from flaxseed (Linum Usitatissimum L.) by high-speed countercurrent chromatography. *J Chromatogr A*, **943**, 299–302.

Degenhardt A, Habben S, Winterhalter P. (2004) Isolation of physiologically active compounds from nutritional beverages by countercurrent chromatography (CCC). In *Chemistry and Flavour of Nutritional Beverages*, Shahidi F, Weerasinghe DK (eds.), ACS Symposium Series, American Chemical Society, Washington, DC, in press.

Degenhardt A, Hofmann S, Knapp H, *et al.* (2000c) Preparative isolation of anthocyanins by high-speed countercurrent chromatography and application of the color activity concept to red wine. *J Agr Food Chem*, **48**, 5812–5818.

Degenhardt A, Knapp H, Winterhalter P. (2000b) Separation and purification of anthocyanins by high-speed countercurrent chromatography and screening for antioxidant activity. *J Agr Food Chem*, **48**, 338–343.

Degenhardt A, Knapp H, Winterhalter P. (2001) Separation of natural food colorants by high-speed countercurrent chromatography. In *Chemistry and Physiology of Selected Food Colorants*, Ames JM, Hofmann T (eds.), ACS Symposium Series 775, American Chemical Society, Washington, DC, 22–42.

Degenhardt A, Winterhalter P. (2001) Isolation and purification of isoflavones from soy flour by high-speed countercurrent chromatography. *Eur Food Res Technol*, **213**, 277–280.

Dix MA, Fairley CJ, Millin DJ, *et al.* (1981) Enzymatic synthesis of tea theaflavin derivatives. *J Sci Food Agr*, **32**, 920–924.

Du QZ, Jiang HY, Ito Y. (2001) Separation of theaflavins of black tea. High-speed counter-current chromatography vs. Sephadex LH-20 gel column chromatography. *J Liq Chromatogr RT*, **24**(15), 2363–2369.

Du YT, Wang X, Wu XD, *et al.* (2005) Keemun black tea extract contains potent fatty acid synthase inhibitors and reduces food intake and body weight of rats via oral administration. *J Enzym Inhib Med Ch*, **20**, 349–356.

Feng Q, Torii Y, Uchida K, *et al.* (2002) Black tea polyphenols: Theaflavins prevent cellular DNA damage by inhibiting oxidative stress and suppressing cytochrome P4501A1 in cell cultures. *J Agr Food Chem*, **50**, 213–220.

Finkel T, Holbrook NJ. (2000) Oxidants, oxidative stress and the biology of ageing, *Nature*, **408**(6809), 239–247.

Furukawa S, Fujita T, Shimabukuro M, *et al.* (2004) Increased oxidative stress in obesity and its impact on metabolic syndrome. *J Clin Invest*, **114**, 1752–1761.

Geissman TA. (1962) The chemistry of flavonoid compounds. In *Pergamon*, Oxford, UK, 468–512.

Gledhill JR, Montgomery MG, Leslie AGW, *et al.* (2007) Mechanism of inhibition of bovine F1-ATPase by resveratrol and related polyphenols. *Proc Natl Acad Sci USA*, **104**, 13632–13637.

Gledhill JR, Walker JE. (2005) Inhibition sites in F1-ATPase from bovine heart mitochondria. *Biochem J*, **386**, 591–598.

Gomes A, Vedasiromoni JR, Das M, *et al.* (1995) Anti-hyperglycemic effect of black tea (Camellia sinensis) in rat. *J Ethnopharmacol*, **45**, 223–226.

Gong YY, Han C, Chen JS. (2000) Effects of tea polyphenols and tea pigment inducing the activity of phase II detoxicating enzymes and on the chemoprevention of liver precancerous lesions. *J Hyg Res*, **29**(3), 159–161.

Goodsall CW, Parry AD, Safford R. (1996) The mechanism of theaflavin oxidation during black tea manufacture. *J Sci Food Agr*, **63**, 435–438.

Green RH. (1994) Inhibition of multiplication of influenza virus by extracts of tea. *Proc Soc Exp Biol Med*, **71**, 84–85.

Gupta S, Chaudhuri T, Ganguly DK, *et al.* (2001) Anticlastogenic effects of black tea (World blend) and its two active polyphenols theaflavins and thearubigins in vivo in Swiss albino mice. *Life Sci*, **69**, 2735–2744.

Gupta S, Hastak K, Afaq F, *et al.* (2004). Essential role of caspases in epigallo-catechin-3-gallate-mediated inhibition of nuclear factor κB and induction of apoptosis. *Oncogene*, **23**, 2507–2522.

Han C, Gong YY. (1999) Experimental studies on the cancer chemoprevention of tea pigments. *J Hyg Res*, **28**(6), 343–348.

Han LK, Takaku T, Li J, *et al.* (1999) Antiobesity action of oolong tea. *Int J Obesity*, **23**, 98–105.

Hasaniya N, Youn K, Xu M, *et al.* (1997) Inhibitory activity of green and black tea in a free radical generating system using 2-amino-3-methylimidazo (4, 5-b) quinoline as substrate. *Jap J Cancer Res*, **88**(6), 553–558.

Hibasami H, Komiya T, Achiwa Y, *et al.* (1998). Black tea theaflavins induce programmed cell death in cultured human stomach cancer cells. *Int Mol Med*, **1**, 725–727.

Higdon JV, Frei B. (2003) Tea catechins and polyphenols: Health effects, metabolism, and antioxidant functions. *Crit Rev Food Sci*, **43**, 89–143.

Huang YW, Liu Y, Dushenkov S, *et al.* (2009) Anti-obesity effects of epigallocatechin-3-gallate, orange peel extract, black tea extract, caffeine and their combinations in a mouse model. *J Funct Foods*, **1**, 304–310.

Hussain SP, Hofseth LJ, Harris CC. (2003) Radical causes of cancer. *Nat Rev Cancer*, **3**(4), 276–285.

Jhoo JW, Lo CY, Li SM, *et al*. (2005) Stability of black tea polyphenol, theaflavin, and identification of theanaphthoquinone as its major radical reaction product. *J Agri Food Chem*, **53**(15), 6146–6150.

Jovanovic SV, Hara Y, Steenken S, *et al*. (1997) Antioxidant potential of theaflavins. A pulse radiolysis study. *J Am Chem Soc*, **119**(23), 5337–5343.

Karmakar S, Das D, Maiti A, *et al*. (2011) Black tea prevents high fat diet-induced non-alcoholic steatohepatitis. *Phytother Res*, **25**, 1073–1081.

Khaled HB, Ghlissi Z, Chtourou Y, *et al*. (2012) Effect of protein hydrolysates from sardinelle (Sardinella aurita) on the oxidative status and blood lipid profile of cholesterol-fed rats. *Food Res Int*, **45**, 60–68.

Khan N, Mukhtar H. (2007) Tea polyphenols for health promotion. *Life Sci*, **81**, 519–533.

Koh WS, Lee SJ, Lee H, *et al*. (1998) Differential effects and transport kinetics of ascorbate derivatives in leuckemic cell lines. *Anticancer Res*, **8**, 2487–2493.

Kuo KL, Weng MS, Chiang CT, *et al*. (2005) Comparative studies on the hypolipidemic and growth suppressive effects of oolong, black, pu-erh, and green tea leaves in rats. *J Agr Food Chem*, **53**, 480–489.

Lafay S, Gil-Izquierdo A. (2008) Bioavailability of phenolic acids. *Phytochem Rev*, **7**, 301–311.

Lambert AJ, Brand MD. (2004) Inhibitors of the quinone-binding site allow rapid superoxide production from mitochondrial NADH:ubiquinone oxidoreductase (Complex I). *J Biol Chem*, **279**, 39414–39420.

Lee DW, Leinung MC, Rozhavskaya-Arena M, *et al*. (2002) Leptin and the treatment of obesity: Its current status. *Eur J Pharmacol*, **440**, 129–139.

Leung LK, Su YL, Chen RY, *et al*. (2001) Theaflavins in black tea and catechins in green tea are equally effective antioxidants. *J Nutr*, **131**(9), 2248–2251.

Lewis JG, Stewart W, Adams DO. (1988) Role of oxygen radicals in induction of DNA damage by metabolite of benzene. *Cancer Res*, **48**(17), 4762–4765.

Lewis JR, Davis AL, Cai Y, *et al*. (1998). Theaflavate B, isotheaflavin-3′-o-gallate and neotheaflavin-3-o-gallate: three polyphenolic pigments from black tea. *Phytochemistry*, **49**(8), 2511–2519.

Li B, Vik SB, Tu YY. (2012) Theaflavins inhibit the ATP synthase and the respiratory chain without increasing superoxide production. *J Nutr Biochem*, **23**, 953–960.

Li MYA. (2001) Comparative study of oral bacteria inhibition of tea polyphenol and tea-pigment. *J Den Prev Treat*, **19**(1), 3–4.

Li W, Wu JX, Tu YY. (2010) Synergistic effects of tea polyphenols and ascorbic acid on human lung adenocarcinoma SPC-A-1 cells. *J Zhejiang Univ-Sci B*, **11**(6), 458–464.

Liang YC, Chen YC, Lin YL, *et al.* (1999) Suppression of extracellular signals and cell proliferation by the black tea polyphenol, thiaflavin-3-3′-digallate. *Carcinogenesis*, **20**, 733–736.

Lin JK, Chen PC, Ho CT, *et al.* (2000) Inhibition of xanthine oxidase and suppression of intracellular reactive oxygen species in HL-60 cells by theaflavin-3,3′-digallate, (–)-epigallocatechin-3-gallate, and propyl gallate. *J Agr Food Chem*, **48**, 2736–2743.

Lin JK, Chen PC, Ho CT, *et al.* (2002) Inhibition of xanthine oxidase and NADPH oxidase by tea polyphenols. *Free Radicals in Food*, **20**, 264–281.

Lin JK, Lin-Shiau SY. (2009) Fermented tea is more effective than unfermented tea in suppressing lipogenesis and obesity. In *Tea and tea products: Chemistry and health promoting properties*, Ho CT, Lin JK, Shahidi F (eds.), Boca Raton, FL: CRC Press.

Lin YL, Tsai SH, Lin-Shiau SY, *et al.* (1999) Theaflavin-3, 3′-digallate from black tea blocks the nitric oxide synthase by downregulating the activation of NF-kappaB in macrophages. *Eur J Pharmacol*, **367**(2-3), 379–388.

Liu JW, Nagao N, Kageyama K, *et al.* (2000) Anti-metastatic effect of an autooxidation-resistant and lipophilic ascorbic acid derivative through inhibition of tumor invasion. *Anticancer Res*, **20**, 113–118.

Lu J, Ho CT, Geetha G, *et al.* (2000) Differential effects of theaflavin monogallates on cell growth, apoptosis, and cox-2 gene expression in cancerous versus normal cells. *Cancer Res*, **60**, 6465–6471.

Lu YP, Lou YR, Xie JG, *et al.* (1997) Inhibitory effect of black tea on the growth of established skin tumors in mice: effects on tumor size, apoptosis, mitosis and bromodeoxyuridine incorporation into DNA. *Carcinogenesis*, **18**, 2163–2169.

Luczaj W, Skrzydlewska E. (2005) Antioxidative properties of black tea. *Prev Med*, **40**, 910–918.

Lung HL, Ip WK, Wong CK, *et al.* (2002) Anti-proliferative and differentiation-inducing activities of the green tea catechin epigallocatechin-3-gallate (EGCG) on the human eosinophilic leukemia EoL-1 cell line. *Life Sci*, **72**(3), 257–268.

Maeda-Yamamoto M, Kawahara H, Tahara N, *et al.* (1999) Effects of tea polyphenols on the invasion and matrix metalloproteinases activities of human fibrosarcoma HT 1080 cells. *J Agr Food Chem*, **47**(6), 2350–2354.

Marchesini G, Moscatiello S, Di Domizio S, *et al.* (2008) Obesity-associated liver disease. *J Clin Endocr M*, **93**, S74–S80.

Matsumoto N, Okushio K, Hara Y. (1998) Effect of black tea polyphenols on plasma lipids in cholesterol-fed rats. *J Nutr Sci Vitaminol (Tokyo)*, **44**, 337–342.

Matsushita K, Ohnishi T, Kaback HR. (1987) NADH-ubiquinone oxidoreductases of the Escherichia coli aerobic respiratory chain. *Biochemistry*, **26**, 7732–7737.

Mayer AM. (1987) Polyphenol oxidases in plants-recent progress. *Phytochemistry*, **26**(1), 11–20.

McArdle NJ, Clifford MN, Ioannides C. (1999) Consumption of tea modulates the urinary excretion of mutagens in rats treated with IQ: role of caffeine. *Mutat Res*, **441**, 191–203.

Miller NJ, Castelluccio C, Tijburg L, *et al.* (1996) The antioxidant properties of theaflavins and their gallate esters-free radical scavengers or metal chelators. *Febs Lett*, **392**(1), 40–44.

Millin DJ, Crispin DJ, Swaine D. (1969) Non-volatile components of black tea and their contribution to the character of the beverage. *J Agr Food Chem*, **17**, 717–122.

Miyamoto S, Martinez GR, Medeiros MHG, *et al.* (2003) Singlet molecular oxygen generated from lipid hydroperoxides by the russell mechanism: studies using O-18-labeled linoleic acid hydroperoxide and monomol light emission measurements. *J Am Chem Soc*, **125**(20), 6172–6179.

Nakayama M, Suzuki K, Toda M, *et al.* (1993) Inhibition of the infectivity of influenza virus by tea polyphenols. *Antiviral Res*, **21**, 289–299.

Nakayama M, Toda M, Okubo S, *et al.* (1990). Inhibition of influenza virus infection by tea. *Lett Appl Microbio*, **11**, 38–40.

Netke SP, Roomi MW, Ivanov V, *et al.* (2003) A specific combination of ascorbic acid, lysine, proline and epigallocatechingallate inhibits proliferation and extracellular matrix invasion of various human cancer cell lines. *Res Commun Pharmacol Toxicol Emerg Drugs*, **11**, 37–50.

Nishiumi S, Bessyo H, Kubo M, *et al.* (2010) Green and black tea suppress hyperglycemia and insulin resistance by retaining the expression of glucose transporter 4 in muscle of high-fat diet-fed C57BL/6J Mice. *J Agr Food Chem*, **58**, 12916–12923.

Nomura M, Ma WY, Chen NY, *et al.* (2000). Inhibition of 12-O-tetradecanoylphorbol-13-acetate-induced NF-kB activation by tea polyphenols, (–)-epigallocatechin gallate and theaflavins, *Carcinogenesis*, **21**(10), 1885–1890.

Ohnishi ST, Shinzawa-Itoh K, Ohta K, *et al.* (2010). New insights into the superoxide generation sites in bovine heart NADH-ubiquinone oxidoreductase

(Complex I): The significance of protein-associated ubiquinone and the dynamic shifting of generation sites between semiflavin and semiquinone radicals. *Biochim Biophys Acta*, **1797**, 1901–1909.

Opie SC, Clifford MN, Robertson A. (1993) The role of (–)-epicatechin and poly-phenol oxidase in the coupled oxidative breakdown of theaflavins. *J Sci Food Agr*, **63**(4), 435–438.

Opie SC, Clifford MN, Robertson A. (1995) The formation of thearubigin-like substances by *in votro* polyphenol oxidase-mediated fermentation of indi-vidual flavan-3-ols. *J Sci Food Agr*, **67**, 501–512.

Opie SC, Robertson A, Clifford MN. (1990) Black tea thearubigins: Their HPLC separation and preparation during in vitro oxidation. *J Sci Food Agr*, **50**(4), 547–562.

Owuor PO, Mutea MJS, Obanda, AM, *et al.* (1986a) Effects of withering on some quality parameters of black tea: CTC Preliminary Results. *Tea*, **7**(2), 13–17.

Owuor PO, Reeves SG. (1986b). Optimizing fermentation time in black tea manufacture. *Food Chem*, **21**, 195–203.

Owuor PO, Reeves SG, Wanyoko JK. (1986c) Correlation of theaflavin content and valuations of Kenyan black teas. *J Sci Food Agr*, **37**, 507–513.

Pan MH, Lin-Shiau SY, Ho CT, *et al.* (2000) Suppression of lipopolysaccharide-induced nuclear factor-kappaB activity by theaflavin-3,3′-digallate from black tea and other polyphenols through down-regulation of IkB kinase activity in macrophages. *Biochem Pharm*, **59**, 357–367.

Powell C, Clifford MN, Opie S, *et al.* (1993) Tea cream formation: the contribu-tion of black tea phenolic pigments determined by HPLC. *J Sci Food Agr*, **63**, 77–80.

Pruidze GN, Mchedlishvili NI, Omiadze NT, *et al.* (2003). Multiple forms of phenol oxidase from Kolkhida tea leaves (Camelia Sinensis) and their role in tea production. *Food Res Int*, **36**, 587–595.

Racotta IS, Leblanc J, Richard D. (1994) The effect of caffeine on food intake in rats: involvement of corticotropin-releasing factor and the sympatho-adrenal system. *Pharmacol Biochem Be*, **48**, 887–892.

Rains TM, Agarwal S, Maki KC. (2011) Antiobesity effects of green tea cate-chins: a mechanistic review. *J Nutr Biochem*, **22**, 1–7.

Ramadan G, El-Beih NM, Abd El-Ghffar EA. (2009) Modulatory effects of black v. green tea aqueous extract on hyperglycaemia, hyperlipidaemia and liver dysfunction in diabetic and obese rat models. *Brit J Nutr*, **102**, 1611–1619.

Ramesh E, Elanchezhian R, Sakthivel M, *et al.* (2008) Epigallocatechin gallate improves serum lipid profile and erythrocyte and cardiac tissue antioxidant parameters in Wistar rats fed an atherogenic diet. *Fund Clin Pharmacol*, **22**, 275–284.

Ricci R, Bevilacqua F. (2012) The potential role of leptin and adiponectin in obesity: A comparative review. *Vet J*, **191**, 292–298.

Rice-Evans CA, Miller NJ, Paganga G. (1997) Antioxidant properties of phenolic compounds. *Trends Plant Sci*, **2**(4), 152–159.

Roberts EAH. (1958) The phenolic substances of manufactured tea. *J Sci Food Agr*, **9**, 212–216.

Roberts EAH. (1962) Assessment of quality in teas by chemical analysis. *Two and A Bud*, **9**, 3–8.

Robertson A. (1983a) Effects of physical and chemical conditions on the in-vitro oxidation of tea catechins. *Phytochemistry*, **22**, 889–896.

Robertson A. (1983b) Effect of catechins concentration on the formation of black tea polyphenols during in vitro oxidation. *Phytochemistry*, **22**, 897–903.

Robertson A, Bendall DS. (1983) Production and HPLC analysis of black tea theaflavins and thearubigins during in vitro oxidation. *Phytochemistry*, **22**, 883–887.

Rompel A, Fisher H, Meiwes D, *et al.* (1999) Substrate specificity of catechol oxidase from Lycopus europaeus and characterizatrion of the bioproducts of enzymic caffeic acid oxidation. *FEBS Lett*, **445**(1), 103–110.

Roomi MW, House D, Eckert-Macksic M, *et al.* (1998) Growth, suppression of malignant leukaemia cell line in vitro by ascorbic acid (vitamin C) and its derivatives. *Cancer Lett*, **122**, 93–99.

Runeckles VC, Tso TC. (1972) Recent advances in phytochemistry. Academic Press, New York. **5**, 247–316.

Sanderson GW, Ranadive AS, Eisenberg LS, *et al.* (1976) Contributions of poly-phenolic compounds to the taste of tea. *ACS Symp Ser*, **26**, 14–16.

Sarkar A, Bhaduri A. (2001) Black tea is a powerful chemopreventor of reactive oxygen and nitrogen species: comparison with its individual catechin con-stituents and green tea. *Biochem Bioph Res Co*, **284**, 173–178.

Schneider I, Kressel G, Meyer A, *et al.* (2011) Lipid lowering effects of oyster mushroom (*Pleurotus ostreatus*) in humans. *J Funct Foods*, **3**, 17–24.

Schwarz M, Hillebrand S, Degenhardt A, *et al.* (2003) Application of high-speed countercurrent chromatography to the large-scale isolation of anthocyanins. *Biochem Eng J*, **14**(3), 179–189.

Shiraki M, Hara Y, Osawa T, *et al.* (1994) Antioxidative and antimutagenic effects of theaflavins from black tea. *Mutat Res*, **323**, 29–34.

Shukla Y, Taneja P. (2002) Anticarcinogenic effect of black tea on pulmonary tumors in Swiss albino mice. *Cancer Lett*, **176**, 137–141.

Shuler LM, Kargi F. (1992) In *Bioprocess Engineering*. Prentice Hall P T R, Englewood Cliff, New Jersey.

Sohn OS, Surace A, Fiala ES, *et al.* (1994) Effects of green and black tea on hepatic xenobiotic metabolizing systems in the male F344 rat. *Xenobiotica*, **24**, 119–127.

Takino Y, Imagawa H, Horikawa H, *et al.* (1964) Studies on the mechanism of the oxidation of tea leaf catechins formation of the reddish/orange pigment and its spectral relationship to some benzotropolone derivatives. *Agr Biol Chem*, **28**, 64–71.

Tan KCB, Shiu SWM, Chu BYM. (1999) Roles of hepatic lipase and cholesteryl ester transfer protein in determining low density lipoprotein subfraction distribution in Chinese patients with non-insulin-dependent diabetes mellitus. *Atherosclerosis*, **145**, 273–278.

Tan KCB, Shiu SWM, Chu BYM. (2001) Effects of gender, hepatic lipase gene polymorphism and type 2 diabetes mellitus on hepatic lipase activity in Chinese. *Atherosclerosis*, **157**, 233–239.

Tian B, Hua YJ. (2005) Concentration dependence of prooxidant and antioxidant effects of aloin and aloe-emodin on DNA. *Food Chem*, **91**(3), 413–418.

Tian G, Zhang TY, Yang FQ, *et al.* (2000) Separation of gallic acid from Cornus officinalis Sieb. et Zucc by high-speed counter-current chromatography. *J Chromatogr*, **886**(1-2), 309–312.

Trevisanato SI, Kim YI. (2000) Tea and health. *Nutr Rev*, **58**(1), 1–10.

Tsushida T, Takeo T. (1981) The hydroxylation of p-cumarate to coffeat by the tea enzyme. *Study of Tea*, **60**, 29–33.

Tu YY, Wu XC. (1992) Mechanism of substitution of polyphenol oxidase during withering of tea leaves. *Proceeding of Tea Science Research*, 119–125.

Tu YY, Xia HL. (2004) Immobilized polyphenol oxidase catalyze purified tea polyphenols into high quality theaflavins, China Patent 02136982.8 ZL.

Tu YY, Xu XQ, Xia HL, *et al.* (2005) Optimization of theaflavins biosynthesis from tea polyphenols using an immobilized enzyme system and response surface methodology. *Biotechnol Lett*, **27**(4), 269–274.

Uchiyama S, Taniguchi Y, Saka A, *et al.* (2011). Prevention of diet-induced obesity by dietary black tea polyphenols extract in vitro and in vivo. *Nutrition*, **27**, 287–292.

Vermeer MA, Mulder TP, Molhuizen HO. (2008) Theaflavins from black tea, especially theaflavin-3-gallate, reduce the incorporation of cholesterol into mixed micelles. *J Agr Food Chem*, **56**, 12031–12036.

Wan XC, Li LX. (2001). Tea pigments and medical functions. *Chinese Natural Product R&D*, **13**(4), 65–70.

Wan XC, Nursten HE, Cai Y, *et al.* (1999) A new type of tea pigment from the chemical oxidation of epicatechin gallate and isolated from tea. *J Sci Food Agr*, **74**(3), 401–408.

Wang J, Xing D. (2002) Detection of vitamin C-induced singlet oxygen formation in oxidized LDL using MCLA as a chemiluminescence probe. *Sheng Wu Hua Xue Yu Sheng Wu Wu li Xue Bao (Shanghai)*, **34**(1), 11–15.

Wang YY, Lin SY, Sheu WHH, *et al.* (2010). Obesity and diabetic hyperglycemia were associated with serum alanine aminotransferase activity in patients with hepatitis B infection. *Metabolism*, **59**, 486–491.

Wu YY, Li W, Xu Y, *et al.* (2011) Evaluation of the antioxidant effects of four main theaflavin derivatives through chemiluminescence and DNA damage analyses. *J Zhejiang Univ-Sci B*, **12**(9), 744–751.

Xie YR, Xu P. (2000) Tea pigment gemfibrozil in treating hyperlipidemia and high uric acid in blood. *Chin J New Drugs Clin Remd*, **19**(1), 25–27.

Xu D, Wang Q, Wei L, *et al.* (1996) Study on the toxicology of theaflavins. *Chinese Journal of Public Health*, **15**(6), 368–370.

Xu F, Zhen YS. (2003) (–)-Epigallocatechin-3-gallate enhances anti-tumor of cytosine arabinoside on HL-60 cells. *Acta Pharmacia Sin*, **24**(2), 163–168.

Xu Y, Jin YX, Wu YY, *et al.* (2010) Isolation and purification of four individual theaflavins using semi-preparative high performance liquid chromatography. *J Liq Chromatogr RT*, **33**(20), 1791–1801.

Yan J, Zhao Y, Suo S, *et al.* (2012) Green tea catechins ameliorate adipose insulin resistance by improving oxidative stress. *Free Radical Biol Med*, **52**, 1648–1657.

Yang CS, Maliakal P, Meng X. (2002). Inhibition of carcinogenesis by tea. *Annu Rev Pharmacol*, **42**, 25–54.

Yang GY, Liao J, Kim K, *et al.* (1998) Inhibition of growth and induction of apoptosis in human cancer cell lines by tea polyphenols. *Carcinogenesis*, **19**(4), 611–616.

Yang MH, Wang CH, Chen HL (2001). Green, oolong and black tea extracts modulate lipid metabolism in hyperlipidemia rats fed high-sucrose diet. *J Nutr Biochem*, **12**, 14–20.

Yang RL, Shi YH, Hao G, *et al.* (2008). Increasing oxidative stress with progressive hyperlipidemia in human: Relation between malondialdehyde and atherogenic index. *J Clin Biochem Nutr*, **43**, 154–158.

Yang XQ, Wang YF, Tu YY. (2003) In *Polyphenols Chemistry*, Shanghai Science and Technology Press, Shanghai.

Yang XQ, Jia ZH, Shen SR, *et al.* (1992) Toxicology test and evaluation of tea polyphenois. *Acta Agricultur AL University Zheijangensis*, **18**(1), 23–29.

Yang ZY, Jie GL, Dong F, *et al.* (2008) Radical-scavenging abilities and antioxidant properties of theaflavins and their gallate esters in H2O2-mediated oxidative damage system in the HPF-1 cells. *Toxicol in Vitro*, **22**(5), 1250–1256.

Yang ZY, Tu YY, Xia HL, *et al.* (2007) Suppression of free-radicals and protection against H2O2-induced oxidative damage in HPF-1 cell by oxidized phenolic compounds present in black tea. *Food Chem*, **105**, 1349–1356.

Yeh CW, Chen WJ, Chiang CT, *et al.* (2003) Suppression of fatty acid synthase in MCF-7 breast cancer cells by tea and tea polyphenols: a possible mechanism for their hypolipidemic effects. *Pharmacogenomics J*, **3**, 267–276.

Yoshida H, Ishikawa T, Hosoai H, *et al.* (1999) Inhibitory effect of tea flavonoids on the ability of cells to oxidize low density lipoprotein. *Biochem Pharmacol*, **58**, 1695–1703.

Yoshino K, Hara Y, Sano M, *et al.* (1994) Antioxidative effects of black tea theaflavins and thearubigin on lipid peroxidation of rat liver homogenates induced by tert-Butylhydroperoxide. *Biol Pharm Bull*, **17**(1), 146–149.

Yun JW. (2010) Possible anti-obesity therapeutics from nature-a review. *Phytochemistry*, **71**, 1625–1641.

Zhang J, Jiang H, Wang Y, *et al.* (2014) Study on the acute toxicity of theaflavins and theaflvains tablet. *Chinese Agricultural Science Bulletin*, **30**(21), 285–288.

Zhao D, Liu ZS, Xi B. (2001) Cloning and alignment of polyphenoloxidase cDNA tea plant. *J Tea Sci*, **21**(2), 94–98.

Chapter 4

Oxidative Stress and Antioxidant Regulation of Tea Polyphenols

Guoliang Jie

Huangshan Maofeng Tea Group Co., Ltd., Anhui Province, China

Baolu Zhao

Institute of Biophysics, Chinese Academy of Sciences, Beijing, China

4.1 Introduction

In the process of metabolism, the human body constantly produces redox reaction, reactive oxygen species (ROS), and reactive nitrogen species (RNS) free radicals, resulting in oxidation and oxidative stress. There are two kinds of antioxidant systems in the body. One is enzyme antioxidant system, including superoxide dismutase (SOD), catalase (CAT), glutathione peroxidase (GSH PX), etc. The other is non-enzymatic antioxidant system, including tea polyphenols, vitamin C, vitamin E, glutathione, α-lipoic acid, carotenoids, trace elements copper, zinc, selenium, etc. Both can remove excess oxidizing substances to balance them and maintain physical health. The balance between the two systems keeps the body healthy. Oxidative stress refers to a state of imbalance between oxidation and anti-oxidation in the body, which tends to oxidize, leading to neutrophil inflammatory infiltration, increased protease secretion, and the production of a large number of oxidative intermediates. Oxidative stress is a

167

negative effect produced by free radicals in the body and is considered to be an important factor leading to aging and disease. An important member of the ROS family is peroxide. ROS includes superoxide anion ($\cdot O_2^-$), hydroxyl radical ($\cdot OH$) and hydrogen peroxide (H_2O_2); RNS includes nitric oxide ($\cdot NO$), nitrogen dioxide ($\cdot NO_2$) and peroxynitrite ($ONOO^-$) [Zhao, 1996a, 1996b].

A large number of studies have shown that oxidative stress increases in neurodegenerative diseases, such as Parkinson's disease (PD) and Alzheimer's disease (AD), and vascular disease such as hypertension and atherosclerosis. In the state of coronary artery disease, oxidative stress is exacerbated by the decrease of vascular extracellular SOD, which is an important protective enzyme against superoxide anion under normal conditions. Oxidative stress plays an important role in pulmonary fibrosis, epilepsy, hypertension, atherosclerosis, and sudden death. Oxidative stress destroys the balance between oxidants and antioxidants, which is the main cause of cell damage. Oxidative stress occurs when there is an imbalance between generation of ROS and inadequate antioxidant defense systems. Oxidative stress can cause cell damage either directly or through altering signaling pathways. Oxidative stress is a unifying mechanism of injury in many types of disease processes, including gastrointestinal diseases. For example, in alcoholic liver disease, ROS has been detected through direct spin-trapping techniques and indirect markers, such as products of lipid peroxidation.

Studies have shown that natural antioxidants, such as tea polyphenols, can reduce the harm caused by oxidative stress. Tea is one of the drinks with the largest consumption in the world and also contains specific polyphenols as we discussed in Chapters 2 and 3. Green tea and black tea are rich in polyphenols, among which epigallocatechin gallate (EGCG) and theaflavin are the most abundant. This chapter discusses oxidative stress and the regulatory effect of tea polyphenols on oxidative stress [Yang *et al.*, 2003].

4.2 Oxidative stress

Oxidative stress refers to the stress response caused by the accumulation of ROS and RNS in the body when the body is subjected to harmful stimulation, the production of oxygen, and NO free radicals exceed the clearance speed in the body, and out of balance. Oxidative stress is defined as a disturbance between the production of ROS and antioxidant defenses. Increasing evidence from research on several diseases show that oxidative stress is associated with the pathogenesis of diabetes, obesity, cancer, ageing, inflammation, neurodegenerative disorders, hypertension, apoptosis, cardiovascular diseases, and heart failure. Based on this research, the emerging concept is that oxidative stress is the "final common pathway" through which risk factors of several diseases exert their deleterious effects. Oxidative stress causes a complex dysregulation of cell metabolism and cell-cell homeostasis.

Increasing evidence have shown that ROS and RNS play important roles in the development of the diseases [Ferrari *et al.*, 2004]. Generation of ROS and RNS are normally tightly regulated. At low and moderate concentrations, ROS/RNS mediate signal transduction cascades involved in a variety of cellular and physiological functions, and are important for defense against infectious agents. In contrast, overproduction of ROS/RNS results in oxidative stress, a deleterious process that causes damages to cell structures including lipids, proteins, and DNA, which can eventually lead to cellular senescence and apoptosis [Afanas'ev, 2005]. Under normal circumstances, biological individuals can always maintain the balance of oxidative stress and ROS and its stability in cell localization. If the balance of oxidative stress and ROS are gradually broken, it may be the beginning of accelerated damage and aging of organisms. If oxidation and antioxidation are difficult to maintain a balance, it may lead to sub-health. In disease and aging, the balance between oxidation and antioxidation is broken (Fig. 4-1). However, it depends on the severity and duration of oxidative stress. Too low or too high ROS and the change of its cell location will affect the normal physiological function of organisms, and then affect individual life span.

Looking around us, we know that we are in an oxidative stress environment. The environment around us is constantly producing free radicals, especially with the intensification of environmental pollution. There are many sources of free radicals in the surrounding environment, mainly including the following sources (Fig. 4-2):

Radiation: all kinds of ionizing radiation and ultraviolet radiation can produce free radicals. Ionizing radiation can make water in organisms to

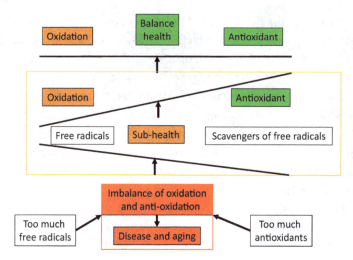

Figure 4-1. The balance between oxidation and anti-oxidation can keep the body healthy.

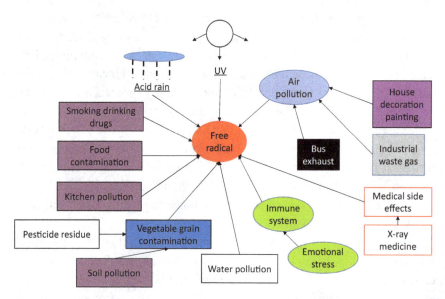

Figure 4-2. Free radicals produced in the environment.

produce hydrated electrons and hydroxyl radicals, and then damage cell components. Ultraviolet light can produce ozone, which in turn produces hydroxyl radicals. Inhaling it can cause lung damage. Radiation (visible light, ultraviolet light, X-ray, γ-ray) can provide the energy required for molecular homolysis. Especially in recent years, due to the destruction of the atmospheric ozone layer by people's excessive development, the ultraviolet ray has increased significantly, which has become the main inducement for people's skin damage and melanoma.

Cigarette smoking: smoking produces a large number of harmful free radicals, which are mainly distributed in smoking tar and smoking smoke. The free radicals in smoking tar are mainly polycyclic aromatic hydrocarbon free radicals. The concentration of free radicals in each cigarette tar is about 6×10^{14} free radicals. The free radicals in smoking cigarette are transient unstable free radicals, which are mainly reactive oxygen free radicals, nitric oxide, nitrogen dioxide free radicals, alkoxy, and alkane free radicals.

Gases emitted by automobile exhaust and coal combustion: the gases released by coal and fuel combustion contain a large number of free radicals, including nitric oxide, nitrogen dioxide free radicals, and aromatic hydrocarbon free radicals, which are important factors of environmental pollution.

Kitchen lampblack and barbecue food: high temperature cooking and barbecue food in the kitchen are the processes of generating free radicals. Kitchen lampblack and barbecue food contain a large number of free radicals, including a large number of lipid peroxidation products in addition to aromatic hydrocarbon free radicals. These substances are very harmful to health.

A large number of studies have shown that not only our environment is oxidative stress, but there is also a lot of oxidative stress in our body. The concept of "oxidative stress" was first put forward by Sohal, an American aging researcher, in 1990 [Chen *et al.*, 2021; Sohal & Allen, 1990]. It refers to the stress response caused by the accumulation of ROS in the body when the body is subjected to harmful stimulation, and the production and clearance of ROS are out of balance. A large number of studies also have shown that most diseases are related to oxidative stress, indicating the universality of oxidative stress. Almost all chapters of this book discuss oxidative stress. Oxidative stress is an intracellular or extracellular state that can lead to the production of ROS, RNS, and their metabolites from chemical or metabolic sources. On the one hand, ROS (such as $\cdot O_2^-$ and H_2O_2) and NO play an important role in cell signal regulation as intracellular second messengers. On the other hand, the reaction between NO and $\cdot O_2^-$ produces $ONOO^-$ and its metabolite $\cdot OH$ with stronger reactivity, resulting in extensive oxidative damage for cells. Many environmental factors and metabolic toxins can induce excessive ROS and RNS, resulting in varying degrees of cytotoxicity. Excessive ROS and RNS will not only affect the permeability of cell membrane, cause the intracellular Ca^{2+} influx and the content of second messenger to increase, and then activate a variety of redox pathways. In addition, it can directly start the apoptotic pathway through mutual feedback regulation with mitochondria. Mitochondria are the organelles sensitive to ROS. Excessive ROS not only affects the normal function of mitochondrial "energy factory", but also increases the production of ROS, releases apoptotic factor cytochrome C, activates caspase protein family, and finally leads to apoptosis [Yuan & Yankner, 2000; Hengartner, 2000; Guo *et al.*, 2005, 2007]. A large number of studies have shown that excessive NO can induce neuronal apoptosis. Under normal physiological conditions, the concentration of NO and superoxide anion in brain cells is very low, so it is difficult to react. In addition, NO molecule is 1000 times smaller than SOD and the diffusion rate is fast. The reaction rate of NO with superoxide anion is 10 times faster than that of SOD. Therefore, most of the superoxide anions produced in cells are captured by NO. Peroxynitrite is a strong oxidant, which can react with sulfhydryl or nitroaromatic amino acids of protein and affect its signal transduction function. Peroxynitrite oxidizes lipids, proteins, and DNA, thereby destroying their function [Cohen *et al.*, 1985; Torreilles *et al.*, 1999]. NO can also react with iron ligands to reduce Fe (III) by binding with NO, and the resulting Fe^{2+} and NO^+ can further react with intracellular sulfhydryl groups to make it nitrous.

4.2.1 *Oxidative stress in nervous system and degenerative various diseases*

Nervous system, such as brain, is a tissue that is relatively prone to oxidative damage in the body. On the one hand, the nervous system is prone to oxidation due to its high oxygen consumption. At the same time, the special anatomical, physiological, and biochemical characteristics of the nervous system make it less tolerant to oxidative damage. This is mainly due to 1) having a large number of easily oxidized substances in the nervous system, such as polyunsaturated fatty acids and catechol ammonia, 2) low levels of antioxidants, such as glutathione, vitamin E, catalase and SOD, 3) endogenous ROS being produced by specific reactions in the brain, 4) some areas having high iron content, such as globus pallidus and substantia nigra, and 5) non-renewable neurons in the central nervous system. Once damaged, they will lead to apoptosis or permanent functional abnormalities. More and more evidence show that oxidative stress caused by ROS and nitrogen stress caused by RNS are involved in the pathological mechanism of neurodegenerative diseases and aging. The use of antioxidants provides an attractive prospect for the prevention and treatment of neurodegenerative diseases. We studied the mechanism of hydroxyl radical induced neuronal apoptosis [Ni *et al.*, 1996]. Oxidative stress, an imbalance toward the pro-oxidant side of the pro-oxidant/antioxidant homeostasis, occurs in several brain neurodegenerative disorders. Among these neurodegenerative brain disorders are those in which protein aggregation is observed, including AD and PD.

In Chapter 5, we will discuss nerve injury and oxidative stress. A large number of studies have shown that oxidative stress damage occurs after nerve poisoning, such as lead, organophosphate, and nerve agent toxicity, nerve injuries such as brain, spinal cord injury, sciatic nerve crush, cavernous, and mitochondria in spinal cord and optic nerves. The role of excessive oxidative stress induction in the pathogenesis of brain aging and aging-related degenerative nervous system diseases has also been confirmed. The oxygen metabolism rate of brain tissue is very high, the content of metal ions such as iron is also very high, and the protective mechanism of antioxidants is relatively lack. This leads to the imbalance of ROS metabolism in brain tissue. In addition, many kinds of cells, such as cortical neurons, cerebellar granule cells, and astrocytes, can produce excessive NO under pathological conditions. Oxygen stress is related to acute and chronic neurodegenerative diseases. Apoptosis plays an important role in the occurrence of neurodegenerative diseases. Apoptosis can be induced by many factors, such as glucocorticoids and removal of nutritional factors. Hydrogen peroxide is a cytotoxic factor of thymocytes and cortical neurons, which can induce apoptosis, while the antioxidants Trolox and N-acetylcysteine can inhibit the apoptosis induced by oxygen stress [Ni *et al.*, 1996]. There is increasing evidence that brain aging and degenerative nervous system diseases are associated with abnormal and excessive neuronal apoptosis [Yuan and Yankner, 2000]. Recent studies have shown that apoptosis is related to the injury and loss of dopaminergic neurons. Apoptosis or programmed cell death is a kind of death regulated by genes, that is, senile death. Apoptosis plays an important role in the development of nervous system and is a normal developmental regulation mechanism.

In Chapter 6, we will discuss AD and oxidative stress. Oxidative stress plays a significant role in the pathogenesis of AD, a devastating disease of the elderly. The brain is more vulnerable than other organs to oxidative stress, and most of the components of neurons (lipids, proteins, and nucleic acids) can be oxidized in AD due to mitochondrial dysfunction, increased metal levels, inflammation, and β-amyloid (Aβ) peptides. Oxidative stress participates in the development of AD by promoting Aβ deposition, tau hyper-phosphorylation, and the subsequent loss of synapses and neurons. The relationship between oxidative stress and AD suggests that oxidative stress is an essential part of the pathological process, and antioxidants may be useful for AD treatment. Aβ induces nerve cells to increase the production of ROS, causes the degradation of mitochondrial function, reduces mitochondrial membrane voltage, and activates caspases, and finally leads to nerve cell apoptosis [Smith *et al.*, 2000]. Oxidative stress promotes the production of pre-amyloid protein Aβ. The expression of the catalytic subunit presenilin protein 1 (PS1) of catabolic enzyme (BACE1) and gama secretase accelerates the generation of Aβ from its precursor APP and forms a vicious circle [Oda *et al.*, 2009, Ansari & Scheff, 2010]. Studies have shown that gene defects in the antioxidant system increase Aβ deposition in the animal brain, while antioxidant intake reduces Aβ deposition and improves the cognitive status of animals. Free radicals peroxidize membrane lipids [Butterfield & Kanski, 2001] and oxidize proteins, resulting in damage of the plasma membrane and crosslinking of cytoskeletal proteins. In addition, free radicals damage RNA [Nurk *et al.*, 2009], nuclear DNA [Gabbita *et al.*, 1998], and DNA. In the brain, the high metabolic rate, the low concentration of glutathione and antioxidant enzyme catalase, and the high proportion of polyunsaturated fatty acids, make the brain tissue particularly susceptible to oxidative damage [Smith *et al.*, 2000].

In Chapter 7, we will discuss Parkinson's disease (PD) and oxidative stress. More and more evidence show that oxidative stress is involved in the pathological process of PD. The excessive production of oxidative stress leads to the depletion of dopaminergic neurons, which is also the main reason for the further damage of residual neurons. Dopamine is a neurotransmitter that can be oxidized by chemical and enzymatic reactions to produce metabolites, ROS and RNS. It was also found that metal ions such as iron ions in substantia nigra of PD patients were overloaded. Excessive ROS, such as hydrogen peroxide, can react with transition metal iron to produce more active species and hydroxyl radicals, increase oxidative stress, deplete cellular antioxidants, and destroy antioxidant defense system. Redox imbalance can block mitochondrial respiratory chain, leading to respiratory failure and energy crisis. In addition, there is feedback between mitochondrial damage and oxidative stress. Finally, dopaminergic neurons died, and the typical clinical characteristics of patients with PD were observed. PD is characterized by nigrostriatal degeneration and may involve oxidative stress α-Synuclein (αS) aggregation, imbalance of redox metal homeostasis, and neurotoxicity. Oxidative stress has been widely believed to be an important pathogenetic mechanism of neuronal apoptosis in PD [Halliwall, 1992]. Overproduction of ROS can lead to oxidative damage in the brain of PD, as shown by increased lipid peroxidation and DNA damage in the substantia nigra. Increased protein oxidation is also apparent in many areas of the brain, whereas substantia nigra is particularly vulnerable [Jenner & Olanow, 1998]. Under physiological conditions, 6-OHDA is rapidly and non-enzymatically oxidized by molecular oxygen to form hydrogen peroxide (H_2O_2) and the corresponding quinone [Soto-Otero *et al.*, 2000]. The former can react with iron (II) to form the reactive and damaging hydroxyl free radical. The latter then undergoes an intramolecular cyclization, followed by a cascade of oxidative reactions, resulting in the formation of an insoluble polymeric pigment related to neuromelanin [Graham *et al.*, 1978]. It has been found that no matter how many factors and changes lead to the pathogenesis of PD, oxidative stress is the most important point. Whether oxidative stress is the cause or result of PD, it is closely related to the pathogenesis and course of PD. However, various stresses *in vivo* and *in vitro*, especially apoptosis induced by oxidative stress, are closely related to the occurrence of degenerative nervous system diseases PD. Experiments show that the apoptosis pathway activated by oxidative stress can lead to the degeneration and damage of dopamine (DA) neurons, and the inhibition of apoptosis pathway can prevent the occurrence of DA neuron degeneration. It is considered that excessive oxidative stress is an important reason for the degeneration of dopamine neurons and the further damage of residual neurons. The levels of α-synuclein or dopamine in endoplasmic reticulum may be related to cellular oxidative stress and PD symptoms.

4.2.2 *Oxidative stress in cardiovascular and cerebrovascular diseases*

In Chapter 8, we will discuss cardiovascular and cerebrovascular diseases and oxidative stress. More and more evidence show that oxidative stress is involved in the pathological process of cardiovascular and cerebrovascular diseases. The incidence and mortality of cardiovascular and cerebrovascular disease are the highest among all disease in China and all over the world and are still increasing with the rising living standards. To find effective treatment and prevention of the diseases, it is important to thoroughly understand the pathobiology of the disease. Oxidative stress is a condition in which oxidant metabolites exert their toxic effect because of an increased production or an altered cellular mechanism of protection. The heart needs oxygen avidly and, although it has powerful defense mechanisms, it is still susceptible to oxidative stress, which occurs, for instance, during post-ischemic reperfusion. Increasing evidence has shown that ROS and RNS play important roles in the initiation and progression of heart disease. Ischemia-reperfusion causes alterations in the defense mechanisms against oxygen free radicals, mainly a reduction in the activity of mitochondrial SOD and a depauperation of tissue content of reduced glutathione. At the same time, production of oxygen free radicals increases in the mitochondria and leukocytes and toxic oxygen metabolite production is exacerbated by re-admission of oxygen during reperfusion. Oxidative stress, in turn, causes oxidation of thiol groups and lipid peroxidation, leading first to reversible damage and eventually to necrosis. There is evidence of oxidative stress during surgical reperfusion of the whole heart, or after thrombolysis, and it is related to transient left ventricular dysfunction or stunning. Recent data have shown a close link between oxidative stress and apoptosis caused by heart failure. Relevant to heart failure is the finding that tumor necrosis factor (TNF-α), which is found increased in failing patients, induces a rapid rise in intracellular ROS and apoptosis. This series of events is not confined to the myocytes but occurs also at the level of endothelium, where TNF-α causes expression of inducible nitric oxide synthase, production of the RNS, and oxidative stress and apoptosis. It is, therefore, possible that the immunological response to heart failure results in endothelial and myocyte dysfunction through oxidative stress mediated apoptosis [Ferrari *et al.*, 2004; Zhao *et al.*, 1989b, 1996c, 1996d, 1997; Zhao and Zhao, 2010].

Stroke and myocardial infarction are one of the most common causes of death and disability in the world. The key factor mediating stroke-related damage is oxidative stress. The ischemic injury behind these diseases is complex, involving complex interactions between many biological functions, including energy metabolism, vascular regulation, oxidative stress, and inflammation. However, reperfusion may cause additional damage to tissues through a process called ischemia/reperfusion injury. There is evidence that ROS produced immediately after acute ischemic stroke increases rapidly, rapidly destroy the antioxidant defense system and cause further tissue damage. These ROS can damage cell macromolecules, leading to autophagy, apoptosis, and necrosis. In addition, the rapid recovery of blood flow increases the level of tissue oxygenation and leads to the generation of second ROS, resulting in reperfusion injury. iNOS was implied in delayed neuron death after brain ischemic damage could increase the protein level of TNF-α and nuclear factor-kappa B (NFκB) and decrease the mRNA level of NOS estimated by western blotting and RT-PCR. Strong evidence about the close relationship between the oxidation of low-density lipoprotein (OxLDL) and atherosclerosis exists as a complex process in which both the protein and the lipids undergo oxidative changes and form complex products. Oxidative stress and LDL oxidation might play a vital role in atherosclerosis, which has been studied for several years [Zhang *et al.*, 2004].

Oxidative stress associated with pathology of atherosclerosis. Endothelial dysfunction is often associated with increased oxidative stress and impaired mitochondrial activity. Oxidative stress would alter endothelial signal transduction and redox-regulated transcription factors to increase vascular endothelial permeability and catalyze leukocyte adhesion. Endothelial dysfunction is often associated with increased oxidative stress [Suganya *et al.*, 2016] and impaired mitochondrial activity [Lum & Roebuck, 2001]. Oxidative stress would alter endothelial signal transduction and redox-regulated transcription factors to increase vascular endothelial permeability and catalyze leukocyte adhesion [Davidson & Duchen, 2007]. Atherosclerosis, includes endothelial dysfunction, low-density lipoprotein oxidation, vascular smooth muscle cell proliferation, inflammatory process by monocytes, macrophages or T lymphocytes, and platelet aggregation. Inflammatory factors and vascular smooth muscle cell proliferation play a vital role in the progression of atherosclerotic plaques. The activation of PPARγ effectively attenuates angiotensin II (AngII)-induced inflammation and intercellular ROS production. Inflammatory process with monocytes, macrophages, and T lymphocytes macrophages play a key role in atherogenesis through their proinflammatory action, which involves the production of Interleukin 1(IL-1) and TNF-α [Cheng *et al.*, 2017].

4.2.3 *Oxidative stress in aging*

In Chapter 9, we will discuss aging and oxidative stress. More and more evidence show that oxidative stress is involved in the pathological process of aging. The core of free radical aging theory is that biological oxidative stress is closely related to aging and life span [Harman, 1956]. Mitochondria is an important organelle of cells, which exists in all human and animal cells except red blood cells. It provides energy for cells, produces ROS that can regulate physiological processes, and participates in the regulation of cell death. Since most ROS are produced in mitochondria, these mitochondrial free radicals (mtROS) are very easy to contact and damage mitochondrial DNA and mitochondrial oxidative phosphorylation system.

As early as 1956, Harman believed that the oxidative damage of mito-chondrial DNA, mitochondrial protein, and phospholipid was the direct cause of aging, which determined the individual life span and put forward the theory of mitochondrial dysfunction [Harman, 1972]. According to Harman's mitochondrial dysfunction theory, cell aging is due to the con-tinuous release of mtROS from mitochondria and the damage of mito-chondrial DNA. Damaged mitochondrial DNA leads to the lack of key enzymes in the electron transport chain and the production of mtROS, which leads to a vicious cycle of ROS and ultimately reduces energy pro-duction [Fariss *et al.*, 2005]. With the deepening of relevant research, a more specific free radical theory of aging (FRTA) was proposed — mito-chondrial free radical theory of aging (mtFRA) [Barja, 2014]. At present, there is a lot of research evidence to support the theory of mitochondrial free radical aging. Through overexpression of mitochondrial targeted catalase (mtCAT), mtROS produced in mitochondria can be effectively eliminated, and the above aging phenotype can be alleviated [Dai *et al.*, 2010]. At the same time, overexpression of mitochondrial targeted cata-lase (mtCAT) in wild-type mice prolongs the life of mice [Dai *et al.*, 2017]. Based on the prediction of mtFRA, the maximum life span of spi-nal thermostatic animals is negatively correlated with the production of mtROS [Lambert *et al.*, 2007]. Many aging related diseases seem to be caused by mtROS overproduction [Dai *et al.*, 2014]. A large number of research evidences support the promoting effect of mitochondrial ROS production on aging, which indicates that ROS has potential toxicity. However, ROS is also an important signal molecule regulating aging. Desjardins *et al.* [2017] studied C elegans and found that both oxidants and antioxidants can prolong or shorten life span, depending on different concentrations, genotypes, and conditions. The experimental results reveal the inverted U-shaped dose-response relationship between ROS level and life span, that is, low-dose ROS, may help to prolong life span. In another recent nematode based study, people increased the concentra-tion of ROS in specific subcellular intervals by using gene manipulation, and found that the location of ROS is very important to determine its impact on life span. A large number of research evidence shows that ROS-or mtROS-induced mitochondrial damage can act as a starting signal and activate several mitochondrial protective pathways, so as to delay aging, inhibit cell death, and possibly prolong life. With the increase of ROS level, it can stimulate the transcription and expression of SIRT3, cause the deacetylation of superoxide dismutase (SOD2) and enhance its antioxidant effect, so as to eliminate excessive mtROS and protect mitochondria [Ansari *et al.*, 2017].

4.2.4 *Oxidative stress and metabolic diseases*

Metabolic diseases include obesity, hyperlipidemia, hyperglycemia, hypertension, hyperuricemia, and diabetes. Studies have shown that metabolic diseases are closely related to oxidative stress.

In Chapter 10, we will discuss obesity and oxidative stress. More and more evidence show that oxidative stress is involved in the pathological process of obesity. Oxidative stress is a key factor linking obesity and its related complications. Systemic and tissue-specific chronic inflammation and oxidative stress are common characteristics of obesity. Obesity itself can induce systemic oxidative stress through various biochemical mechanisms, such as superoxide production by NADPH oxidase, oxidative phosphorylation, automatic oxidation of glyceraldehyde, activation of protein kinase C, polyol, and hexosamine pathway. Other factors of oxidative stress in obese patients include hyperleptinemia, low antioxidant defense, chronic inflammation, and postprandial ROS production. Oxidative stress and inflammation are central mediators of obesity and molecular mechanisms of macronutrients-mediated oxidative stress and inflammation. In addition, other molecules promote atherosclerotic inflammation. Once the body fat accumulates too much, it will secrete a large number of cytokines [Grundy, 2004] that inhibit the function of fat and muscle tissue. These cytokines mainly include free fatty acids (FFA) and inflammatory factors (such as TNF-α), ROS etc. [Chung *et al.*, 2020]. Cytokines can act directly on adipocytes in the form of autocrine and paracrine to produce insulin resistance and disorder glucose and lipid metabolism [Holguin *et al.*, 2019]. Obese patients secrete more free fatty acids than normal people, which is the product of triglyceride decomposition by adipocytes [Fernández-Sánchez *et al.*, 2011]. Once the fat is stored too much, the visceral adipose tissue secretes more FFA than the peripheral adipose tissue [Flashner *et al.*, 2020]. Free fatty acids will lead to the inhibition of carbohydrate oxidation in muscle tissue [Fernández-Sánchez *et al.*, 2011]. Free fatty acids can inhibit the phosphorylation of insulin receptor, thereby inhibiting the insulin signaling pathway [Becerril *et al.*, 2019]. The secretion of inflammatory factors such as IL-6 and TNF-α also increased significantly in obese patients. They can induce muscle insulin resistance, damage vascular endothelial cells, and lead to atherosclerosis [Ridker & Morrow, 2003]. More importantly, inflammatory factors can act directly on adipocytes in the form of autocrine and paracrine, inhibiting insulin signaling pathway and PPARγ [Miles *et al.*, 1997; Yan *et al.*, 2013]. Inflammatory factors can also increase the level of ROS in adipocytes, leading to oxidative stress [Furukawa *et al.*, 2004]. Adipocytes that accumulate excess fat increase the activity of NADPH oxidase and synthesize a large amount of ROS [Sonta *et al.*, 2004]. Oxidative stress of adipocytes not only leads to increased phosphorylated amino terminal protein kinase (P-JNK) levels, inhibits protein kinase B (AKT) phosphorylation, the insulin signaling pathway, and type 2 diabetes mellitus [Houstis *et al.*, 2006], but also secretes ROS into the blood by adipocytes, leading to systemic oxidative stress and damage to islets β cells .

In Chapter 11, we will discuss diabetes and oxidative stress. More and more evidence show that oxidative stress is involved in the pathological process of diabetes. The role of oxidative stress in the pathogenesis of insulin resistance and beta-cell dysfunction need to be discussed. ROS and oxidative stress play key roles in the occurrence, development, and chronic complications of diabetes. Hyperglycemia induced ROS can reduce islets β cell insulin secretion, which lowers insulin sensitivity and signal transduction in insulin sensitive tissues and changes endothelial cell function. Recent studies have shown that oxidative stress and excessive fatty acids can reduce glucose stimulated insulin secretion, inhibit insulin gene expression, and lead to β-cell death. The occurrence of diabetes and its complications is related to oxidative stress and low chronic inflammation. Oxidative stress is an imbalance between cellular oxidants and antioxidant system, which is the result of excessive production of free radicals and related ROS. High glucose increases the markers of chronic inflammation, promotes ROS production, and oxidative stress injury, leading to diabetic complications, including vascular dysfunction. In addition, the increase of ROS level will reduce insulin secretion and damage the signal transduction of insulin sensitivity and insulin responsive tissues. Therefore, the appropriate treatment of high glucose and inhibition of excessive production of ROS is crucial for delaying the occurrence and development of diabetes and preventing subsequent complications. The over nourished adipocytes themselves will be in a state of oxidative stress, produce insulin resistance, and secrete a large amount of ROS into the blood. Inflammatory factors and ROS can directly act on β cells. Therefore, type I diabetes and type II diabetes, especially type II diabetes, are closely related to over-nutrition and obesity, which can lead to obesity [Zhou & Grill, 1994, 1995; Mason *et al.*, 1999]. In addition, recent studies have found that diabetes are related to nitric oxide metabolism. Gene expression analysis suggests that ROS levels are increased in both models, this was later confirmed through measures of cellular redox state. ROS have previously been proposed to be involved in insulin resistance. In cell culture using six treatments designed to alter ROS levels, all ameliorated insulin resistance were in varying degrees. One of these treatments was tested in obese, insulin-resistant mice and was shown to improve insulin sensitivity and glucose homeostasis. This findings suggest that increased ROS levels are an important trigger for insulin resistance in numerous settings [Jacqueminet *et al.*, 2000; Briaud *et al.*, 2001; Houstis *et al.*, 2006]. In a diabetic milieu, high levels of ROS are induced. ROS generated in macrophage activation and function in diabetes. This contributes to the vascular complications of diabetes. Recent studies have shown that ROS formation is exacerbated in diabetic monocytes and macrophages due to a glycolytic metabolic shift. Macrophages are important players in the progression of diabetes and promote inflammation through the release of pro-inflammatory cytokines and proteases. Since ROS is an important mediator for the activation of pro-inflammatory signaling pathways, obesity and hyperglycemia-induced ROS production may favor induction of pro-inflammatory macrophages during diabetes onset and progression. ROS induces mitogen activated protein kinase (MAPK) and NFκB signaling and interferes with macrophage differentiation via epigenetic (re) programming. ROS production links metabolism and inflammation in diabetes and its complications. ROS contributes to the crosstalk between macrophages and endothelial cells in diabetic complications [Rendra *et al.*, 2019].

4.2.5 *Oxidative stress and bacteria and viruses*

In Chapter 12, we will discuss bacteria and viruses and oxidative stress. More and more evidence show that oxidative stress is involved in the pathological process caused by bacteria and viruses. Studies have shown that both bacteria and viruses can cause fever and severe inflammation and oxidative stress. Oxidative stress defines a condition in which the pro-oxidant-antioxidant balance in the cell is disturbed, resulting in DNA hydroxylation, protein denaturation, lipid peroxidation, and apoptosis, ultimately compromising cells' viability.

Studies have shown that various bacteria can cause fever and severe inflammation and oxidative stress. It was previously reported that in mice susceptible to intestinal precancerous lesions (APCMin/+) and knockout of anti-inflammatory factor (IL-10-/-), the mice model is susceptible to colorectal cancer, pro-inflammatory response, which is related to the occurrence of early tumors [Wu *et al.*, 2009]. The contribution of *Streptococcus bovis* to colorectal cancer is related to the increased expression of pro-inflammatory genes, such as IL-1, IL-8, and COX-2. In addition, APCMin/+ and IL-10-/- mice infected with *Clostridium nucleatum* and *Enterococcus faecalis*, respectively, increased immune cell infiltration and pro-inflammatory cytokines such as TNF-α, IL-1β, IL-6, and IL-8. *Escherichia coli* (*E. coli*) is one of the most characteristic bacteria associated with inflammatory bowel disease. It is reported that in patients with inflammatory bowel disease, adhesive and invasive *E. coli* colonize the intestinal mucosa abnormally [Darfeuille-Michaud *et al.*, 2004]. It showed that *E. coli* associated with colorectal cancer can induce the expression of pro-inflammatory gene COX-2 in macrophages, supporting the regulation of inflammation by bacteria during the occurrence of colorectal cancer [Raisch *et al.*, 2015].

Studies shown that various viruses such as human immunodeficiency virus (HIV) and COVID-19 can cause fever and severe inflammation and oxidative stress. Human immunodeficiency virus HIV, is the last stage of human immunodeficiency virus infection. The HIV attacks and gradually destroys the human immune system, leaving the host unprotected when infected. Several chronic diseases, including HIV infection, are associated with oxidative stress. In addition to HIV itself, some antiretroviral drugs can also increase oxidative stress and reduce virus replication. Antiretroviral therapy based on integrase inhibitors can reduce the oxidative stress caused by HIV infection, which may be a good treatment choice for HIV infected people [De Bruyne *et al.*, 1999].

COVID-19 viruses cause respiratory inflammation augment ROS production in host cells. Factors leading to inflammation and immune response include NO/ROS ratio, activation of activation (M1) macrophages injury. Normal NO/ROS balance is essential for normal vascular function. ROSs, such as superoxide species, also act as host defense and are induced during stress, such as viral infection. Excessive ROS (such as virus overload) activates M1 macrophages, recruits neutrophils, and promotes the production of peroxynitrite together with NO, so as to effectively respond to the invading virus, accompanied by endothelial dysfunction, osmotic vascular, and lipid membrane peroxidation. Macrophages produce a large number of highly reactive nitrogen and oxygen derived molecular species and pro-inflammatory cytokines, such as interleukin (IL)-2, IL-6, IL-8, interferon IFN-α/β, and tumor necrosis factor TNF-α. They neutralize invading organisms but also exacerbate vascular damage. In addition, inflammation induced platelet activation (which can be alleviated by NO) can lead to increased coagulation, which is an important consequence of the disease. Recently, evidence showing that COVID-19 destroys erythrocyte hemoglobin has been reported, which may lead to significant oxidative stress. Subsequent hemolysis can lead to anemia, but more importantly, it can exacerbate the inflammatory process. Acellular hemoglobin clears coagulation regulators such as endothelial NO, while the release of pro-inflammatory heme and iron activates platelets. Enhanced coagulation and slow blood flow lead to systemic hypoxia in oxygen sensitive organs such as kidney. Under the promotion of NO, macrophage response produces inflammatory cascade and cytokine storm, resulting in the formation of cell debris and edema, which is manifested as respiratory distress syndrome in the lung [Banu *et al.*, 2020]. The high neutrophil to lymphocyte ratio observed in critically ill patients with COVID-19 is associated with excessive levels of ROS, which promote a cascade of biological events that drive pathological host responses. ROS induce tissue damage, thrombosis, and red blood cell dysfunction, which contribute to COVID-19 disease severity. Decreased expression of SOD3 in the lungs of elderly COVID-19 patients correlates with disease severity [Laforge *et al.*, 2020]. Anti-inflammatory molecules can potentially reduce the damage of ROS to cells and restoring the balance of NO/ ROS can reduce the damage of re-oxygenation and reperfusion to cells [Alvarez *et al.*, 2020]. ROS changes vascular tension by increasing intracellular calcium concentration and reducing the bioavailability of NO. How NO level and bioavailability affect COVID-19 patients are closely related to their unique characteristics [Ricciardolo *et al.*, 2020].

4.2.6 *Oxidative stress and immune and inflammatory diseases*

In Chapter 13, we will discuss immune and inflammatory diseases and oxidative stress. More and more evidence show that oxidative stress is involved in the pathological process of immune and inflammatory diseases. A large number of studies have shown that inflammation causes oxidative stress and oxidative stress plays important functions for body health. ROS and RNS free radicals produced by fever and inflammation react to produce peroxynitrite ions, which further leads to myocardial oxidative stress injury. However, in some immune system diseases, ROS and RNS free radicals and oxidative stress play the opposite role in body. It can damage invading microorganisms and play a key role in preventing diseases and ensuring physical health.

Autoimmune diseases are the kind of diseases that seriously endanger human health, such as lupus erythematosus and rheumatoid arthritis. These kind of diseases are also related to inflammation. Naturally, these diseases are closely related to oxygen free radicals and oxidative stress. Oxygen free radicals produced by activated macrophages can change the antigenic behavior of immunoglobulins, further activate macrophages, and produce inflammation and oxidative stress in animals [Theofilopoulous & Dixon, 1982]. Ascorbic acid in synovial fluid and plasma of rheumatic patients is significantly lower than that of normal people, and mostly exists in the form of oxidative dehydroascorbic acid, which may be caused by the oxidation of oxygen free radicals and hypochlorous acid produced by activated neutrophils, and oxidative stress injury [Axford, 1987]. Patients with spontaneous hemochromatosis and iron excess are often accompanied by arthritis. Peroxidation injury can make ferritin and hemo-siderin release iron, lead to free radical reaction and lipid peroxidation, thereby resulting in oxidative stress injury [Giordano, 1984].

Inflammation is a protective reaction after the body is invaded by external microorganisms. Phagocytes play an important role in the inflammatory reaction. During the inflammatory reaction, they phago-cytize bacteria, activate by stimulation, produce respiratory burst, and release a large number of reactive oxygen free radicals and various enzymes. These products will not only kill invaders, but also generate oxidative stress and damage normal body tissues. In recent years, people have paid special attention to the study of oxygen free radicals produced by stimulated or sensitized phagocytes. Phagocytes produce respiratory burst during phagocytosis or stimulated, consume oxygen, activate hex-ose phosphorylation branch, and release superoxide anion free radical, hydrogen peroxide, and singlet oxygen. Using ESR spin trapping and chemiluminescence technique, we directly captured ROS and NO radicals produced by respiratory burst of human polymer macrophage leukocytes (PMN) stimulated by cancer promoter phorbol-1,2-myristate-1,3-acetate (PMA) [Zhao *et al.*, 1989a, 1996a, 1996b; Li *et al.*, 1990, 1996; Yang *et al.*, 1991]. Lymphocytes can also produce respiratory burst and oxygen radicals as polymorphonuclear leukocytes [Zhao *et al.*, 1990].

4.2.7 *Oxidative stress and cancer*

In Chapter 14, we will discuss cancer and oxidative stress. More and more evidence show that oxidative stress is involved in the pathological process of cancers. Oxidative stress is complex for cancer. Oxidative stress can cause cell and DNA damage and lead to cancer. It can also kill cancer cells and play a role in cancer treatment. Many perturbations, such as increased oxidative stress, pathogen infection, and inflammation, promote the accu-mulation of DNA mutations and ultimately lead to carcinogenesis. ROS are induced through a variety of endogenous and exogenous sources. Overwhelming antioxidant and DNA repair mechanisms in the cell by ROS may result in oxidative stress and oxidative damage to the cell. This resulting oxidative stress can damage critical cellular macromolecules and/or modulate gene expression pathways. Oxidative damage resulting from ROS generation can participate in all stages of the cancer process. An association of ROS generation and human cancer induction has been shown. It appears that oxidative stress may both cause as well as modify the cancer process. Growing evidence supports a role of ROS-induced generation of oxidative stress in these epigenetic processes. Oxidative stress is a physiological state where high levels of ROS and free radicals are generated. Several signaling pathways associated with carcinogenesis can additionally control ROS generation and regulate ROS downstream mechanisms. Cancer initiation may be modulated by the nutrition-mediated elevation in ROS levels, which can stimulate cancer initiation by triggering DNA mutations, damage, and pro-oncogenic signaling. ROS can also promote tumor formation by inducing DNA mutations and pro-oncogenic signaling pathways. Inflammation and oxidative stress are common and co-substantial pathological processes accompanying, promoting, and even initiating numerous cancers. Oxidative stress can activate a variety of transcription factors including NF-κB, AP-1, p53, HIF-1α, PPAR-γ, β-catenin/Wnt, and Nrf2. All these complex and opposing interactions between the canonical WNT/β-catenin pathway and PPARγ appear to be fairly common in inflammation, oxidative stress, and cancers [Vallée & Lecarpentier, 2018]. Oxidative stress is caused by an imbalance between ROS production and a biological system's ability to readily detoxify these reactive intermediates or easily repair the resulting damage. Excessive exposure of skin to UV radiation triggers the generation of oxidative stress, inflammation, immunosuppression, apoptosis, matrix-metalloproteases production, and DNA mutations, leading to the onset of photo ageing and photo-carcinogenesis. At the molecular level, these changes occur via activation of several protein kinases as well as transcription pathways, formation of ROS, and release of cytokines, interleukins, and prostaglandins together [Garg *et al.*, 2020]. Oxidative stress has been shown to be involved in the multistage process of human carcinogenesis via many different mechanisms. Amongst them are ROS-induced oxidative modifications on major cellular macromolecules like DNA, proteins, and lipids with the resulting byproducts being involved in the pathophysiology of human malignant and pre-malignant lesions [Hanafi *et al.*, 2012]. ESR technology is also used to study the singlet oxygen, hydroxyl radical, superoxide anion radical, and hematoporphyrin anion radical produced by Yangzhou hematoporphyrin under light. The study on the photosensitive mechanism of hypocrellin A and hypocrellin B has also obtained very meaningful results [Zhang *et al.*, 1986, 1988, 1991].

4.2.8 *Oxidative stress and radiation damage*

In Chapter 15, we will discuss radiation damage and oxidative stress. More and more evidence show that oxidative stress is involved in the pathological process of radiation damages. The detrimental effects of various radiation are mostly mediated via the overproduction of ROS, especially the hydroxyl radical (•OH). Radiation causes damage to irradiated tissues and also tissues that do not receive direct irradiation through a phenomenon called out-of-field effects. This damage through signals such as inflammatory responses can be transmitted to unirradiated cells/tissues and causes many effects such as oxidative damage. The radio-protective and anti-inflammatory effects of antioxidants have been demonstrated in various studies. UVA is particularly important in causing immunosuppression in both humans and mice, and UV lipid peroxidation-induced prostaglandin production and UV activation of nitric oxide synthase is important mediators of this event. Other immunosuppressive events are likely to be initiated by UV oxidative stress. Antioxidants have also been shown to reduce photo-carcinogenesis [Halliday, 2005]. Gamma radiation-induced damage also stimulated generation of ROS in marrow cells [Kondo *et al.*, 2010]. ROS produced by ultraviolet radiation, oxidizers, or metabolic processes can damage cells and initiate pro-inflammatory cascades. Ionizing radiation was responsible for augmentation of hepatic oxidative stress in terms of lipid peroxidation and depletion of endogenous antioxidant enzymes. Radiation induced the activation of stress-activated protein kinase/c-Jun NH2-terminal kinase (SAPK/JNK)-mediated apoptotic pathway and deactivation of the NF-E2-related factor 2 (Nrf2)-mediated redox signaling pathway [Khan *et al.*, 2015].

4.2.9 *Oxidative stress and osteoporosis and anemia*

In Chapter 16, we will discuss osteoporosis, anemia, and oxidative stress. More and more evidence show that oxidative stress is involved in the pathological process of osteoporosis and anemia. Oxidative stress has been implicated as a causative factor in many disease states, possibly including the diminished bone mineral density in osteoporosis. Hind-limb suspension (HLS) induces reduction of bone mineral density in tibiae while preserving bone structure in tibiae and mechanical strength in femurs. HLS-induced oxidative stress is marked by reduced malondialdehyde content and increased total sulfhydryl content in femurs. In cultured MC3T3-E1 cells, modeled microgravity-induced reactive oxygen species (ROS) formation enhances osteoblastic differentiation. In cultured RAW264.7 cells, modeled microgravity-induced ROS formation attenuates osteoclastogenesis. In addition, microgravity-induced vitamin D receptor (VDR) expression is observed in femurs of rats exposed to HLS and MC3T3-E1 cells exposed to modeled microgravity [Xin *et al.*, 2015]. HLS induces reduction of bone mineral density, ultimate load, stiffness, and energy in femur and lumbar vertebra. HLS also induces augmentation of malondialdehyde content and peroxynitrite content and reduction of total sulfhydryl content in femur and lumbar vertebra. In cultured MC3T3-E1 cells, modeled microgravity-induced ROS formation reduces osteoblastic differentiation, increases ratio of receptor activator of nuclear factor kappa B ligand to osteoprotegerin, induces nitric oxide synthetase upregulation and Erk1/2 phosphorylation. In cultured RAW264.7, modeled microgravity-induced ROS formation leads to osteoclastic differentiation and osteoclastogenesis [Sun *et al.*, 2013].

Human erythrocytes are organelle-free cells packaged with iron-containing hemoglobin, specializing in the transport of oxygen. The oxidative status of each patient is affected by multiple internal and external factors, including genetic makeup, health conditions, nutrition, physical activity, age, and the environment (e.g., air pollution, radiation). In addition, oxidative stress is influenced by the clinical manifestations of the disease (unpaired globin chains, iron overload, anemia, etc.). A paper summarized the role of oxidative stress in Thalassemia [Fibach & Dana, 2019]. Intracellular production of ROS, H_2O_2, and $\cdot O_2^-$ treatment with vitamin E dialyzer effectively reduced the 8-OHdG content in leukocyte DNA and suppressed intracellular ROS production of granulocytes [Tarng *et al.*, 2000]. The vast majority of patients with chronic kidney disease (CKD) seem to be iron-deficient. Iron deficiency frequently complicates anemia in patients with chronic kidney disease, and ferrous iron cation is a co-factor that is needed for hydroxyl radical production, which can promote cytotoxicity and tissue injury. Prescription of intravenous iron may exacerbate oxidative stress and hence, endothelial dysfunction, inflammation, and progression of cardiovascular disease. Correction of anemia represents an effective approach to reduce oxidative stress and, consequently, cardiovascular risk [Garneata, 2008].

4.2.10 *Oxidative stress and corruption and preservative process*

In Chapter 17, we will discuss corruption and preservative process and oxidative stress. More and more evidence show that oxidative stress is involved in the process of corruption and preservations. Oxidative stress caused by ROS and RNS play an important role in the food spoilage. Metabolic processes in muscle tissue *in vivo* result in the production of ROS and oxidative compounds including superoxide anions and NO. All bacteria were able to consume dissolved oxygen and all strains showed significant growth enhancement in the presence of heme, indicating respiratory activity and possibly accelerated spoilage. Lipofuscin is one of the indicators of oxidative stress. A study strongly suggest that high oxidative stress, especially with a significant accumulation of lipofuscin, is associated with high oxidative stress [Hasegawa *et al.*, 2021]. A paper studied the relationship between oxygen concentration, shear force, and protein oxidation in modified atmosphere packaged pork. Pork loins were stored at 5°C for 14 days in modified atmosphere packaging contained 0–80% O_2, 20% CO_2, and balanced with N_2. The results showed that shear force and thiobarbituric acid-reactive substances (TBARS) values increased with increasing oxygen concentration. Protein oxidation when measured as loss of free thiol groups, was greater in meat packaged under oxygen (20–80%). Myosin heavy chain (MHC) cross-linking, another marker of protein oxidation, was greater in 80% oxygen than 0% and 20% oxygen. In oil emulsion and cooked-meat homogenates, ferrous iron and hemoglobin had strong pro-oxidant effects, but ferritin became pro-oxidant only when ascorbate was present. In raw meat, homogenates ferrous had strong pro-oxidant effects even in the presence of $\bullet O_2^-$, or H_2O_2 [Ahn & Kim, 1998]. The influence of nitric oxide in the development of meat tenderness has been studied through postmortem manipulation and also through *in vivo* studies. NO is an oxidant although the effects of NO on effector proteins can be distinguished from a direct oxidation reaction. It is proposed that postmortem meat tenderization is influenced by skeletal muscle's release of NO pre-slaughter and the oxidation of proteases postmortem. Progress has been made, including studies by manipulating the NO levels in muscle cells, suggesting possible effects in the pre-slaughter and post-slaughter environment. It is speculated that NO and protein S-nitrosylation may be involved in muscle to meat conversion through the regulation of postmortem biochemical pathways including glycolysis, Ca^{2+} release, proteolysis, and apoptosis. Nitric oxide is also an antioxidant in processed meat and nitric oxide as a modulator of transmetallisation reactions in meat seems possible. NO and NO donor NOR-3 have dual effect on μ-calpain activity, autolysis, and proteolytic ability [Li *et al.*, 2014; Hou *et al.*, 2020].

4.2.11 *Oxidative stress and cigarette smoke*

In Chapter 18, we will discuss cigarette smoke and oxidative stress. More and more evidences show that oxidative stress is involved in the pathological process of diseases caused by cigarette smoke. A large number of studies have shown that cigarette smoke can cause oxidative stress. Cigarette smoking is a very complex physical and chemical process that can produce thousands of substances, including a large number of free radicals. These free radicals are distributed in the tar and gas phase of cigarette smoking and play different roles in the pathogenesis. In this complex redox process, electrons are easily converted between these substances to form a variety of stable and unstable products, free radicals, which are distributed in the gas phase of cigarette smoking and tar [Pryor 1983, Zhao *et al.*, 1990]. Most of the free radicals in the gas phase of cigarette smoking are transient unstable free radicals. The initial combustion of cigarettes produces nitric oxide, which is not highly reactive with most organic substances, but it is easy to react with oxygen in the flue gas stream to produce nitrogen dioxide, which is easy to react with olefins in the flue gas to produce alkyl free radicals. Alkyl radicals are very easy to oxidize to form alkoxy radicals. Using ESR spectra technique, they measured the radicals in cigarette smoking gas-phase [Zhao *et al.*, 1990; Church & Pryor 1985; Halpern & Knieper, 1985; Yan *et al.*, 1991a, 1991b]. A major toxic target of cigarette smoking is lung injury and the resulting, bronchitis and lung cancer. Due to cigarette smoking, gaseous substances are in direct contact with lung cells. Another way to cause lung injury is that cigarette smoking can cause macrophages to gather and activate in the lung, produce respiratory burst, release a large number of oxygen free radicals, and cause greater damage to lung cells. We studied the effect of cigarette smoking gaseous substances on macrophage respiratory burst, and found that cigarette smoking gaseous substances can directly affect the process of macrophage respiratory burst, and can also induce macrophage respiratory burst by oxidizing cell membrane. Smoking stimulates macrophages to release oxygen free radicals and leads to oxidative stress in two ways, one is direct and the other is indirect [Yan *et al.*, 1991b; Yang *et al.*, 1992, 1993a, 1993b, 1993c]. The results suggest that 1) O_2^-. is probably generated in the upper and lower respiratory tract lining fluid when they come in contact with cigarette smoke, 2) such generated O_2^-. can primarily impair PMN capabilities to generate ROS, and 3) these effects may contribute to the pathogenesis of cigarette smoke-related lung diseases [Tsuchiya *et al.*, 1992; Tsuchimy *et al.*, 1993].

4.3 Regulative effects of tea polyphenols on oxidative stress

Tea polyphenols are widely considered as a kind of excellent antioxidant agents. Tea polyphenols and their derivatives contribute to the beneficial effects ascribed to tea. A large number of studies have shown that tea polyphenols can inhibit or regulate oxidative stress. Tea catechins are ROS scavengers and metal ion chelators, whereas their indirect antioxidant activities comprise of induction of antioxidant enzymes, inhibition of pro-oxidant enzymes, and production of the phase II detoxification enzymes and antioxidant enzymes. Tea catechins and polyphenols are effective scavengers of ROS and may also function indirectly as antioxidants through their effects on transcription factors and enzyme activities. The fact that catechins are rapidly and extensively metabolized emphasizes the importance of demonstrating their antioxidant activity *in vivo*. In addition, tea polyphenols have also been observed a potent pro-oxidant capacity, which directly leads to the effects of the green tea polyphenols as they relate to cancer prevention. Such pro-oxidant effects appear to be responsible for the induction of apoptosis in tumor cells. These pro-oxidant effects may also induce endogenous antioxidant systems in normal tissues that offer protection against carcinogenic insult [Lambert & Elias, 2010].

In Chapters 2 and 3, the antioxidant effects of green tea polyphenols and polymerized tea polyphenols and theaflavins in black tea and other tea were introduced, respectively. Next, we will discuss more antioxidant effects of green tea polyphenols and polymerized tea polymerized tea polyphenols in Pu'er tea.

4.3.1 *Scavenging effect of green tea polyphenols on oxygen free radicals*

Oxygen free radicals are the main cause of oxidative stress. A large number of studies have shown that tea polyphenols can scavenge oxygen free radicals produced by different systems [Shen *et al.*, 1993; Yang *et al.*, 1994; Zhao *et al.*, 1989c]. Oxygen free radicals produced by polymorphonuclear leukocytes (PMNs) play an important role in the process of immune killing but they also damage normal cells. Oxygen free radicals produced by this system are often used to test the scavenging effect of antioxidants on oxygen free radicals produced by cell system. We use ESR spin trapping technology to capture the oxygen free radicals produced by the respiratory burst of human PMN stimulated by cancer promoter PMA. The results show that tea polyphenols almost completely eliminate the oxygen free radicals produced by this system, which is more effective than vitamin C and other antioxidants. In the superoxide anion radical system produced by riboflavin/EDTA system under light, the scavenging rate of tea polyphenols and other antioxidants to superoxide anion radical is similar to that of vitamin C [Shen *et al.*, 1993; Yang *et al.*, 1994; Zhao *et al.*, 1989c].

Tea polyphenols mainly contain four monomer components: EGCG, ECG, EC, and EGC (Fig. 4-3). They have different scavenging effects on oxygen free radicals. It was found that green tea polyphenols with hydroxyl groups have stronger antioxidant capacity. The scavenging ability of these polyphenols to oxygen free radicals decreases according to the following relationship: EGCG > ECG > EC > EGC. It was also found that several tea polyphenols had synergistic effects on scavenging oxygen free radicals. The scavenging effect of two combinations on superoxide anion radical is greater than that of one component [Shen *et al.*, 1993; Yang *et al.*, 1994; Zhao *et al.*, 1989c]. In the antioxidant process of green tea polyphenols, the first reduction potential should play a decisive role. The level of this potential directly affects the antioxidant capacity of the polyphenols. Our study

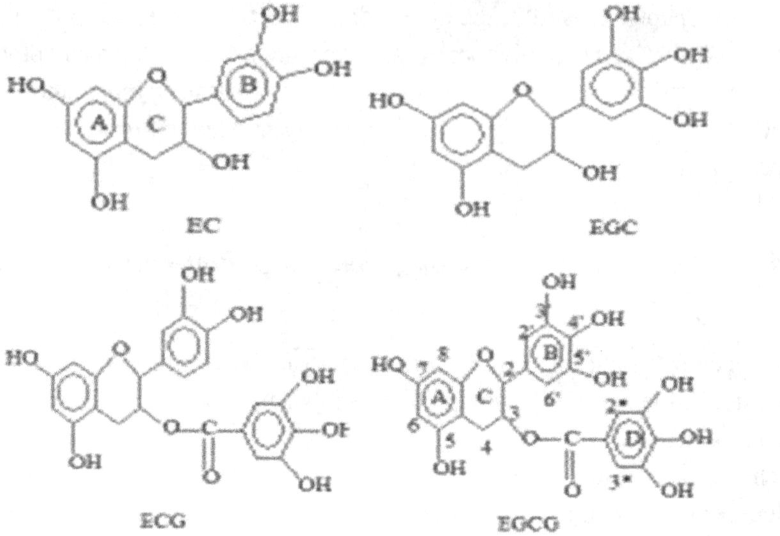

Figure 4-3.   The structures of green tea polyphenols.

Figure 4-4.   Tea polyphenols antioxidant cycle.

found that EGCG has the highest first reduction potential among the poly-phenols and its antioxidant capacity is enhanced [Shen *et al.*, 1993, Yang *et al.*, 1994; Zhao *et al.*, 1989c]. The maximum scavenging rate of super-oxide anion radical by EGCG determined by chemiluminescence is about 98%. We calculated the stoichiometric factors of the reaction of EGCG with oxygen radicals. The reaction rate constant of EGCG with superoxide anion radical determined by chemiluminescence method is $k = 7.71 \times 10^{-6}$ L. mM S^{-1}, stoichiometric factor is 5.98. It shows that each EGCG mole-cule can capture and scavenge six superoxide anion radicals [Shen *et al.*, 1993; Yang *et al.*, 1994].

From the above results, it can be seen that these tea polyphenols have a cycle in the process of antioxidation, and they are a synergistic cycle. The intensity of this cycle is related to the composition and proportion of polyphenols. This cycle is shown in Figure 4-4.

4.3.2 *Inhibitory effect of green tea polyphenols on peroxynitrite oxidation activity*

NO has important biological functions *in vivo*. For example, it can prevent platelet aggregation. It is endothelial cell relaxation factor (EDRF) but it is still a free radical. Therefore, it has active chemical properties, strong reactivity, cytotoxicity, and plays an important role in the process of immune killing. Recent studies have found that when cells produce NO, they also produce superoxide anion free radicals ant same time. NO and superoxide anion radical have a very high reaction rate constant ($6.4 \times 10^9 mol/L^{-1}s^{-1}$) and react to form peroxynitrite ($ONOO^-$). This is a highly oxidizing substance, which can oxidize the thio-hydrogen groups of cell membrane lipids and proteins, leading to cell damage and disease. Peroxynitrite is relatively stable under alkaline conditions. Once protonated, it decomposes immediately to produce hydroxyl and NO_2 free radicals. Experiments show that such reactions occur in many pathological states, so the research on peroxynitrite is very interesting. Many components of $ONOO^-$ damaged cells are through the mechanism of free radicals. The free radicals produced by $ONOO^-$ were directly captured by ESR spin trapping technology and spectrum analysis and computer simulation. Polymorphonuclear leukocytes, endothelial cells, and nerve cells can produce superoxide anion and NO free radical at the same time under some pathological conditions. Considering the fast disproportionation reaction of superoxide anion catalyzed by enzyme *in vivo* ($10^9 mol/L^{-1}s^{-1}$) and the short life of hydroxyl radical ($10^{-6}s$), it is difficult to reach the biological target molecule nearby its position. The rate constants of NO and $\bullet O_2^-$ reactions are of the same order of magnitude ($6.7 \times 10^9 mol/L^{-1}s^{-1}$), and they are produced in the same cell at the same time. Therefore, they should first react to produce $ONOO^-$, and its half-life is 1.9 seconds at pH 7.4, which allows it to diffuse to several cells and can reach the position that superoxide anion and hydroxyl cannot reach. Therefore, under various pathological conditions, the cytotoxicity of NO and oxygen free radicals may be mainly realized through $ONOO^-$ and many cell components attacked and damaged by $ONOO^-$ are through free radical mechanism. We have found tea polyphenols could scavenge the free radicals and the inhibition of $ONOO^-$ oxidation activity [Zhao *et al.*, 1996e].

4.3.3 *Inhibitive effect of green tea polyphenols on lipid peroxidation and scavenging effect on lipid free radical in brain synaptosomes*

The oxygen free radicals produced by the reaction of enzyme system and non-enzyme system attack the polyunsaturated fatty acids in biological cell membrane phospholipids and cause lipid peroxidation. Lipid peroxidation can eventually cause cell metabolism, dysfunction, and even death. In the field of central nervous system, researchers pay more and more attention to the relationship between oxygen free radical production, lipid peroxidation, and central nervous system injury. Therefore, it is of great significance to find antioxidants to reduce lipid peroxidation and brain injury. We studied and compared the inhibitory effects of four tea polyphenol monomers on iron induced lipid peroxidation in brain synaptosomes by spin trapping method [Guo *et al.*, 1996].

1. Inhibitory effects of EGCG, ECG, EGC, and EC on lipid peroxidation injury of brain synaptosomes

We measured the inhibitory effects of four tea polyphenol monomers on 2-thiobarbituric (TBA) reactants formed during lipid peroxidation of brain synaptosomes caused by iron ion (Fe^{2+}/Fe^{3+} system). With the increase of the concentrations of EGCG, ECG, EGC, and EC, the formation amount of TBA reactants gradually decreased and the inhibition rate gradually increased, with a good dose-response relationship. Their IC50 were 0.35 mmol/L, 0.24 mmol/L, 0.19 mmol/L, and 0.11 mmol/L respectively, indicating that their inhibitory effect on lipid peroxidation injury of brain synaptosomes induced by iron ions was enhanced in the following order: EGCG > ECG > GTP > EGC > EC.

2. Scavenging effects of four monomers of tea polyphenols on lipid free radicals

After Fe^{2+}/Fe^{3+} system initiated lipid peroxidation in brain synaptosomes, the IC50 of ECG, EGCG, EC and EGC of scavenging effect of tea polyphenols on lipid free radicals were 7.31mg/ml, 14.9 mg/ml, 22.14 mg/ml, and 59.28 mg/ml, respectively. Therefore, their ability to scavenge lipid free radicals was in this order: ECG > EGCG > EC > EGC.

The scavenging effect of tea polyphenols on lipid free radicals reflects the scavenging effect of tea polyphenols on lipid free radicals produced in the process of lipid peroxidation. It includes both lipid free radicals produced in the initiation process of lipid peroxidation of brain synaptosomes induced by iron ions and lipid free radicals produced in the chain reaction process of lipid peroxidation. The results given by it reflect the action mechanism of tea polyphenols with free radicals. The inhibitory effect of tea polyphenols on lipid peroxidation of brain synaptosomes measured by TBA method reflects the inhibitory effect of tea polyphenols on the formation of TBA reactants, the final product of lipid peroxidation. The results given by it mainly reflect the overall effect and final result of antioxidation and inhibition of free radical induced oxidative stress by tea polyphenols. Many problems are involved in explaining the differences between them, including the complexation of tea polyphenols to iron ions, the scavenging effect of tea polyphenols on hydroxyl radicals and lipid radicals, and the stability of semiquinone radicals formed after the reaction of tea polyphenols with free radicals.

4.3.4 *Scavenging effects of different isomers of tea polyphenols on reactive oxygen species*

Not only do tea polyphenols with different structures have different antioxidant properties and scavenging effects on different free radicals, but tea polyphenol isomers with the same structure but different isomers may also have different scavenging effects on free radicals. Here, we selected three pairs of tea polyphenol isomers: EC–(+)-C, EGC–GC, EGCG–GCG, with the same structure to study (Fig. 4-5), and found that they do have different antioxidant properties, especially lower concentrations have different scavenging effects on different free radicals [Guo *et al.*, 1996, 1999].

1. Scavenging effect of tea polyphenol isomers on superoxide anion

O_2^- was produced by EDTA/light riboflavin system, which was studied by ESR spin trapping method. After adding the six isomers of tea polyphenols, the ESR signal is different with their concentration. Taking monomer EGCG as an example, we analyze these change processes in detail. When the concentration of EGCG increased from 0.01 mmol/L to 0.02 mmol/L,

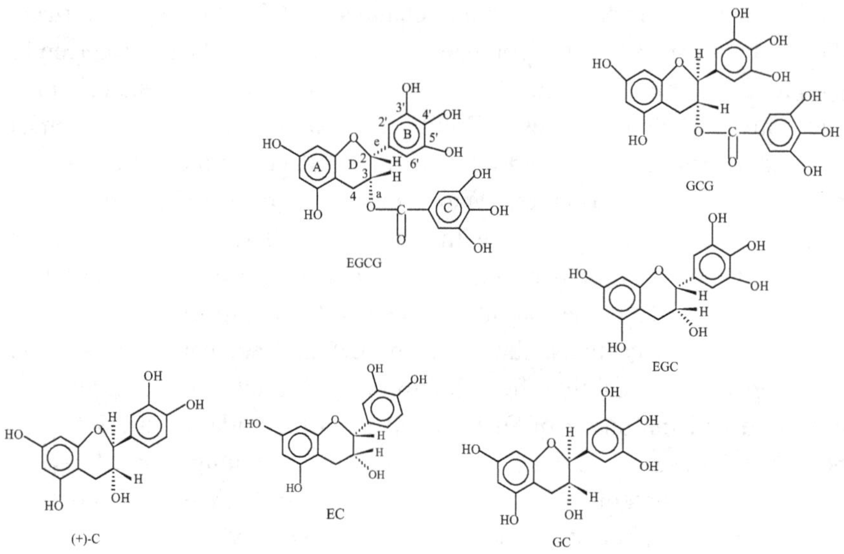

Figure 4-5.   The structures of green tea polyphenol isomers.

the scavenging rate of O_2^- increased, which were 64.9% and 76.2% respectively. The scavenging effect of the six isomers of tea polyphenols at these two concentrations increased in the following order: EC<(+)-C < EGC < GC < EGCG < GCG, that is, the scavenging effect of GCG and EGCG with one more anoic acid group (CH (OH) COO) at position 3 was stronger than that of GC, EGC, EC and (+)-C. The scavenging effect of EGC and GC with one more hydroxyl group at position 5 is stronger than that of EC and (+)-C. It was also found that the difference of their spatial structure had an impact on their scavenging effect. The scavenging effects of GCG, GC, and (+)-C were better than EGCG, EGC, and EC respectively, and the difference was more significant at low concentration.

The experimental results also found that the relationship between the concentration of GCG, GC, and EGC and the ESR signal was consistent with that of the above EGCG, that is, when their concentration was 0.02 mmol/L, there was $\bullet O_2^-$, the hydroxyl radical generated by the reaction of GCG (or GC or EGC) with $\bullet O_2^-$, and the corresponding tea polyphenol isomer radical. When the concentration was 0.2 mmol/ L and 1.0 mmol/L, the concentration of the corresponding tea polyphenol isomer radical and hydroxyl radical reached the maximum respectively, and then, decrease as the isomer concentration continues to increase. However, the results of EC and (+)-C are different from those of the above four monomers. When their concentration increases to 0.2mmol/l, three free radicals appear at the same time. When the concentration is 1.0mmol/l and 2.0mmol/l, the concentrations of EC and (+)-C radicals and hydroxyl radicals reach the maximum respectively, and then decrease with the continuous increase of monomer concentration.

These show that the scavenging process of different tea polyphenol isomers on superoxide anion free radicals produced by lipid peroxidation at different concentrations is complex. It needs deep and careful research to understand the reaction mechanism.

2. Scavenging effect of tea polyphenol isomers on singlet oxygen

Hematoporphyrin was used to produce singlet oxygen $^1O_2$ under light. The concentration of singlet oxygen $^1O_2$ produced by hematoporphyrin under light increases with time. When it was illuminated for 12 minutes, its concentration reached the maximum, then tends to be stable after 15 minutes. The kinetic process of singlet oxygen produced by hematoporphyrin under light is changed by adding six isomers of tea polyphenols (taking GC as an example). When the GC concentration is 0.1 mmol/l, it shortens the time to reach the maximum and decreases the maximum. With the increase of GC concentration, the production rate of singlet oxygen decreases, the time to reach the maximum gradually moves forward and the maximum decreases significantly. The effects of other tea polyphenol isomers were similar. When GCG and EC were at higher or lower concentrations there was no significant difference between GC and EGCG about the scavenging ability of singlet oxygen. When their concentration decreased to 0.1 mmol/l, the scavenging effect of GCG on singlet oxygen was better than that of EGCG. These results suggest that when they are at high concentration, the difference of spatial structure has no obvious effect on their ability to scavenge singlet oxygen, while at low concentration, their spatial structure has an effect on their ability to scavenge singlet oxygen, and the scavenging effect of GCG is better than that of EGCG. Similarly, at high concentration, the spatial structure difference of stereoisomers GC and EGC and (+) — C had no effect on their ability to scavenge singlet oxygen, while at low concentration, their spatial structure has an effect on their ability to scavenge singlet oxygen, and the scavenging effects of GC and (+) — C are better than EGC and EC, respectively. The results indicated that the tea polyphenol monomer has a stronger ability to scavenge singlet oxygen and their scavenging effect on singlet oxygen increases in the following order: EC {(+) — C} < EGC (GC) < EGCG (GCG).

3. Scavenging effect of tea polyphenol isomers on alkane free radicals produced by the decomposition of AAPH

2,2′-Azobis(2-amidinopropane)dihydrochloride;2,2′-Azobis(isobutyramidine)dihydrochloride (AAPH) is a common water-soluble alkane radical generator. At 37°C, it can produce alkane radicals in the carbon center. Six isomers of tea polyphenols can scavenge alkane free radicals produced by the decomposition of AAPH. It was found that the concentration of these isomers in the range of 0.01–0.10 mmol/l had a good scavenging effect on the alkane free radicals produced by the decomposition of AAPH in a concentration dependent manner. The scavenging effect increased in the following order: EC < (+) — C < EGC < GC < EGCG < GCG. It was also found that the difference of their spatial structure had an impact on their scavenging effect.

4.3.5 *Molecular mechanism of scavenging oxygen free radicals by tea polyphenols*

In order to deeply understand the scavenging mechanism and antioxidant mechanism of tea polyphenols on oxygen free radicals, we studied the complexing effect of four monomers of tea polyphenols on iron ions, the scavenging effect of hydroxyl free radicals and lipid free radicals in iron free ion system, and the stability of semiquinone free radicals generated after their reaction with free radicals. Based on the structure of tea polyphenols, the reactive active sites of four monomers of tea polyphenols to scavenge oxygen free radicals were deeply analyzed [Guo e*t al*., 1996, 1999; Zhao *et al*., 2001].

1. The complexation ratio of tea polyphenols to Fe (III) was determined by molar ratio method

Iron ion is an important factor in inducing lipid peroxidation, so the difference of iron ion complexation between the four monomers of tea polyphenols need be considered. Here, we first check the complex molecular ratio of four monomers of tea polyphenols to iron ions quantitatively determined by molar ratio method. The results are as follows:

$$EGC: Fe^{2+}/Fe^{3+} = 3:2; EGCG\ Fe^{2+}/Fe^{3+} = 2:1; ECG\ Fe^{2+}/Fe^{3+} = 2:1;$$
$$EC\ Fe^{2+}/Fe^{3+} = 3:1$$

This result shows that the complexation of tea polyphenols to iron ions plays a certain role in its antioxidant effect, and the order of their complexation strength to iron ions is EGC > EGCG = ECG > EC. However, the protective effect of tea polyphenols to iron ions on lipid peroxidation of brain synaptosomes is not the only factor. Next, we check the scavenging effect of tea polyphenols on hydroxyl radicals.

2. Scavenging effect of tea polyphenol monomer on hydroxyl radical produced by light hydrogen peroxide system

Another important factor of oxidative stress that initiates lipid peroxidation is hydroxyl radical. So, we studied the hydroxyl radical produced by light hydrogen peroxide without involvement of iron ions. The scavenging effect of tea polyphenols on hydroxyl can be directly tested. When the concentration of ECG is 0.75 mmol/L, it can remove 46.5% hydroxyl, EC can remove 19.1% hydroxyl, and the other three substances cannot remove hydroxyl at this concentration. When the concentration of EGCG increases to 1.5 mmol/L, it can remove 61.1% hydroxyl, and 3.75 mmol/L can remove 84.2% hydroxyl but EGC did not show the ability to scavenge hydroxyl radicals. Therefore, in this system, the order of scavenging hydroxyl radical by tea polyphenol monomers is: ECG > EC > EGCG > EGC. This result is consistent with their results of scavenging lipid peroxidation of brain synaptosomes to produce lipid free radicals. It shows that although the lipid free radicals produced by lipid peroxidation of brain synaptosomes are induced by iron ions, hydroxyl free radicals also play a certain role. When we consider their inhibitory effect on lipid peroxidation, we must consider other factors.

3. Scavenging of lipid free radicals produced by lipoxygenase system by tea polyphenol monomers EGCG, ECG, EGC and EC

Lipoxygenase acts on lecithin to produce lipid free radicals without iron ion involved, which can directly to be used to detect the scavenging effect of tea polyphenols on lipid free radicals. The IC50 values of EGCG, ECG, EGC, and EC for scavenging lipid free radicals are 0.37 mmol/L, 0.46 mmol/L, 0.27 mmol/L, and 0.30 mmol/L respectively, so their ability to scavenge lipid free radicals is as follows: ECG > EGCG > EGC > EC. The overall results are consistent with the inhibitory effect of tea polyphenols on lipid peroxidation of brain synaptosomes but inconsistent with the second half of the inhibitory effect. In order to fully explain the inhibitory effect of tea polyphenols on lipid peroxidation of brain synaptosomes, other factors need to be considered.

4. Stability of free radicals formed by the reaction of tea polyphenol monomers with free radicals

Tea polyphenols react with oxygen free radicals to form semiquinone free radicals. The stability of semiquinone free radicals is an important factor for the ability of tea polyphenols to scavenge oxygen free radicals and inhibit lipid peroxidation. The changes of semiquinone radicals formed by the reaction of four monomers of tea polyphenols with oxygen radicals with time were directly observed by ESR, in which the stability of semiquinone radicals formed by EGCG, ECG, and EC but EGC was unstable. This result is basically consistent with the result that tea polyphenols inhibit the lipid peroxidation of brain synaptosomes, indicating that the stability of semiquinone free radicals formed by the reaction between tea polyphenols and oxygen free radicals plays a major role in the overall effect of inhibiting the formation of final products by lipid peroxidation [Guo *et al.*, 1996].

These results show that the above four aspects play an important role in protecting the mechanism of lipid peroxidation injury. The total effect is that EGCG and ECG are better than EGC and EC.

4.3.6 *Quantum chemical basis of scavenging oxygen free radicals by tea polyphenols*

Quantum chemistry is a subject that studies chemical problems based on the basic principles and methods of quantum mechanics. It can calculate the molecular mechanism from the microscopic point of view on the characteristics and laws of electronic bonding and bonding characteristics of molecular electronic junction. We discussed why tea polyphenols can scavenge oxygen free radicals and the reactive active centers of oxygen free radicals from quantum chemistry perspective, that is, the results of quantum chemical calculation of the molecular structure and molecular orbital of tea polyphenols, and compare them with vitamin E. Thus, the reaction sites of tea polyphenols in inhibiting oxidative stress and scavenging free radicals can be studied [Zhao *et al.*, 1992, 1996e].

1. The best conformation of catechin in aqueous solution
The molecular orbital approximation method is used for quantum chemistry calculation and results show that the best conformation of catechin in aqueous solution is that the benzhydrinfuran of catechin is located on a plane, and the other phenol ring is perpendicular to this plane.

2. Deriving the structural characteristics of catechins from the eigenvector of the highest occupied orbit
The structural characteristics of a molecule can be known from their various eigenvector assignments. The electrons of Px and Py usually form S bond, while the electrons of Pz form P bond. In the plane of benzhydrinfuran, the electrons on the carbon atom and oxygen atom 7, 11, and 13 of the benzene ring are mainly concentrated in the Pz orbital, which means that the benzene ring in this plane, the oxygen atom of the two phenolic hydroxyl groups connected to it, and the oxygen atom on the pyrrole ring form a large P bond. The two phenolic hydroxyl groups are not real phenolic hydroxyl groups, but in the intermediate form of phenolic hydroxyl and quinone. The inclusion of the oxygen atom on the Py ran ring in this large p-bond increases the composition of quinone and makes it easier for electrons to leave this large p-bond. The other phenol ring is perpendicular to the plane of benzhydrinfuran, so the Px electrons of carbon atoms 15, 16, 17, 18, and 19 and oxygen atoms 22 and 23 participate in the formation of a large bond, but smaller than that of benzhydrinfuran ring. The two phenolic hydroxyls on this ring also contain quinone components, but less than the two phenolic hydroxyls on benzhydrinfuran, and the degree of electron delocalization is also smaller (Fig. 4-6).

3. Electron multiple distribution and charge difference between phenol hydroxyl oxygen atom and hydrogen atom on catechin ring
The greater the multiple distribution and electrostatic charge difference between two atoms, the stronger the bond between two atoms. The O-H

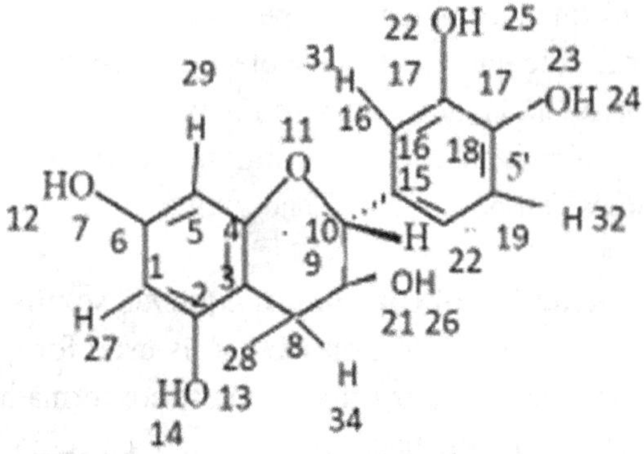

Figure 4-6.   The structure of catechin.

bond of the two phenolic hydroxyl groups on the benzhydrinfuran ring is stronger than that of the two phenolic hydroxyl groups on the benzhydrin-furan ring, which also shows that when catechin reacts with oxygen radi-cal, the hydrogen atoms of the two phenolic hydroxyl groups on the benzhydrinfuran ring are easier to be removed than the hydrogen atoms of the two hydroxyl groups on the phenol ring.

The above results show that when tea polyphenols inhibit oxidative stress to scavenge free radicals, the reaction site on the hydrogen atom of two phenol hydroxyl groups on the benzohydrofluorane ring is more likely than on the hydrogen atom of two hydroxyl groups on the phenol ring.

4.3.7 *Inhibitory effect of tea polyphenols on oxidative stress in DNA damage*

DNA is a major cellular component of ROS attack, which will directly lead to mutation and tumor. The bases of DNA damage and oxidation, especially 8-oxo-7,8-dihydro-2'-deoxyguanosine (8-oxodGuo), are effec-tive and main biomarkers for the above two methods to detect DNA oxida-tive damage *in vivo* and *in vitro*. However, while these methods provide information about the final DNA oxidation products, they cannot provide some information during the oxidation process, especially at the begin-ning of the oxidation reaction. The generation of chemiluminescence may be one of the earliest reactions of DNA molecules to oxidative stress before other changes can be detected. Electron spin resonance (ESR) can directly provide some information about free radicals that cause DNA oxidative damage. In order to study the protective mechanism of tea polyphenols on oxidative damage to DNA, as a model for studying oxidative damage to DNA, the copper ion 1,10-phenanthroline complex (two chelating molecules 1,10-phenanthroline (OP) chelate a copper ion to form [(OP)$_2$Cu^{2+}]) is widely used as a chemical nuclease to study the structure of DNA [Nie *et al.*, 2001]. (OP)$_2$Cu^{2+} reacts with hydrogen peroxide and ascorbic acid to produce chemiluminescence, with maximum absorption at 460–480nm, which is mainly the chemiluminescence generated by superoxide anion and mercapto radical. When DNA was added into the system, the chemiluminescence showed a significant lag and the maximum absorption blue shifted to 400–420 nm, which was mainly caused by the carbonyl radical produced by the modification of guanine by DNA oxidative damage. It was found that this chemiluminescence was proportional to the concentration of DNA (0.2–1.2 µg/ml) and hydrogen peroxide, and was related to the concentration of ascorbic acid, Cu^{2+} and pH value. In order to study the protective mechanism of tea polyphenols on oxidative damage DNA, we used Cu^{2+} chelator (1,10-phenanthroline (OP) (OP)$_2$Cu^{2+} as the model of DNA damage induced by chemical nuclease, and studied the protective mechanism of four tea polyphenol monomers on oxidative damage DNA by ESR and chemiluminescence technology [Nie *et al.*, 2001].

1. Protective effect of tea polyphenols on DNA against oxidation detected by chemiluminescence

When four kinds of tea polyphenol monomers were added to the Cu^{2+} chelator 1,10-phenanthroline (OP) (OP)$_2$Cu^{2+} as the model of DNA system, it was found that the chemiluminescence decreased significantly. The concentration dependence of four tea polyphenol monomers on chemiluminescence produced by DNA oxidative damage. The analysis results show that the order of their protective effects of DNA against oxidative damage is: ECG > EC > EGCG > EGC. The IC50 of their chemiluminescence inhibition on DNA oxidative damage were 2.6 µg/ml, 3.8 µg/ml, 6.0 µg/ml, and 11.5 µg/ml, respectively.

2. Protective effect of tea polyphenols on DNA against oxidation detected by ESR

In order to determine what free radicals cause DNA damage, we measured the generated free radicals by ESR spin trapping technique. The results showed that the spectrum of sulfhydryl radicals was captured. It was also found that the order of ESR signal scavenging effect of adding four tea polyphenol monomers into the system was ECG > EC > EGCG > EGC. This proves that the sulfhydryl radical is the main cause of DNA damage in the reaction system of (OP) $2Cu^{2+}$ with hydrogen peroxide and ascorbic acid.

4.3.8 *Inhibitory effect of tea polyphenols on oxidative stress in animal experiments*

In addition to the above-mentioned studies on the inhibitory effect of tea polyphenols on oxidative stress and scavenging on oxygen free radical in chemical and cellular systems, there are also many reports on the inhibitory effect of tea polyphenols on oxidative stress and scavenging effect on oxygen free radical in animal systems.

1. Antioxidant effect of green tea polyphenols in animals

A paper studied the ability to scavenge the free radicals was compared in mouse liver mitochondria. Mitochondrial swelling, mitochondrial membrane potential (MMP), cardiolipin peroxidation, and respiratory chain complex (RCC) I-V activities were also evaluated as the index of the mitochondrial oxidative damage. The study supports evidence that the tea polyphenols extract exhibited significant protection against oxidative damage on mitochondrial. Furthermore, the effect of the tea polyphenols on antioxidant ability in zebrafish embryo was evaluated. After the tea polyphenols pretreatment for one (1) day, zebrafish embryos showed a significantly higher survival rate as well as heart rate when facing the oxidative stress [Gao *et al.*, 2020]. This was accompanied by enhanced lipid and protein oxidation and compromised antioxidant defenses associated with increased expression of the oxidative stress markers 4-hydroxynonenal (4-HNE), anti-hexanoyl lysine, dibromotyrosine, and 8-hydroxy 2-deoxy-guanosine (8-OHdG). Dietary administration of tea polyphenon effectively reduced preneoplastic and neoplastic lesions, modulation of xenobiotic-metabolizing enzymes, and amelioration of oxidative stress.

A paper studied the inhibitory effect of green tea extract and tea polyphenols in the mouse model of chronic fatigue syndrome. It was found that treatment with green tea polyphenols (25 or 50 mg/kg, i.p.) and catechin (50 or 100 mg/kg, i.p.) for seven days significantly reversed the increased lipid peroxidation levels and decreased glutathione levels in mouse whole-brain homogenate [Singal *et al.*, 2005]. A study demonstrated that tea polyphenols, alter the production of ROS, glutathione metabolism, lipid peroxidation, and protein oxidation under *in vitro* conditions. It also demonstrated that EGCG and black tea polyphenols affected redox metabolism under cell culture conditions. Induction of apoptosis was observed, after the treatment with tea polyphenols, as shown by increased DNA breakdown and activation of the apoptotic markers, cytochrome c, caspase 3, and poly-(ADP-ribose) polymerase. These results may have implications in determining the chemopreventive and therapeutic use of tea polyphenols *in vivo* [Raza & John, 2008].

Green tea components have been shown to affect the peroxisome proliferator-activated receptors (PPARs) signaling pathways. We found that the malondialdehyde (MDA) level in the high-fat diet group was significantly higher than that in the control group ($p < 0.05$). Decreased PPAR-α and sirtuin 3 (SIRT3) expression, and increased manganese superoxide dismutase (MnSOD) acetylation levels were also detected in the high-fat diet group ($p < 0.05$). GTP treatment upregulated SIRT3 and PPARα expression, increased the PPARα mRNA level, reduced the MnSOD acetylation level, and decreased MDA production in rats fed a high-fat diet ($p < 0.05$). The reduced oxidative stress detected in kidney tissues after green tea polyphenol treatment was partly due to the higher SIRT3 expression, which was likely mediated by PPARα [Yan *et al.*, 2013; Yang *et al.*, 2015].

2. Antioxidant effect of theaflavins and thearubigins in animals

Oolong, black, and dark teas contain abundant black tea polyphenols modulate xenobiotic-metabolizing enzymes, oxidative stress, and adduct formation in a rat [Murugan *et al.*, 2008]. Vlack tea polyphenols can reduce body weight, alleviate metabolic syndrome, and prevent diabetes and cardiovascular diseases in animal models and humans. Due to strong antioxidant property, black tea inhibits the development of various cancers by regulating oxidative damage of biomolecules, endogenous antioxidants, and pathways of mutagen and transcription of antioxidant gene pool. Tea polyphenols appear as direct antioxidants by scavenging reactive oxygen/nitrogen species, chelating transition metals, and inhibiting lipid, protein, and DNA oxidations. Intra-gastric administration of 7,12-dimethylbenz[a]anthracene (DMBA) induced adenocarcinomas that showed enhanced activities of phase I carcinogen activation and phase II detoxification enzymes with increased lipid and protein oxidation and decrease in antioxidant status. Dietary administration of polyphenol effectively suppressed the incidence of mammary tumors as evidenced by modulation of xenobiotic-metabolizing enzymes and oxidant-antioxidant status, inhibition of cell proliferation and angiogenesis, and induction of apoptosis [Kumaraguruparan *et al.*, 2007].

4.3.9 *Scavenging effect of Pu'er tea extracts on free radicals and their protective effect on oxidative damage in human fibroblast cells*

Pu'er tea is a post-fermentation tea with unique flavor and multiple health benefits. Pu'er tea displays cholesterol-lowering properties, but the underlying mechanism has not been elucidated. The theaflavins is one of the most active and abundant pigments in Pu'er tea. Pu'er tea is rich in multiple active constitute such as flavonoids, catechins, phenolic acids, flavanols polymer, purine alkaloids, and hydrolysable tannin as a microbial-fermented tea. Pu'er tea has a variety of pharmacologically activities, such as anti-hyper-lipidemic, anti-diabetic, anti-oxidative, anti-tumor, anti-bacterial, anti-inflammatory, and anti-viral effects [Gu *et al.*, 2017]. We successively extracted the Pu'er tea with acetone, water, chloroform, ethyl acetate, and n-butanol, and the extracts were then isolated by column chromatography. Our study demonstrates that the Pu'er tea ethyl acetate extract, n-butanol extract, and their fractions had scavenging activity on superoxide anion and hydroxyl radical. Fractions 2 and 8 from the ethyl acetate extract and fractions 2, 4, and 5 from the n-butanol extract showed protective effects against hydrogen peroxide-induced damage in human fibroblast HPF-1 cells and increased the cells' viability under normal cell culture conditions. In addition, it was found that these fractions, except fraction 5 from the n-butanol extract, decreased the accumulation of intracellular ROS in hydrogen peroxide-induced HPF-1 cells. Interestingly, the antioxidant effect of fraction 8 from the ethyl acetate extract on the above four systems was much stronger than that of the typical green tea EGCG, but there were almost no monomeric polyphenols, theaflavins, and gallic acid in fraction 8 [Jie *et al.*, 2006].

1. Scavenging effect on hydroxyl radicals

The scavenging effects of Pu'er tea fractions on the •OH were evaluated by means of Fenton-type reaction. Fractions PEF2, PEF5, PEF6, PEF7, PEF8, PBF2, PBF3, PBF4, and PBF5 showed a good linear relationship between their scavenging percent and the logarithm of their concentration. On the basis of the comparison among the IC50 values of each fraction, the five fractions of Pu'er tea ethyl acetate extract and the four fractions of butyl alcohol extract exhibit higher antioxidant potency than that of EGCG and even higher than that of ascorbic acid. PEF8 shows the strongest scavenging effect on the hydrogen radicals.

2. Scavenging effect on superoxide anions

The scavenging effects of Pu-erh tea fractions on $•O_2^-$ were evaluated by means of pyrogallol autoxidation. That upon linear regression analysis of their scavenging percent and sample concentration, fractions PEF2, PEF5, PEF6, PEF8, PBF3, and PBF5 had a good linear relationship between their two parameters. Like the scavenging effect on HO•, the Pu-erh tea ethyl acetate extract's fractions also exhibit stronger scavenging effects on $•O_2^-$ than that of EGCG and even stronger than that of ascorbic acid. A little difference was shown in the fractions PEF7, PBF2, and PBF4. PBF3, unlike its scavenging effect on •OH, showed a lower scavenging effect on $•O_2^-$ even than that of ascorbic acid. Consistent with its scavenging effect on •OH, PEF8 also showed the strongest scavenging effect on the $•O_2^-$.

3. Decreased H2O2-induced accumulation of ROS in HPF-1 cells

ROS is the main factor that causes oxidative stress, which results in decreasing cell viability. The level of DCF fluorescence is an indicator of ROS production. After treatment with different doses of H_2O_2 in the HPF-1 cells, the level of ROS in the cells increased dose-dependently. After treatment with 600 μM H_2O_2 for one hour, the DCF fluorescence intensity increased about 70–80% in comparison with that of the negative control. The increase in the DCF fluorescence intensity was eliminated partly when the cells were co-treated with different concentrations of EGCG or PEF8. The fluorescence intensity of cells treated with EGCG and PEF8 at a concentration of 1.6 μg/mL was decreased by about 31% and 33%, respectively. When the cells were co-treated with PEF8, the decrease in fluorescence intensity was dose-dependent. In addition, even with no exposure to H_2O_2, treatment with EGCG or PEF8 at concentrations of 0.2–1.6 μg/mL for one hour resulted in decreased DCF fluorescence intensity in the HPF-1cells.

4. Ameliorated H2O2-induced loss of HPF-1 cell viability

Treatment with 600 μM H_2O_2 for 24 hours decreased the viability of HPF-1 cells about 30–35% relative to the negative control. After pretreatment with PEF2, PEF8, PBF2, PBF4, and PBF5 at different concentrations (0.6 μg/mL, 1.2 μg/mL, 2.4 μg/mL, 4.8 μg/mL, or 9.6 μg/mL), the cell viability was almost dose-dependently ameliorated. PEF2, at the concentration of 4.8 μg/mL, showed the best protective effect on the damaged HPF-1 cell. The cell viability after pretreatment with PEF8 gradually increased in accordance with the concentration at 9.6 μg/mL, the viability was about 18.9% higher than that of the negative control group. The cell viability after pretreatment with PBF2 and PBF5 also significantly increased at different concentrations. PBF4 showed the strongest protective effect at the concentration of 2.4 μg/mL, then the increased viability decreased in accordance with the concentration. In addition, we also checked the cell viability when the cells were treated with Pu'er fractions alone. The cells that were treated with PEF2 and PBF4 at concentrations of 0.6–9.6 μg/mL had no improvements in viability. But after treatment with 4.8 μg/mL PEF8, the increased viability was about 18.9% higher that of the negative control. PBF2 and PBF5 can distinctly increase the cell viability at all five concentrations of the experiment.

The above results show that Pu'er tea can not only directly scavenge ROS, but also has a stronger scavenging effect than EGCG. In addition, it has a significant inhibitory effect on the oxidative stress induced by hydrogen peroxide and a good protective effect on cell damage.

4.3.10 *Effects of tea polyphenols on the activities of antioxidant enzymes in the leaves of wheat seedlings under salt stress*

In addition to the antioxidant effects of tea polyphenols in cell, animal, and human experiments discussed above, there was also a report on the antioxidant effects of tea polyphenols on plant systems. It was found that tea polyphenols could regulate oxidative stress in plant system. This paper studied the effects of tea polyphenols on the activities of antioxidant enzymes in the leaves of wheat seedlings under salt stress. NaCl stress alone inhibited the seedling growth, increased sodium content and ROS accumulation, but reduced potassium (K) and calcium levels at different culture times, thus resulting in the oxidative damage to the leaves. The 25 mg L^{-1} or 100 mg L^{-1} tea polyphenols treatment alone led to the significant increases of $\bullet O_2^-$ and H_2O_2 generation, tea polyphenols-treated leaves exhibited the reduction of relative electrical conductivity and no change of malondialdehyde content. In addition, the activities and gene expression of superoxide dismutase, catalase, and peroxidase as well as diamine oxidase and polyamine oxidase were changed to different degrees due to NaCl or tea polyphenols treatment alone. Further study showed that the presence of tea polyphenols promoted the growth, increased K^+ and Ca^{2+} contents, and reduced $\bullet O_2^-$ and H_2O_2 accumulation in salt-stressed wheat seedlings [Zhang *et al.*, 2012].

4.4 Clinical implications of oxidative stress and antioxidant therapy of tea polyphenols

The above extensive and in-depth research on oxidative stress and the inhibitory effect of tea polyphenols on oxidative stress in theories, cell, and animal systems provide a solid foundation for clinical application. In addition, epidemiological investigation also found that drinking tea and tea polyphenols can also be used to treat a variety of diseases.

For example, a statistical analysis on green tea and esophageal cancer in 2020, including 16 studies, explored the relationship between green tea intake and esophageal cancer risk through meta-analysis. In Asian and non-Asian countries, male and female case-control studies have shown that green tea can be used as a preventive factor for esophageal cancer. Another study found that the consumption of green tea and the number of years of drinking green tea were negatively correlated with the risk of breast cancer. In China, the number of people who drink green tea suffering from breast cancer decreased by about 23.5% compared with the control group. It is believed that drinking green tea may have a positive effect on reducing the risk of breast cancer, especially in the case of long-term and high dose of green tea drinking [Abe & Inoue, 2020].

Many studies reported statistical analysis results related to green tea and cardiovascular diseases. Two prospective studies showed that 90,914 Japanese aged 40 to 69 years old between 1990 and 1994 had a negative correlation between green tea intake and all-cause mortality and cardio-vascular mortality. Three statistical analyses on green tea consumption and cardiovascular disease mortality support a risk reduction of 18–33% [Zhang *et al.*, 2010; Tang *et al.*, 2015].

Many analysis randomized controlled trials reported that green tea was negatively correlated with fasting blood glucose (FBG) and glyco-sylated hemoglobin (HbA1c). The Japanese Cancer Risk Assessment Collaborative Cohort Study found that the RR of women was 0.49 (95% CI 0.30–0.79), while the RR of men was 0.91 (95% CI 0.55–1.52) for consumption green tea ≥6 cups/day. Four statistical analyses of green tea showed a decrease in body weight or body mass index. Four statistical analyses on the relationship between green tea and blood pressure reported that both systolic and diastolic blood pressure decreased [Liu *et al.*, 2013; Vazquez *et al.*, 2017; Hibi *et al.*, 2018].

Human epidemiology and animal data show that drinking tea can reduce the incidence rate of AD and PD. Epidemiological studies have found that the incidence rate of PD in Asia is 5–10 times lower than that in western countries, which may be related to the widespread drinking of green tea in Asian countries [Pan *et al.*, 2003].

The relationship between tea drinking and bone mineral density (BMD) in elderly British women was studied. BMD of lumbar spine, femoral neck, trochanter, and Ward triangle were measured in 1,256 free living women aged 65–76 years in Cambridge, England. Research shows that compared with those who do not drink tea, those who drink tea for a long time have higher bone mineral density in lumbar vertebrae, femoral neck, and Ward triangle. Those who drank black tea, oolong tea, or green tea for more than 10 years had the highest bone density in their hips. Those who drank tea had significantly higher mean BMD measurements in lumbar spine (0.033 g/cm; $p = 0.03$), greater trochanter (0.028 g/cm; $p = 0.004$), and Ward's triangle (0.025 g/cm) [Hegarty *et al.*, 2000]. Another random survey was conducted on 623 postmenopausal women of Han nationality in Fuzhou China, and 593 women met the requirements after completing the test. The greater trochanter bone density (0.807 ± 0.117 kg/cm^2) in the tea drinking group was higher than that in the non-tea drinking group (0.778 ± 0.117 kg/cm^2, $p = 0.042$,), and the Ward's triangle bone density (0.676 ± 0.130 kg/cm^2) in the tea drinking group was higher than that in the non-tea drinking group (0.643 ± 0.138 kg/cm^2, $p = 0.022$). Comparison of the incidence of postmenopausal fracture: 13.39% in the tea drinking group and 12.06% in the non-tea drinking group, $p > 0.05$, with no statistically significant difference [Wang *et al.*, 2014].

Green tea polyphenols, the most interesting constituent of green tea leaves, have been shown to have both pro-oxidant and antioxidant properties. Both pro-oxidant and antioxidant properties are expected to contribute to modulation of oxidative stress response under ideal optimal dosage regimens. Postprandial oxidative stress is attenuated when dietary antioxidants are supplied together with a meal rich in oxidized or oxidizable lipids. Ingestion of dietary polyphenols from tea, improves endothelial dysfunction and lowers the susceptibility of LDL lipids to oxidation. Tea polyphenols affect endothelial function not solely as antioxidants but also as modulatory signaling molecules. Exposure to a low concentration of a pro-oxidant prior to exposure to oxidative stress induces the expression of genes that code for proteins that induce adaptation in a subsequent oxidative stress. On the contrary, exposure to an antioxidant concurrently with exposure to the oxidative stress affords protection through free radical scavenging or through other indirect antioxidant mechanisms. In any case, the optimal conditions that afford protection from oxidative stress should be defined for any substance with redox properties. Green tea polyphenols, naturally occurring substances, should to be an ideal option for the modulation of oxidative stress response. In inflammatory bowel disease, oxidative stress has been postulated to play a role in disease initiation and progression, and antioxidant therapy, such as green tea polyphenols and gene therapy with SOD, has a markedly attenuated disease. Tea polyphenols supplementation has been used with some success in the treatment of chronic pancreatitis. Tea polyphenols could reduce oxidative stress in neural damage and mitochondrial dysfunction to neuronal death, cell swelling, and brain edema in ischemia [Dryden *et al.*, 2005; Yiannakopoulou, 2013].

A paper elucidated the interaction between green tea and EGCG, its main polyphenol and oral peroxidases. It was found elderly trained people who drink green tea for three months, have a higher level of oral peroxidases activity compared to non-drinkers. Addition of green tea and black tea infusions and EGCG to saliva, resulted in a sharp rise of oral peroxidases activity ($p = 0.009$), respectively. Also, following green tea infusion mouth rinsing, a rise of oral peroxidases activity was observed ($p = 0.159$). These results may be of great clinical importance, as tea consumer's oral epithelium may have better protection against the deleterious effects of hydroxyl radicals, produced by hydrogen peroxides in the presence of metal ions [Narotzki *et al.*, 2013].

The inhibitory effects of green tea polyphenols on oxidative stress have extensively been demonstrated in different models of neurotoxicity. A study demonstrated that white tea extracts protect striatal cell lines against oxidative stress-mediated cell death. The effects of white tea on protection of striatal cell cultures are likely associated with the antioxidant properties of white tea components since neuronal cell loss induced by non-oxidative insults such as D1 dopamine receptor activation cannot be prevented by pre-treatment with white tea. This results indicated that six hours isoflurane anesthesia induced cognitive impairment in early three days, meanwhile, the hippocampus SOD declined in step. Green tea derived polyphenols 25mg/kg per day effectively mitigated isodlurane-induced declines of SOD [Song *et al.*, 2019]. A paper evaluated the role of heme oxygenase 1 (HO-1) induction by EGCG and the transcriptional mechanisms involved. The results demonstrated that EGCG induced HO-1 expression in cultured neurons, possibly by activation of the transcription factor Nrf2, and by this mechanism was able to protect against oxidative stress-induced cell death [Romeo *et al.*, 2009].

Clinically, the quantitative determination of oxidative stress is a representative biomarker. 8-hydroxydeoxyguanosine (8-OHdG) is a sensitive marker of DNA damage, which is formed by a hydroxyl group attached to the 8th carbon of guanine. Modulation of urinary excretion of green tea polyphenols and oxidative DNA damage biomarker, 8-OHdG, were assessed in urine samples collected from a randomized, double-blinded, and placebo-controlled phase IIa chemoprevention trial with green tea polyphenols in 124 individuals. In green tea polyphenols-treated groups, EGC and EC levels displayed significant and dose-dependent increases in both the 500 mg group and 1000 mg group ($p < 0.05$). At the end of the three months' intervention, 8-OHdG levels decreased significantly in both GTP-treated groups ($p = 0.007$) [Luo *et al.*, 2006]. Acrylamide neurotoxicity in humans is a significant public health issue attracting wide attention. Acrylamide-treated PC12 cells pretreated with various concentrations of EGCG for 24 hours had increased viability and acetylcholinesterase activity and reduced apoptosis and necrosis compared to cells exposed to acrylamide alone. EGCG reduced the expression of bax mRNA, decreased cytochrome c release, reduced intracellular calcium levels, inactivated caspase 3, and increased mitochondrial membrane potential, suggesting that EGCG prevents acrylamide-induced apoptosis through a mitochondrial-mediated pathway. In addition, EGCG inhibited the formation of reactive oxygen species and lipid peroxidation while enhancing superoxide dismutase activity and glutathione levels, thereby reducing oxidative stress [He *et al.*, 2017].

The above research results show that tea polyphenols can prevent and treat some diseases through antioxidant effects in population experiments and clinical studies.

4.5 Conclusion

From above discussion, we can clearly find that oxidative stress plays an important role in biological system, especially in some disease states. Tea polyphenols can regulate oxidative stress and protect the health of the body. Tea polyphenols regulate oxidative stress by scavenging ROS and

RNS, chelating redox active transition metal ions and playing the role of antioxidants in human body. They may also inhibit redox sensitive transcription factors and nuclear factors by 1) having κB and activator protein-1 indirectly play the role of antioxidants, 2) inhibiting "oxidant" enzymes, such as inducible nitric oxide synthase, lipoxygenase, cyclooxygenase, and xanthine oxidase, 3) inducing phase II and antioxidant enzymes such as glutathione S-transferase and superoxide dismutase, and 4) having tea polyphenols to not only act as antioxidants, but also as signal molecules to regulate endothelial function. Prior to exposure to oxidative stress, exposure to low concentrations of pro-oxidants can induce the expression of genes encoding proteins, thereby inducing adaptation in subsequent oxidative stress. However, simultaneous exposure to antioxidants and oxidative stress can provide protection by scavenging free radicals or other indirect antioxidant mechanisms.

In the following chapters of this book, the antioxidant effects of tea polyphenols in different systems and their effects on health will be discussed in more detail.

References

Abe SK, Inoue M. (2020) Green tea and cancer and cardiometabolic diseases: a review of the current epidemiological evidence. *Eur J Clin Nutr*, **75**(6), 865–876.

Afanas'ev IB. (2005) On mechanism of superoxide signaling under physiological and pathophysiological conditions. *Med Hypotheses*, **64**, 127–129.

Ahn DU, Kim SM. (1998) Prooxidant effects of ferrous iron, hemoglobin, and ferritin in oil emulsion and cooked-meat homogenates are different from those in raw-meat homogenates. *Poult Sci*, **77**(2), 348–355.

Alvarez RA, Berra L, Galdwin MT. (2020) Home nitric oxide therapy for COVID-19. *Am J Respir Crit Care Med*, **202**, 16–20.

Ansari MA, Scheff SW. (2010) Oxidative stress in the progression of Alzheimer disease in the frontal cortex. *J Neuropathol Exp Neurol*, **69**(2), 155–167.

Ansari A, Rahman MS, Saha SK, *et al.* (2017) Function of the SIRT3 mitochondrial deacetylase in cellular physiology, cancer, and neurodegenerative disease. *Aging Cell*, **16**(1), 4–16.

Axford JS. (1987) Reduced B-cell galactosyltransferase activity in rhematoid arthrits. *Lanecet II*, 1486–1492.

Banu NSS, Panikar LR, Leal AR. (2020) Protective role of ACE2 and its down-regulation in SARS-CoV-2 infection leading to Macrophage Activation Syndrome: therapeutic implications. *Life Sci*, **256**, 117905.

Barja G. (2014) The mitochondrial free radical theory of aging. *Prog Mol Biol Transl Sci*, **127**, 1–27.

Becerril S, Rodríguez A, Catalán V, *et al.* (2019) Functional relationship between leptin and nitric oxide in metabolism. *Nutrients*, **11**(9), 2129.

Briaud I, Harmon JS, Kelpe CL, Segu VB, Poitout V. (2001) Lipotoxicity of the pancreatic β-cell is associated with glucose-dependent esterification of fatty acids into neutral lipids. *Diabetes*, **50**, 315–321.

Butterfield DA, Kanski J. (2001) Brain protein oxidation in age-related neurode-generative disorders that are associated with aggregated proteins. *Mech Ageing Dev*, **122**, 945–962.

Chen X, Yang C, Jiang G. (2021) Research progress on skin photoaging and oxidative stress. *Postepy Dermatol Alergol*, **38**(6), 931–936.

Cheng YC, Sheen JM, Hu WL, Hung YC. (2017) Polyphenols and oxidative stress in atherosclerosis-related ischemic heart disease and stroke. *Oxid Med Cell Longev*, **2017**, 8526438.

Chung KH, Chiou HY, Chang JS, Chen YH. (2020) Associations of nitric oxide with obesity and psychological traits among children and adolescents in Taiwan. *Pediatr Obes*, **15**(3), e12593.

Church DF, Pryor WA. (1985) Free-radical chemistry of cigarette smoke and its toxicological implications. *Environ Health Perspect*, **64**, 111–123.

Cohen G, Pasik, P, Cohen B, *et al.* (1985) Pargyline and deprenyl prevent the neurotoxicity of1-methyl-4-phenyl-1,2,3,6-tetrahydropyridine (MPTP) in monkeys. *Eur J Pharmacol*, **106**, 209–210.

Dai DF, Chen T, Wanagat, *et al.* (2010) Age-dependent cardiomyopathy in mito-chondrial mutator mice is attenuated by overexpression of catalase targeted to mitochondria. *Aging Cell*, **9**(4), 536–544.

Dai DF, Chiao YA, Martin GM, *et al.* (2017) Mitochondrial-targeted catalase: Extended longevity and the roles in various disease models. *Prog Mol Biol Transl Sci*, **146**, 203–241.

Dai DF, Chiao YA, Marcinek DJ, *et al.* (2014) Mitochondrial oxidative stress in aging and healthspan. *Longev Healthspan*, **3**, 6.

Darfeuille-Michaud A, Boudeau J, Bulois P, *et al.* (2004) High prevalence of adherent-invasive Escherichia coli associated with ileal mucosa in Crohn's disease. *Gastroenterology*, **127**,412–421.

Davidson SM, Duchen MR. (2007) Endothelial mitochondria: Contributing to vascular function and disease. *Circ Res*, **100**(8), 1128–1141.

De Bruyne T, Pieters L, Witvrouw M, De Clercq E, Vanden Berghe D, Vlietinck AJ. (1999) Biological evaluation of proanthocyanidin dimmers and related polyphenols. *J Nat Prod*, 62, 954–958.

Desjardins D, Cacho-Valadez B, Liu JL, *et al.* (2017) Antioxidants reveal an inverted U-shaped dose-response relationship between reactive oxygen species levels and the rate of aging in Caenorhabditis elegans. *Aging Cell*, **16**(1), 104–112.

Dryden GW Jr, Deaciuc I, Arteel G, McClain CJ. (2005) Clinical implications of oxidative stress and antioxidant therapy. *Curr Gastroenterol Rep*, **7**(4), 308–316.

Fariss MW, Chan CB, Patel M, *et al.* (2005) Role of mitochondria in toxic oxidative stress. *Mol Interv*, **5**(2), 94.

Fernandez-Sanchez A, Madrigal-Santillan E, Bautista M, *et al.* (2011) Inflammation, oxidative stress, and obesity. *Int J Mol Sci*, **12**(5), 3117–3132.

Ferrari R, Guardigli G, Mele D, Percoco GF, Ceconi C, Curello S. (2004) Oxidative stress during myocardial ischaemia and heart failure. *Curr Pharm Des*, 10, 1699–1711.

Fibach E, Dana M. (2019) Oxidative stress in beta-thalassemia. *Mol Diagn Ther*, **23**(2), 245–261.

Flashner BM, Rifas-Shiman SL, Oken E, *et al.* (2020) Obesity, sedentary lifestyle, and exhaled nitric oxide in an early adolescent cohort. *Pediatr Pulmonol*, **55**(2), 503–509.

Furukawa S, Fujita T, Shimabukuro M, *et al.* (2004) Increased oxidative stress in obesity and its impact on metabolic syndrome. *J Clin Invest*, **114**, 1752–1761.

Gao T, Shi Y, Xue Y, Yan F, Huang D, Wu Y, Weng Z. (2020) Polyphenol extract from superheated steam processed tea waste attenuates the oxidative damage in vivo and in vitro. *J Food Biochem*, **44**(1), e13096.

Gabbita SP, Lovell MA, Markesbery WR. (1998) Increased nuclear DNA oxidation in the brain in Alzheimer's disease. *J Neurochem*, **71**, 2034–2040.

Garg C, Sharma H, Garg M. (2020) Skin photo-protection with phytochemicals against photo-oxidative stress, photo-carcinogenesis, signal transduction pathways and extracellular matrix remodeling-An overview. *Ageing Res Rev*, **62**, 101127.

Garneata L. (2008) Intravenous iron, inflammation, and oxidative stress: is iron a friend or an enemy of uremic patients? *J Ren Nutr*, **18**(1), 40–45.

Giordano N. (1984) Increased storege of iron and anaemia in rheumatoid arthritis: usefulness of deferrioxamine. *Br Med J*, **289**, 961–970.

Graham D, Tiffany SM, Bell WR Jr, Gutknecht WF. (1978) Autoxidation versus covalent binding of quinines as the mechanism of toxicity of dopamine,

6-hydroxydopamine, and related compounds toward C1300 neuroblastoma cells in vitro. *Mol Pharmacol*, **14**, 644–653.

Grundy SM. (2004) Obesity, metabolic syndrome, and cardiovascular disease. *J Clin Endocrinol Metab*, **89**, 2595–2600.

Gu XP, Pan B, Wu Z, Zhao YF, Tu PF, Zheng J. (2017) Progress in research for pharmacological effects of Pu-erh tea. *Zhongguo Zhong Yao Za Zhi*, **42**(11), 2038–2041.

Guo Q, Zhao BL, Li MF, Shen SR, Xin WJ. (1996) Studies on protective mechanisms of four components of green tea polyphenols (GTP) against lipid peroxidation in synaptosomes. *Biochim Biophys Acta*, **1304**, 210–222.

Guo Q, Zhao B-L, Hou J-W, Xin W-J. (1999) ESR study on the structure-antioxidant activity relationship of tea catechins and their epimers. *Biochim Biophys Acta*, **1427**, 13–23.

Guo SH, Bezard E, Zhao BL. (2005) Protective effect of green tea polyphenols on the SH-SY5Y cells against 6-OHDA induced apoptosis through ROS-pathway. *Free Rad Biol Med*, **39**, 682–695.

Guo SH, Bezard E, Zhao BL. (2007) Protective effects of green tea polyphenols in the 6-OHDA rat model of Parkinson's disease through inhibition of ROS-pathway. *Biol Psychiatry*, **62**(12), 1353–1362.

Halliwall B. (1992) Reactive oxygen species and the central nervous system. *J Neurochem*, **59**, 1609–1623.

Halliday GM. (2005) Inflammation, gene mutation and photoimmunosuppression in response to UVR-induced oxidative damage contributes to photocarcinogenesis, *Mutat Res*, **571**(1-2), 107–120.

Halpern A, Knieper J. (1985) Spin trappings of radicals in gasephase tobacco smoke. In *Proceedings of the seventeenth International Symposium on Free Radicals*, Colorado, 306.

Hanafi R, Anestopoulos I, Voulgaridou GP, *et al.* (2012) Oxidative stress based-biomarkers in oral carcinogenesis: how far have we gone? *Curr Mol Med*, **12**(6), 698–703.

Harman D. (1956) Aging: a theory based on free radical and radiation chemistry. *J Gerontol*, **11**(3), 298–300.

Harman D. (1972) The biologic clock: the mitochondria? *J Am Geriatr Soc*, **20**(4), 145–147.

Hasegawa Y, Kawasaki T, Maeda N, Yamada M, Takahashi N, Watanabe T, Iwasaki T. (2021) Accumulation of lipofuscin in broiler chicken with wooden breast. *Anim Sci J*, **92**(1), e13517.

Hegarty VM, May HM, Khaw KT. (2000) Tea drinking and bone mineral density in older women. *Am J Clin Nutr*, **71**, 1003–1007.

He Y, Tan D, Mi Y, Bai B, Jiang D, Zhou X, Ji S. (2017) Effect of epigallocate-chin-3-gallate on acrylamide-induced oxidative stress and apoptosis in PC12 cells. *Hum Exp Toxicol*, **36**(10), 1087–1099.

Hengartner MO. (2000) The biochemistry of apoptosis. *Nature*, **407**(6805), 770–776.

Hibi M, Takase H, Iwasaki M, Osaki N, Katsuragi Y. (2018) Efficacy of tea cate-chin-rich beverages to reduce abdominal adiposity and metabolic syndrome risks in obese and overweight subjects: a pooled analysis of 6 human trials. *Nutr Res*, **55**, 1–10.

Holguin F, Grasemann H, Sharma S, *et al.* (2019) L-Citrulline increases nitric oxide and improves control in obese asthmatics. *JCI Insight*, **4**(24), e131733. polycystic ovary syndrome. *J Clin Endocrinol Metab*, **91**, 336–340.

Hou Q, Zhang CY, Zhang WG, Liu R, Tang H, Zhou GH. (2020) Role of protein S-nitrosylation in regulating beef tenderness. *Food Chem*, **306**, 125616.

Houstis N, Rosen ED, Lander ES (2006) Reactive oxygen species have a causal role in multiple forms of insulin resistance. *Nature*, **440**, 944–948.

Jacqueminet S, Briaud I, Rouault C, Reach G, Poitout V (2000) Inhibition of insulin gene expression by long-term exposure of pancreatic β-cells to pal-mitate is dependent on the presence of a stimulatory glucose concentration. *Metabolism*, **49**, 532–536.

Jenner P, Olanow CW. (1998) Understanding cell death in Parkinson's disease. *Ann Neurol*, **44**(Suppl 1), S72–S84.

Jie G, Lin Z, Zhang L, Lv H, He P, Zhao B-L. (2006) Free radical scavenging effect of Pu-erh tea extracts and their protective effect on oxidative damage in human fibroblast cells. *J Agric Food Chem*, **54**, 8058–8064.

Khan A, Manna K, Das DK, *et al.* (2015) Gossypetin ameliorates ionizing radia-tion-induced oxidative stress in mice liver--a molecular approach. *Free Radic Res*, **49**(10), 1173–1186.

Kondo H, Yumoto K, Alwood JS, Mojarrab R, Wang A, Almeida EA, Searby ND, Limoli CL, Globus RK. (2010) Oxidative stress and gamma radiation-induced cancellous bone loss with musculoskeletal disuse. *J Appl Physiol*, **108**(1), 152–161.

Kumaraguruparan R, Seshagiri PB, Hara Y, Nagini S. (2007) Chemoprevention of rat mammary carcinogenesis by black tea polyphenols: modulation of xenobiotic-metabolizing enzymes, oxidative stress, cell proliferation, apop-tosis, and angiogenesis. *Mol Carcinog*, **46**(9), 797–806.

Laforge M, Elbim C, Frère C, *et al.* (2020) Tissue damage from neutrophil-induced oxidative stress in COVID-19. *Nature Reviews Immunology*, **20**(9), 515–516.

Lambert AJ, Boysen HM, Buckingham JA, *et al.* (2007) Low rates of hydrogen peroxide production by isolated heart mitochondria associate with long maximum lifespan in vertebrate homeotherms. *Aging Cell*, **6**(5), 607–618.

Lambert JD, Elias RJ. (2010) The antioxidant and pro-oxidant activities of green tea polyphenols: a role in cancer prevention. *Arch Biochem Biophys*, **501**(1), 65–72

Li H-T, Zhao B-L, Hou J-W, Xin W-J. (1996) Two peak kinetic curve of chemiluninencence in phorbol stimulated macrophage. *Biochem Biophys Res Commn*, **223**, 311–314.

Li X-J, Zhao B-L, Hou J-W, Xin W-J. (1990) Active oxygen radicals produced by leukocytes of malignant lymphoma, *Chinese Medical J*, **103**, 899–905.

Li YP, Liu R, Zhang WG, Fu QQ, Liu N, Zhou GH. (2014) Effect of nitric oxide on mu-calpain activation, protein proteolysis, and protein oxidation of pork during post-mortem aging. *J Agric Food Chem*, **62**(25), 5972–5977.

Liu K, Zhou R, Wang B, Chen K, Shi LY, Zhu JD, *et al.* (2013) Effect of green tea on glucose control and insulin sensitivity: a meta-analysis of 17 randomized controlled trials. *Am J Clin Nutr*, **98**, 340–348.

Lum H, Roebuck KA. (2001) Oxidant stress and endothelial cell dysfunction. *Am J Physiol — Cell Physiol*, **280**(4), C719–C741.

Luo H, Tang L, Tang M, Billam M, Huang T, Yu J, Wei Z, Liang Y, Wang K, Zhang ZQ, Zhang L, Wang JS. (2006) Phase IIa chemoprevention trial of green tea polyphenols in high-risk individuals of liver cancer: modulation of urinary excretion of green tea polyphenols and 8-hydroxydeoxyguanosine. *Carcinogenesis*, **27**(2): 262–268.

Mason TM, Goh T, Tchipashvili V, Sandhu H, Gupta N, Lewis GF, Giacca A. (1999) Prolonged elevation of plasma free fatty acids desensitizes the insulin secretory response to glucose in vivo in rats. *Diabetes*, **48**, 524–530.

Miles PD, Romeo OM, Higo K, Cohen A, Rafaat K, Olefsky JM. (1997) TNF-alpha-induced insulin resistance in vivo and its prevention by troglitazone. *Diabetes*, **46**, 1678–1683.

Murugan RS, Uchida K, Hara Y, Nagini S. (2008) Black tea polyphenols modulate xenobiotic-metabolizing enzymes, oxidative stress and adduct formation in a rat hepatocarcinogenesis model. *Free Radic Res*, **42**(10), 873–884.

Narotzki B, Levy Y, Aizenbud D, Reznick AZ. (2013) Green tea and its major polyphenol EGCG increase the activity of oral peroxidases. *Adv Exp Med Biol*, **756**, 99–104.

Ni Y, Zhao B-L, Hou J, Xin W (1996) Ginkgo biloba extract protection of brain neurons from damage induced by free radicals. In *Proceedings of the international symposium on natural antioxidants molecular mechanisms and health effects*, Packer L, Traber MG, Xin W (eds.), AOCS Press Champaign Illinois.

Ni YC, Zhao BL, Hou JW, Xin WJ. (1996) Protection of cerebellar neuron by Ginkgo-biloba extract against apoptosis induced by hydroxyl radicals. *Neuron Science Letter*, **214**, 115–118.

Nie GJ, Jin C-F, Cao Y-L, Shen S-R, Zhao B-L. (2002a) Distinct effects of tea catechins on 6-hydroxydopamine-induced apoptosis in PC12 cells. *Arch Biochem Biophys*, **397**, 84–90.

Nie GJ, Cao YL, Zhao B-L. (2002b) Protective effects of green tea polyphenols and their major component, (-)-epigallocatechin-3-gallate (EGCG), on 6-hydroxyldopamine-induced apoptosis in PC12 cells. *Redox Report*, **7**, 170–177.

Nie GJ, Wei TT, Shen SR, Zhao BL. (2001) Polyphenol protection of DNA against damage, *Meth Enzym*, **335**, 232–244.

Nurk E, Refsum H, Drevon CA, Tell GS, Nygaard HA, Engedal K, Smith AD. (2009) Intake of flavonoid-rich wine, tea, and chocolate by elderly men and women is associated with better cognitive test performance. *J Nutr*, **139**, 120–127.

Oda A, Tamaoka A, Araki W. (2009) Oxidative stress up-regulates presenilin 1 in lipid rafts in neuronal cells. *J Neurosci Res*, **88**(5), 1137–1145.

Pan T, Jankovic J, Le W. (2003) Potential therapeutic properties of green tea polyphenols in Parkinson's disease. *Drugs Aging*, **20**(10), 711–721.

Pryor WA. (1983) An electron spin resonence study of mainstream and sidestream cigarette smoke: nature of the free radicals in gas-phase smoke. *Environ Health Perspect*, **47**, 345–355.

Raisch J, Rolhion N, Dubois A, Darfeuille-Michaud A, Bringer MA. (2015) Intracellular colon cancer-associated Escherichia coli promote protumoral activities of human macrophages by inducing sustained COX-2 expression. *Lab Invest*, **95**, 296–307.

Raza H, John A. (2008) In vitro effects of tea polyphenols on redox metabolism, oxidative stress, and apoptosis in PC12 cells. *Ann N Y Acad Sci*, **1138**, 358–365.

Rendra E, Riabov V, Mossel DM, Sevastyanova T, Harmsen MC, Kzhyshkowska J. (2019) Reactive oxygen species (ROS) in macrophage activation and function in diabetes. *Immunobiology*, **224**(2), 242–253.

Ricciardolo FL, Bertolini F, Carriero V, Högman M. (2020) Nitric oxide's physiologic effects and potential as a therapeutic agent against COVID-19. *J Breath Res*, **15**(1), 014001.

Ridker PM, Morrow DA. (2003) C-reactive protein, inflammation, and coronary risk. *Cardiol Clin*, **21**, 315–325.

Romeo L, Intrieri M, D'Agata V, Mangano NG, Oriani G, Ontario ML, Scapagnini G. (2009) The major green tea polyphenol, (-)-epigallocatechin-3-gallate, induces heme oxygenase in rat neurons and acts as an effective neuroprotective agent against oxidative stress. *J Am Coll Nutr*, **28** Suppl, 492S–499S.

Shen S, Yang X, Yang F, Zhao B-L, Xin W. (1993) Synergistic enhancement effect of catechins on antioxidation. *Tea Sci*, **13**(2), 141–146.

Singal A, Kaur S, Tirkey N, Chopra K. (2005) Green tea extract and catechin ameliorate chronic fatigue-induced oxidative stress in mice. *J Med Food*, **8**(1), 47–52.

Smith MA, Rottkamp CA, Nunomura A, Raina AK, Perry G. (2000) Oxidative stress in Alzheimer's disease. *Biochim Biophys Acta*, **1502**(1) 139–144.

Sonta T, Inoguchi T, Tsubouchi H, *et al.* (2004) Evidence for contribution of vascular NAD(P)H oxidase to increased oxidative stress in animal models of diabetes and obesity. *Free Radic Biol Med*, **37**, 115–123.

Song Y, Li X, Gong X, Zhao X, Ma Z, Xia T, Gu X. (2019) Green tea polyphenols improve isoflurane-induced cognitive impairment via modulating oxidative stress. *J Nutr Biochem*, **73**, 108213.

Soto-Otero R, Méndez-Alvarez E, Hermida-Ameijeiras A, Muñoz-Patiño AM, Labandeira-Garcia JL. (2000) Autoxidation and neurotoxicity of 6-hydrodopamine in the presence of some antioxidants: potential implication in relation to the pathogenesis of Parkinson's disease. *J Neurochem*, **74**, 1605–1612.

Suganya N, Bhakkiyalakshmi E, Sarada DV, Ramkumar KM. (2016) Reversibility of endothelial dysfunction in diabetes: role of polyphenols. *Br J Nutr*, **116**(2), 223–246.

Sun Y, Shuang F, Chen DM, Zhou RB. (2013) Treatment of hydrogen molecule abates oxidative stress and alleviates bone loss induced by modeled microgravity in rats. *Osteoporos Int*, **24**(3), 969–978.

Tang J, Zheng JS, Fang L, Jin Y, Cai W, Li D. (2015) Tea consumption and mortality of all cancers, CVD and all causes: a meta-analysis of eighteen prospective cohort studies. *Br J Nutr*, **114**,673–83.

Tarng DC, Huang TP, Liu TY, Chen HW, Sung YJ, Wei YH. (2000) Effect of vitamin E-bonded membrane on the 8-hydroxy 2'-deoxyguanosine level in leukocyte DNA of hemodialysis patients. *Kidney Int*, **58**(2), 790–799.

Theofilopoulous AN, Dixon FB. (1982) Autoimmunone diseases: Immunology and etiopathogenesis. *Am J Pathol*, **108**, 321–330.

Torreilles FO, Salman-Tabcheh S, Guerin MC, Torreilles J. (1999) Neurodegenerative disorders: the role of peroxynitrite. *Brain Res Rev*, **30**, 153–163.

Tsuchiya M, Thompson DF, Suzuki YJ, Cross CE, Packer L. (1992) Superoxide formed from cigarette smoke impairs polymorphonuclear leukocyte active oxygen generation activity. *Arch Biochem Biophys*, **299**(1), 30–37.

Tsuchimy M, Suzuki Y, Cross CE, Packer L. (1993) Superoxide generation by cigarette smoke damages the respiratory bust and induces physical change in the membrane order and water organization of inflammatory cells. *Ann N Y Acad Sci*, **686**, 39–52.

Vallée A, Lecarpentier Y. (2018) Crosstalk between peroxisome proliferator-activated receptor gamma and the canonical WNT/beta-catenin pathway in chronic inflammation and oxidative stress during carcinogenesis. *Front Immunol*, **9**, 745.

Vazquez Cisneros LC, Lopez-Uriarte P, Lopez-Espinoza A, Navarro Meza M, Espinoza-Gallardo AC, Guzman Aburto MB. (2017) Effects of green tea and its epigallocatechin (EGCG) content on body weight and fat mass in humans: a systematic review. *Nutr Hosp*, **34**, 731–737.

Wang G, Liu G, Zhao H, Zhang F, Li S, Chen Y, Zhang Z. (2014) Study of the relationship between tea drinking and bone mineral density in postmenopausal Han women. *Cell Biochem Biophys*, **70**(2), 1289–1293.

Wu S, Rhee KJ, Albesiano E, Rabizadeh S, Wu X, Yen HR, Huso DL, Brancati FL, Wick E, McAllister F, *et al.* (2009) A human colonic commensal promotes colon tumorigenesis via activation of T helper type 17 T cell responses. *Nat Med*, **15**, 1016–1022.

Xin M, Yang Y, Zhang D, Wang J, Chen S, Zhou D. (2015) Attenuation of hindlimb suspension-induced bone loss by curcumin is associated with reduced oxidative stress and increased vitamin D receptor expression. *Osteoporos Int*, **26**(11), 2665–2676.

Yan J, Zhao Y, Zhao B. (2013) Green tea catechins prevent obesity through modulation of peroxisome proliferator-activated receptors. *Sci China Life Sci*, **56**, 804–810.

Yan LJ, Zhao BL, Xin WJ. (1991a) Experimental studies on smoke aspects of toxicological effects of gas phase cigarette smoke. Research *Chem Interm*, **16**, 15–24.

Yan LJ, Zhao BL, Li X-J, Xin WJ. (1991b) ESR was used to study the effect of cigarette smoke on polymorphonuclear leukocyte respiratory burst. *J Environ Sci (China)*, **11**, 79–82.

Yan, LJ, Zhao BL, Xin WJ. (1991c) Study on the physical properties of biofilm caused by smoking. *J Biophys*, **7**, 5–9.

Yang FJ, Zhao,BL, Xin,WJ. (1991) The activity of NADPH oxidase to produce O2 — was studied by chemiluminescence. *J Biophys*, **7**, 530–538.

Yang FJ, Zhao BL, Xin WJ. (1992) ESR spin trapping method was used to study the effect of liposomes treated with smoking smoke on O2 — production by rat granulocytes. *J Biophys*, **8**, 659–663.

Yang FJ, Zhao BL, Xin WJ. (1993a) ESR spectroscopic study on lipid peroxidation of rat liver microsomes induced by smoking smoke. *Environ Chem*, **12**, 117–125.

Yang FJ, Zhao BL, Xin WJ. (1993b) ESR study on the relationship between lipid peroxidation induced by cigarette gaseous substances and granulocyte respiratory burst. *J Environ Sci (China)*, **13**, 355–359.

Yang FJ, Zhao BL, Ren X-J, Xin,WJ. (1993c) ESR study of tea polyphenols inhibiting lipid free radical production in rat liver microsomes stimulated by smoking gaseous substances. *J Biophysics*, **9**, 468–471.

Yang H, Zuo XZ, Tian C, He DL, Yi WJ, Chen Z, Zhang PW, Ding SB, Ying CJ. (2015) Green tea polyphenols attenuate high-fat diet-induced renal oxidative stress through SIRT3-dependent deacetylation. *Biomed Environ Sci*, **28**(6), 455–459.

Yang X, Shen S, Hou J, Zhao B-L. (1994) Scavenging mechanism of epigallocatechin gallate on reactive oxygen free radicals. *Chinese J Pharmacol*, **15**(4), 350–353.

Yang X, Wang Y, Chen L. (2003) Chemistry of tea polyphenols. Shanghai scientific and technology Press.

Yiannakopoulou ECh. (2013) Targeting oxidative stress response by green tea polyphenols: clinical implications. *Free Radic Res*, **247**(9), 667–671.

Yuan J, Yankner B. (2000) Apoptosis in the nervous system. *Nature*, **407**, 802–809.

Zhang D-L, Yin J-J, Zhao B-L. (2004) Oral administration of Crataegus extraction protects against ischemia/reperfusion brain damage in the Mongolian gerbils. *J Neur Chem*, **90**, 211–219.

Zhang J, Huang N, Zhao B-L, Li, Xin W. (1988) Effects of photosensitization of hematoporphyrin derivatives on lipid kinetics and phase diagram of artificial membrane. *Scientific bulletin*, **33**, 1258–1260.

Zhang J, Huang N, Zhao B-L, Chen L, Xin W. (1986) ESR Study on free radical capture by DMPO in hematoporphyrin biological photosensitive system. *J Chem*, **44**, 627–630.

Zhang X, Albanes D, Beeson WL, van den Brandt PA, Buring JE, Flood A, *et al.* (2010) Risk of colon cancer and coffee, tea, and sugar-sweetened soft drink intake: pooled analysis of prospective cohort studies. *J Natl Cancer Inst*, **102**(11), 771–783.

Zhang Y, Zhang Q, Zhao B-L, *et al.* (1991) Scavenging effect of Promethazine on semiquinone free radicals in rat myocardium induced by adriamycin. *Chinese J Pharmacol*, **12**, 20–28.

Zhang Y, Li G, Si L, Liu N, Gao T, Yang Y. (2012) Effects of tea polyphenols on the activities of antioxidant enzymes and the expression of related gene in the leaves of wheat seedlings under salt stress. *Environ Sci Pollut Res Int*, **28**(46), 65447–65461.

Zhao B-L, Duan S-J, Xin W-J. (1990) Lymphocytes can produce respiratory burst and oxygen radicals as polymorphonuclear leukocytes. *Cell Biophys*, **17**, 205.

Zhao B-L, Guo Q, Xin W-J. (2001) Free radical scavenging by green tea polyphenols. *Method Enzym*, **335**, 217–231.

Zhao BL, Li XJ, Xin WJ. (1989a) ESR study on oxygen consumption during the respiratory burst of human polymophonuclear leukocytes. *Cell Biol Intern Report*, **13**, 317–325.

Zhao BL, Xin WJ, Yang WD, Zh, HL. (1989b) Direct measurement of active oxygen free radicals from ischemia-reperfusion rabbit myocardium. *Chin Sci Bull*, **34**, 780–787.

Zhao BL, Li XJ, He RG, Cheng SJ, Xin WJ. (1989c) Scavenging effect of extracts of green tea and natural antioxidants on active oxygen radicals. *Cell Biophys*, **14**, 175–181.

Zhao BL, Liu SL, Chen RS, Xin WJ. (1992) Scavenging effect of catechin on free radicals studied by molecular orbital calculation. *Acta Pharmacol Sinica,* **13**, 9–14.

Zhao B-L, Shen, J-G, Li, M,Xin, W-J. (1997) Study on NO free radicals generated from ischemia-reperfused heart and macrophage. *Chinese J Magn Reson*, **14**, 99–106.

Zhao BL, Wang JC, Hou J, Xin WJ. (1996a) NO and superoxide anion radicals produced by polymorphonuclear leukocytes mainly form ONOO-. *Chinese Science*, **26**, 406–413.

Zhao BL, Wang JC, Hou JW, Xin WJ (1996b) Studied the nitric oxide free radicals generated from polymorphonuclear leukocytes (PMN) stimulated by phobol myristate (PMA). *Cell Biol Intern*, **20**, 343–350.

Zhao BL, Shen JG, Li M, Xin WJ. (1996c) Synergic effect of NO and oxygen free radicals in ischemia-reperfusion rabbit myocardium. *Sci China*, **26**, 331–338.

Zhao BL, Shen JG, Li M, Xin WJ. (1996d) Scavenging effect of Chinonin on NO and oxygen free radicals generated from ischemia reperfusion myocadium. *Biochim Biophys Acta Biomembr*, **1317**, 131–137.

Zhao BL, Wang J, Hou J, Xin W. (1996e) Scavenging effect of tea polyphenols on methyl radical produced by peroxynitrite oxidation dimethyl sulfoxide. *Sci Bull*, **41**, 925–927.

Zhao B-L, Yan L-J, Hou J-W, Xin W-J. (1990) Electron spin resonance spin trapping of gaseous free radicals in smoking. *Chin J Med*, **70**, 386–391.

Zhao Y, Zhao B-L. (2010) Protective effect of natural antioxidant on heart against ischemis-reperfusion damage. *Curr Pharm Biotechnol*, **11**(8), 868–874.

Zhou YP, Grill VE. (1994) Long-term exposure of rat pancreatic islets to fatty acids inhibits glucose-induced insulin secretion and biosynthesis through a glucose fatty acid cycle. *J Clin Invest*, **93**, 870–876.

Zhou YP, Grill V. (1995) Long term exposure to fatty acids and ketones inhibits B-cell functions in human pancreatic islets of Langerhans. *J Clin Endocrinol Metab*, **80**, 1584–1590.

Chapter 5

Protective Effects of Tea Polyphenols on Nerves

Baolu Zhao

Institute of Biophysics, Chinese Academy of Sciences, Beijing, China

5.1 Introduction

The nervous system is the human headquarters and the regulatory system that plays a leading role in the body. It is mainly composed of nerve tissue, which is divided into central nervous system and peripheral nervous system. The central nervous system includes brain and spinal cord, and the peripheral nervous system includes brain nerve and spinal nerve. Under its direct or indirect regulation and control, the functions of various organs and various physiological processes of the human body are interrelated and interact and cooperate closely so as to make the human body a complete and unified organism and maintain normal life activities. The nervous system can feel the changes of the external environment and constantly make rapid and perfect adjustments to various functions in the body so as to make the human body adapt to the changes of the internal and external environment. The human nervous systems are highly developed. The cerebral cortex has not only evolved into the highest center of regulation and control, but also evolved into an organ capable of thinking activities. Human beings can not only adapt to the environment, but also understand and transform the world.

At present, there are more than 300 million people suffering from stroke, Alzheimer's disease (AD), Parkinson's disease (PD), heart disease, hyperlipidemia, hyperglycemia, and hypertension, which seriously affect people's health. These diseases can damage the nervous system. The incidence rate and mortality rate of cardiovascular and cerebrovascular diseases are highest. In recent years, the incidence rate of stroke has increased dramatically. If the treatment is not timely, it will leave sequela and even lead to death, which will bring enormous pressure and burden to the family and society. This will be a severe medical and social problem faced by people. How to prevent and treat neurological related diseases is not only the responsibility of medical workers, but also an arduous task in scientific research and health. Therefore, it is necessary to find effective methods and drugs to protect the health of the nervous system and prevent and treat nervous system diseases. The oxygen free radicals produced by the reaction of enzyme system and non-enzyme system attack the polyunsaturated fatty acids in cell membrane phospholipids and cause lipid peroxidation. Lipid peroxidation can eventually cause cell metabolism, dysfunction, and even death. In the field of central nervous system, researchers pay more and more attention to the relationship between oxygen free radical production, lipid peroxidation, and central nervous system injury. There is a relative lack of preventive mechanism against oxygen free radical injury in the brain, almost no catalase, glutathione peroxidase, and GSH, and the content of vitamin E is also very low. In addition, the brain also contains a large number of easily oxidized polyunsaturated fatty acids and active metals such as iron and copper, and iron mediated lipid peroxidation is the key factor causing cell damage.

Many studies have shown that natural antioxidant tea polyphenols can chelate metal ions, scavenge free radicals and antioxidant, protect the nervous system, reduce and inhibit the nervous system damage caused by oxidative stress, activate nerves, and prevent and treat neurodegenerative diseases. However, tea polyphenols must be at the appropriate concentration in the appropriate environment to have such a role. If they are not used properly, there are also some reports that tea polyphenols have neurotoxic effects. This chapter will discuss this issue in detail.

5.2 Nerve injury and oxidative stress

Nervous system, such as the brain, is a tissue that is relatively prone to oxidative damage in the body. On the one hand, the nervous system is prone to oxidation due to its high oxygen consumption. On the other hand, the special anatomical, physiological, and biochemical characteristics of the brain make it less tolerant to oxidative damage. This is mainly due to 1) the large number of easily oxidized substances in the nervous system, such as polyunsaturated fatty acids and catechol ammonia, 2) low levels of antioxidants, such as glutathione, vitamin E, catalase and SOD, 3) endogenous ROS, which is produced by specific reactions in the nervous system, 4) some areas having high iron content, such as globus pallidus and substantia nigra, and 5) non-renewable neurons in the central nervous system. Once damaged, they will lead to apoptosis or permanent functional abnormalities. More and more evidence show that oxidative stress is caused by reactive oxygen substances (ROS) and nitrogen stress caused by reactive nitrogen substances (RNS) and some metal ions and toxic substances, and some nerve injuries are involved in the pathological mechanism of neurodegenerative diseases and aging. Although several mechanisms have been proposed to explain the lead-induced toxicity, the known mechanisms of lead toxicity are incapable of explaining some of the toxic effects of lead [Tian & Lowrence, 1995]. Many nerve injuries caused by oxidative stress led to complex neuronal apoptosis pathways.

Oxidative stress plays a central role in neuronal injury and cell death in acute and chronic pathological conditions. The cellular responses to oxidative stress embrace changes in mitochondria and other organelles, notably endoplasmic reticulum, and can lead to a number of cell death paradigms, which cover a spectrum from apoptosis to necrosis and include autophagy. In AD, and other pathologies including PD, protein aggregation provides further cellular stresses that can initiate or feed into the pathways to cell death engendered by oxidative stress. Specific attention is paid here to mitochondrial dysfunction and programmed cell death, and the diverse modes of cell death mediated by mitochondria under oxidative stress [Higgins *et al.*, 2010]. Many studies have shown that excessive ROS and RNS, some metal ions, toxic substances, and nerve injuries can induce neuronal apoptosis and oxidative stress [Ni *et al.*, 1996]. Therefore, it needs in-depth research and discussion.

5.2.1 *Toxic substances induce neuronal injuries and oxidative stress*

Studies have shown that oxidative stress damage occurs after nerve poisoning, such as lead, organophosphate.

1. Oxidative stress caused by Pb^{2+} in Cells

Lead (Pb) is one of the first discovered and most widely used metals in human history and therefore, it is one of the metals most commonly encountered in the environment [Shotyk *et al.*, 1998]. Its continued release into the environment as an exhaust emission product, as well as widespread industrial use, has made lead a serious threat to human health [Juberg *et al.*, 1997]. Pregnant women, children, and inhabitants of large cities are at risk of lead intoxication, and lead poisoning is a serious occupational disease in some industries. Exposure to low-level of lead has been associated with behavioral abnormalities, learning impairment, decreased hearing, and impaired cognitive functions in humans and in experimental animals [Cory-Slechta & Pound, 1995; Lidsky & Schneider, 2003].

A growing amount of evidence indicates that cellular damage mediated by ROS may be involved in the pathology associated with lead intoxication [Hermes-Lima1 *et al.*, 1991; Bechara *et al.*, 1993]. The malondialdehyde levels in blood were strongly correlated with lead concentration in blood of exposed workers [Jiun & Hsien, 1994]. In erythrocytes from workers exposed occupationally to lead, the activities of the antioxidant enzymes, superoxide dismutase (SOD), and glutathione peroxidase, were remarkably higher than that in non-exposed workers [Monteiro *et al.*, 1985]. Lead exposure decreased defense capacity of sperm to the oxidative stress and elevated the ROS generation and reduced sperm motility and oocyte penetration capability [Hsu *et al.*, 1998]. It also demonstrated that lead increased prooxidant/antioxidant ratio in a (dose) concentration-dependent manner in lead-treated CHO cells and rats [Gurer *et al.*, 1999]. The results suggest that antioxidants might play an important role in the treatment of lead poisoning.

Recent studies have shown that lead causes oxidative stress by inducing the generation of ROS and reducing the antioxidant defense system of cells, which suggests that antioxidants may play an important role in the treatment of lead poisoning [Chen *et al.*, 2004]. We have studied that lead causes oxidative stress in nerve cell lines PC12; it was found that lead significantly decreased reduced glutathione (GSH)/oxidative glutathione (GSSG) and protein sulphydryl groups (PSH)/glutathione-protein mixed disulphide (GSSP) ratio, as well as glutathione reductase activities in a concentration-dependent manner, which would render cells more susceptible to oxidative damage (Fig. 5-1) [Chen *et al.*, 2003].

Cells (5×10^5) were exposed for an additional 24 hours to either sodium acetate. The supernatant obtained was used for the assay of intracellular GSH (non-protein thiols) and 100 μM Pb^{2+} and the cell pellet was resuspended in 180 μl PBS-EDTA and 20 μl 0.1 M NaOH and kept under shaking for the specified times for the GSSP measurement.

Glutathione reductase catalyzes the reduction of glutathione disulfide (GSSG) to glutathione (GSH) and thereby serves as a critical function in

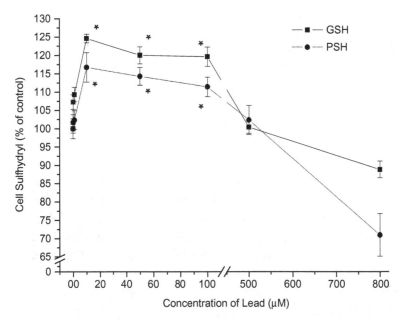

Figure 5-1. Effect of Pb^{2+} on cell sulfhydryl and disulphide in PC12 cells.

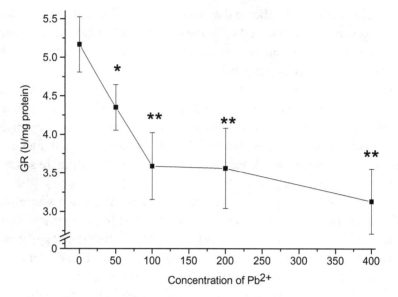

Figure 5-2. The intracellular glutathione reductase activity in lead-treated PC12 cells.

cellular defenses against injury by ROS and other oxidants. In order to evaluate the role of glutathione reductase in lead toxicity, PC12 cells were treated for 24 hours with different concentrations of lead. It was found that the treatment strongly decreased intracellular glutathione reductase activities in a concentration-dependent manner: $100\mu M$ lead was found to reduce about 30% of the total glutathione reductase activities after treatment for 24 hours, whereas the total glutathione reductase activities decreased a little further over the range of concentration 100–400 μM (Fig. 5-2) [Chen *et al.*, 2002].

Glutathione reductase activity was determined by monitoring the oxidation of NADPH at 340 nm. Glutathione reductase mixture contained, in a final volume of 1 ml, 500 μl potassium phosphate buffer, 0.2 M, pH 7.0 with 2 mM EDTA, 50 μl of a 2 mM NADPH solution dissolved in 10 mM HCl Tris, pH 7.0, 100 μM Pb^{2+} distilled water and 100 μl of freshly isolated cellular extracts.

We also studied the oxidative stress caused by lead toxicity to HepG2 cells of liver cancer tissue. Lipid peroxidation was assayed by determining the production rate of thiobarbituric acid reactive substances (TBARS)

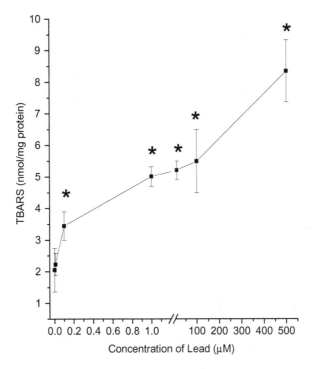

Figure 5-3. Effects of different concentration of Pb^{2+} on TBARS formation in HepG2 Cell. HepG2 cells were cultured with various concentration of Pb^{2+} for 24 hours and lipid peroxidation level was determined. *, significant difference from control by ANOVA, $p < 0.05$.

and was expressed as malondialdehide (MDA) equivalents. The membrane fluidity characteristics were estimated from the line width and shape of the ESR spectra. Lower order and faster motion means higher membrane fluidity. The order of membrane hydrocarbon chains is described by the order parameter (S). The study showed that exposure to Pb^{2+} reduced cell viability and stimulated lipid peroxidation of cell membrane (Fig. 5-3) and lead exposure reduces the fluidity of the polar surface of the cell membrane. (Fig. 5-4) [Chen *et al.*, 2002].

We studied the oxidative stress in PC12 cells exposed to lead. The experimental results showed that lead decreased the viability of PC12 cells and induced a rapid increase $[Ca^{2+}]$, followed by the accumulation of reactive oxygen species (ROS) and the reduction of mitochondrial

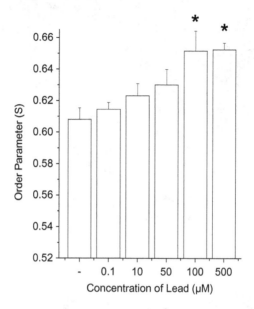

Figure 5-4.   Effect of different concentration of Pb^{2+} on order parameters (S) of HepG2 cell membrane. HepG2 cells were cultured with various concentration of Pb^{2+}. *, significant difference from control by ANOVA, $p < 0.05$.

membrane potential (MMP). ROS formation was detected with a fluorescent spectrometer using a non-fluorescent compound, 2′,7′-dichlorofluorescin diacetate (DCFH2-DA). Once inside the cell, the de-esterified product becomes the fluorescent compound, 2′,7′-dichlorofluorescin on oxidation by ROS, and the fluorescent signal is proportional to ROS production. The changes in MMP were estimated using a fluorescent cationic dye Rh123, which accumulates in mitochondria as a direct function of the membrane potential and is released upon membrane depolarization.

As the cytotoxic effect of Pb^{2+} is significant appearance at high concentration, the following experiment was conducted in the cells exposed to 100–500 μM Pb^{2+}. It was found that MMP, intracellular ROS, and calcium concentration were significantly different in PC12 cells exposed to 100 μM Pb^{2+}. When the cells were treated with 100 μM Pb^{2+}, significant increase of intracellular calcium was found after six hours ($p < 0.01$). When cells were treated for 12 hours, intracellular ROS formation significantly increased about 30% ($p < 0.05$). While in the case of 24 hours, more than ~34.5% of the cells were found to have decreased MMP as evidenced by the lower Rh123 fluorescence intensity ($p < 0.05$). The results indicated both intracellular ROS formation and MMP decrease were later than Ca^{2+}-increase and the decrease of MMP was later than the formation of ROS in the cells exposed to 100 μM Pb^{2+}.

Lead is a highly neurotoxic environmental pollutant. Recently, lead induced oxidative stress in liver, kidneys, brain, and other organs has been postulated to be one of the possible mechanisms of lead induced toxic effects [Gurer *et al.*, 1999, 2000]. Lead increased the pro-oxidant/antioxidant ratio in a concentration-dependent manner in lead-treated Chinese hamster ovary cells and rats [Gurer *et al.*, 1999, 2000]. An increased lead concentration has shown to be accompanied by increased lipid peroxidation in the rat brain [Shafiq-ur-Rehman, 1995]. Increased contents of brain thiobarbituric acid reactive substances accompanied by altered antioxidant defense systems were confirmed by Adanaylo and Oteiza [Adanaylo & Oteiza, 1999]. Several recent studies confirmed the possible involvement of reactive oxygen species (ROS) in lead induced toxicity [Gurer *et al.*, 2000]. Therefore, we believe that antioxidants should be considered as a component of an effective treatment for lead poisoning.

2. The oxidative stress induced by organophosphate and nerve agent toxicity

Organophosphate nerve agents exert their toxicity through inhibition of acetylcholinesterase. The excessive stimulation of cholinergic receptors rapidly causes neuronal damage, seizures, death, and long-term neurological impairment in those that survive. A study has shown that oxidative stress occurs in models of subacute, acute, and chronic exposure to organophosphate agents. This study found alterations in mitochondrial function and increased free radical-mediated injury, such as lipid peroxidation. Owing to the lethality of organophosphorus agents and the growing risk they pose, medical interventions that prevent organophosphate toxicity and the delayed injury response are much needed. Understanding the sources, mechanisms, and pathological consequences of organophosphate-induced oxidative stress can lead to the development of rational therapies for treating toxic exposures to organophosphorus agents [Pearson & Patel, 2016].

5.2.2 *Nerve injuries induce oxidative stress*

Nerve injury includes multiple nerve injuries, such as brain injury, spinal cord, sciatic nerve crush, sciatic nerve crush, spinal cord, cavernous nerve, and glaucoma injury, and some neurological diseases, such as AD, PD, and optic nerve disease. All these injuries can cause oxidative stress. Since AD, PD, and oxidative stress will be discussed in later chapters, they will not be discussed in detail here.

1. Brain injury induced oxidative stress

Intracerebral hemorrhage (ICH) is a cerebrovascular disease with high mortality and morbidity. A study investigated the therapeutic effects and mechanisms of melatonin on the secondary brain injury (SBI) after ICH. The results of the *in vivo* study showed that severe brain edema and behavior disorders induced by ICH. Indicators of blood-brain barrier (BBB) integrity, DNA damage, inflammation, oxidative stress, apoptosis, and mitochondria damage showed a significant increase after ICH. Meanwhile, decreased expression levels of antioxidant indicators was induced by ICH. Microscopically, it showed that increased numbers of ICH induced apoptotic cells. Inflammation and DNA damage indicators exhibited an identical pattern compared to those above. Additionally, the *in vitro* study demonstrated that the apoptotic neurons and the mitochondrial membrane potential was induced by OxyHb [Tian & Lowrence, 1995]. These results suggest that oxidative stress caused by brain injury.

Autophagy is a highly controlled lysosome-mediated function in eukaryotic cells to eliminate damaged or aged long-lived proteins and organelles. It is required for restoring cellular homeostasis in cell survival under multiple stresses. Autophagy is known to be a double-edged sword because too much activation or inhibition of autophagy can disrupt homeostatic degradation of protein and organelles within the brain and play a role in neuronal cell death. Many factors affect autophagy flux function in the brain such as oxidative stress and aging. Newly emerged research indicates that altered autophagy flux functionality is involved in neurodegeneration of the aged brain, chronic neurological diseases, and after traumatic and ischemic brain injuries. In search of identifying neuroprotective agents that may reduce oxidative stress and stimulate autophagy, one particular neuroprotective agent, antioxidants, docosahexaenoic acid, presents unique functions in reducing endoplasmic reticulum (ER) and oxidative stress and modulating autophagy [Yin *et al.*, 2017].

2. Spinal cord injury associated oxidative stress

Spinal cord injury (SCI) is a devastating event, leading to the progression of chronic neuropathic pain syndrome. The pathophysiology of spinal cord injury (SCI) is characterized by an initial primary injury followed by secondary deterioration. Its pathophysiology comprises acute and chronic phases and incorporates a cascade of destructive events such as ischemia, oxidative stress, inflammatory events, apoptotic pathways, and locomotor dysfunctions. Oxidative stress is considered a hallmark injury of SCI. ROS and oxidative stress have significant roles in the pathophysiology of SCI. Although the etiology and pathogenesis of SCI remain to be fully understood, it has been suggested that ROS and oxidative stress have a significant role in the pathophysiology of SCI.

Endoplasmic reticulum ROS and oxidative stress play an important role in the pathophysiology of SCI. Mice exhibited exacerbated locomotor function, mechanical, and thermal hypersensitivity in the hind paws after SCI. In addition, peripheral nerve injury-related biomarkers were up-regulated in SCI mice. Further, spinal cord injury enhanced oxidative stress markers but suppressed anti-oxidants, such as superoxide dismutase-1 (SOD1), NAD(P)H: quinone oxidoreductase-1 (NQO-1), heme oxygenase-1 (HO-1), and nuclear factor E2-related factor 2 (Nrf2) in the injured spinal cord. Triggering receptor expressed on myeloid cells 1 (TREM1) is an innate immune receptor expressed on neutrophils and monocytes/macrophages. The results suggested that TREM1 was highly expressed in the spinal cord tissues of WT mice after SCI. TREM1-knockout (TREM1^{KO}) TREMKO mice exhibited improved locomotor function, mechanical, and thermal hypersensitivity in the hind paws after SCI. In addition, peripheral nerve injury-related biomarkers were down-regulated by TREM1KO in SCI mice. Thus, alleviating oxidative stress may be an effective strategy for therapeutic intervention of SCI [Li *et al.*, 2019; Jia *et al.*, 2012].

3. Oxidative stress and sciatic nerve crush injury

Peripheral nerve injury is a debilitating condition that may lead to partial or complete motor, sensory, and autonomic function loss and lacks effective therapy until date. Studies have shown that oxidative stress is closely related to sciatic nerve crush injury. A compression injury was induced in the sciatic nerve of the right leg in the mice. Oxidative stress and blood cell count were measured by biochemistry and hematological analyses in this work. It was found that the muscle mass attenuates total oxidant status and the total antioxidant capacity of the biological system decreased. Several studies found that antioxidants preparation (curcumin, isoquertrin, strychnos nuxvomical L. seed) promotes functional recovery and attenuates oxidative stress in a mouse model of sciatic nerve crush injury [Caillaud *et al.*, 2018; Qiu *et al.*, 2019; Razzaq *et al.*, 2020].

4. Oxidative stress in cavernous nerve injury

A study investigated the functional and morphological changes in the corpus cavernosum after cavernous nerve (CN) injury or neurectomy. The bilateral cavernous nerve injury and the neurectomy groups exhibited decreases in erectile response and increases in apoptosis and oxidative stress, compared with the Sham group. The results suggested that cavernous nerve injury damaged erectile function and significantly reduced corporal apoptosis and oxidative stress by inhibiting the Akt/Bad/Bax/caspase-3 and Nrf2/Keap-1 pathways [Wang *et al.*, 2019]. Another study was to investigate potential antioxidant and anti-fibrotic in a cavernosal nerve injury (CNI)-induced erectile dysfunction (ED) rat model. Superoxide dismutase (SOD), catalase (CAT), malondialdehyde (MDA), and prolidase levels were measured in serum to evaluate antioxidant. Histopathological examination of tissues revealed that the highest fibrosis rate was in the group of CNI^+ distilled water (66.84%). Collagen 1 and 3, alpha and beta actin, and fibronectin levels were significantly different among groups ($p < 0.05$). Differences between plasma SOD, CAT, MDA, and prolidase were also significant among those groups ($p < 0.05$) [Resim *et al.*, 2020].

Another study investigated the role of oxidative stress in surgical cavernous nerve (CN) injury in a rat model. Eighty-four males' oxidative stress were evaluated by malondialdehyde levels, SOD activities, and glutathione peroxidase (GPX) activities in the serum. Erectile function and the number of myelinated axons of CNs and NADPH-diaphorase-positive nerve fibers statistically decreased between groups, from sham to crush to transection. Both nerve-injury groups showed increased oxidative stress markers at early time points, with the transection group showing greater oxidative stress than the crushed group and values normalizing to sham levels by week 12. GPX expression and NT-3 levels in penile tissue were in concordance with the results of SOD and GPX. These results show that oxidative stress plays an important role in injured CNs, and different methods of CN injury can lead to different degrees of oxidative stress in a rat model [Wang *et al.*, 2015].

5.2.3 *Oxidative stress in optic nerve*

Traumatic optic neuropathy (TON) is commonly associated with head trauma, and thus, is a known comorbidity of traumatic brain injury. TON has not received much attention in basic research despite being associated with permanent vision loss, color blindness, and loss of visual fields. Secondary degeneration contributes substantially to structural and functional deficits following traumatic injury to the central nervous system. Oxidative stress is a feature of secondary degeneration, contributing to reactive species and resultant oxidized products. A study was designed to identify contributors to and consequences of oxidative stress in a white matter tract vulnerable to secondary degeneration. Partial dorsal transection of the optic nerve was used to model secondary degeneration in ventral nerve unaffected by the primary injury. ROS and/or RNS increased at one, three, and seven days after injury, in ventral optic nerve. ROS increased when linked to infiltrating microglia/macrophages in dorsal optic nerve. Similarly, immunoreactivity for glutathione peroxidase and heme oxygenase-1 increased in ventral on at three and seven days after injury, respectively. Despite increased antioxidant immunoreactivity, DNA oxidation was evident from one day, lipid oxidation at three days, and protein nitration at seven days after injury. Nitrosative and oxidative damage was particularly evident in oligodendrocytes. The incidence of mitochondrial autophagic profiles was also significantly increased from three days. Despite modest increases in antioxidant enzymes, increased reactive species are accompanied by oxidative and nitrosative damage to DNA, lipid, and protein, associated with increasing abnormal mitochondria, which together may contribute to the deficits of secondary degeneration [O'Hare Doig *et al.*, 2014]. Traumatic optic neuropathy, in the context of traumatic brain injury and mechanisms, may be involved in the ongoing optic nerve degeneration. It focus particularly on endoplasmic reticulum and redox stress processes because of the overlapping presence of these degenerative mechanisms in both traumatic brain injury and various retinopathies, even though these stress pathways have not yet been used to explain retinal degeneration in a model of traumatic optic neuropathy [Cansler & Evanson, 2020]. Future research needs to uncover whether endoplasmic reticulum and redox stress function independently or whether one precedes the other.

Glaucomatous subjects might have a genetic predisposition, rendering them more susceptible to reactive oxygen species-induced damage. The perturbation of the pro-oxidant/antioxidant balance can lead to increased oxidative damage, especially when the first line of antioxidant defense weakens with age. Chronic changes in the composition of factors present in aqueous or vitreous humor may induce alterations both in trabecular cells and in cells of the optic nerve head. Free radicals and reactive oxygen species are able to affect the cellularity of the human trabecular meshwork (HTM). These findings suggest that intraocular pressure increase, which characterizes most glaucomas, is related to oxidative and degenerative processes affecting the HTM and, more specifically, its endothelial cells. It is likely that specific genetic factors contribute to both the elevation of IOP and susceptibility of the optic nerve/retinal ganglion cells (RGCs) to degeneration. Thus, oxidative stress plays a fundamental role during the arising of glaucoma-associated lesions, first in the HTM and then when the balance between nitric oxide and endothelins is broken in neuronal cell. Vascular damage and hypoxia, often associated with glaucoma, lead to apoptosis of RGCs and may also contribute to the induction of oxidative damage to the HTM. These findings support the hypothesis that oxidative damage is an important step in the pathogenesis of primary open-angle glaucoma and might be a relevant target for both prevention and therapy [Saccà & Izzotti, 2008]. This supports the theory that glaucomatous damage is the pathophysiological consequence of oxidative stress.

5.2.4 *Oxygen stress and neurodegenerative diseases*

Studies have found that oxidative stress plays an important role in neuro-degenerative diseases AD and PD. Since Chapters 6 and 7 of this book will be discussed in detail, only a brief discussion will be made here.

1. Oxygen stress and AD

We also found that overexpression of wild-type human Aβ decreased iron content and increased oxidative stress in neuroblastoma SH-SY5Y cells. We found that the overexpression of wild-type human AβPP695 decreased the iron content and increased the oxidative stress in neuroblastoma SH-SY5Y cells. The catalase activity of stably transfected cells overexpressing wild-type AβPP695 (AβPP cells) was significantly lower than that of the control cells. Intracellular reactive oxygen species (ROS) generation and calcium levels significantly increased in AβPP cells compared to control cells. The mitochondrial membrane potential of AβPP cells was significantly lower than that of the control cells. Moreover, iron treatment decreased ROS and calcium levels and increased cell viability of AβPP cells. The iron deficiency in AβPP cells may contribute to the pathogenesis of AD [Wan *et al.*, 2012].

2. Oxygen stress and PD

A large number of studies have shown that PD can damage the nervous system, especially the memory system. The accumulation of free radical damage products in PD brain is related to the central nervous system and leads to changes in the antioxidant defense of peripheral tissues [Padurariu *et al.*, 2010]. Our study shows that the 6-OHDA-induced PD rat model led to the injury of the substantia nigra through the ROS-RNS pathway, decrease in mitochondrial membrane potential, and increase in accumulation of ROS and of intracellular free Ca^{2+}. Furthermore, 6-OHDA-induced nitric oxide increased the overexpression of nNOS and iNOS and increased the level of protein-bound 3-nitrotyrosine (3-NT), lipid peroxidation, nitrite/nitrate content, inducible nitric oxide synthase, and protein-bound 3-nitro-tyrosine [Guo *et al.*, 2005, 2007]. The accumulation of iron in MPTP induced neurodegeneration is related to a nitric oxide dependent mechanism, leading to significant degradation of iron regulatory proteins through ubiquitination, resulting in these nerve injuries in mice and rats, iron and α accumulation of synuclein in the substantia nigra compacta (SNPC) [Mandel *et al.*, 2004].

5.2.5 *Oxygen stress and neuronal apoptosis*

Apoptosis plays an important role in the occurrence of neurodegenerative diseases. Oxidative stress can mediate apoptosis which is related to acute and chronic neurodegenerative diseases. Apoptosis can be induced by many factors, such as glucocorticoids and removal of nutritional factors. Hydrogen peroxide is a cytotoxic factor of thymocytes and cortical neurons, which can induce apoptosis, while the antioxidants Trolox and N-acetylcysteine can inhibit the apoptosis induced by oxygen stress. The use of antioxidants provides an attractive prospect for the prevention and treatment of neurodegenerative diseases. We studied the mechanism of hydroxyl radical induced apoptosis of cerebellar neuron [Ni *et al.*, 1996].

1. Cell morphology

Rat cerebellar nerve cells were treated with 50 mmol/L hydrogen peroxide and 100 mmol/L ferrous sulfate and cultured at $37^{\circ}C$ and 5% CO_2 for eight hours. The control cells were cultured without hydrogen peroxide and ferrous sulfate. Under microscope and electron microscope, it was found that the chromatin of the cells treated with hydroxyl radical was concentrated, showing the morphological characteristics of apoptotic cells. Nuclear chromatin is distributed along the nuclear membrane and forms apoptotic bodies, while the chromatin of control cells and cells pre-treated with antioxidants was evenly distributed in the nucleus, and the nuclear membrane is complete, indicating that hydroxyl free radicals can induce apoptosis.

2. DNA fragments induced by hydroxyl radical

The appearance of DNA fragments is an obvious sign of apoptosis. DNA analysis of brain nerve cells extracted after being treated with 50 mmol/L H_2O_2 and 100 mmol/L Fe^{2+} for six or eight hours showed obvious electrophoretic ladder bands of apoptotic cells. After the cells are treated with hydroxyl radical, the delay of electrophoretic DNA ladder distribution indicates that the DNA fragment is produced by the direct action of hydroxyl radical on the cell DNA. It also includes a series of changes leading to the formation of DNA fragment and cell suicide death. Oxygen stress is easy to react with membrane polyunsaturated fatty acids and cholesterol, causing membrane lipid peroxidation before DNA damage. Hydrogen peroxide induced apoptosis can be inhibited by Trolox (vitamin E water-soluble analogue, membrane protective agent).

3. Quantitative analysis of hydroxyl radical induced apoptosis by flow cytometry

Flow cytometric DNA analysis estimates the proportion of apoptotic cells according to the fact that the DNA content of apoptotic cells is lower than that of diploid cells. The percentage of normal cells (DNA content is diploid or polyploid) and apoptotic cells (lower than diploid) is calculated according to DNA content. In normal serum medium, less than 7% of cells undergo apoptosis. About 21% of the cells in the low serum medium had apoptosis. Adding 50 mmol/L H_2O_2 and 100 mmol/L Fe^{2+} to the low serum medium could make more than 41% of the cells apoptosis. The quantitative analysis of DNA fragments of cerebellar nerve cells were treated with 50 mmol/L H_2O_2 and 100 mmol/L Fe^{2+} at different times and showed that the DNA fragments of cerebellar nerve cells increased significantly with the extension of action time. After eight hours, 55% of the cell DNA changed into fragment DNA, indicating that the DNA fragment induced by hydroxyl radical has a time-dependent relationship.

4. Tetrazolium assay (MTT) was used to quantitatively determine the apoptosis induced by hydroxyl radical

MTT method was used to detect the ability of MTT to be reduced to methyllunar replacement. It mainly reflects the integrity of electron transport chain and is an indirect early indicator of cell death. After the cells were treated with 50 mmol/L H_2O_2 and 100 mmol/L Fe^{2+} at different times, the reduction of MTT decreased significantly with the increase of action time. After eight hours, only 77% MTT was reduced compared with the control. The results showed that hydroxyl radical induced cell death in a time-dependent manner.

5. Effect of hydroxyl radical on bcl-2 mRNA level of primary cultured rat cerebellar neurons

Bcl-2 is the most common regulator of apoptosis and overexpression of Bcl-2 can inhibit apoptosis. Hydroxyl radicals may induce apoptosis by antagonizing the activity of Bcl-2. Therefore, the level of bcl-2 mRNA in primary rat cerebellar neurons treated with hydroxyl radical was measured. After the cells were treated with hydroxyl radical for 1.5 hours, the level of Bcl-2 mRNA was significantly reduced.

6. Effect of hydroxyl radical on Fos level of primary rat cerebellar neurons

It has been reported that FOS is related to apoptosis. Transcription factor AP-1 containing FOS heterodimer is involved in a variety of cellular processes, such as cell proliferation, differentiation, and apoptosis. The results showed that when the cells were treated with hydroxyl radical, Fos protein increased 1.5 times. Immediate early genes, such as FOS encoding transcription regulators, may play a central role in transforming extracellular signals into long-lasting intracellular signals. FOS protein can directly induce apoptosis, while hydrogen peroxide can induce the expression of fos gene. Conversely, antioxidants can reduce the expression of c-fos mRNA. Overexpression of Bcl-2 can inhibit apoptosis induced by c-fos. The ability of FOS to bind DNA is regulated by redox. Our results show that when the cells were treated with hydroxyl radical, FOS protein increased 1.5 times which indicates that hydroxyl radicals can induce apoptosis by regulating gene expression by FOS.

7. Effect of hydroxyl radical on lipid peroxidation

The metabolites of lipid peroxidation have been proved to be related to apoptosis, so hydroxyl free radicals may induce apoptosis through lipid peroxidation. When cerebellar neurons were treated with 50 mmol/L H_2O_2 and 100 mmol/L $FeSO_4$, it was found that the level of lipid peroxidation increased significantly with the extension of action time. After eight hours, TBARS was 1.2 times higher than that of the control group, indicating that the lipid peroxidation induced by hydroxyl radical is time-dependent. The results here suggest that hydroxyl radicals induce lipid peroxidation and mediate the signal transduction process of membrane surface receptors.

8. Changes of sulfhydryl binding sites of membrane proteins induced by hydroxyl radicals

Membrane proteins have two sulfhydryl binding sites, one on the surface of the protein (W) and the second located in the deep layer of the three-dimensional structure of the protein (S). The experimental results showed that the S/W ratio of apoptotic nerve cells induced by hydroxyl radical increased, that is, the second sulfhydryl in the deep layer of the three-dimensional structure of the protein increased. The above results show that the sulfhydryl groups on the surface of many membrane proteins in apoptotic cells are oxidized or shifted to the deep layer of the three-dimensional structure of proteins, which may be partly due to the changes of cell membrane proteins, which may change the function of receptors and activate the signal pathway of apoptosis.

In conclusion, hydroxyl free radicals can induce apoptosis of cerebellar neurons, accompanied by the decrease of Bcl-2, the increase of FOS, the increase of lipid peroxidation level, and the change of sulfhydryl binding site of membrane protein. Hydroxyl radicals can induce apoptosis through different signal pathways, which form intracellular signal networks and interact through ROS and surrounding redox reactions.

5.2.6 *NO induces apoptosis*

Recent studies on a variety of systems have found that NO can induce apoptosis of neuron. Activation of glutamate receptor (NMDA) products NO and O_2^- then formed different concentrations of $ONOO^-$ which induces apoptosis or necrosis in a time and concentration-dependent manner. These may be the result of direct damage of NO free radical to cells or the result of activating the expression of tumor suppressor gene p53. High or low concentrations of NMDA stimulant, NO donors, or a small amount of short-term exposure of peroxynitrite can induce apoptosis of cerebral cortical cells. On the contrary, long-term high concentration exposure can produce necrosis, it shows that the initial injury intensity determines whether nerve cells are apoptotic or necrotic [Zhang & Zhao, 2003].

NO or its oxidized form can directly destroy DNA molecules and cause DNA single strand breaks. NO can modify or inhibit key enzymes in the metabolic process, resulting in cytotoxicity. NO significantly inhibited the activities of complex I and complex II of the mitochondrial respiratory chain, combined with cytochrome C oxidase, competed with oxygen, reversibly inhibited the respiratory chain, blocked the production process of mitochondria, and did not specifically cause the release of calcium in mitochondria. NMDA receptor, one of the subtypes of glutamate receptor, includes a transmembrane calcium channel. When overactivated by glutamate, it will cause a large amount of calcium influx, activate NO synthase-dependent on calcium and calmodulin, synthesize a large amount of NO and spread to peripheral nerve cells, trigger guanylate cyclase activity, mediate a series of biochemical reactions through cGMP, and finally lead to cell apoptosis [Zhang & Zhao, 2003].

NO can destroy the antioxidant mechanism in cells, leading to the inactivation of glutathione peroxidase, the decrease of reducing glutathione, and the inhibition of mitochondrial SOD, resulting in the increase of intracellular ROS and the aggravation of oxidative damage. NO regulates the translation of ferritin mRNA by binding to the center of specific protein, and finally affects the balance of iron ions in cells. In addition, NO can inhibit the activity of nucleotide cyclooxygenase, DNA synthesis, and cell division. Using primary cultured rat cerebellar nerve cells, we studied the possible pathway and mechanism of NO induced neuronal apoptosis. The results showed that NO could lead to cerebellar nerve cell apoptosis, and antioxidants had a protective effect. It was also found that peroxynitrite-induced apoptosis could also be blocked by antioxidants. NO donor sodium nitroprusside or S-nitrosoglutathione could lead to cell death. It showed that the bulge of nerve cells was significantly reduced, the cell body was shrunk and became blurred, some cells broke off the wall and floated and the cell density became smaller, and the chromatin of the cells treated with NO donor was concentrated and some nuclei were broken. The DNA showed characteristic ladder bands on agarose electrophoresis and the characteristics of apoptosis. Treatment with antioxidant and NO scavenger can significantly protect the cells [Zhang & Zhao, 2003].

5.3 Protective effect of tea polyphenols on nerves

Above discussion suggest that oxidative stress resulting in ROS and RNS generation and inflammation play a pivotal role in the age-associated cognitive decline and neuronal loss in neurodegenerative diseases, nerve injuries, and neurodegenerative diseases. Human epidemiological and new animal data suggest that tea drinking may decrease neurodegenerative diseases and phenolic compounds have received increasing interest because of the numerous studies. There is a growing recognition that tea polyphenolic catechins exert a protective role in neurodegeneration. The neuroprotective effect has been long established in animal models of neurological disorders. Tea polyphenol EGCG has been shown to improve age-related cognitive decline and protect against cerebral ischemia/reperfusion injuries and brain inflammation and neuronal damage in experimental autoimmune encephalomyelitis [Zhao, 2012]. These are discussed in detail below.

5.3.1 *Protective effect of tea polyphenols on nerve against neurodegenerative diseases*

Epidemiological and animal data suggest that tea drinking may decrease the incidence of neurodegenerative diseases and their major constituent, EGCG, has diverse pharmacological activities such as for the treatment of neurodegenerative diseases. Here, we will focus on PD, AD, amyotrophic lateral sclerosis, brain ischemia and reperfusion, autoimmune encephalomyelitis, Huntington's disease, lipid peroxidation of brain synaptosomes, and summarize the current knowledge on neuroprotective effects of tea polyphenols and their molecular mechanisms responsible for the neuroprotection in various models of neurodegenerative and neural injury.

1. Protective effect of tea polyphenols on nerves against PD

Cumulative evidence suggests that tea drinking is associated with a lower risk of PD. Many human epidemiological and animal studies have shown that green tea polyphenols may promote health, reduce disease occurrence, and may prevent PD. Tens of thousands of people have found that drinking three cups of green tea a day can reduce the incidence rate of PD [Zhao, 2020].

Levites's group demonstrated the neuroprotective property of green tea extract and ECG in the MPTP-treated mice model of PD [Levites *et al.*, 2001]. Our study indicated protective effects of green tea polyphenols in the 6-OHDA rat model of PD through inhibition of ROS-NOS pathway [Guo *et al.*, 2005, 2007]. Tea catechins has the ability to modulate L-DOPA methylation and to protect nervous against oxidative hippocampal injury. In addition, tea (+)-catechin strongly reduces glutamate-induced oxidative cytotoxicity in HT22 mouse hippocampal neurons *in vitro* through inactivation of the nuclear factor-κB signaling pathway. Tea catechin is a dietary polyphenolic that may have beneficial effects in L-DOPA-based treatment of PD patients by inhibiting L-DOPA methylation plus reducing oxidative neurodegeneration [Guo *et al.*, 2005, 2007; Kang *et al.*, 2013].

The accumulation of iron in MPTP-induced neurodegeneration has been linked to nitric oxide-dependent mechanism, resulting in degradation of prominent iron regulatory proteins by ubiquitination. Green tea catechin polyphenol EGCG are neuroprotective against these neurotoxins in mice and rats, preventing the accumulation of iron and alpha-synuclein in dense part of substantia nigra (SNpc). EGCG may play a role in the growth of PC12 cells, where it stimulates survival-promoting pathways [Mandel *et al.*, 2004]. Brain penetrating property of polyphenols, as well as their antioxidant and iron-chelating properties may make such compounds an important class of drugs to be developed for treatment of neurodegenerative diseases where oxidative stress has been implicated [Levites *et al.*, 2001].

2. Protective effect of tea polyphenols on nerves against AD

The component is antioxidant, green tea can prevent hippocampal neuron apoptosis and improve cognitive function by inhibiting JNK/ MLCK pathway [Fernando *et al.*, 2017]. It was reported that long-term (26 weeks) administration of green tea polyphenols (0.5% green tea polyphenols in water) could prevent amyloidosis β-induced cognitive impairment in rats. In addition to preventing cognitive impairment, lipid peroxides and ROS in hippocampus and plasma were more than 20% lower than those in the control group [Haque *et al.*, 2008]. We found preventing effects of natural antioxidants tea polyphenols on neurodegenerative diseases AD and PD [Zhao, 2005, 2009; Zhao and Zhao, 2012]. We studied the pathogenic mechanism of iron in AD and the regulatory effect of tea polyphenol EGCG on iron imbalance and results show that tea polyphenol EGCG could reduce the oxidative damage of AD cells by complexing too much iron in efforts to protect AD cells. EGCG reduced the content of Aβ in APPsw cells, ROS, and intracellular calcium, and improved the mitochondrial membrane potential. Tea polyphenol EGCG may have a preventive and therapeutic effect on AD by chelating excessive iron in the iron pool of APPsw cells [Wan *et al.*, 2012]. Additionally, it was demonstrated that a similar role of EGCG in streptozotocin induced dementia in rats [Biasibetti *et al.*, 2013]. The free radical scavenging activity of tea polyphenols [Sang *et al.*, 2003] and the chelating properties of metal iron may contribute to these antioxidant effects [Seeram *et al.*, 2006]. Metal ions such as copper (II) and iron (III) can be chelated by tea polyphenols, and iron chelation reduces the production of ROS by inhibiting the Fenton reaction [Weinreb *et al.*, 2009]. Copper (II) and iron (III) ions also accumulate in the brain of AD patients [Ward *et al.*, 2009]. The above results show that drinking tea and its effective component tea polyphenols have preventive and therapeutic effects on AD.

3. Protective effect of tea polyphenols on nerves against amyotrophic lateral sclerosis

Long-term administration of the preparation of green tea catechins EGCG was demonstrated to improve spatial cognition learning ability in rats. There is a report on the effect of EGCG on a model of amyotrophic lateral sclerosis *in vivo*. This study evaluated the effect of EGCG on amyotrophic lateral sclerosis (ALS) model mice with the human G93A mutated Cu/Zn-superoxide dismutase (SOD1) gene. The treatment of more than 2.9 mg EGCG/g body weight significantly prolonged the symptom onset and life span, preserved more survival signals, and attenuated death signals [Koh *et al.*, 2006]. Another study showed that long-term administration of green tea catechins improves spatial cognition learning ability in rats. They also had lower plasma concentrations of lipid peroxides and greater plasma ferric-reducing anti-oxidation power than controls. Furthermore, rats that administered EGCG had lower hippocampus ROS concentrations than controls. This improvement in spatial cognitive learning ability is due to the anti-oxidative activity of green tea catechins

[Haque *et al.*, 2006]. These data suggest that EGCG could be a potential therapeutic candidate for model of amyotrophic lateral sclerosis as a disease-modifying agent.

4. Protective effect of tea polyphenols on nerves against brain ischemia and reperfusion

Oxygen free radical injury plays an important role in neuronal damage induced by brain ischemia and reperfusion. A study examined whether EGCG would reduce neuronal damage after transient global ischemia in the gerbils because EGCG has a potent antioxidant property as a green tea polyphenol. EGCG at the dose of 10 mg/kg failed to reduce hippocampal neuronal damage. Moreover, EGCG when administered at the dose of 25 mg/kg or 50 mg/kg, significantly reduced hippocampal neuronal damage in a dose-dependent manner ($p < 0.001$). The results show that the green tea polyphenol, EGCG, has a neuroprotective effect against neuronal damage following global ischemia in the gerbils [Lee *et al.*, 2000]. Another study examined the effect of ad libitum oral-administration of (-) catechin solution on ischemia-reperfusion-induced cell death of hippocampal CA1 in the gerbil. When (-)catechin solution instead of drinking water was orally administered ad libitum for two weeks, dose-dependent protection against neuronal death followed by transient ischemia and reperfusion was observed [Inanami *et al.*, 1998].

5. Protective effect of tea polyphenols on nerves against autoimmune encephalomyelitis

A study showed that green tea EGCG had neuroprotection in autoimmune encephalomyelitis. It shows that the major green tea constituent, EGCG, dramatically suppresses experimental autoimmune encephalomyelitis induced by proteolipid protein 139–151. EGCG reduced clinical severity when given at initiation or after the onset of experimental autoimmune encephalomyelitis by both limiting brain inflammation and reducing neuronal damage. In orally-treated mice, tea polyphenols abrogated proliferation and TNF-α production of encephalitogenic T cells. In human myelin-specific CD4$^+$ T cells, cell cycle arrest was induced, down-regulating the cyclin-dependent kinase 4. Since its structure implicates additional anti-oxidative properties, EGCG was capable of protecting against neuronal injury in living brain tissue induced by N-methyl-D-aspartate. Thus, a natural green tea constituent may open a new therapeutic avenue for young disabled adults with inflammatory brain disease by combining, on one hand, anti-inflammatory and, on the other hand, neuroprotective capacities [Aktas *et al.*, 2004]. Another study showed that green tea polyphenol EGCG had protective effects against hippocampal neuronal damage after transient global ischemia in gerbils. Histological assessment showed that EGCG significantly reduced infarct volume in comparison to ischemia + saline. In addition, EGCG significantly reduced total citrulline/30 min/mg protein and inducible nitric oxide synthase (iNOS) activity in comparison to ischemia + saline control citrulline/30 min/mg protein for total NOS and iNOS activity, respectively, iNOS protein expression was also reduced. In contrast, EGCG significantly increased endothelial and neuronal NOS protein expression compared with ischemia controls. EGCG also significantly preserved mitochondrial energetics (complex I-V) and citrate synthase activity [Sutherland *et al.*, 2004]. The neuroprotective effects of EGCG are, in part, due to modulation of NOS isoforms and preservation of mitochondrial complex activity and integrity. Therefore, the neuroprotective effects of EGCG are not exclusively due to its antioxidant effects but involve more complex signal transduction mechanisms. Green tea constituent may open a new therapeutic avenue for disabled adults with inflammatory brain disease by combining anti-inflammatory and neuroprotective capacities.

6. Protective effect of tea polyphenols on nerves against Huntington's disease

Huntington's disease (HD) is a progressive neurodegenerative disorder for which only symptomatic treatments of limited effectiveness are available. Preventing early misfolding steps and thereby aggregation of the polyglutamine (poly Q)-containing protein huntingtin (htt) in neurons of patients may represent an attractive therapeutic strategy to postpone the onset and progression of HD. From a biomedical point of view, the scavenging activity of reactive oxygen species (ROS) makes them a potential tool for the treatment of neurodegenerative diseases including Huntington's disease, dementia, and amyotrophic lateral sclerosis (ALS). Extracts derived from tea show multi-therapeutic effects by cooperatively acting on different biochemical pathways. A paper demonstrate that the green tea polyphenol EGCG potently inhibits the aggregation of mutant huntingtin exon 1 protein in a dose-dependent manner. EGCG also significantly reduced poly Q-mediated huntingtin protein aggregation and cytotoxicity in a yeast model of HD. When EGCG was fed to transgenic HD flies overexpressing a pathogenic huntingtin exon 1 protein, photoreceptor degeneration and motor function improved. These results indicate that modulators of huntingtin exon 1 misfolding and oligomerization like EGCG are likely to reduce poly Q-mediated toxicity *in vivo* [Ehrnhoefer *et al.*, 2006]. Tea polyphenols EGCG may provide the basis for the development of a novel pharmacotherapy for HD and related poly Q disorders.

7. Drinking tea reduces the risk of depression

Drinking tea helps to treat neurodegenerative diseases, and depression is a common symptom. A sample of the Finnish general population (n = 2011) was investigated. Those who reported drinking tea daily were less depressed than others. Those who drank five or more cups of tea a day had no depression. Fifteen observational studies (nine cross-sectional studies and six prospective studies) of beverage consumption and depression, included 20,572 cases of depression among 347,691 participants. The inverse association with coffee or tea consumption and the positive association with soft drink consumption for risk of depression did not vary by gender, country, high consumption category, and adjustment factors such as alcohol, smoking and physical activity [Kang *et al.*, 2018]. Eleven studies with 13 reports were eligible for inclusion in the meta-analysis (22,817 participants with 4,743 cases of depression). Eight reports were included in the dose-response analysis of tea consumption and depression risk (10,600 participants with 2,107 cases). There was a linear association between tea consumption and the risk of depression, with an increment of three cups/day in tea consumption associated with a decrease in the risk of depression of 37% [Dong *et al.*, 2015]. This study investigated these associations in a sample of the Finnish general population (n = 2011). Those who reported drinking tea daily were less depressed than the others [Hintikka *et al.*, 2005].

In conclusion, an inverse relationship between daily tea drinking and the risk of being depressed was found in a relatively large general population sample. Nevertheless, the underlying mechanisms are unresolved and further studies are needed.

5.3.2 *Protective effect of tea catechins on neurons against toxic substances*

A large number of studies have shown that tea polyphenols can prevent nerve damage caused by toxic substances, such as lead, acrylamide, methamphetamine, and tetanus toxin.

1. Protective effect of tea catechins on lipid peroxidation in PC12 cells caused by Pb^{2+}

As a kind of excellent scavenger of free radicals and chelator of heavy metals, tea catechins have protective effects on oxidative stress caused by lead treatment in cell systems. Tea catechins, as a kind of efficient free radical scavenger, could improve lead-induced oxidative damage in cell systems. We found that lead significantly reduced GSH/oxidative glutathione (GSSG) and protein PSH/ GSSP ratio (Fig. 5-5), as well as

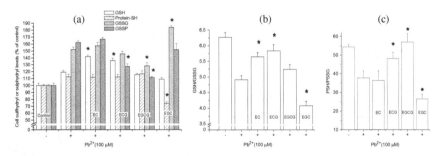

Figure 5-5. Effect of tea catechins on intracellular sulfhydryl and sulphydryl levels of lead-exposed PC12 cells. After the initial 24-hour attachment period, PC12 cells were exposed to both Pb^{2+} (100 μM) and different concentrations of tea catechins for an additional 24 hours and intracellular sulfhydryl and sulphydryl levels were determined by fluorometry assay. Data are expressed as a percentage of the untreated control ± S.E., n = 5. *, significant difference from the corresponding control in the presence of 100 μM Pb^{2+} by ANOVA, $p < 0.05$. The concentrations of tea catechins were 100 μM.

glutathione reductase activities in a concentration-dependent manner, which would render cells more susceptible to oxidative damage. Both EC and ECG supplementation resulted in increased GSH/GSSG ratio and glutathione reductase activities. The ECG or EGCG treatment significantly decreased the glutathione-protein mixed disulphide levels and increased intracellular PSH/GSSP ratio in lead-exposed PC12 cells. The results suggested that the role of tea catechins for lead-exposed PC12 cells was related to its ability to regulate intracellular thiol status and glutathione reductase activities (Fig. 5-6) [Chen *et al.*, 2004].

We also studied whether tea catechins had any protective effects against lead-induced ROS formation, mitochondrial dysfunction, and calcium dysregulation in PC12 cells. The experimental results showed that lead decreased PC12 cell viability and induced a rapid elevation of $[Ca^{2+}]$,

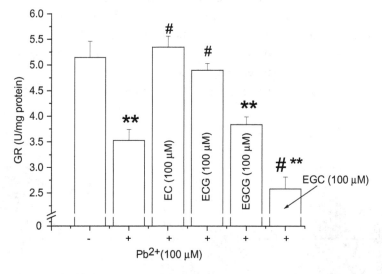

Figure 5-6. Effects of catechins on intracellular glutathione reductase activity in lead treated PC12 cells. Glutathione reductase activity was determined by monitoring the oxidation of NADPH at 340 nm. The absorbance rate of the reaction mixture at 340 nm was measured after incubation at 37°C for five minutes. One unit of glutathione reductase activity is defined as the amount of enzyme reducing 1 μmol of GSSG per minute and mg of soluble protein. *, $p < 0.05$ and **, $p < 0.01$ *vs* the control in the absence of Pb^{2+}; #, $p < 0.05$ *vs* the control in the presence of 100 μM Pb^{2+}.

which was followed by an accumulation of ROS and a decrease of mito-chondrial membrane potential (MMP). Treatment by tea catechins signifi-cantly increased cell viability, decreased intracellular Ca^{2+} levels and ROS formation, and improved MMP in PC12 cells exposed to lead. The gal-loylated catechins showed a greater effect on ROS formation and mito-chondrial dysfunction than that of nongalloylated catechins, which was similar to the result of their scavenging ability on free radical [Chen *et al.*, 2003].

We designed an experiment to elucidate if tea catechins have any pro-tective effects on lipid peroxidation damage in lead-exposed HepG2 cells. Exposure of HepG2 cells to Pb^{2+} decreased cell viability and stimulated lipid peroxidation of cell membranes as measured by thioburbituric acid reaction. Results showed that tea catechins treatment significantly increased cell viability, decreased lipid peroxidation levels and protected cell membrane fluidity in lead-exposed hepG2 cells in a concentration-dependent manner (Figs. 5-7, 5-8, 5-9). The galloylated catechins showed stronger effect than that of nongalloylated catechins. Co-treatment with EGCG and EC, ECG and EGCG showed synergistically protective effects. The results suggest that tea catechins supplementation may have a role to play in modulating oxidative stress in lead-exposed HepG2 cells [Chen *et al.*, 2002].

Above results suggest that tea catechins supplementation may play a role for modulating oxidative stress in PC12 and HepG2 cells exposed to lead.

2. The protective effect of EGCG on nerves against acrylamide-induced apoptosis in PC12 cells

Acrylamide (ACR) is a neurotoxic industrial chemical intermediate, which is also present in food and water. This study investigated the poten-tial neuroprotective effects of EGCG, the most abundant polyphenolic compound in green tea, in PC12 cells treated with ACR. ACR-treated PC12 cells pretreated with various concentrations of EGCG increased viability and acetylcholinesterase activity and reduced apoptosis and necrosis compared to cells exposed to ACR alone. EGCG reduced the expression of bax mRNA, decreased cytochrome c release, reduced intra-cellular calcium levels, inactivated caspase 3, and increased mitochondrial

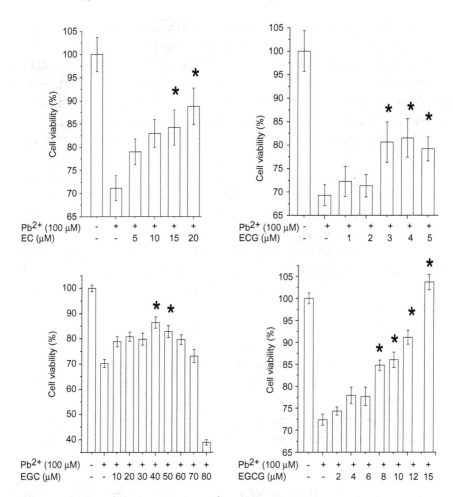

Figure 5-7. Effect of tea catechins on Pb^{2+}-induced cell toxicity. After the initial 24-hour attachment period, HepG2 cells were exposed to both Pb^{2+} (100 μM) and different concentration of tea catechins for an additional 24 hours and cell viability was determined by MTT assay. Data are expressed as a percentage of the untreated control ± S.E., n = 7. *, significant difference from control in the presence of 100 μM Pb^{2+} by ANOVA, $p < 0.05$.

membrane potential, suggesting that EGCG prevents ACR-induced apoptosis through a mitochondrial-mediated pathway. In addition, EGCG inhibited the formation of reactive oxygen species and lipid peroxidation while enhancing superoxide dismutase activity and glutathione levels, thereby reducing oxidative stress. The results indicate that pretreatment of

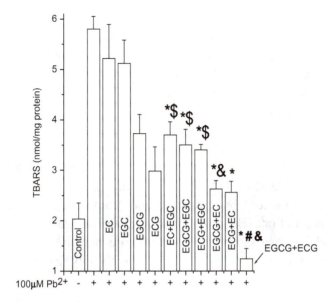

Figure 5-8. Effect of tea catechins on TBARS formation in HepG2 cells treated by Pb^{2+}. HepG2 cells were cultured with various concentration of tea catechins and 100 μM Pb^{2+} for 24 hours and lipid peroxidation level was determined. *, significant difference from control in the presence of 100 μM Pb^{2+} by ANOVA, $p < 0.05$.

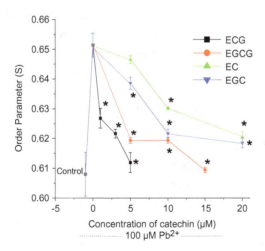

Figure 5-9. Effects of tea catechins on Order Parameters (S) of HepG2 cells exposed to Pb^{2+}. HepG2 cells were cultured with various concentrations of tea catechins and 100 μM Pb^{2+} for 24 hours and Order Parameter (S) was determined. *, significant difference from control in the presence of 100 μM Pb^{2+} by ANOVA, $p < 0.05$.

PC12 cells with EGCG attenuates ACR-induced apoptosis by reducing oxidative stress. [He *et al.*, 2017a]. Another study investigated the neuro-protective effects of EGCG, the most abundant polyphenol in green tea, on ACR-treated rat brain. EGCG increased acetylcholinesterase (AChE) activity and the rate of Nissl-positive cells in ACR-treated rats. Senescence-associated β-galactosidase (SA-β-gal) staining indicated that EGCG attenuated ACR-induced senescence. Tumor necrosis factor alpha (TNF-α), iNOS, and cyclooxygenase 2 (COX-2) protein expression indicated that EGCG inhibited ACR-induced inflammation. EGCG promoted brain regeneration in ACR-treated rats. The results suggest that EGCG can attenuate ACR-induced brain damage and promote regeneration in the cerebral cortex of rats [He *et al.*, 2017b]. It is also shown that tea polyphenols could increase glutathione S-transferase activity and remarkably inhibit acrylamide toxicity in mice by tea polyphenols dietary constituents [Xie *et al.*, 2008].

Therefore EGCG may alleviate ACR-related nerve injury and drinking green tea may reduce nerve injury induced by ACR.

3. Tea polyphenols attenuate methamphetamine-induced neuronal damage

Methamphetamine (METH) acts strongly on the nervous system and damages neurons, and is known to cause neurodegenerative diseases such as AD and PD. METH is an illegal psychoactive substance that is abused worldwide, and repeated exposure to METH could form mass free radicals and induce neuronal apoptosis. It has been reported that free radicals generated by METH treatment can oxidize DNA and hence, produce strand breaks. Additionally, tea polyphenols could protect PC12 cells against METH-induced cell viability loss, reactive oxide species and nitric oxide production, and mitochondrial dysfunction and suppress METH-induced apoptosis. Furthermore, tea polyphenols could increase the anti-oxidant capacities and expressions of phosphorylation of ataxia telangiectasia and checkpoint kinase 2 and then attenuate DNA damage via activating the DNA repair signaling pathway. Tea polyphenols exert bioactivities through antioxidant-related mechanisms [Ru *et al.*, 2005]. In addition, it was also observed that EC abrogated the activation of ERK, p38, and inhibited the expression of CHOP and DR4. EC also reduced

METH-induced ROS accumulation and changes of mitochondrial membrane potential (MMP) [Kang *et al.*, 2019].

These studies showed that tea polyphenols can reduce the damage caused by toxic substances such as methamphetamine.

4. Antagonistic effect of epigallocatechin-3-gallate on neurotoxicity induced by formaldehyde

The toxicity of formaldehyde (FA) has always been of great concern, particularly since its use is very extensive and unavoidable. After 14 days of exposure to 3 mg/m^3 formaldehyde, mice exhibited significant cognitive impairment. In the FA group, a significant increase in iNOS level compared with the control group was observed. The reduced GSH level was significantly decreased. The levels of IL-1β, TNF-α, and Caspase-3 were obviously raised, while H&E and Nissl staining illustrated significant neuronal damage. After administering EGCG as a protective agent, all the above observed changes were reversed, and the protective effect of EGCG became gradually evident in the 20–500 mg/kg range. Immunohistochemistry results showed that EGCG could activate the Nrf2 signaling pathway, thus alleviating the oxidative damage caused by formaldehyde [Huang *et al.*, 2019].

Above study show that tea polyphenols can reduce the brain and nerve damage caused by toxic substances such as formaldehyde.

5. Black tea extract protects nerves against the neurotoxin

Botulinum neurotoxin type A evoked twitches. Thearubigin fraction mixed with the toxin protected against the *in vivo* paralytic effect of the toxin. Thearubigin fraction had no protective effect on other toxins, such as tetrodotoxin and saxitoxin. The specific binding of [^{125}I] tetanus toxin to rat cerebrocortical synaptosomes was inhibited by mixing iodinated toxin with thearubigin fraction. Thearubigin fraction counteracts the effect of tetanus toxin by binding with toxin and also suggest that this fraction may be able to apply for prophylaxis of tetanus [Satoh *et al.*, 2001]. Another study examined the effects of thearubigin fraction extracted from a black tea infusion for neuromuscular blocking action on tetanus toxin in mouse phrenic nerve-diaphragm preparations and on the binding of this toxin to the synaptosomal membrane preparations of rat cerebral cortices. The interaction between tetanus toxin and thearubigin fraction was also investigated. Tetanus toxin abolished indirect twitches in mouse phrenic nerve-diaphragm preparations. Mixing iodinated toxin with thearubigin fraction inhibited the specific binding of tetanus toxin to the synaptosomal membrane preparation. The effects of thearubigin fraction were dose-dependent. [Satoh *et al.*, 2002].

6. Protective effects of tea polyphenols for brain synaptosomes

Synaptosomes include presynaptic membranes, synaptic spaces, and post-synaptic membranes that are not on the synaptosomes. Synaptosomes belong to the branches at the end of axons of nerve cells, which also refer to synaptosomes at the end of axons of nerve cells that are connected with the next nerve cell to transmit excitation or inhibition through it, synaptic gap, and postsynaptic membrane. As synaptosomes contain a lot of unsaturated phospholipids, they are very easy to be oxidized and damaged. We studied the protective mechanisms of four components of green tea polyphenols against lipid peroxidation in synaptosomes and compared the protective effects of four components of green tea polyphenols — EGCG, ECG, EGC, and EC against iron-induced lipid peroxidation in synaptosomes. With the increase of the concentrations of EGCG, ECG, EGC, and EC, the formation amount of TBA reactants gradually decreased and the inhibition rate gradually increased, with a good dose-response relationship. Their protective effect on lipid peroxidation injury of brain synaptosomes induced by iron ions was enhanced in the following order: EGCG > ECG > GTP > EGC > EC. We compared the scavenging effects of four monomers of tea polyphenols on lipid free radicals. The ability to scavenge lipid free radicals in the lipid peroxidation of brain synaptosomes caused by Fe^{2+}/Fe^{3+} was in this order: ECG > EGCG > EC > EGC. Taking the content of TBA reactants and the end product of lipid peroxidation as the index, their ability to inhibit the formation of TBA reactants increased with the order of EC, EGC, ECG, and EGCG. Although their effects in these three experiments are different in order, it can be seen that the effects of EGCG and ECG are better than EC and EGC for synaptosomes [Guo *et al.*, 1996, 1999]. Treatment with tea polyphenol could significantly decrease the increased production of synaptosomal ROS and thus reduced the deteriorative ROS-sensitive Na^+, K^+-ATPase, and Mg^{2+}-ATPase activity [Chen *et al.*, 2008]. The behavioral and neurotoxic effects of intracerebroventricular injection of NMDA were ameliorated by treatment with tea polyphenol alone [Chang-Mu *et al.*, 2010].

These findings provide useful information about the potential application tea polyphenols in preventing clinical excitotoxic injury such as brain trauma, brain ischemia, epilepsy, PD, and AD.

5.3.3 *The protective effect of tea polyphenols on neuronal nerve against neuronal injury*

Many studies have shown that tea polyphenols have protective effects on nerves against various brain and peripheral nerve injuries, caused by microbiota, nerve crush injury etc.

1. The protective effect of tea polyphenols on brain nerve against intestinal microbiota

Cerebral nerves have become a hot topic of research, and studies have marked the importance of microbial flora and tea polyphenols in protecting cerebral nerves. Tea polyphenols can regulate the composition of human intestinal flora and be bio-transformed by the intestinal microbiota, resulting in relative metabolites, which prevent nerve damage, promote neuro-cognition, and increase resistance to oxidative stress. Tea polyphenols play vital roles in maintaining the steady status of intestinal flora and protecting cerebral nerve damage [Zhang *et al.*, 2021a]. A study investigated the effects of tea polyphenols on learning improvement. In the step-down avoidance test and the Morris water maze test, it showed that tea polyphenols was effective in reversing memory impairment in scopolamine-treated mice. In addition, tea polyphenols significantly improved cellular uptake of cationic peptide P1 and anionic peptide P2 by brain capillary endothelial cells. [Sheng *et al.*, 2020]. Lots of research have shown that tea polyphenols have possible therapeutic effect against high fat diet induced obesity, which is closely related to the gut flora of the host. A better understanding on the interactions of tea polyphenols with intestinal microbiota and their implication for cellular signal conditioning mechanism will lead us to better evaluate the contribution of the microbial metabolites of tea polyphenols, as well as the regulation of intestinal bacterial diversity and abundance for host health. Tea polyphenols is the main functional substances in tea. After the bio-transformation by intestinal flora, the metabolites of tea polyphenols have positive effects on the health of the host. Understanding the interactions of tea polyphenols with intestinal microbiota and their implication for cellular signal conditioning mechanism that will enable us to better study the two-way effects of tea polyphenols and intestinal microbiota on host health improvement. The dysfunction of the microbiota-gut-brain axis is the main pathological basis of depression, and its abnormality may be the direct cause and potential influencing factor of psychiatric disorders. The interrelationship between tea polyphenols and intestinal microorganisms prevent psychiatric disorders by modulating host intestinal microorganisms [Sun *et al.*, 2021]. Neuroscience studies are primarily focused on discerning the functional mechanism of the nervous system. Tea polyphenols play a vital role in maintaining the steady status of intestinal flora and protecting cerebral nerve damage. An in-depth understanding of the tea polyphenols and intestinal microbiota interaction, its implication on cerebral nerve protection, and the associated underlying mechanism will allow us to expand the therapeutic applications of tea polyphenols [Zhang *et al.*, 2021b]. Tea polyphenols can act indirectly on the central nervous system by affecting the "microflora-gut-brain axis", in which the microbiota and its composition represent a factor that determines brain health. Bidirectional communication between the intestinal microflora and the brain (microbe-gut-brain axis) occurs through a variety of pathways, including the vagus nerve, immune system, neuroendocrine pathways, and bacteria-derived metabolites. This axis has been shown to influence neurotransmission and behavior, which is usually associated with neuropsychiatric disorders.

Tea polyphenols and their metabolites may provide benefits by restoring the imbalance of intestinal microbiota and that tea polyphenols are metabolized by intestinal flora, to provide a new idea for tea polyphenols to play a neuroprotective role by regulating intestinal flora.

2. Protective effects of tea catechin on nerves against peripheral nerve injuries

Studies showed that EGCG, a major metabolite of tea catechin, which is the principle bioactive compound in green tea, had protective effect on peripheral nerve injury. Oxidative stress and large amounts of nitric oxide (NO) have been implicated in the pathophysiology of neuronal injury and neurodegenerative disease. Recent studies have shown that tea polyphenol EGCG has potent antioxidant effects against free radical-mediated lipid peroxidation in ischemia-induced neuronal damage. EGCG could attenuate neuronal expression of NADPH-d/nNOS in the motor neurons of the lower brainstem following peripheral nerve crush. In rats treated with high dosages of EGCG (25 or 50 mg/kg), NADPH-d/nNOS reactivity and cell death of the motor neurons were significantly decreased. EGCG can reduce NADPH-d/nNOS reactivity and thus may enhance motor neuron survival time following peripheral nerve injury [Wei *et al.*, 2011]. A study showed that the daily consumption, 25 mg/kg EGCG, and 50 mg/kg EGCG groups demonstrated statistically significant decrease in lipid peroxidation levels and particularly, the daily consumption and the 25 mg/kg EGCG group showed a statistical reduction of degeneration and edema histologically [Yildirim *et al.*, 2015]. This study shows that tea polyphenols and its derivatives have a protective effect on peripheral nerve injury.

3. Green tea polyphenols protect nerves against sciatic nerve crush injury

Hip sciatic nerve injury is one of the most difficult and most effective injuries in peripheral nerve injury. This study examined the capacity of the major polyphenolic green tea extract, EGCG, to suppress oxidative stress, stimulate the recovery, and prompt the regeneration of sciatic nerve after crush injury. After sciatic nerve crush injury, the EGCG-treated animals showed significantly better recovery of foot position and toe spread and 50% greater improvement in motor recovery than the saline-treated animals. EGCG displayed an early hopping response at the beginning of the third week post-injury. EGCG group also showed a significant reduction in mechanical allodynia and hyperalgesia latencies and significant improvement in recovery from nociception deficits in both heat withdrawal and tail flick withdrawal latencies compared with the crush group. EGCG significantly increased the mean cross-sectional area of axons, myelin thickness in the sciatic nerve, myelin basic protein concentration and gene expression in both the injured sciatic nerve and spinal cord, and fiber diameter to axon diameter ratio and myelin thickness to axon diameter ratio at week 2 after sciatic nerve injury. EGCG-treated rats showed significantly less increase of isoprostanes than saline-treated animals and showed full recovery of total antioxidant capacity levels by day 14 after nerve injury. In spinal cord tissue analysis, EGCG-treated animals showed induced glutathione reductase and suppressed induction of heme oxygenase 1 gene expression compared with non-treated animals [Renno *et al.*, 2017]. Another study found that EGCG attenuates functional deficits and morphological alterations by diminishing apoptotic gene overexpression in skeletal muscles after sciatic nerve crush injury. Semi-quantitative desmin immunohistochemistry revealed intense staining in the saline-treated injured animals, whereas EGCG treatment decreased the desmin immunoreactivity back to sham control levels. EGCG treatment induced a significant anti-apoptotic effect in injured muscle tissues by normalizing the Bax/Bcl-2 ratio back to baseline levels and inhibiting overexpression of the p53 apoptotic gene at days 3 and 7 post-surgery. EGCG enhances functional recovery, protects muscle fibers from cellular death by activating anti-apoptotic signaling pathway, and improves morphological recovery in skeletal muscle after nerve injuries [Renno *et al.*, 2012]. These findings suggest that EGCG can be used as an adjunctive therapeutic remedy for nerve injury. However, further investigations are needed to establish the antioxidative mechanism involved in the regenerative process after nerve injury.

4. Green tea polyphenols protect nerves against thermal hyperalgesia after spinal cord injury

Spinal cord injury (SCI) is a debilitating condition which is characterized by an extended secondary injury due to the presence of inflammatory local milieu. This study aimed to assess the tea polyphenol EGCG treatment effects on thermal hyperalgesia, spinal cord gliosis, and modulation of Ras homologue gene family member A (RhoA), fatty acid synthase (FASN), and tumor necrosis factor alpha (TNF-α) expression after spinal cord contusion in mice. Within a short time, EGCG treatment reduced significantly thermal hyperalgesia but had no effect on locomotor recovery in spinal cord injured mice. Furthermore, EGCG treatment downregulated the TNF-α proteins expression and decreased astro and microglia reactivity in spinal cord [Álvarez-Pérez *et al.*, 2016]. A study evaluated the beneficial effects of EGCG on recovery from SCI. The results demonstrated that EGCG-treated rats displayed a superior behavioral performance in a flat beam test, higher axonal sprouting, and positive remodelation of glial scar. Cytokine analysis revealed a reduction in inflammatory factors IL-6, IL2, MIP1α, and RANTES levels on days 1 and 3, and an upregulation of IL-4, IL-12p70, and TNFα 1 day following SCI in EGCG-treated rats. Treatment with EGCG was effective in decreasing the nuclear translocation of subunit p65 (RelA) of the NF-κB dimer, and therefore canonical NF-κB pathway attenuation. A significant increase in the gene expression of growth factors (FGF2 and VEGF), was noted in the spinal cord of EGCG-treated rats. Further, EGCG influenced expression of M1 and M2 macrophage markers. [Machova Urdzikova *et al.*, 2017]. These results demonstrated that within a short time, EGCG treatment reduces thermal hyperalgesia and gliosis via FASN and RhoA pathway, causing a decrease in cytokines in spinal cord, suggesting that the therapeutic value of EGCG in SCI .

5. Neuroprotection of green tea catechins on surgical menopause-induced overactive bladder

Bladder and urinary tract dysfunction is caused by damage to the central nervous system or peripheral nerves controlling urination. A rat model of ovariectomy-induced voiding dysfunction has been established, which mimicked the urge incontinence in postmenopausal women. Previous studies have identified strong anti-inflammatory/antioxidant properties of green tea and its associated polyphenols. The aim of this study was to evaluate whether the green tea extract, EGCG, could prevent an ovariectomy-induced overactive bladder. Long-term ovariectomy significantly increased non-voiding contractions, whereas treatment with EGCG significantly attenuated the frequency of non-voiding contractions. Ovariectomy significantly decreased the numbers of neuro-filament and increased M2 and M3 of muscarinic cholinergic receptor (MChR) protein and mRNA expressions. Treatment with EGCG restored the amount of neuro-filament staining and decreased M2 and M3 MChR protein and mRNA overexpressions [Juan *et al.*, 2012; Ikeguch *et al.*, 2003]. A study investigated oxidative stress induced by metabolic syndrome (MetS) and bilateral ovariectomy (OVX), and elucidated the mechanism underlying the protective effect of

EGCG (10 umol/kg/day) on bladder overactivity. Metabolic syndrome and MetS + OVX rats showed overexpression of inflammatory and fibrosis markers (1.7–3.8-fold of control). EGCG pre-treatment alleviated storage dysfunction and protected the bladders from metabolic syndrome and OVX-induced interstitial fibrosis changes. Moreover, OVX exacerbated metabolic syndrome related bladder apoptosis (2.3–4.5-fold of control; 1.8–2.6-fold of Mets group), enhanced oxidative stress markers (3.6–4.3-fold of control; 1.8–2.2-fold of Mets group), and mitochondrial enzyme complexes subunits (1.8–3.7-fold of control; 1.5–3.4-fold of Mets group). EGCG pretreatment alleviated bladder apoptosis, attenuated oxidative stress, and reduced the mitochondrial and endoplasmic reticulum apoptotic signals [Lee *et al.*, 2018].

These studies confirmed that EGCG prevented ovariectomy-induced bladder dysfunction through neuroprotective effects in a dose-dependent fashion.

6. Green tea polyphenols prevent herbicide-induced neurotoxicity

Herbicide pollutes soil and rivers, causing threats to the ecosystem. Studies found that acetochlor exposure could damage multiple organs and tissues in fish and mammal. Tea polyphenols, a natural antioxidant that extracted from tea, has been widely used in food and feed additions. After herbicide poisoning, serious patients may suffer from myocardial damage, blood pressure drop, acute renal failure, and other symptoms, and at the same time, they may be accompanied by neurological manifestations of different degrees, such as dizziness, headache, hallucination, coma, and convulsion. Atrazine is a broad-spectrum herbicide in wide use around the world. However, atrazine is neurotoxic and can cause cell death in dopaminergic neurons, leading to neurodegenerative disorders. Autophagy is the basic cellular catabolic process involving the degradation of proteins and damaged organelles. Studies have shown that certain plant compounds can induce autophagy and prevent neuronal cell death. This prompted us to investigate plant compounds that might reduce the neurotoxic effects of atrazine. Green tea polyphenols had the strongest activity against atrazine-induced neuronal apoptosis. Atrazine reduced the expression of hydroxylase, whereas green tea polyphenols increased hydroxylase expression. In addition, atrazine inhibited autophagy, whereas green tea polyphenols induced autophagy through the accumulation of LC3-II and decreased expression of p62; this effect was abolished by 3-methyl-adenine (3-MA) [Li *et al.*, 2017]. A study showed that herbicide exposure modified antioxidant enzyme activities, induced oxidative stress, resulting in the decline of mitochondrial membrane potential (MMP) and ATP levels, enhanced glycolysis and lactate accumulation, and triggered apoptosis and necroptosis in cells. However, tea polyphenols could inhibit CYP450s expression, activate Nrf2 pathway, enhance antioxidant capacity, and further effectively alleviate herbicide induced cell death. This proved that herbicide exposure triggered mitochondrial damage and lactate accumulation-mediated apoptosis and necroptosis through CYP450s/ROS/MAPK/NF-κB pathway [Zhao *et al.*, 2022].

This study shows that green tea polyphenols activate hydroxylase-dependent autophagy and some signals, reducing and protecting against herbicide induced neuronal apoptosis.

7. EGCG protects motor neurons following peripheral nerve injury

Oxidative stress and large amounts of NO have been implicated in the pathophysiology of neuronal injury and neurodegenerative disease. Recent studies have shown that EGCG, one of the green tea polyphenols, has potent antioxidant effects against free radical-mediated lipid peroxidation in ischemia-induced neuronal damage. The purpose of this study was to examine whether EGCG would attenuate neuronal expression of NADPH-d/nNOS in the motor neurons of the lower brainstem following peripheral nerve crush. In rats treated with high dosages of EGCG (25 mg/kg or 50 mg/kg), NADPH-d/nNOS reactivity and cell death of the motor neurons were significantly decreased [Wei *et al.*, 2011]. Studies also showed that protection of motor neuron by EGCG is associated with regulating glutamate level in organotypic culture of rat spinal cord. In this model, EGCG blocked glutamate excitotoxicity caused by threohy-droxyaspartate, an inhibitor of glutamate transporter. This property of EGCG may be not due to its intrinsic antioxidative activity, because another antioxidant could not regulate glutamate level under the same condition [Yu *et al.*, 2010]. These evidence indicated that EGCG can reduce NADPH-d/nNOS reactivity and thus, may enhance motor neuron survival time following peripheral nerve injury.

8. Green tea polyphenols protect transplantation of peripheral nerve

Nerve transplantation treatment refers to the transplantation of neural stem cells, which is mainly used to treat neurodegenerative changes and neutralize central nervous system damage. The neural stem cells entering the body can be directed to the pathological site of the nervous system, to survive, proliferate, and differentiate into neurons or glial cells in the pathological site of the host. Green tea polyphenols have a buffering effect and are the best conditions for storing peripheral nerves. The nerve segments was assessed by vital staining (calcein-AM/ethidium homodimer), by electron microscopy, and by genomic studies before transplantation. Allogeneic-transplanted peripheral nerve segments preserved for one month in green tea polyphenol solution at 4°C could regenerate nerves in rodents, thus demonstrating the same extent of nerve regeneration as isogeneic fresh nerve grafts. Green tea polyphenols can protect nerve tissue from ischemic damage for one month [Nakayama *et al.*, 2010; Ikeguchi *et al.*, 2003].

Above studies show successful nerve regeneration in the green tea polyphenol-treated nerve allografts when transplanted, indicating that polyphenols can protect nerve tissue from ischemic damage for one month.

9. Green tea polyphenols protect spinal cord neurons

Spinal nerve is an important nervous system of human body. Green tea polyphenols are strong antioxidants and can reduce free radical damage. The neuroprotective potential of tea polyphenols on spinal cord neurons against oxidative damage in spinal cord neurons was investigated. Results indicate that green tea polyphenols play a protective role in spinal cord neurons under oxidative stress [Zhao *et al.*, 2014]. EGCG has beneficial effects on the neuropathic pain alleviation. Chemokine fractalkine has been suggested as an important signal during neuropathic pain development. It participates in the migration and activation of leukocytes, especially phagocytes and lymphocytes, and at the same time, it also shows adhesion and mediates intercellular adhesion. This study aimed to investigate whether chemokine fractalkine expression may be modulated by EGCG treatment reducing hyperalgesia in chronic constriction injured mice. Results revealed that EGCG treatment significantly reduced thermal hyperalgesia in sciatic nerve-injured mice at short time, and this antihyperalgesic effect was associated with a down-regulation of chemokine fractalkine protein expression in the spinal cord. However, EGCG treatment did not affect the chemokine fractalkine transcription [Bosch-Mola *et al.*, 2017]. The results suggest a new role of EGCG-treatment in an experimental model of neuropathic pain as a mediator of nociceptive signaling crosstalk between neurons and glial cells in the dorsal horn of the spinal cord.

In the model of SCI rats, it was found that the water content and the BSCB permeability were decreased by bone marrow stromal cells (BMSCs) and GTPs treatment, and their combination had a synergistic effect. Further, the motor function of rats was also greatly improved by BMSCs and GTPs administration. After treated by the combination of BMSCs and GTPs, SCI rats showed the up-regulated expression of tight junction (TJ) associated proteins claudin-5, occluding, and ZO-1 by increased expression levels of claudin-5, occludin, and ZO-1 were the most obvious in the spinal cord microvessels using immunohistochemistry assay. This shows that the combination of BMSCs and GTPs could decrease the BSCB permeability by up-regulating protein expression levels of claudin-5, occludin, and ZO-1. In addition, after BMSCs and GTPs administration, the results of Western blot and enzyme-linked immunosorbent assay (ELISA) revealed a significant decrease in protein expression level and the activation of nuclear factor-κB (NF-κB) p65. The results indicated that the combination of BMSCs and GTPs could improve motor function after SCI, which might be correlated with improvements in BSCB integrity, and that NF-κB might be involved in the modulating process [Yu *et al.*, 2015].

These studies show that EGCG is the most abundant component in tea catechins and the main biological activity of green tea extracts, such as antioxidant, anti-inflammatory, and anti-apoptotic effects. Tea polyphenols have protective effects on many kinds of nerves, such as the brain and peripheral nerve injury caused by anti-microbial population and nerve crush injury.

5.3.4 *EGCG can active the neurons*

Some studies have shown that tea polyphenols can activate nerves such as cerebral cortex, perivascular nerves, activate neural pathways, and inhibit neuronal cell necrosis.

1. EGCG facilitates glutamate release via activation of protein kinase C in rat cerebral cortex

Central glutamatergic activity is crucial to cognitive function. EGCG has been reported to improve cognitive decline. A study investigated the effect of EGCG on the release of endogenous glutamate using nerve terminals purified from rat cerebral cortex. Results showed that the release of glutamate evoked by 4-aminopyridine (4AP) was facilitated by EGCG in a concentration-dependent manner, and this effect resulted from an enhancement of vesicular exocytosis and not from an increase in Ca^{2+}-independent efflux via glutamate transporter. The effect of EGCG on cytoplasmic free Ca^{2+} concentration (Ca^{2+}) revealed that the facilitation of glutamate release could be attributed to an increase in Ca^{2+} influx through N- and P/Q-type voltage-dependent Ca^{2+} channels. EGCG-mediated facilitation of 4AP-evoked glutamate release was significantly prevented in synaptosomes pre-treated with a combination of the N- and P/Q-type Ca^{2+} channel blockers. Additionally, inhibition of protein kinase C (PKC) by treatment with PKC inhibitor Ro318220 significantly reduced the facilitatory effect of EGCG on 4AP-evoked glutamate release and phosphorylation of PKC or its presynaptic target myristoylated alanine-rich C kinase substrate (MARCKS) [Chou *et al.*, 2007]. Systolic, diastolic, and mean blood pressure levels in the spontaneously early aged hypertensive group (SHR). SHR-EGCG were reduced compared to the SHR. The percentage of neural cell deaths, the levels of cytosolic Endonuclease G, cytosolic AIF (Caspase-independent apoptotic pathway), Fas, Fas Ligand, FADD, Caspase-8 (Fas-mediated apoptotic pathway), t-Bid, Bax/Bcl-2, Bak/Bcl-xL, cytosolic Cytochrome C, Apaf-1, Caspase-9 (Mitochondrial-mediated apoptotic pathway), and Caspase-3 (Fas-mediated and Mitochondria-mediated apoptotic pathways) were increased in the hypertensive group relative to control Wistar-Kyoto group and reduced in SHR-EGCG relative to hypertensive group. In contrast, the levels of Bcl-2, Bcl-xL, p-Bad, 14-3-3, Bcl-2/Bax, Bcl-xL/Bak, p-Bad/Bad (Bcl-2 family-related pro-survival pathway), Sirt1, p-PI3K/PI3K, and p-AKT/AKT (Sirt1/PI3K/AKT-related pro-survival pathway), were reduced in SHR relative Wistar-Kyoto group and enhanced in SHR-EGCG relative to SHR [Hsieh *et al.*, 2021].

These results suggest that EGCG effects a facilitation of glutamate release from glutamatergic terminals by positively modulating N- and P/Q-type Ca^{2+} channel activation through a signaling cascade involving PKC. In this EGCG/PKC signaling cascade facilitating glutamate release, the regulation of cytoskeleton dynamics was also indicated to be involved by the disruption of cytoskeleton organization with cytochalasin D occluded from the EGCG-mediated facilitation of 4AP-evoked glutamate. Green tea flavonoid EGCG might prevent neural apoptotic pathways and activate neural survival pathways, providing therapeutic effects on early aged hypertension-induced neural apoptosis.

2. EGCG induces activation of channels in perivascular nerves

There are nerves around the blood vessels. If there is a disease, there may be pain, limb weakness, and other symptoms. Vasculitis peripheral neuropathy can occur in vasculitis associated with connective tissue diseases, such as rheumatoid arthritis, systemic lupus erythematosus, Sjogren's syndrome, and sarcoidosis, and can also occur in primary vasculitis. Green tea polyphenol EGCG promotes vasodilation and reduces blood pressure. Recent reports suggested that EGCG can activate heterologously expressed in mouse and zebrafish TRPA1 channels. Activation of TRPA1 in sensory neurons triggers the release of calcitonin gene-related peptide (CGRP), a potent vasodilator. CGRP-containing (CGRPergic) sensory nerves contribute to EGCG-induced reduction in vascular resistance. This study demonstrated that intravenous infusion of EGCG elevated the plasma level of CGRP in mice, an effect that was attenuated by TRPA1 channel blocker A-967079. EGCG-induced increase in mesenteric artery blood flow and reduction in mean arterial pressure were reversed by A-967079, CGRP receptor antagonist CGRP8-37, and CGRP depletion in perivascular nerves. Moreover, EGCG stimulated TRPA1-dependent intracellular Ca^{2+} elevation and CGRP release in a differentiated rat embryonic dorsal root ganglion/mouse neuroblastoma hybrid cell line [Peixoto-Neves *et al.*, 2019]. These results suggest that EGCG-induced activation of TRPA1 channels in perivascular CGRPergic nerves decreases vascular resistance via Ca^{2+}-dependent exocytosis of CGRP.

3. The inhibitive effect of EGCG on neuronal cell necrosis

There are many situations that lead to the phenomenon of nerve necrosis, most of which are caused by nerve necrosis after injury without timely treatment. Once the nerve is necrotic, there is no possibility of recovery like other tissues in the human body. A study indicated the inhibition effect of EGCG and green tea extract on neuronal necroptosis based on necroptosis in male. It found that there are significant inhibition effect of EGCG on neuronal cell necroptosis ($p < 0.05$), which means both EGCG and green tea extract decrease the number of neuron cell necroptosis. EGCG will decrease neuron cell necroptosis starting from the dose of 20 mg/kg BW/day. EGCG 30 mg/kg BW/day produces the best result compared to other doses [Syaharani *et al.*, 2021]. Another study focused on specific identification of proteins involved in the neurorescue activity of the green tea polyphenol EGCG in a progressive model of neuronal death, induced by long-term serum deprivation of human neuroblastoma SH-SY5Y cells. It was found that the neurorescue action of EGCG was associated with its iron chelating properties and its ability to regulate metabolic energy balance and affect cell morphology. The results suggested the notion that the diverse molecular signaling pathways participating in the neurorescue activity of EGCG rendered this multifunctional compound [Weinreb *et al.*, 2008]. Above results indicate the diverse molecular signaling pathways participating in the neurorescue activity of tea polyphenols as potential agent to reduce risk of various neurodegenerative diseases.

5.4 Green tea polyphenols enhance neurotoxicity

Most study results show that tea polyphenols are antioxidants, which can protect oxidative damage and apoptosis, while other studies have found that tea polyphenols may promote oxidation and apoptosis. Green tea polyphenols are usually expected as potent chemo-preventive agents due to their ability of scavenging free radicals and chelating metal ions. However, not all actions of green tea polyphenols are necessarily beneficial in some special environments.

5.4.1 *Green tea polyphenols enhance sodium nitroprusside-induced neurotoxicity in human neuroblastoma SH-SY5Y cells*

Oxidative stress is a main mediator in nitric oxide (NO)-induced neurotoxicity and has been implicated in the pathogenesis of many neurodegenerative disorders. Our study demonstrated that higher-concentration green tea ployphenols significantly enhanced the neurotoxicity by treatment of sodium nitroprusside (SNP), a nitric oxide donor (Fig. 5-10). SNP induced apoptosis in human neuroblastoma SH-SY5Y cells in a concentration and time-dependent manner, whereas treatment with green tea polyphenols alone had no effect on cell viability. Pre-treatment with lower dose green tea polyphenols (50 μM and 100 μM) had only slightly deleterious effect in the presence of SNP, while higher dose green tea polyphenols (200 μM and 500 μM) synergistically damaged the cells severely. We further compared the effects of four catechins on SNP induced toxicity and found that the toxic effects of tea catechins connected comparatively to their activity. The possible reactions between GTPs or catechins with SNP were then examined by ESR technique. Adducts with free radical property were captured after mixing GTPs or catechins with SNP and the amounts of this adducts were comparative to the deleterious effects of catechins with NO, suggesting that in the presence of NO, GTPs may act as pro-oxidant by generation of ROS (Fig. 5-11). Thus, much consideration for safety should be paid when designing catechins as therapeutic reagents or nutrition supplements. Mechanisms of the synergistic neurotoxicity of NO and green tea polyphenols were further studied, and it was found that oxidative stress mediated the toxicity of NO and NO/GTPs co-treatment. SNP treatment caused loss of mitochondrial membrane potential, decrease of intracellular GSH, and accumulation of reactive oxygen species. Oxidative-related signal cascade was involved in the regulation of neuronal apoptosis. The intracellular levels of NFκB, p53, and c-Jun were up-regulated, while Bcl-2 expression was down-regulated. Low concentrations GTPs co-treatment improved the cellular redox status and attenuated SNP induced up-regulation of p53 and c-Jun. While high concentrations GTPs

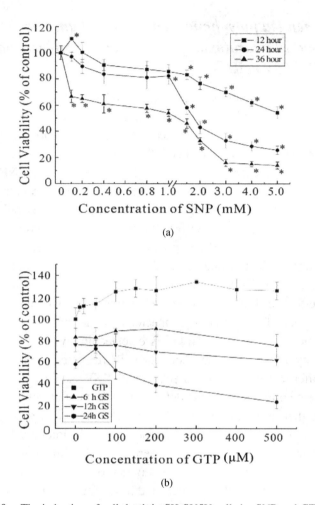

(a)

(b)

Figure 5-10.   The induction of cell death in SH-SY5Y cells by SNP and GTP. (a) Cells were incubated in drug-free medium or medium containing different concentrations of SNP for indicated time. It shows a concentration- and time-dependent loss of viability after SNP challenge. (b) Cells were incubated with different concentrations of green tea polyphenols in the presence (GS) or absence (GTP) of 1.5 mM SNP for 6 hours, 12 hours, and 24 hours, respectively. The cell viability was estimated by MTT assay. Data are expressed as percentage of the untreated control $\pm$ SD, n = 8. *, $p < 0.05$ significant difference from control by ANOVA; #, $p < 0.05$ in comparison with the SNP treated cells of same incubation time.

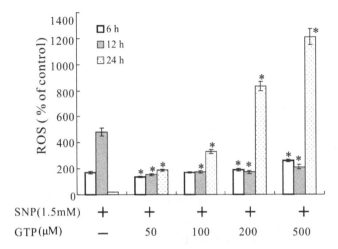

Figure 5-11. Effects of green tea polyphenols on reactive oxygen species (ROS) generation in SH-SY5Y cells exposed to SNP. Cells were exposed to 1.5 mM SNP without or with different concentration of green tea polyphenols for indicated time. Data are expressed as percentage of the untreated control ± SD, n = 6. There are significant difference between control group and all treated groups. *, $p < 0.05$ in comparison with the SNP treated cells by ANOVA.

exacerbated NO-induced oxidative stress and augment the expression of NFκB, p53, and c-Jun, resulting in further down-regulation of Bcl-2 and apoptosis. The regulation of GTPs on apoptotic signal cascade showed a positive correlation with cellular redox status, suggesting that GTPs regulate the signal transduction pathway via modulation of oxidative stress. [Zhang & Zhao, 2003].

5.4.2 *Green tea polyphenols potentiate the action of nerve growth factor to induce neuritogenesis: Possible role of ROS*

Exogenously administered nerve growth factor (NGF) repairs injured axons but it does not cross the blood-brain barrier. Thus, agents that could potentiate the neuritogenic ability of endogenous NGF would be of great utility in treating neurological injuries. Using the PC12 cell model, the results showed that unfractionated green tea polyphenols at low concentrations (0.1 µg/ml) potentiate the ability of low concentrations of NGF (2 ng/ml) to induce neuritogenesis at a level comparable to that induced by optimally high concentrations of NGF (50 ng/ml) alone. Among the polyphenols present in green tea polyphenols, EGCG alone appreciably potentiated NGF-induced neurite outgrowth. Although other polyphenols present in green tea polyphenols, particularly epigallocatechin and epicatechin, lack this activity, they synergistically promoted this action of EGCG. Green tea polyphenols also induced an activation of extracellular signal-regulated kinases (ERKs). An inhibitor of the ERK pathway, blocked the expression of GAP-43. K252a, an inhibitor of TrkA-associated tyrosine kinase, partially blocked the expression of these genes and ERK activation. These results show for the first time that green tea polyphenols potentiates NGF-induced neuritogenesis, likely through the involvement of sublethal levels of reactive oxygen species, and suggest that unfractionated green tea polyphenols is more effective in this respect than its fractionated polyphenols [Gundimeda *et al.*, 2010]. A synergistic interaction was observed between GTP constituents, where epigallocatechin and epicatechin, both individually lacking this activity, promoted the action of EGCG. GTP-induced potentiation of brain-derived neurotrophic factor (BDNF) action required the cell-surface associated 67 kDa laminin receptor (67LR) to which EGCG binds with high affinity. A cell-permeable catalase abolished GTP/EGCG-induced potentiation of BDNF action, suggesting the possible involvement of H_2O_2 in the potentiation. Consistently, exogenous sublethal concentrations of H_2O_2, added as a bolus dose (5 µM) or more effectively through a steady-state generation (1 µM), potentiated brain-derived neurotrophic factor action. Collectively, these results suggest that EGCG, dependent on 67 LR and H_2O_2, potentiates the neuritogenic action of brain-derived neurotrophic factor. Intriguingly, this effect requires only submicromolar concentrations of EGCG. This is significant as extremely low concentrations of polyphenols are believed to reach the brain after drinking green tea [Gundimeda *et al.*, 2014].

5.5 Conclusion

The above results show that the nervous system is a fragile system which is very vulnerable to oxidative stress injury, resulting in nerve injury, and neurodegenerative diseases. Drinking tea and its effective component, natural antioxidant tea polyphenols, have antioxidant capacity, can inhibit oxidative stress, prevent its damage to the nervous system, and prevent and treat neurodegenerative diseases. However, in the future, more rigorous clinical trials should be carried out to verify its role in order to benefit the majority of patients. At the same time, it should also be studied in what concentration and under what conditions tea polyphenols have a protective effect on nerves, and under what conditions they may cause nerve damage so as to avoid it.

References

Adanaylo VN, Oteiza PI. (1999) Lead intoxication: Antioxidant defenses and oxidative damage in rat brain. *Toxicology*, **135**, 77–85.

Aktas O, Prozorovski T, Smorodchenko A, Savaskan NE, Lauster R, Kloetzel PM, Infante-Duarte C, Brocke S, Zipp F. (2004) Green tea epigallocatechin-3-gallate mediates T cellular NF-kappa B inhibition and exerts neuroprotection in autoimmune encephalomyelitis. *J Immunol*, **173**, 5794–5800.

Álvarez-Pérez B, Homs J, Bosch-Mola M, Puig T, Reina F, Verdú E, Boadas-Vaello P. (2016) Epigallocatechin-3-gallate treatment reduces thermal hyperalgesia after spinal cord injury by down-regulating RhoA expression in mice. *Eur J Pain*, **20**(3), 341–352.

Bechara EJH, Medeiros MHG, Monteiro HP, Her-mesLima M, Pereira B, Demasi M, Costa CA, Ab-dalla DSP, Onuki J, Wendel CMA, Di Mascio P. (1993) A free radical hypothesis of lead poisoning and inborn porphyrias associated with 5-aminolevulinic acid overload. *Quim Nova,* **16**, 385–392.

Biasibetti R, Tramontina AC, Costa AP, Dutra MF, Quincozes-Santos A, Nardin P, Bernardi CL, Wartchow KM, Lunardi PS, Goncalves CA. (2013) Green tea (–)epigallocatechín-3-gallate reverses oxidative stress and reduces acetylcholinesterase activity in a streptozotocin-induced model of dementia. *Behav Brain Res*, **236**, 186–193.

Bosch-Mola M, Homs J, Álvarez-Pérez B, Puig T, Reina F, Verdú E, Boadas-Vaello P. (2017) (-)-epigallocatechin-3-gallate antihyperalgesic effect associates with reduced CX3CL1 chemokine expression in spinal cord. *Phytother Res*, **31**(2), 340–344.

Caillaud M, Chantemargue B, Richard L, *et al.* (2018) Local low dose curcumin treatment improves functional recovery and remyelination in a rat model of sciatic nerve crush through inhibition of oxidative stress. *Neuropharmacology*, **139**, 98–116.

Cansler SM, Evanson NK. (2020) Connecting endoplasmic reticulum and oxidative stress to retinal degeneration, TBI, and traumatic optic neuropathy. *J Neurosci Res*, **98**(3), 571–574.

Chang-Mu C, Jen-Kun L, Shing-Hwa L, Shoei-Yn LS. (2010) Characterization of neurotoxic effects of NMDA and the novel neuroprotection by phytopolyphenols in mice. *Behav Neurosci*, **124**(4), 541–553.

Chen CM, Lin JK, Liu SH, Lin-Shiau SY. (2008) Novel regimen through combination of memantine and tea polyphenol for neuroprotection against brain excitotoxicity. *J Neurosci Res*, **86**(12), 2696–2704.

Chen L-J, Yang X-Q, Jiao H-L, Zhao B-L. (2002) Tea catechins protect against lead-induced cytotoxicity, lipid peroxidation, and membrane fluidity in HepG2 cells. *Toxcol Sci*, **69**, 149–156.

Chen L-J, Yang X-Q, Jiao H-L, Zhao B-L. (2003) Tea Catechins protect against lead-induced ROS formation, mitochondrial dysfunction and calcium dysregulation in PC 12 cells. *Chem Res Toxiol*, **16**, 1155–1161.

Chen LJ, Yang XQi, Jiao HL, Zhao BL. (2004) Effect of tea catechins on the change of glutathione levels caused by Pb^{++} in PC12 cells. *Chem Res Toxiol*, **17**, 922–928.

Chou CW, Huang WJ, Tien LT, Wang SJ. (2007) (-)-Epigallocatechin gallate, the most active polyphenolic catechin in green tea, presynaptically facilitates Ca2+-dependent glutamate release via activation of protein kinase C in rat cerebral cortex. *Synapse*, **61**(11), 889–902.

Cory-Slechta DA, Pound JG. (1995) Lead neurotoxicity. In *Handbook of Neurotoxicology*, Chang LW, Dyer RS (eds.), New York: Dekker, 61–89.

Dong X, Yang C, Cao S, Gan Y, Sun H, Gong Y, Yang H, Yin X, Lu Z. (2015) Tea consumption and the risk of depression: a meta-analysis of observational studies. *Aust N Z J Psychiatry*, **49**(4), 334 –345.

Ehrnhoefer DE, Duennwald M, Markovic P, Wacker JL, Engemann S, Roark M, Legleiter J, Marsh JL, Thompson LM, Lindquist S, Muchowski PJ, Wanker EE. (2006) Green tea (-)-epigallocatechin-gallate modulates early events in huntingtin misfolding and reduces toxicity in Huntington's disease models. *Hum Mol Genet*, **15**(18), 2743–2751.

Fernando WMADB, Somaratne G, Goozee KG, *et al.* (2017) Diabetes and alzheimer's disease: Can tea phytochemicals play a role in prevention? *J Alzheimers Dis*, **59**(2), 481–501.

Gundimeda U, McNeill TH, Schiffman JE, Hinton DR, Gopalakrishna R. (2010) Green tea polyphenols potentiate the action of nerve growth factor to induce neuritogenesis: possible role of reactive oxygen species. *J Neurosci Res*, **88**(16), 3644–3655.

Gundimeda U, McNeill TH, Fan TK, Deng R, Rayudu D, Chen Z, Cadenas E, Gopalakrishna R. (2014) Green tea catechins potentiate the neuritogenic action of brain-derived neurotrophic factor: role of 67-kDa laminin receptor and hydrogen peroxide. *Biochem Biophys Res Commun*, **445**(1), 218–224.

Guo Q, Zhao BL, Li MF, Shen SR, Xin WJ. (1996) Studies on protective mechanisms of four components of green tea polyphenols (GTP) against lipid peroxidation in synaptosomes. *Biochem Biophys Acta*, **1304**, 210–222.

Guo Q, Zhao B-L, Hou J-W, Xin W-J. (1999) ESR study on the structure-antioxidant activiity relationship of tea catechins and their epimers. *Bichim Biophys Acta*, **1427**(1),13–23.

Guo S, Bezard E, Zhao B-L. (2005) Protective effect of green tea polyphenols on the SH-SY5Y cells against 6-OHDA induced apoptosis through ROS-NO pathway. *Free Rad Biol Med*, **39**, 682–695.

Guo S, Yan J, Bezard E, Yang T, Yang X, Zhao B-L. (2007) Protective effects of green tea polyphenols in the 6-OHDA rat model of Parkinson's disease through inhibition of ROS-NO pathway. *Biol Psychiatry*, **62**, 1353–1362.

Gurer H, Ozgunes H, Oztezcan S, Ercal N. (1999) Antioxidant role of α-lipoic acid in lead toxicity. *Free Radic Biol Med*, **27**, 75-81.

Gurer H, Nuran E. (2000) Can antioxidants be beneficial in the treatment of lead poisoning? *Free Rad Biol Med*, **29**, 927-945.

Haque AM, Hashimoto M, Katakura M, Tanabe Y, Hara Y, Shido O. (2006) Long-term administration of green tea catechins improves spatial cognition learning ability in rats. *J Nutr*, **136**, 1043–1047.

Haque AM, Hashimoto M, Katakura M, Hara Y, Shido O. (2008) Green tea catechins prevent cognitive deficits caused by Aβ1–40 in rats. *J Nutr Biochem*, **19**, 619–626.

Hermes-Lima M, Pereira B, Bechara EJH. (1991) Are free radicals involved in lead poisoning? *Xenobiotica*, **21**, 1085–1090.

He Y, Tan D, Mi Y, Zhou Q, Ji S. (2017a) Epigallocatechin-3-gallate attenuates cerebral cortex damage and promotes brain regeneration in acrylamide-treated rats. *Food Funct*, **8**(6), 2275–2282.

He Y, Tan D, Mi Y, Bai B, Jiang D, Zhou X, Ji S. (2017b) Effect of epigallocatechin-3-gallate on acrylamide-induced oxidative stress and apoptosis in PC12 cells. *Hum Exp Toxicol*, **36**(10), 1087–1099.

Higgins GC, Beart PM, Shin YS, Chen MJ, Cheung NS, Nagley P. (2010) Oxidative stress: Emerging mitochondrial and cellular themes and variations in neuronal injury. *J Alzheimers Dis*, **20**(Suppl 2), S453–S473.

Hintikka J, Tolmunen T, Honkalampi K, Haatainen K, Koivumaa-Honkanen H, Tanskanen A, Viinamäki H. (2005) Daily tea drinking is associated with a low level of depressive symptoms in the Finnish general population. *Eur J Epidemiol*, **20**(4), 359–363.

Hsieh MH, Cui ZY, Yang AL, Nhu NT, Ting SY, Yu SH, Cheng YJ, Lin YY, Wu XB, Lee SD. (2021) Cerebral cortex apoptosis in early aged hypertension: effects of epigallocatechin-3-gallate. *Front Aging Neurosci*, **13**, 705304.

Hsu P-C, Liu M-Y, Hsu C-C, *et al.* (1998) Effects of vitamin E and/or C on reactive oxygen species-related lead toxicity in the rat sperm. *Toxicology*, **128**, 169–179.

Huang J, Lu Y, Zhang B, Yang S, Zhang Q, Cui H, Lu X, Zhao Y, Yang X, Li R. (2019) Antagonistic effect of epigallocatechin-3-gallate on neurotoxicity induced by formaldehyde. *Toxicology*, **412**, 29–36.

Ikeguchi R, Kakinoki R, Okamoto T, Matsumoto T, Hyon SH, Nakamura T. (2003) Successful storage of peripheral nerve before transplantation using green tea polyphenol: an experimental study in rats. *Exp Neurol*, **184**(2), 688–696.

Inanami O, Watanabe Y, Syuto B, Nakano M, Tsuji M, Kuwabara M. (1998) Oral administration of (–) catechin protects against ischemia–reperfusion-induced neuronal death in the gerbil. *Free Radical Res*, **29**, 359–365.

Jia Z, Zhu H, Li J, Wang X, Misra H, Li Y. (2012) Oxidative stress in spinal cord injury and antioxidant-based intervention. *Spinal Cord*, **50**(4), 264–274.

Jiun YS, Hsien LT. (1994) Lipid peroxidation in workers exposed to lead. *Archiv Environ Health*, **49**, 256–259.

Juan Y-S, Chuang S-M, Long C-Y, Chen C-H, Levin RM, Liu K-M, Huang C-H. (2012) Neuroprotection of green tea catechins on surgical menopause-induced overactive bladder in a rat model. *Menopause*, **19**(3), 346–354.

Juberg DR, Kleiman CF, Kwon SC. (1997) Position paper of the American council on science and health. *Ecotoxicol Environ Saf*, **38**, 162–180.

Kang D, Kim Y, Je Y. (2018) Non-alcoholic beverage consumption and risk of depression: epidemiological evidence from observational studies. *Eur J Clin Nutr*, **72**(11), 1506–1516.

Kang KS, Yamabe N, Wen Y, Fukui M, Zhu BT. (2013) Beneficial effects of natural phenolics on levodopa methylation and oxidative neurodegeneration. *Brain Res*, **1497**, 1–14.

Kang Y, Lee JH, Seo YH, Jang JH, Jeong CH, Lee S, Jeong GS, Park B. (2019) Epicatechin prevents methamphetamine-induced neuronal cell death via inhibition of ER stress. *Biomol Ther (Seoul)*, **27**(2), 145–151.

Koh SH, Lee SM, Kim HY, Lee KY, Lee YJ, Kim HT, Kim J, Kim MH, Hwang MS, Song C, Yang KW, Lee KW, Kim SH, Kim OH. (2006) The effect of epigallocatechin gallate on suppressing disease progression of ALS model mice. *Neurosci Lett*, **395**, 103–107.

Lee S, Suh S, Kim S. (2000) Protective effects of the green tea polyphenol (-)-epigallocatechin gallate against hippocampal neuronal damage after transient global ischemia in gerbils. *Neurosci Lett*, **287**, 191–194.

Lee YL, Lin KL, Wu BN, Chuang SM, Wu WJ, Lee YC, Ho WT, Juan YS. (2018) Epigallocatechin-3-gallate alleviates bladder overactivity in a rat model with metabolic syndrome and ovarian hormone deficiency through mitochondria apoptosis pathways. *Sci Rep*, **8**(1), 5358.

Levites Y, Weinreb O, Maor G, Youdim MBH, Mandel S. (2001) Green tea polyphenol epigallocatechin-3-gallate prevents MPTP induced dopaminergic neurodegeneration. *J Neurochem*, **78**, 1073–1082.

Lidsky TI, Schneider JS. (2003) Lead neurotoxicity in children: basic mechanisms and clinical correlates. *Brain*, **126**(Pt 1), 5–19.

Li P, Ma K, Wu HY, Wu YP, Li BX. (2017) Isoflavones induce BEX2-dependent autophagy to prevent ATR-induced neurotoxicity in SH-SY5Y cells. *Cell Physiol Biochem*, **43**(5), 1866 –1879.

Li Z, Wu F, Xu D, Zhi Z, Xu G. (2019) Inhibition of TREM1 reduces inflammation and oxidative stress after spinal cord injury (SCI) associated with HO-1 expressions. *Biomed Pharmacother*, **109**, 2014–2021.

Machin A, Syaharani R, Susilo I, Hamdan M, Fauziah D, Purwanto DA. (2021) The effect of Camellia sinensis (green tea) with its active compound EGCG on neuronal cell necroptosis in Rattus norvegicus middle cerebral artery occlusion (MCAO) model. *J Basic Clin Physiol Pharmacol*, **32**(4), 527–531.

Machova Urdzikova L, Ruzicka J, Karova K, Kloudova A, Svobodova B, Amin A, Dubisova J, Schmidt M, Kubinova S, Jhanwar-Uniyal M, Jendelova P. (2017) A green tea polyphenol epigallocatechin-3-gallate enhances neuroregeneration after spinal cord injury by altering levels of inflammatory cytokines. *Neuropharmacology*, **126**, 213–223.

Mandel S, Maor G, Youdim MB. (2004) Iron and alpha-synuclein in the substantia nigra of MPTP-treated mice: effect of neuroprotective drugs R-apomorphine and green tea polyphenol (-)-epigallocatechin-3-gallate. *J Mol Neurosci*, **24**(3), 401–416.

Monteiro HP, Abdalla DSP, Arcuri AS, Bechara EJH. (1985) Oxygen toxicity related to exposure to lead. *Clin Chem*, **31**, 1673–1676.

Nakayama K, Kakinoki R, Ikeguchi R, Yamakawa T, Ohta S, Fujita S, Noguchi T, Duncan SF, Hyon SH, Nakamura T. (2010) Storage and allogeneic transplantation of peripheral nerve using a green tea polyphenol solution in a canine model. *J Brachial Plex Peripher Nerve Inj*, **5**, 17.

Ni YC, Zhao BL, Hou JW, Xin WJ. (1996) Protection of cerebellar neuron by Ginkgo-biloba extract against apoptosis induced by hydroxyl radicals. *Neuronsci Letter*, **214**, 115–118.

O'Hare Doig RL, Bartlett CA, Maghzal GJ, Lam M, Archer M, Stocker R, Fitzgerald M. (2014) Reactive species and oxidative stress in optic nerve vulnerable to secondary degeneration. *Exp Neurol*, **261**, 136–146.

Padurariu MA, Ciobica A, Hritcu L, Stoica B, Bild W, C Stefanescu. (2010) Changes of some oxidative stress markers in the serum of patients with mild cognitive impairment and Alzheimer's disease. *Neurosci Lett*, **469**(1), 6–10.

Pearson JN, Patel M. (2016) The role of oxidative stress in organophosphate and nerve agent toxicity. *Ann N Y Acad Sci*, **1378**(1), 17–24.

Peixoto-Neves D, Soni H, Adebiyi A. (2019) CGR pergic nerve TRPA1 channels contribute to epigallocatechin gallate-induced neurogenic vasodilation. *ACS Chem Neurosci*, **10**(1), 216–220.

Qiu J, Yang X, Wang L, *et al.* (2019) Isoquercitrin promotes peripheral nerve regeneration through inhibiting oxidative stress following sciatic crush injury in mice. *Ann Transl Med*, **7**(22), 680.

Razzaq A, Hussain G, Rasul A, Xu J, Zhang Q, Malik SA, Anwar H, Aziz N, Braidy N, de Aguilar JG, Wei W, Li J, Li X. (2020) Strychnos nux-vomica L. seed preparation promotes functional recovery and attenuates oxidative stress in a mouse model of sciatic nerve crush injury. *BMC Complement Med Ther*, **20**(1), 181.

Renno WM, Benov L, Khan KM. (2017) Possible role of antioxidative capacity of (-)-epigallocatechin-3-gallate treatment in morphological and neurobehavioral recovery after sciatic nerve crush injury. *J Neurosurg Spine*, **27**(5), 593–613.

Renno WM, Al-Maghrebi M, Al-Banaw A. (2012) (-)-Epigallocatechin-3-gallate (EGCG) attenuates functional deficits and morphological alterations by diminishing apoptotic gene overexpression in skeletal muscles after sciatic nerve crush injury. *Naunyn Schmiedebergs Arch Pharmacol*, **385**(8), 807–822.

Resim S, Koluş E, Barut O, Kucukdurmaz F, Bahar AY, Dagli H. (2020) Ziziphus jujube ameliorated cavernosal oxidative stress and fibrotic processes in cavernous nerve injury-induced erectile dysfunction in a rat model. *Andrologia*, **52**(7), e13632.

Ru Q, Xiong Q, Tian X, Chen L, Zhou M, Li Y, Li C. (2005) Tea polyphenols attenuate methamphetamine-induced neuronal damage in PC12 cells by alleviating oxidative stress and promoting DNA repair. *Curr Top Med Chem*, **5**(7), 721–736.

Saccà SC, Izzotti A. (2008) Oxidative stress and glaucoma: injury in the anterior segment of the eye. *Prog Brain Res*, **173**, 385–407.

Sang S, Tian S, Wang H, Stark RE, Rosen RT, Yang CS, Ho CT. (2003) Chemical studies of the antioxidant mechanism of tea catechins: Radical reaction products of epicatechin with peroxyl radicals. *Bioorg Med Chem*, **11**, 3371–3378.

Satoh E, Ishii T, Shimizu Y, Sawamura S, Nishimura M. (2001) Black tea extract, thearubigin fraction, counteracts the effect of tetanus toxin in mice. *Exp Biol Med (Maywood)*, **226**(6), 577–580.

Satoh E, Ishii T, Shimizu Y, Sawamura S, Nishimura M. (2002) A mechanism of the thearubigin fraction of black tea (Camellia sinensis) extract protecting against the effect of tetanus toxin. *J Toxicol Sci*, **27**(5), 441–417.

Seeram NP, Henning SM, Niu Y, Lee R, Scheuller HS, Heber D. (2006) Catechin and caffeine content of green tea dietary supplements and correlation with antioxidant capacity. *J Agric Food Chem*, **54**, 1599–1603

Shafiq-ur-Rehman S, Rehman S, Chandra O, Abdulla M. (1995) Evaluation of malondialdehyde as an index of lead damage in rat brain homogenates. *Biometals*, **8**, 275–279.

Sheng J, Yang X, Liu Q, Luo H, Yin X, Liang M, Liu W, Lan X, Wan J, Yang X. (2020) Coadministration with tea polyphenols enhances the neuroprotective effect of defatted walnut meal hydrolysate against scopolamine-induced learning and memory deficits in mice. *J Agric Food Chem*, **68**(3), 751–758.

Shotyk W, Weiss D, Appleby G, Cheburkin AK, Gloor RFM, Kramers JD, Reese S, Van Der Knaap WO. (1998) History of atmospheric lead deposition since 12, 370(14) C yr BP from a peat bog, Jura mountains. Switzerland: Science, 281.

Sun Q, Cheng L, Zhang X, Wu Z, Weng P. (2021) The interaction between tea polyphenols and host intestinal microorganisms: an effective way to prevent psychiatric disorders. *Food Funct*, **12**(3), 952–962.

Sutherland BA, Shaw OM, Clarkson AN, Jackson DM, Sammut IA, Appleton I. (2004) Neuroprotective effects of (-)-epigallocatechin gallate after hypoxia-ischemia-induced brain damage: novel mechanisms of action. *FASEB J*, **19**: 258–260.

Syaharani R, Susilo I, Hamdan M, Fauziah D, Purwanto DA. (2021) The effect of Camellia sinensis (green tea) with its active compound EGCG on neuronal cell necroptosis in Rattus norvegicus middle cerebral artery occlusion (MCAO) model. *J Basic Clin Physiol Pharmacol*, **32**(4), 527–531.

Tian L, Lowrence D. (1995) Lead inhibits nitric oxide production in vitro by murine splenic macrophages. *Toxicol Appl Pharmacol*, **132**, 156–163.

Torres JD, Dueik V, Carré D, Bouchon P. (2019) Effect of the addition of soluble dietary fiber and green tea polyphenols on acrylamide formation and in vitro starch digestibility in baked starchy matrices. *Molecules*, **24**(20), 3674.

Wan L, Nie G, Zhang J, Zhao B (2012) Overexpression of human wild-type amyloid-β protein precursor decreases the iron content and increases the oxidative stress of neuroblastoma SH-SY5Y cells. *Journal of Alzheimer's Disease*, **30**, 523–530.

Wang Y, Meng XH, Zhang QJ, Wang YM, Chen C, Wang YC, Zhou X, Ji CJ, Song NH. (2019) Losartan improves erectile function through suppression of corporal apoptosis and oxidative stress in rats with cavernous nerve injury. *Asian J Androl*, **21**(5), 452–459.

Wang H, Ding XG, Li SW, Zheng H, Zheng XM, Navin S, Li L, Wang XH. (2015) Role of oxidative stress in surgical cavernous nerve injury in a rat model. *J Neurosci Res*, **93**(6), 922–929.

Ward R, Zucca FA, Duyn JH, Crichton RR, Zecca L. (2014) The role of iron in brain ageing and neurodegenerative disorders. *Lancet Neurol.* **13**, 1045–1060.

Wei IH, Tu HC, Huang CC, Tsai MH, Tseng CY, Shieh JY. (2011) (-)-Epigallocatechin gallate attenuates NADPH-d/nNOS expression in motor neurons of rats following peripheral nerve injury. *BMC Neurosci*, **12**, 52.

Weinreb O, Amit T, Youdim MB. (2008) The application of proteomic for studying the neurorescue activity of the polyphenol (-)-epigallocatechin-3-gallate. *Arch Biochem Biophys*, **476**(2), 152–160.

Weinreb O, Amit T, Mandel S, Youdim MB. (2009) Neuroprotective molecular mechanisms of (−)-epigallocatechin-3-gallate: a reflective outcome of its antioxidant, iron chelating and neuritogenic properties. *Genes Nutr*, **4**, 283–296.

Wei IH, Tu HC, Huang CC, Tsai MH, Tseng CY, Shieh JY. (2011) (-)-Epigallocatechin gallate attenuates NADPH-d/nNOS expression in motor neurons of rats following peripheral nerve injury. *BMC Neurosci*, **1**(12), 52.

Xie Q, Liu Y, sun H, Liu Y, Ding X, Fu D, Liu K, Du X, Jia G. (2008) Inhibition of acrylamide toxicity in mice by three dietary constituents. *J Agric Food Chem*, **56**(15), 6054–6060.

Yildirim AE, Dalgic A, Divanlioglu D, Akdag R, Cetinalp NE, Alagoz F, Helvacioglu F, Take G, Guvenc Y, Koksal I, Belen AD. (2015) Biochemical and histopathological effects of catechin on experimental peripheral nerve injuries. *Turk Neurosurg*, **25**(3), 453–460.

Yin Y, Sun G, Li E, Kiselyov K, Sun D. (2017) ER stress and impaired autophagy flux in neuronal degeneration and brain injury. *Ageing Res Rev*, **34**, 3–14.

Yu J, Jia Y, Guo Y, Chang G, Duan W, Sun M, Li B, Li C. (2010) Epigallocatechin-3-gallate protects motor neurons and regulates glutamate level. *FEBS Lett*, **584**(13), 2921–2925.

Yu DS, Liu LB, Cao Y, Wang YS, Bi YL, Wei ZJ, Tong SM, Lv G, Mei XF. (2015) Combining bone marrow stromal cells with green tea polyphenols attenuates the blood-spinal cord barrier permeability in rats with compression spinal cord injury. *J Mol Neurosci*, **56**(2), 388–396.

Zhang Y, Zhao B. (2003) Green tea polyphenols enhance sodium nitroprusside induced neurotoxicity in human neuroblastoma SH-SY5Y cells. *J Neur Chem*, **86**, 1189–1200.

Zhang Y, Cheng L, Zhang X. (2021a) Interactions of tea polyphenols with intestinal microbiota and their effects on cerebral nerves. *J Food Biochem*, **45**(1), e13575.

Zhang Z, Zhang Y, Li J, Fu C, Zhang X. (2021b) The neuroprotective effect of tea polyphenols on the regulation of intestinal flora. *Molecules*, **26**(12), 3692.

Zhao B-L. (2005) Natural antioxidant for neurodegenerative diseases. *Mol Neurobiol*, **31**, 283–293.

Zhao B-L. (2009) Natural antioxidants protect neurons in alzheimer's disease and parkinson's disease. *Neurochem Res*, **34**, 630–638.

Zhao B-L. (2012) Natural antioxidant green tea polyphenols and health. *Acta Biophys Sin (Shanghai)*, **28**, 26–36.

Zhao B-L. (2020) The pros and cons of drinking tea. *Tradit Med Mod Med*, **3**, 1–12.

Zhao X, Shi X, Liu Q, Li X. (2022) Tea polyphenols alleviates acetochlor-induced apoptosis and necroptosis via ROS/MAPK/NF-κB signaling in Ctenopharyngodon idellus kidney cells. *Aquat Toxicol*, **246**, 106153.

Zhao Y, Zhao B. (2012) Oxidative stress, natural antioxidants protect neurons against alzheimer's disease. *Front Biosci*, **15**, 454–461.

Zhao J, Fang S, Yuan Y, Guo Z, Zeng J, Guo Y, Tang P, Mei X. (2014) Green tea polyphenols protect spinal cord neurons against hydrogen peroxide-induced oxidative stress. *Neural Regen Res*, **9**(14), 1379–1385.

Chapter 6

Tea Polyphenols and Alzheimer's Disease

Baolu Zhao

Institute of Biophysics, Chinese Academy of Sciences, Beijing, China

6.1 Introduction

The latest census shows that there are 130 million elderly people over the age of 60 in China, which will reach 439 million by 2050, accounting for 1/4 of the total population. The number of people suffering from Alzheimer's disease (AD) and Parkinson's disease (PD) was more than 8 million, reaching more than 15 million in 2010. China has entered an aging society ahead of time. According to the survey, the incidence rate of AD among people over 65 years old in China is 4.8%, 11.5% over 75 years old, and 30% above 85 years old. According to this speed of development, AD will become the biggest health risk for the elderly in China and even the world. However, the pathogenesis of AD is not completely clear, and there is no effective drug. Neurofilament winding, senile plaques, and amyloid Aβ precipitation generally appear in the brain of patients with AD [Querfurth & LaFerla, 2010]. At present, neurotoxicity related to amyloid and tau protein, changes in cholinergic neurotransmission, oxidative stress, and changes in calcium homeostasis are considered to be the key factors of AD. Amyloid is composed of amyloid precursor protein (APP) made by enzyme treatment. Studying the pathogenesis of

AD and finding effective methods to treat AD have attracted extensive attention, and it is also the bounden task of scientific researchers and medical workers. There is evidence that oxidative stress and NO are closely related to the pathogenesis of AD. Research shows that drinking tea and antioxidant tea polyphenols have great benefits for AD. The prevention and treatment of AD by tea polyphenols have a certain role and play a regulatory role in this process. This chapter discusses the research and problems about oxidative stress and tea polyphenols in this regard.

6.2 Alzheimer's disease

AD is a progressive neurodegenerative disease, a devastating disease of the elderly. Those who come on before the age of 65 are called premature aging syndrome; those who come on after the age of 65 are called AD. It is the most common cause of dementia (it is estimated that about 60% to 80% of cases are caused by AD). As it was discovered by a German doctor, Alzheimer, in 1906, it was therefore named Alzheimer's disease. AD is one of the most common neurodegenerative diseases, which mostly occurs in the elderly. The patients gradually suffer from a series of high-level neurological functions such as cognition and movement, and finally lose the ability to take care of themselves [Querfurth & LaFerla, 2010]. AD is a latent progressive neurodegenerative disease. Clinically, it is characterized by memory impairment, aphasia, apraxia, agnosia, impairment of visuospatial skills, executive dysfunction, personality and behavior changes, and the etiology is complex and caused by many factors (including biological and psychosocial factors). The main pathological features of AD are senile plaques, neurofilament winding, and neuron loss in the patient's brain. The main manifestations are the decline of cognitive function, mental symptoms and behavioral disorders, and the gradual decline of daily living ability. According to the deterioration of cognitive ability and physical function, it can be divided into three periods. Mild dementia: Manifested by memory loss and prominent forgetting of recent events; Moderate dementia: Severe impairment of distant and near memory; Severe dementia stage: Patient needs to be completely dependent on the caregiver and suffers severe memory loss with only fragments of memory, is unable to take care of themselves in daily life, and is incontinence of urine, silence, and experiences rigidity of limbs. The physical examination shows that the pyramidal tract sign is positive, and there are primitive reflexes such as strong grip, groping, and sucking. Eventually coma, usually died of infection and other complications.

An important feature of AD is the presence of senile plaques in brain cells. Senile plaques are mainly formed by beta amyloid deposition [Ansari & Scheff, 2010]. Aβ is generated from transmembrane amyloid precursor (APP) digested by β secretase and γ secretase in sequence. According to different cutting sites of γ secretase, Aβ with different lengths can be produced; among them, Aβ 40 with a length of 40 amino acids and Aβ 42 with 42 amino acids are the main forms of Aβ, while Aβ1–42 is more oligomerized and more toxic [Ankner & Lu, 2009]. A large number of studies have shown that Aβ-induced neurotoxicity is closely related to the degradation and loss of neurons, and the loss of neurons is the reason for a series of clinical symptoms of AD. Therefore, the control of Aβ neurotoxicity is an important target in the prevention and treatment of AD. Amyloid Aβ was observed in older people. The corresponding increase Aβ in production may be amyloid in the brain. Tau protein is a highly soluble protein related to microtubule structure and function, and its increase is also related to AD pathology. Microtubules are involved in neuronal growth and axon transport. In AD, hyper-phosphorylated tau protein was observed, resulting in abnormal aggregation [Querfurth & LaFerla, 2010; Ankner & Lu, 2009].

Another important feature of AD is neurofibrillary tangles (NFT), and the main component of NFT is tau hyper-phosphorylation. Tau protein hyper-phosphorylation is considered to be an important factor in the pathogenesis of AD, which can produce cytotoxicity and mediate neuronal apoptosis. Tau protein lost its function of stabilizing microtubules after hyper-phosphorylation and formed NFT deposits, which affected axon transport mediated by microtubules and even led to axonal degeneration at the distal end [Pîrşcoveanu *et al.*, 2017].

Increased oxidative stress and neuronal inflammation are also associated with neuronal dysfunction and neurodegeneration. Oxidative stress is caused by the imbalance between reactive oxygen species (ROS), reactive nitrogen species (RNS), and antioxidants. The oxidative stress will increase with age in the brain. Mitochondria play an important role in maintaining the balance between ROS and antioxidants. In addition, metal imbalance also play an important role in AD brain. In addition, many studies have shown that tea polyphenols are found in tea and tea can prevent and treat AD. This chapter will discuss the research in this area in depth.

6.3 Oxidative stress and AD

Oxidative stress plays a significant role in the pathogenesis of AD. The brain is more vulnerable than other organs to oxidative stress, and most of the components of neurons (lipids, proteins, and nucleic acids) can be oxidized in AD due to mitochondrial dysfunction, increased metal levels, inflammation, and β-amyloid (Aβ) peptides. Oxidative stress participates in the development of AD by promoting Aβ deposition, tau hyper-phosphorylation, and the subsequent loss of synapses and neurons. The relationship between oxidative stress and AD suggests that oxidative stress is an essential part of the pathological process, and antioxidants may be useful for AD treatment. Many studies have shown that oxidative stress plays an important role in Aβ-induced cytotoxicity [Smith *et al.*, 2000]. Aβ induces nerve cells to increase the production of ROS, causes the degradation of mitochondrial function, reduces mitochondrial membrane voltage, and activates caspases, and finally leads to nerve cell apoptosis [Smith *et al.*, 2000]. Oxidative stress promotes the production of pre-amyloid protein Aβ. The expression of the catalytic subunit presenilin protein 1 (PS1) of catabolic enzyme (BACE1) and gama secretase, accelerates the generation of Aβ from its precursor APP and forms a vicious circle [Oda *et al.*, 2009, Ansari & Scheff, 2010]. Studies have shown that gene defects in the antioxidant system increase Aβ deposition in the animal brain, while antioxidant intake reduces Aβ deposition and improves the cognitive status of animals [Dumont *et al.*, 2009].

Undoubtedly, extra free radical results in injury of the biological body. Free radicals peroxidize membrane lipids [Butterfield & Kanski, 2001] and oxidize proteins [Stadtman, 1990], resulting in damage of the plasma membrane and crosslinking of cytoeskeletal proteins. In addition, free radicals damage RNA [Nurk *et al.*, 2009], nuclear DNA [Gabbita *et al.*, 1998], and DNA [Mecocci *et al.*, 1994]. In the brain, the high metabolic rate, the low concentration of glutathione and antioxidant enzyme catalase, and the high proportion of polyunsaturated fatty acids, make the brain tissue particularly susceptible to oxidative damage [Smith *et al.*, 1998]. Oxidative stress, an imbalance toward the pro-oxidant side of the pro-oxidant/antioxidant homeostasis and protein aggregation occurs in several brain neurodegenerative disorders.

6.3.1 *Reactive oxygen and reactive nitrogen free radicals cause oxidative stress in the brain of AD*

Oxidative stress is a state caused by the imbalance between oxidant production and endogenous antioxidant defense system when oxidant production exceeds the scavenging capacity of antioxidant defense system in the brain of AD. During oxidative stress, reactive oxygen species (ROS) and reactive nitrogen species (RNS) react with proteins, nucleic acids, and lipids in cells, damaging their functions, causing progressive neuron cell damage, and finally leading to brain cell death [Stadtman, 1990]. Vascular dysfunction and decreased cerebral blood flow are linked to Alzheimer's disease (AD). Loss of endothelial nitric oxide (NO) and oxidative stress in human cerebrovascular endothelium increase expression of amyloid precursor protein (APP) and enhance production of the Aβ peptide, suggesting that loss of endothelial NO contributes to AD pathology. Long-term activation of the stress response results in neuronal oxidative stress via reactive oxygen and nitrogen species generation, contributing to the development of depression. Stress-induced depression shares a high comorbidity with other neurological conditions including Alzheimer's disease (AD) and dementia, often appearing as one of the earliest observable symptoms in these diseases. Furthermore, stress and/or depression appear to exacerbate cognitive impairment in the context of AD associated with dysfunctional catecholaminergic signaling. Given there are a number of homologous pathways involved in the pathophysiology of depression and AD, stress-induced perturbations in oxidative stress, and particularly NO signaling, contribute to neurodegeneration. So, dysregulation of stress systems and nitric oxide signaling underlies neuronal dysfunction in AD [Spiers *et al.*, 2019, Roberts *et al.*, 2016]. The brain is particularly vulnerable to oxidative stress because of its high metabolic rate, high content of polyunsaturated fatty acids and transition metals, and relatively low levels of antioxidants. More and more evidence show that in addition to the established pathology of senile plaques, the presence of extensive oxidative stress is also a feature of AD brain [Pratico, 2008a, 2008b]. The increase of oxidative stress and free radical production in AD is based on the following facts: Protein oxidation, manifested as the increase of protein carbonyl and 3-nitrotyrosine levels, as well as markers of DNA and RNA oxidative damage, for example 8-hydroxydeoxyguanosine (8OHdG) and 8-hydroxyguanosine, are very prominent in AD brain [Beal, 2002; Butterfield & Kanski, 2001]. Lipid peroxidation products such as malondialdehyde (MDA), 4-hydroxynona-nal, and F2 iso-prostaglandin are also increased in multiple brain regions and cerebrospinal fluid (CSF) of patients with AD or mild cognitive impair-ment (MCI) [Pratico & Sung, 2004; Williams *et al.*, 2006]. In addition, the accumulation of free radical injury products in AD brain is combined with the changes of antioxidant defense in central nervous system and peripheral tissues [Padurariu *et al.*, 2010]. The expression of superoxide dismutase (SOD) is abnormally increased in neuropathy in the brain of AD patients, which may be an adaptive response to the increased oxidative damage in these regions. In contrast, most studies have shown that the activity of anti-oxidant enzymes (including SOD and catalase) in the brain of AD patients is reduced [Padurariu *et al.*, 2010]. The difference between the expression and activity of antioxidant enzymes may reflect the redistribution of anti-oxidant enzymes in neuropathy or enzyme inactivation due to oxidation [Omar *et al.*, 1999]. In addition, in AD brains, the increase of lipid and protein oxidative damage and the decrease of glutathione and antioxidant enzyme activities are more limited to synapses and related to the severity of the disease, indicating that oxidative stress is involved in AD related synaptic loss [Ansari & Scheff, 2010]. Many of the above studies have shown that oxidative stress in AD is elevated, which is considered to be an intermediate state between normal aging and dementia, suggesting that oxidative stress injury in AD may occur before the onset of the disease. These evidence show that oxidative stress is the earliest change in the pathogenesis of AD. Oxidative stress has become one of the important fac-tors in the pathogenesis of AD, the mechanism of the change of redox bal-ance, and the source of free radicals may cause by mitochondrial. Mitochondrial dysfunction β-amyloid mediated processes, transition metal accumulation, and microglia activation are considered to play an important role in redox imbalance [Smith *et al.*, 2000].

Accumulated data demonstrated that oxidative damage occurs in the AD brain. Amyloid-β (Aβ) peptide has been proven to produce hydrogen peroxide (H_2O_2) through metal ion reduction, with concomitant release of thiobarbituric acid-reactive substances (TBARS), a process probably mediated by the formation of hydroxyl radicals [Huang *et al.*, 1999]. The cytotoxicity of Aβ fibrils is also implicated as an oxidative mechanism. Aβ-fibrils-induced H_2O_2 was detected by several laboratories [Markesbery & Carney, 1999]. There is considerable evidence consistent with the importance of oxidative stress in the pathology of AD [Varadarajan, 2000]. Evidence supporting the notion of free-radical oxidative stress in the AD brain includes increased redox-active metal ions in the AD brain, increased lipid peroxidation detected by decreased levels of polyunsaturated fatty acids, and increased levels of the lipid peroxidation products, acrolein, TBARS, isoprostanes, and neuroprostanes, increased protein oxidation, increased oxidation of DNA and RNA, and decreased activity of oxidative prone enzymes, such as glutamine synthetase (GS).

Oxidation of proteins is normally caused by free radicals, and this process, from a chemical thermodynamics standpoint, is an exothermic event. Oxidative reactions of peptides are mediated mainly by the hydroxyl radical (•OH). There are two possible oxidative pathways that can occur: backbone oxidation and side-chain oxidation. Backbone oxidation is initiated by carbon abstraction of hydrogen by the free radical, leading to the formation of a carbon-centered radical. In the presence of oxygen, this radical is converted to a peroxyl radical. This can lead to the formation of an alkoxyl radical and subsequent hydroxylation of the peptide backbone. The oxidation of amino acid side chains greatly depends on their structure. An important oxidative process with profound functional and structural consequences involves the irreversible nitration of tyrosine residues by peroxynitrite ($ONOO^-$) [Beckman, 1996]. The levels of protein oxidation in membrane systems can be indirectly monitored by the use of electron paramagnetic resonance (EPR) spin-labeling techniques [Xin *et al.*, 1984]. The changes in protein conformation resulting from protein oxidation may be related with the mechanism of AD.

Understanding these mechanisms related with ROS and RNS free radicals may provide new therapeutic targets for the prevention and intervention of AD.

6.3.2 *Metal ion metabolic homeostasis and oxidative stress in AD patients*

Many studies have shown that there is a close relationship between the destruction of metal stasis and AD. One mechanism of Aβ accumulation may be due to the disorder of metal homeostasis in AD brain [Strausak *et al.*, 2001]. Aβ peptide is the main component of senile plaque amyloid core, derived from APP and secreted into extracellular space. High concentrations of copper, zinc, and iron were found in amyloid deposits in AD brain. When Aβ is under mild acidic conditions, Cu^{2+} can rapidly precipitate peptide and zinc can rapidly precipitate peptide at low physiological (submicromolar) concentrations [Bush *et al.*, 1994]. An age-related binding β peptide containing excess brain metals (copper, iron, and zinc) can induce the precipitation of β peptide into metal rich plaque [Bush, 2002]. We have studied the steady state failure of iron and copper, oxidative stress β-amyloid (Aβ), amyloid precursor protein (APP), iron regulatory protein (IRP), and divalent metal transporter 1 (DMT1). It is found that iron and copper overload may be closely related to oxidative stress injury in the later stage of AD, and iron and copper deficiency may be closely related to the early onset of AD. Natural antioxidants can protect AD by regulating iron and copper homeostasis. The destruction of metal homeostasis directly causes oxidative stress [Zhao & Wan, 2012].

It was observed that Aβ increased the levels of iron content and oxidative stress in SH-SY5Y cells, overexpressing the Swedish mutant form of human β-amyloid precursor protein (APPsw) and in *Caenorhabditis elegans* Aβ-expressing strain CL2006. Intracellular iron and calcium levels and ROS and nitric oxide generation significantly increased in APPsw cells compared to control cells. The activity of superoxide dismutase (SOD) and the antioxidant levels of APPsw cells were significantly lower than those of control cells. Moreover, iron treatment decreased cell viability and mitochondrial membrane potential and aggravated oxidative stress damage as well as the release of Aβ1–40 from the APPsw cells. The iron homeostasis disruption in APPsw cells is probably associated with elevated expression of the iron transporter divalent metal transporter 1 but not transferrin receptor. Furthermore, the *C. elegans* with Aβ-expression had increased iron accumulation [Wan *et al.*, 2011]. We also found that the overexpression of wild-type human AβPP695 decreased the iron content and increased the oxidative stress in neuroblastoma SH-SY5Y cells. The catalase activity of stably transfected cells overexpressing wild-type AβPP695 (AβPP cells) was significantly lower than that of the control cells. Intracellular ROS generation and calcium levels significantly increased in AβPP695 cells compared to control cells. The mitochondrial membrane potential of AβPP695 cells was significantly lower than that of the control cells. Moreover, iron treatment decreased ROS and calcium levels and increased cell viability of AβPP695 cells. The iron deficiency in AβPP695 cells may contribute to the pathogenesis of AD [Wan *et al.*, 2011]. In aggregate, these results demonstrate that Aβ accumulation in neuronal cells correlated with neuronal iron homeostasis disruption and probably contributed to the pathogenesis of AD. Reactive β metal will lead to ROS, and ROS plays a key role in the life cycle [Multhaup *et al.*, 2002]. Cu^{2+} acts on β peptides as a cofactor to promote the process of oxidative stress. Double electrons can be transferred to oxygen (O_2), resulting in H_2O_2 β-bound copper and reduction of Cu^{2+} to Cu^+ [Opazo *et al.*, 2002]. Then, Cu^+ reacts with H_2O_2 to generate hydroxyl radicals (Fenton type reaction) [Lynch *et al.*, 2000]. These ROS directly induce oxidative damage in AD brain.

We studied the effects of antioxidant nicotine on metal homeostasis in hippocampus and cortex of APPV717I (APP London mutant) transgenic mice. After nicotine treatment, the metal contents of copper and zinc in senile plaques and nerve fibers decreased significantly. After nicotine treatment, the distribution density of copper and zinc in hippocampal CA1 region was also reduced. We used overexpression of Swedish mutant human APP (APPsw) SH-SY5Y cells to further study the mechanism of nicotine mediated metal homeostasis effects. Nicotine treatment reduces the concentration of intracellular copper and weakens the absorption of intracellular copper β, which promotes mediated neurotoxicity through the addition of copper, and these effects are independent of the activation of nicotinic acetylcholine receptor. In addition, the levels of copper and zinc in senile plaques and nerve endings of nicotine treated mice were significantly reduced by about 10–20% compared with the control group treated with sucrose. At the same time, the distribution of copper and zinc were observed in hippocampal CA1 region. Copper is enriched in pyramidal neuron layer, while zinc is relatively dispersed. After nicotine treatment, the metal content of copper or zinc decreased significantly, especially in the pyramidal neuron layer. ROS are mainly produced by transition metals (especially copper) catalysis; oxidative stress plays a key role in the pathogenesis of AD. In this study, metal level and Aβ toxicity are manifested in three aspects: 1) the effect of metal alone or in combination with APP and Aβ on cell viability 2) Aβ-induced neurotoxicity related to oxidative stress, indicated by ROS, and 3) when APPsw cells express APP and produce Aβ. So, compared to APPsw cells expressing only APP, Cu^{2+} and APP Cu^{2+} can catalyze more ROS production [Zhang *et al.*, 2006a]. Nicotine decreases homeostasis β-amyloidosis by regulating metal. These data suggest that nicotine reduces β-amyloidosis, and is partly mediated by regulating the balance of metal internal environment [Liu *et al.*, 2007].

We found that two subtypes of divalent metal transporter 1 (DMT1), DMT1-ire and DMT1 –non-ire, were associated AD with Aβ in postmortem cerebral plaques co-location. Using APP/ PS1 transgenic mouse model, we found that the levels of DMT1-IRE and DMT1 non-IRE in cortex and hippocampus were significantly higher than those in wild-type control group (Fig. 6-1, 6-2). DMT1-IRE and DMT1 non-IRE have cell type specificity and subcellular distribution specificity. We used SH-SY5Y cell line stably overexpressing human APP Swedish mutation (APPsw) as a cell model to further verify that DMT1 may be involved in APP processing and Aβ-proposed mechanism of secretion. We found that overexpression of APPsw resulted in increased expression levels of DMT1-ire and DMT1 non-IRE in SH-SY5Y cells. Interestingly, endogenous DMT1 is silenced by RNA interference, thereby reducing divalent ion influx, resulting in APP expression and Aβ decreased secretion. These findings suggest that DMT1 plays a key role in ion mediated neuropathy of AD [Zheng *et al.*, 2009]. We investigated the interaction and toxicity of Aβ1–42 and copper in the Aβ1–42 transgenic *Caenorhabditis elegans* worm model CL2006. Our data show that the paralysis behavior of CL2006 worms significantly deteriorated after exposure to 10^{-3} mol/L copper ions. However, the paralysis behavior was dramatically attenuated with exposure to 10^{-4} mol/L copper ions. The exogenous copper treatment also

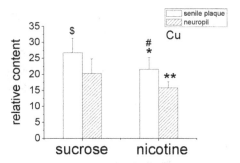

Figure 6-1. The copper level in the senile plaques and their distribution in the hippocampus. The relative levels of copper in senile plaque and neuropil of the mice from the sucrose and nicotine treated groups, respectively. The result is expressed in mean ± SEM, n = 4. Statistical analysis was done by ANOVA. *, $p < 0.05$ comparing plaque (sucrose) *vs* neuropil (sucrose).

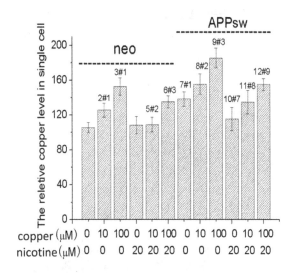

Figure 6-2. The copper level in the single in the SH-SY5Y cells detected by synchrotron radiation X-ray fluorescence analysis. The SH-SY5Y cells were washed with 0.1M Tris-Cl and put on the maylar film. The cross section of the beam focused on a single cell and collected the fluorescence signal. The result is expressed in mean ± SEM, n = 10. Statistical analysis was done by ANOVA. #, $p < 0.05$ compared with nicotine (0).

partially changed the homeostatic balance of zinc, manganese, and iron. Our data suggest that the accumulation of ROS was responsible for the paralysis induced by Aβ and copper in CL2006. The ROS generation induced by Aβ and copper appear to be through *sod-1*, *prdx-2*, *skn-1*, *hsp-60*, and *hsp-16.2* genes [Luo *et al.*, 2011]. Our data suggest that the accumulation of ROS was responsible for the paralysis induced by Aβ and copper in CL2006. Drug blockade of DMT1 may provide a new therapeutic strategy for AD.

The relative ROS generation in neo and APPsw cells after copper and antioxidants fermented papaya preparation (FPP) treatment were measured by the fluorescence method and the results are showed in Figure 6-3. Without copper treatment, the ROS level in neo and APPsw cells did not show any significant difference. The 100 μM copper increased ROS generation by about 110% in APPsw cells. This increasing in the DCF fluorescence intensity was eliminated by about 63% when the APPsw cells were post-treated with FPP. The 100 μM copper increased ROS production only by about 23% in neo cell and FPP administration lowered the

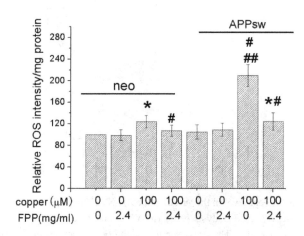

| copper (μM) | 0 | 0 | 100 | 100 | 0 | 0 | 100 | 100 |
| FPP(mg/ml) | 0 | 2.4 | 0 | 2.4 | 0 | 2.4 | 0 | 2.4 |

Figure 6-3. ROS generation in the neo and APPsw cells after copper and antioxidant FPP treatment was measured using DCF-DA. *, $p < 0.05$ comparing neo copper(0) *vs* FPP(0); #, $p < 0.05$ comparing neo copper(100) *vs* FPP(0); ##, $p < 0.05$ comparing APPsw copper(0) *vs* FPP(0); *#, $p < 0.05$ comparing APPsw copper(100) *vs* FPP(0). The result is expressed in mean ± SEM, n = 4. Statistical analysis was done by ANOVA.

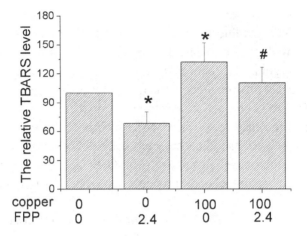

Figure 6-4. The relative content of TBARS in the cells treated by FPP and copper. The result is expressed in mean ± SEM, n = 4. Statistical analysis was done by ANOVA. *, $p < 0.05$ comparing copper(0) *vs* FPP(0); #, $p < 0.05$ comparing copper(100) *vs* FPP(0).

ROS level significantly. The TBARS concentration, an end product of lipid peroxidation in the APPsw cells, is showed in Figure 6-4. FPP treatment alone decreased the lipid peroxide level by about 30% compared with control group. Administration of 100 μM copper significantly increased the TBARS level about 31%. After FPP administration, the increased TBARS level was reversed [Zhang *et al.*, 2018].

Understanding these mechanisms related with toxicity of metals to brain nerves may provide new therapeutic targets for the prevention and intervention of AD.

6.3.3 *Mitochondrial dysfunction and oxidative stress in AD*

Oxidative modifications of cellular components can disrupt the integrity of membranes and alter the function of essential proteins, leading to the disruption of ion homeostasis, mitochondria dysfunction, and eventually the activation of apoptotic pathway and neuronal cell death [Mattson, 1997]. Mitochondrial respiratory chain is a major site of ROS production in the cell [Caspersen *et al.*, 2005]. In fact, in isolated mitochondria, Aβ could cause oxidative injury to mitochondrial membrane, disrupted lipid polarity and protein mobility, and inhibited key enzymes of mitochondria respiratory chain, leading to increased mitochondrial membrane permeability and cytochrome c release [Casley *et al.*, 2002]. Consistently, the data from transgenic mice showed that the presence of Aβ in mitochondria was associated with impaired mitochondrial metabolism and increased mitochondrial ROS production. Furthermore, oxidative stress mediated by Aβ accumulation may result in modification and damages of important cellular components including enzymes critical for antioxidant defense. Manganese superoxide dismutase (MnSOD), a primary antioxidant enzyme protecting mitochondria against superoxide, was found to be a target of nitration and inactivation in a double homozygous knock-in mouse model expressing APP and PS-1 mutants. The decreased activity of antioxidant defense enzymes such as MnSOD may further increase ROS levels and compromise mitochondria function, contributing to the loss of mitochondrial membrane potential and eventually caspase activation and apoptosis [Anantharaman *et al.*, 2006].

Aβ was also shown to alter other cellular protective mechanisms against oxidative stress. Uncoupling proteins (UCPs) are a family of mitochondrial anion carrier proteins that are located on the inner mitochondrial membrane with a diverse physiological functions [Rousset *et al.*, 2004]. It has been demonstrated that UCP-2 and UCP-3 can be activated by ROS or products of lipid peroxidation to diminish proton motive force, reduce mitochondrial membrane potential and ATP production, cause mitochondria uncoupling, and decrease of ROS generation from mitochondria [Echtay, 2007]. Therefore, the expression and activation of UCPs are considered to be a protective mechanism in response to oxidative stress. This protective mechanism appears dysfunctional in AD brains where the expression of UCP-2, UCP-4, and UCP-5 is significantly decreased [de la Monte & Wands, 2006]. In SH-SY5Y neuroblastoma cells overexpressing APP or APP mutant, it was found that the upregulation of UCP2 and UCP4 protein levels in response to the exposure of superoxide was abrogated; the mechanisms may be that Aβ accumulation may lead to irreversible cellular alterations that render the cell more susceptible to oxidative stress. Moreover, the UCP2-dependent and UCP4-dependent upregulation of mitochondrial free calcium in response to superoxide treatment was found to be diminished in cells overexpressing

APP or APP mutant, indicating that the Abeta accumulation may be associated with a dysfunction of mitochondria as the reserve pool of intracellular calcium that leads to an increased cell sensitivity to the loss of calcium homeostasis [Wu *et al.*, 2009].

We found that antioxidant nicotine inhibited N-methyl-4-phenylpyridine (MPP^+) and calcium-induced mitochondria high amplitude swelling and cytochrome c release from intact mitochondria. Intra-mitochondria redox state was also maintained by nicotine, which could be attributed to an attenuation of mitochondria permeability transition. Further investigation revealed that nicotine did not prevent MPP^+ or calcium-induced mitochondria membrane potential loss, but instead decreased the electron leak at the site of respiratory chain complex I. In the presence of mecamylamine hydrochloride, a non-selective nicotinic acetylcholine receptor inhibitor, nicotine significantly postponed mitochondria swelling and cytochrome c release induced by a mixture of neurotoxins (MPP^+ and 6-hydroxydopamine) in SH-SY5Y cells, suggesting that there is a receptor-independent nicotine-mediated neuroprotective effect of nicotine [Xie *et al.*, 2005].

Taken together, Aβ toxicity, oxidative stress, and mitochondria dysfunction appear to be interlinked in the pathogenesis of AD. These results show that interaction of oxidative stress generated from mitochondrial respiratory chain together with antioxidant effects should be considered in the neuroprotective diseases.

6.3.4 *Aβ toxicity and oxidative stress induce AD*

A great deal of research has suggest the presence of oxidative stress in Aβ-induced neurotoxicity. *In vitro* experiments using cell models showed that Aβ treatment could increase the levels of hydrogen peroxide and lipid peroxides [Behl *et al.*, 1994]. Consistently, in various AD transgenic mouse models carrying mutants of APP and PS-1, increased hydrogen peroxide and nitric oxide production as well as elevated oxidative modifications of proteins and lipids correlated the age-associated Aβ accumulation, further confirming that Aβ promotes oxidative stress [Manczak *et al.*, 2006]. In hippocampal neuronal cell cultures, the induction of ROS by soluble Aβ oligomers required activation of N-methyl-d-aspartate

(NMDA) receptor and was associated with a rapid increase in neuronal calcium levels, suggesting the possible role of soluble Aβ oligomers as proximal neurotoxins and the involvement of oxidative stress in the synaptic impairment and neuronal loss induced by soluble Aβ oligomers [De Felice *et al.*, 2007]. In addition to mediating Aβ-induced cytotoxity, numerous studies have indicated oxidative stress in the increased Aβ production in AD. It was demonstrated that defects in antioxidant defense system caused elevated oxidative stress and significantly increased Aβ deposition in transgenic mice overexpressing APP mutant [Nishida *et al.*, 2006], while dietary antioxidants such as curcumin lowered the elevation of oxidized proteins and decreased brain Aβ levels and Aβ plaque burden [Lim *et al.*, 2001]. In line with these findings, it was recently reported that overexpression of MnSOD in transgenic mice overexpressing APP mutant decreased protein oxidation and increased antioxidant defense capability in brains while reducing Aβ plaque burden and restoring the memory deficit [Dumont *et al.*, 2009]. Meanwhile, Aβ was found to be localized to mitochondria in AD patients as well as transgenic mice and neuroblastoma cells stably expressing human mutant APP [Caspersen *et al.*, 2005]. Moreover, the increased Aβ deposition and its associated earlier onset and more severe cognitive dysfunction induced by the defect in antioxidant defense system could be ameliorated by antioxidant supplementation [Nishida *et al.*, 2006]. Studies on how oxidative stress enhances Aβ production have pointed to a direct relationship between oxidative stress and activation of BACE and gama-secretase, enzymes critical for generation of Aβ from APP [Oda *et al.*, 2009]. Moreover, it was found that the induction of BACE1 and PS1 expression and the activation of γ-secretase by oxidative stress was dependent on the activation of *c-jun* N-terminal kinase (JNK)/*c-jun* pathway, a major cell signaling cascade that is stimulated by oxidative stress [Shen *et al.*, 2008]. In fact, the promoter and 5' untranslated region of BACE gene contain binding sites for multiple transcription factors, including the redox-sensitive activator protein (AP)1 and nuclear factor (NF)-kappa B, activation of which by oxidative stress may in turn enhances BACE expression [Sambamurti *et al.*, 2004]. In AD brains, both the activation of JNK signaling cascade [Lagalwar *et al.*, 2006] and the elevation of BACE1 and PS1 expression/activity have been detected [Matsui *et al.*, 2007], thus it is possible that the increased oxidative stress in AD brain may initiate the activation of a cascade of redox-sensitive cell signal pathways including JNK, which promotes the expression of BACE1 and PS1, eventually enhancing the production of Aβ and deterioration of cognitive function. As JNK has also been implicated in Aβ-induced neuronal apoptosis, pharmacological inhibition of the redox-sensitive signaling pathways such as JNK may reduce Aβ accumulation as well as inhibit neuronal apoptosis [Yao *et al.*, 2005].

These evidence suggest that the enhancement of Aβ production/ plaque formation by oxidative stress is important for the initiation and development of AD. It is possible that the increased oxidative stress in AD brain may initiate the activation of a cascade of redox-sensitive cell signal pathways including JNK, which promotes the expression of BACE1 and PS1, eventually enhancing the production of Aβ and deterioration of cognitive function.

6.3.5 *Tao protein, neurofibrillary tangles, and oxidative stress*

AD is most commonly characterized by neurofibrillary tangles (NFTs) composed of Tau protein. Although pathological hallmarks of AD are senile plaques, neurofibrillary tangles, and neuronal degeneration, which are associated with increased oxidative stress, synaptic loss is an early event in the pathogenesis of AD. The involvement of major kinases such as mitogen-activated protein kinase (MAPK), extracellular receptor kinase (ERK), calmodulin-dependent protein kinase (CaMKII), glycogen synthase-3β (GSK-3β), cAMP response element-binding protein (CREB), and calcineurin is dynamically associated with oxidative stress-mediated abnormal hyper-phosphorylation of tau and suggests that alteration of these kinases could exclusively be involved in the pathogenesis of AD. N-methyl-D-aspartate (NMDA) receptor (NMDAR) activation and Aβ toxicity alter the synapse function, which is also associated with protein phosphatase (PP) inhibition and tau hyper-phosphorylation (two main events of AD). The oxidative stress is involved in synapse dysfunction. Oxidative stress and free radical generation in the brain, along with excitotoxicity, leads to neuronal cell death. It is inferred from several studies that excitotoxicity, free radical generation, and altered synaptic function encouraged by oxidative stress are associated with AD pathology. NMDARs maintain neuronal excitability, Ca^{2+} influx, and memory formation through mechanisms of synaptic plasticity. The mechanism of the synapse redox stress is associated with NMDARs altered expression [Karmat *et al.*, 2016]. Oxidative stress mediated through NMDAR and their interaction with other molecules might be a driving force for tau hyper-phosphorylation and synapse dysfunction. Thus, understanding the oxidative stress mechanism and degenerating synapses is crucial for the development of therapeutic strategies designed to prevent AD pathogenesis.

Antioxidant α-Lipoic acid (LA) has been found to stabilize the cognitive function of AD patients, and animal study findings have confirmed its anti-amyloidogenic properties. It was found that LA supplementation effectively inhibited the hyper-phosphorylation of tau at several AD-related sites, accompanied by reduced cognitive decline in P301S tau transgenic mice. Furthermore, it found that LA not only inhibited the activity of calpain1, which has been associated with Tao pathway development and neurodegeneration via modulating the activity of several kinases, but also significantly decreased the calcium content of brain tissue in LA-treated mice. Various modes of neural cell death was screened in the brain tissue of LA-treated mice. It found that caspase-dependent apoptosis was potently inhibited, whereas autophagy did not show significant changes after LA supplementation. Interestingly, tau-induced iron overload, lipid peroxidation, and inflammation, which are involved in ferroptosis, were significantly blocked by LA administration [Zhang *et al.*, 2018].

These results provide compelling evidence that LA plays a role in inhibiting tau hyper-phosphorylation and neuronal loss, including ferroptosis, through several pathways such as oxidative stress etc.

AD and other neurodegenerative diseases are characterized by abnormal and prominent protein aggregation in the brain, partially due to deficiency in protein clearance. It was found that aged tree shrews presented an increased number of activated microglia containing ferritin, but microglia labeled with Iba1 with a dystrophic phenotype was more abundant in aged individuals. With aging, oxidative damage to RNA (8OHG) increased significantly in all hippocampal regions, while tau hyper-phosphorylation (AT100) was enhanced in DG, CA3, and SUB in aged animals. Phagocytic inclusions of 8OHG and AT100-damaged cells were observed in activated M2 microglia in old and aged animals [Rodriguez-Callejas *et al.*, 2020].

Above evidence have demonstrated that the accumulation of Abeta and neurofibrillary tangles (NFTs) composed of tau protein promote oxidative stress, which mediates Abeta and neurofibrillary tangles (NFTs) composed of tau protein-induced neurotoxicity and further enhances Abeta and neurofibrillary tangles (NFTs) composed of tau protein production, forming a vicious cycle in AD pathogenesis. Regardless a primary or secondary event, oxidative stress is an important factor contributing to the development of AD, thus, removal of ROS or prevention of their formation may inhibit the progression of AD.

6.4 Preventive and therapeutic effects of tea polyphenols on AD

Cumulative evidence suggests that tea drinking is associated with a lower risk of AD. Many human epidemiological and animal studies have shown that green tea polyphenols may promote health, reduce disease occurrence, and may prevent AD and other neurodegenerative diseases. The incidence rate of AD in Asia is 5–10 times lower than that in western countries, which may be related to green tea consumption in Asian countries. Oxidative stress of ROS and inflammation plays a key role in neurodegenerative diseases, supporting the clinical use of free radical scavenger transition metal green tea and tea polyphenols, as iron and copper chelating agents and non-vitamin natural antioxidants are considered to be a preventive neuroprotective agent for brain aging and are potential neuroprotective agents in neurodegenerative diseases such as AD. Tens of thousands of people have found that drinking three cups of green tea a day can reduce the incidence rate of AD. Green tea can prevent hippocampal neuron apoptosis and improve cognitive function by inhibiting JNK / MLCK pathway [Fernando *et al.*, 2017].

With the increase of aging population, the incidence rate of AD has increased significantly worldwide. Unless effective preventive and/or therapeutic strategies are formulated, the incidence rate of will continue to rise by an increase of over 100%. Many drugs have been approved for the treatment of AD, however, they have little benefit and cause a variety of side effects [Eskelinen *et al.*, 2009]. Therefore, there is an urgent need for new AD prevention and treatment strategies with better curative effect and fewer side effects. Oxidative stress is one of the earliest changes in the pathogenesis of AD, which is related to Aβ-induced synaptic dysfunction and neuronal loss. The accumulation of Aβ stimulates oxidative stress, which leads to the enhancement of the production of Aβ, thus forming a vicious circle and promoting the occurrence and progress of AD, which indicates that the prevention of oxidative stress may delay the onset of AD and slow down the disease progress of AD patients.

Many studies have shown that tea polyphenols can reduce the neuro-toxicity induced by Aβ, such as oxidative stress, mitochondrial dysfunc-tion, and neuronal apoptosis. Since the pathogenesis of AD involves a variety of molecular events, the neuroprotective effect of natural antioxi-dants also seems to involve a variety of mechanisms. In addition to reduc-ing oxidative stress, preventing apoptosis, and promoting neurogenesis, natural antioxidants can also inhibit the accumulation of Aβ, restore cal-cium homeostasis and reduce transition metal overload in the brain. The anti-inflammatory properties of many antioxidants are also related to their neuroprotective mechanisms. Most evidence suggests that supplementing natural antioxidants or eating foods rich in natural antioxidants may be beneficial to the elderly at risk of AD and AD patients with minimal side effects. Foods such as fresh fruits, vegetables, and tea are associated with reducing the risk of high serum cholesterol and glucose intolerance. It has been proven that high serum cholesterol and glucose intolerance are asso-ciated with an increased risk of AD and dementia [Ide *et al.*, 2018].

"Modern" medicine and pharmacology require an effective medical drug with a single compound for a specific disease. This seems very sci-entific but usually has unavoidable side effects. For example, the chemical therapy to patients can totally damage the immunological ability leading to early death rather than non-treatment. In contrast, natural antioxidant drugs not only can cure the disease but also can enhance the immunologi-cal ability and health for the patient, although they usually have several compounds or a mixture. For the degenerative disease such as AD, natural antioxidant drugs are suitable drugs because the pathogenesis of these diseases is complex with many targets and pathways. These effects are more evidence when the clinic trial is for long-term treatment. The author's

studies on the preventing effects of natural antioxidants, such as tea polyphenols and flavonoids on neurodegenerative diseases, can be found in
these references [Zhao, 2005, 2009; Zhao & Zhao, 2012].

6.4.1 *Prevention and treatment using tea polyphenols for AD in epidemiological investigation*

Taking green tea as the research object, the relationship between green tea
intake and dementia, AD, mild cognitive impairment, or cognitive impairment was systematically observed and studied. Articles registered as of 23
August 2018 in the PubMed database on tea and cognition were searched.
Finally, one cohort study and three cross-sectional studies supported the
positive effects of green tea intake. A cohort study and a cross-sectional
study reported some positive effects. These results seem to support the
hypothesis that green tea intake may reduce the risk of dementia, AD, mild
cognitive impairment, or cognitive impairment [Kakutani *et al.*, 2019]. In
Japan, authors conducted a cohort study in the elderly and conducted an
epidemiological analysis of the correlation between the frequency of drinking green tea and cognitive function assessed by mini mental state examination. Among 1,003 Japanese aged > 70, those who drank more green tea
had a significantly reduced prevalence of cognitive impairment [Kuriyama
et al., 2006]. It has been reported that the same trend has been observed in
the drinking of tea by the elderly population in China for more than 60
years [Gu *et al.*, 2017]. In the past 55 years or more, the consumption of
green tea has been negatively correlated with the prevalence of cognitive
impairment [Ng *et al.*, 2008]. A largest longitudinal study was conducted
in 2012. The study included more than 7,000 Chinese elderly over the age
of 80, including a seven-year follow-up. The cognitive function of tea
drinkers was higher than that of non-tea drinkers at all time points [Feng
et al., 2012]. Tea consumption was also associated with better cognitive
performance in community-living Chinese older adults in Singapore. The
protective effect of tea consumption on cognitive function was not limited
to a particular type of tea [Feng *et al.*, 2010]. Other studies in Norway and
Singapore have also demonstrated a similar association [Nurk *et al.*, 2009].
However, one study showed that there was no significant correlation
between green tea consumption and cognitive impairment [Shen *et al.*,

2015], and another study showed that there were gender differences in the scope of influence. In the latter study, compared with the control group, a significant association was observed only in men who drank green tea before and after, but not in women [Huang *et al.*, 2009]. A study conducted in the United States showed gender differences. A total of 1,438 participants assessed cognitive function 1–2 years (median 16 months) after baseline assessment. Even after adjusting for confounding variables, higher levels of tea consumption were significantly associated with lower prevalence of cognitive decline [Arab *et al.*, 2011]. Although the observations are still somewhat uncertain, Liu *et al.* recently published the results of a meta-analysis. A total of 48,435 people were included. It was found that tea drinking was significantly negatively correlated with the risk of cognitive impairment. When layering by tea type, only green tea was found to be associated with this, because green tea has high EGCG content. The authors also evaluated the dose-dependent effect of green tea consumption and observed a linear relationship of 100–500 ml/day [Liu *et al.*, 2017]. It is also worth noting that a meta-analysis shows that tea consumption is dose-dependent.

Above epidemiological investigation suggests the prevention and treatment effects of tea polyphenols on AD. There are some different results, which may be caused by the different consumed amount of tea. More studies are needed to confirm these results.

6.4.2 *Molecular mechanism of tea polyphenols in prevention and treatment AD*

There are many reports about the mechanism that drinking tea can prevent AD, most of which focus on the tea polyphenol antioxidant effect, scavenging free radicals, iron chelating characteristics, inhibition of inflammation, regulation of cell survival/death genes and induction of neuronal activity by mitochondrial function, Aβ deposits, and signal transduction pathway. There are not only animal experiments, but also molecular experiments. Many studies on the molecular mechanism of the effect of tea polyphenols on AD have been carried out *in vivo* and *in vitro*, as well as in silicon wafer [Ali *et al.*, 2016].

1. Tea polyphenol antioxidant effect, scavenging free radicals in AD

We studied the pathogenic mechanism of iron in AD and the regulatory effect of tea polyphenol EGCG on ROS free radicals. The results show that EGCG reduced the content of Aβ in APPsw cells, ROS, and intracellular calcium, and improve the mitochondrial membrane potential. Tea polyphenol EGCG may have a preventive and therapeutic effect on AD by scavenging ROS in APPsw cells (Fig. 6-5) [Wan *et al.*, 2011, 2012]. Also, it demonstrated a related role to EGCG in another paper. The author evaluated a similar role of EGCG in streptozotocin-induced dementia in rats. One month after oral administration of EGCG (10 mg/kg/day) the cognitive impairment assessed by Morris water maze was reversed, ROS levels and NO production (based on nitrates in the hippocampus) were significantly reduced [Biasibetti *et al.*, 2013]. Tea polyphenols have scavenging activity on free radical [Sang *et al.*, 2003]. In fact, lipid peroxide, protein, and oxidized DNA are increased in AD patients, and the antioxidant effects of tea polyphenols may help prevent AD. [Praticò *et al.*, 2008b; Kim *et al.*, 2015].

Tea polyphenol EGCG may have a preventive and therapeutic effect on AD by their scavenging ROS and antioxidant effects.

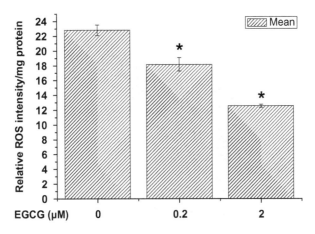

Figure 6-5. Effects of different concentrations of EGCG on ROS content in APPsw cells. *, $p < 0.05$ compared without EGCG. The results are expressed as mean ± standard error, n = 4.

2. The regulatory effect of tea polyphenol EGCG on iron imbalance in AD

The imbalance of iron in the body is closely related to the occurrence of AD. The important thing is that iron overload leads to nerve cell damage and AD related diseases. In addition, if iron deficiency leads to anemia and hypoxia, it is also related to AD. We studied the pathogenic mechanism of iron in AD and the regulatory effect of tea polyphenol EGCG on iron imbalance. The results show that tea polyphenol EGCG can reduce the oxidative damage of AD cells by complexing too much iron, so as to protect AD cells. APPsw of transferred Aβ into SH-SY5Y cells were treated with different concentrations of EGCG for 48 hours and its iron content in the cell iron pool was measured. After treatment with 20 mM EGCG, the iron content in the cell iron pool significantly reduced. Tea polyphenol EGCG may have a preventive and therapeutic effect on AD by chelating excessive iron in the iron pool of APPsw cells (Fig. 6-6) [Wan *et al.*, 2011, 2012]. The chelating properties of EGCG for metal iron may contribute to these antioxidant effects [Seeram *et al.*, 2006]. Metal ions such as copper (II) and iron (III) can be chelated by tea polyphenols, and iron chelation reduces the production of ROS by inhibiting

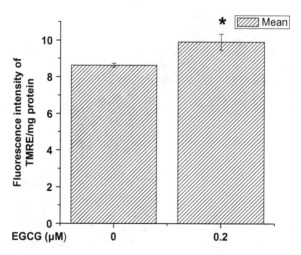

Figure 6-6.   Effect of EGCG on membrane potential of APPsw cells. *, $p < 0.05$ compared without EGCG. The results are expressed as mean ± standard error, n = 4.

the Fenton reaction [Weinreb *et al.*, 2009]. Copper (II) and iron (III) ions also accumulate in the brain of AD patients [Ward *et al.*, 2014].

These studies show that tea polyphenols EGCG may have a preventive and therapeutic effect on AD by chelating excessive iron in the iron pool and reduce oxidative stress in peripheral and brain tissues and may inhibit behavioral changes associated with cognitive impairment.

3. Inhibition effect of tea polyphenols on Aβ accumulation in brain
Studies have also shown that tea polyphenols can prevent amyloid β plaque formation and enhances cognitive function and may therefore be helpful in the treatment of patients with AD or dementia. Recent studies suggest that green tea flavonoids may be used for the prevention and treatment of a variety of neurodegenerative diseases by inhibition effect of tea polyphenols on Aβ accumulation in brain. Recently, long-term administration of green tea polyphenol or EGCG preparation was demonstrated to improve spatial cognition learning ability in rats and reduce cerebral amyloidosis in Alzheimer's transgenic mice, respectively [Haque *et al.*, 2006]. A study reported that EGCG, the main polyphenolic constituent of green tea, reduces Aβ generation in both murine neuron-like cells (N2a) transfected with the human "Swedish" mutant amyloid precursor protein (APP) and in primary neurons derived from Swedish mutant APP-overexpressing mice (APPsw). In concert with these observations, it found that EGCG markedly promotes cleavage of the alpha-C-terminal fragment of APP and elevates the N-terminal APP cleavage product, soluble APP-alpha. These cleavage events are associated with elevated alpha-secretase activity and enhanced hydrolysis of tumor necrosis factor alpha-converting enzyme, a primary candidate alpha-secretase. As a validation of these findings *in vivo*, we treated APPsw transgenic mice overproducing Aβ with EGCG and found decreased Aβ levels and plaques associated with promotion of the non-amyloidogenic alpha-secretase proteolytic pathway. GCG, the main tea polyphenol constituent of green tea, was reported to reduce Aβ generation in both N2a overexpressing human Swedish mutant APP and in primary neurons derived from Tg2576 through promotion of the non-amyloidogenic alpha-secretase proteolytic pathway. EGCG reduces Aβ generation in both murine neuron-like cells (N2a) transfected with the human "Swedish" mutant amyloid precursor

protein (APP) and in primary neurons derived from Swedish mutant APP-overexpressing mice (Tg APPsw line 2576). In concert with these observations, it was found that EGCG markedly promoted cleavage of the α-C-terminal fragment of APP and elevates the N-terminal APP cleavage product, soluble APP-α. These cleavage events are associated with elevated α-secretase activity and enhanced hydrolysis of tumor necrosis factor α-converting enzyme, a primary candidate α-secretase. As a validation of these findings *in vivo*, Tg APPsw transgenic mice overproducing Aβ were treated with EGCG and found decreased Aβ levels and plaques associated with promotion of the non-amyloidogenic α-secretase proteolytic pathway [Rezai-Zadeh *et al.*, 2005]. It was reported that long-term (26 weeks) administration of green tea polyphenols (0.5% green tea polyphenols in water) could prevent Aβ induced cognitive impairment in rats. In addition to preventing cognitive impairment, lipid peroxides and ROS in hippocampus and plasma were more than 20% lower than those in the control group [Haque *et al.*, 2008]. We treated APPsw of SH-SY5Y cells transferred into Aβ with different concentrations of EGCG for 48 hours and it was found that Aβ 1-42 content in APPsw cell culture medium significantly reduced (Fig. 6-7) [Wan *et al.*, 2011, 2012].

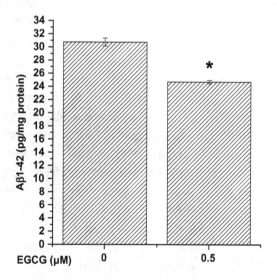

Figure 6-7.   Effect of EGCG on Aβ 1-42 content in APPsw cell culture medium. *, $p <$ 0.05 compared without EGCG. The results are expressed as mean ± standard error, n = 4.

Green tea polyphenol EGCG exerts a beneficial role on reducing brain Aβ levels, resulting in mitigation of cerebral amyloidosis in a mouse model of AD. EGCG seems to accomplish this by modulating amyloid precursor protein (APP) processing, resulting in enhanced cleavage of the α-COOH-terminal fragment (α-CTF) of APP and corresponding elevation of the NH2-terminal APP product, soluble APP-α (sAPP-α). These beneficial effects were associated with increased α-secretase cleavage activity, but no significant alteration in β- or α-secretase activities. To gain insight into the molecular mechanism whereby EGCG modulates APP processing, a study evaluated the involvement of three candidate α-secretaseenzymes, α-disintegrinand metalloprotease (ADAM) 9, 10, or 17, in EGCG-induced non-amyloidogenic APP metabolism. Results showed that EGCG treatment of N2a cells stably transfected with "Swedish" mutant human APP (SweAPP N2a cells) led to markedly elevated active (~60 kDa mature form) ADAM10 protein. Elevation of active ADAM10 correlates with increased α-CTF cleavage, and elevated sAPP-α. To specifically test the contribution ofADAM10 to non-amyloidogenic APP metabolism. Results show that ADAM10 (but not ADAM9 or ADAM17) is critical for EGCG-mediated α-secretase cleavage activity [Obregon *et al.*, 2006]. ADAM10 activation is necessary for EGCG promotion of non-amyloidogenic (α-secretase cleavage) APP processing. Thus, ADAM10 represents an important pharmaco-therapeutic target for the treatment of cerebral amyloidosis in AD. Interestingly, it was reported that oral co-treatment with fish oil and EGCG led to a synergetic effect on the inhibition of cerebral Aβ deposits in Tg2576 mice. As the comparable effective dose of EGCG in humans may exceed clinical convenience and/or safety, the study provides a solution by co-treatment of EGCG and fish oil, which enhanced bioavailability of EGCG, allowing supplementation with moderate dose to achieve significant therapeutic effects [Zhao, 2012].

These data raise the possibility that for inhibition effect of tea polyphenols on Aβ accumulation in brain, EGCG may provide effective prophylaxis for AD.

4. Preventive effects of tea polyphenols on AD by inhibition of related diseases

A number of studies have examined the role of dietary patterns on late-life cognition, with accumulating evidence that the combinations of foods and nutrients may act synergistically to provide stronger benefit than those conferred by individual dietary components. Higher adherence to the Mediterranean dietary pattern has been associated with decreased cognitive decline and incident AD. Another dietary pattern with neuroprotective actions is the Dietary Approach to Stop Hypertension (DASH). The combination of these two dietary patterns has been associated with slower rates of cognitive decline and significant reduction in incident AD. Studies have found that obesity, hyperlipidemia, hypertension, cardiovascular disease and stroke, and diabetes are associated with the risk of AD. Several epidemiological studies suggest that the regular consumption of foods and beverages rich in flavonoids is associated with a reduction in the risk of several pathological conditions ranging from hypertension to coronary heart disease, stroke, and dementia. The impairment of endothelial function is directly related to ageing and an association between decreased cerebral perfusion and dementia has been shown to exist [Dominguez & Barbagallo, 2018; Ghosh & Scheepens, 2009]. The flavonoid phytochemicals in tea, namely polyphenol or catechin, provide potential benefits for reducing the risk of diabetes and AD by targeting common risk factors, including obesity, hyperlipidemia, hypertension, cardiovascular disease, and stroke. In addition, polyphenols found in tea have important antioxidant properties and natural properties that can regulate intracellular neuronal signal transduction pathway and mitochondrial function. Although diabetes is considered to be a different disease, it has recently been considered an important factor in the risk of dementia. Some studies have mentioned type 3 diabetes, a disease caused by insulin resistance in the brain. Interestingly, AD and diabetes have common basic pathological processes, common risk factors, and important intervention pathways [Fernando *et al.*, 2017].

The tea polyphenols may provide potential benefits for reducing the risk of diabetes and AD by targeting common risk factors, including obesity, hyperlipidemia, hypertension, cardiovascular disease, and stroke.

5. Preventive effects of tea polyphenols on AD by inhibition of inflammation

It has been found that brain nerve injury caused by inflammation is an important factor in AD. The inflammatory process produces a large number of ROS and RNS free radicals, leading to oxidative stress damage.

Increased oxidative stress and neuronal inflammation are also associated with neuronal dysfunction and neurodegeneration. Tea polyphenols also has anti-inflammatory properties, which may also be the basis of its mechanism of action on AD. Neuronal injury leads to the secretion of pro-inflammatory factors (such as cytokines and cytotoxic factors), which triggers neuronal death [Morales *et al.*, 2014]. In a study using mice injected with lipopolysaccharide, it was demonstrated that EGCG given in advance (1.5 mg/kg and 3 mg/kg for 3 weeks) can prevent lipopolysaccharide-induced memory damage and inhibit the increase of cytokines and inflammatory proteins in untreated control group [Lee *et al.*, 2013]. Another *in vitro* study on BV-2 microglia showed that the reactions related to lipopolysaccharide-induced inflammation (including nitric oxide production, cyclooxygenase-2 expression, and inducible nitric oxide synthase expression) were inhibited by EGCG [Wu *et al.*, 2012].

Above studies suggest that green tea flavonoids may be used for the prevention and treatment of a variety of neurodegenerative diseases AD by inhibition of inflammation.

6. Preventive effects of tea polyphenols on AD through signal transduction pathway

Many factors leading to AD can activate multiple signaling molecules such as PKC, Bax NF-kB, and ERK through multiple signaling pathways. Protein kinase C (PKC)-related mechanisms may also contribute to the effect of tea polyphenols on AD. PKC plays an important role in cell survival and soluble non-toxic amyloid β (SAPP) in generation [Alkon *et al.*, 2007]. Several isozymes, including PKCα and ε, activated α-secretory enzymes that directly led to the cleavage of amyloid APP into non-toxic Aβ. *In vitro* and *in vivo* studies published in a paper showed that low concentration EGCG (1–5μM) stimulated human neuroblastoma and PC12 cells to produce SAPP. Compared with control treated animals, oral EGCG (2 mg/kg/day) for two weeks increased PKCα and ε in mouse hippocampus [Levites *et al.*, 2003].

Other possible mechanisms of tea polyphenols have also been reported. Studies related to acetylcholinesterase inhibition using tea polyphenols were conducted [Kaur *et al.*, 2008; Kim *et al.*, 2004]. The effects of green tea extract on learning, memory, behavior, and acetylcholinesterase activity in young and old male rats were studied and the passive avoidance test was used for evaluation. The learning and memory abilities of aged Wister rats treated with green tea extract (0.5%) for eight weeks significantly improved compared with control young rats, the acetylcholinesterase activity in the brain of aged rats treated was reduced [Kaur *et al.,* 2008]. Furthermore, when 0.2% was given to mice through diet tea polyphenols, scopolamine induced amnesia was reversed. With the change of behavior, tea polyphenols significantly inhibited the activity of acetylcholinesterase [Kim *et al.*, 2004]. In addition to these *in vitro* and *in vivo* studies, the electronic docking of tea polyphenols and cholinesterase was also studied [Srividhya *et al.*, 2012]. Co-treated green tea extract (10–50 μg/ml) dose-dependently attenuated Aβ (25–35) (50 μM)-induced cell death, intracellular ROS levels, and 8-oxodG formation, in addition to p53, Bax, and caspase-3 expression, but upregulated Bcl-2. Furthermore, green tea extract prevented the Aβ (25-35)-induced activations of the NF-kB, ERK, and p38 MAP kinase pathways. Inhibitory effect of green tea extract on amyloid β induced PC12 cell death by inhibiting the activation of NF-kB and ERK/p38 MAP kinase pathways through antioxidant mechanisms [Lee *et al.*, 2005].

Above discussion suggests that the preventive and therapeutic effects of tea polyphenols on AD are multi-target, including antioxidant effect, free radical scavenging, iron chelating characteristics, the inhibition of hydroxyl-dopamine, signal transduction pathway, regulation of cell survival/death genes, and induction of neuronal activity by mitochondrial function. Studies have also shown that tea polyphenols can prevent amyloid β. The formation of plaque enhances cognitive function, so it may be helpful to treat patients with AD or dementia. Therefore, the use of tea polyphenols as multi-target drugs is of great significance for the prevention and treatment of AD and PD.

6.5 Preventive and therapeutic effects of L-theanine on AD

L-theanine is an important component of tea and a very good natural antioxidant. L-theanine, which has become one of the new favorites of natural health products in the international market in recent years, is known as a "natural sedative". L-theanine, also known as N-ethyl-γ-Glutamine, is a unique amino acid in tea. L-theanine was first discovered from green tea by Sakato, a Japanese scholar [Sakato, 1949], and then Casimir and Tsushida isolated it from mushroom and tea plum [Casimir & Jadot,1960; Tsushida & Takeo, 1984]. L-theanine is one of the main flavor substances of green tea. It has a special fresh taste and can alleviate the astringency of tea [Arai *et al.*, 1989]. Its content is positively correlated with the quality of tea, with a correlation coefficient of 0.787–0.876. It is one of the important indexes to evaluate the quality of green tea [Wang, 1980]. L-theanine accounts for 1.0–2.0% of the dry weight of tea [Ekborg-Ott *et al.*, 1997], and plays an important role in tea plants. It is the existence and storage form of soluble nitrogen in tea plants, and it is also an important precursor for the formation of catechin ring during catechin biosynthesis [Kito *et al.*, 1968]. L-theanine is a major amino acid derivative component in green tea that is also widely used as a food additive to reduce anxiety [Finger *et al.*, 1992]. It is also suggested that L-theanine may be involved in cognitive performance [Bryan, 2008]. As a natural antagonist of glutamate, L-theanine can inhibit the re-uptake of glutamate from the synaptic cleft and block the glutamate receptors in the hippocampus [Kakuda *et al.*, 2002]. Many biological activities of L-theanine, such as inhibiting caffeine-induced excitement and calming nerves, have been widely recognized. In recent years, with the further study of its mechanism, L-theanine has been found to have neuroprotective effect. L-theanine can prevent and treat AD through various ways and mechanisms, such as inhibiting NMDA receptor induced excitotoxicity, reducing a β level, inhibiting hippocampal long-term enhancement and memory damage, inhibiting age-induced D-galactose induced brain damage, protecting mitochondrial function, inhibiting hippocampal long-term enhancement and memory damage, reducing Aβ 42 level, and antioxidant. The preventive and therapeutic effects of L-theanine on AD and its mechanism are briefly discussed below.

L-theanine can prevent and treat AD through many ways and mechanisms, such as inhibition of NMDA receptor induced excitotoxicity, reduction of Aβ levels, inhibition of hippocampal long-term enhancement and memory damage, inhibiting advanced glycation end products (AGEs) formation d-galactose-induced brain damage, protection of mitochondrial function, inhibition of hippocampal long-term enhancement and memory damage, and reduced Abeta42 levels, antioxidant.

6.5.1 *Inhibition effect of L-theanine on NMDA receptor-induced excitotoxicity*

Excitatory glutamatergic neurotransmission via N-methyl-d-aspartate receptor (NMDAR) is critical for synaptic plasticity and survival of neurons. However, excessive NMDAR activity causes excitotoxicity and promotes cell death, underlying a potential mechanism of neurodegeneration occurred in Alzheimer's disease (AD). Studies indicate that the distinct outcomes of NMDAR-mediated responses are induced by regionalized receptor activities, followed by different downstream signaling pathways. The activation of synaptic NMDARs initiates plasticity and stimulates cell survival. In contrast, the activation of extrasynaptic NMDARs promotes cell death and thus contributes to the etiology of AD [Wang & Reddy, 2017]. Over-stimulation of NMDA (N-methyl-D-aspartate) subtype of L-glutamate receptors, which causes calcium influx, increases ROS production, and triggers Aβ-induced neuronal death, also accelerates Aβ production [Parameshwaran *et al.*, 2008]. NMDA receptors not only play an important physiological role in the development of nervous system, such as regulating the survival of neurons, regulating the development of dendrites and axons, and participating in the formation of synaptic plasticity, but also play a key role in the formation of neuronal circuits. Some data show that NMDA receptors are important receptors in the process of learning and memory recipient. However, a lot of experimental evidence shows that Aβ and NMDA receptor-induced excitotoxicity play an important role in the pathological process of AD, and the over activation of NMDA receptor will promote Aβ. It then causes the release of glutamate, resulting in the death of neurons.

We found that L-theanine, like NMDA receptor inhibitor and NO synthase inhibitor, attenuated the decrease of cell viability and apoptosis of APPsw cells induced by glutamate (Fig. 6-8). L-theanine pretreatment significantly inhibited the increase of calcium level in APPsw cells (Fig. 6-9). After L-theanine pretreatment, the content of Aβ in APPsw cells decreased and the amount secreted by APPsw cells into the culture medium β1-40 increased (Fig. 6-10). It can also significantly reduce the content of ROS and internal calcium, increase the protein expression of nNOS and iNOS (Fig. 6-11), significantly inhibit the up regulation of p-JNK and Caspase-3 expression in cells caused by glutamate, improve

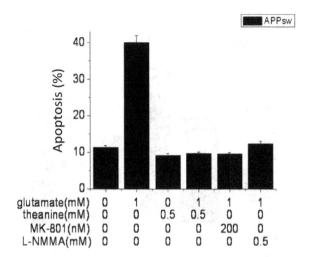

Figure 6-8. Effects of L-glutamate, L-theanine, MK801 (the inhibitor of NMDA receptor), and L-NMMA (the inhibitor of iNOS) on apoptosis ratio in APPsw cells. Untreated cells, L-glutamate (1 mmol) treated, L-theanine (0.5 mmol) treated, cells pretreated with L-theanine (0.5 mmol) or MK-801 (200 nM) or L-NMMA (0.5 mmol), then exposed to L-glutamate. The mean ± SEM for three independent experiments. The cells were treated with different compounds and examined by flow cytometry.

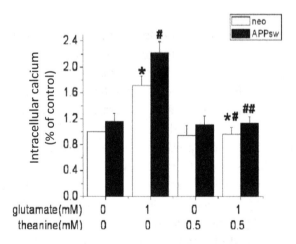

Figure 6-9. Intracellular calcium levels in neo and APPsw cells were detected by Fluo-3AM fluorescence. The intracellular [Ca2+]i was measured after neo or APPsw exposed to L-glutamate or L-theanine with probe Fluo-3AM. The result is expressed in mean ± SEM, $n = 4$. *, $p < 0.05$ comparing neo L-glutamate 0 *vs* L-theanine 0; #, $p < 0.05$ comparing APPsw L-glutamate 0 *vs* L-theanine 0; *#, $p < 0.05$ comparing neo L-glutamate 1 *vs* L-theanine 0; ##, $p < 0.05$ comparing APPsw L-glutamate 1 *vs* L-theanine 0.

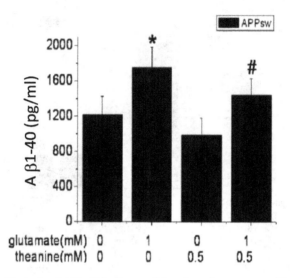

Figure 6-10.   Contents of Aβ1-40 in the conditioned media were quantified by ELISA. The effect of L-glutamate and L-theanine on Aβ secretion of the APPsw cells was measured by a sensitive fluorescence-based sandwich ELISA assay using a kit. The result is expressed in mean ± SEM, n = 4. *, $p < 0.05$ comparing L-glutamate 0 *vs* L-theanine 0; #, $p < 0.05$ comparing L-glutamate 1 *vs* L-theanine 0.

the mitochondrial membrane potential, and reduce the apoptosis of APPsw cells (Fig. 6-12). According to above findings, it is possible that L-theanine may have protective effects against neurotoxicity induced by L-glutamate and Aβ. Results from studying the neuroprotective effects of L-theanine using an AD cell model overexpressing APPsw (APPsw) support this hypothesis [Di *et al.*, 2010].

In this cell model, the expression of APPsw renders cells more susceptible to glutamate-induced excitotoxicity. It was found that the cell viability was decreased by L-glutamate treatment, which was significantly improved by L-theanine. Meanwhile, apoptosis and caspase-3 activation induced by L-glutamate was suppressed by L-theanine. Further, L-theanine ameliorated glutamate-induced apoptosis in a way similar to that of the NMDA receptor inhibitor MK-801 and the NOS inhibitor L-NMMA, indicating that L-theanine may protect APPsw cells from glutamate-induced apoptosis via inhibition of NMDA receptor over-activation, NO overproduction, and the related pathways. In fact, excessive

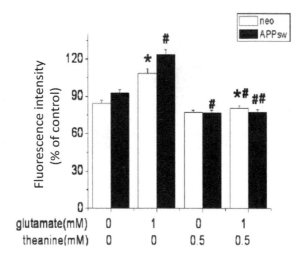

Figure 6-11. The relative content of nitric oxide generated in the cells treated by L-glutamate or L-theanine. Nitric oxide was measured using DAF-FM DA as a probe to detect its fluorescence intensity. The result is expressed in mean ± SEM of a ratio between fluorescence intensity and protein content, n = 4. *, $p < 0.05$ comparing neo L-glutamate 0 *vs* L-theanine 0; #, $p < 0.05$ comparing APPsw L-glutamate 0 *vs* L-theanine 0; *#, $p < 0.05$ comparing neo L-glutamate 1 *vs* L-theanine 0; ##, $p < 0.05$ comparing APPsw L-glutamate 1 *vs* L-theanine 0.

NO formation can be caused by stimulation of the NMDA receptor [Hynd *et al.*, 2004]. In APPsw cells, L-glutamate significantly increased the generation of NO, while pretreatment of cells with L-theanine prevented the increase of NO production, an effect likely resulted from the down-regulation of iNOS and neuronal nitric oxide synthase (nNOS) protein levels by L-theanine. Over-activation of the NMDA receptor by L-glutamate stimulation may cause increase of intracellular calcium and disturbance of Ca^{2+} homeostasis that has been indicated in the neuronal loss seen in AD [Hynd *et al.*, 2004]. Pretreatment of L-theanine signifi-cantly prevented the elevation of intracellular calcium level and the dis-turbance of Ca^{2+} homeostasis induced by L-glutamate stimulation. L-theanine treatment also inhibited the increase of Abeta secretion induced by L-glutamate [Hynd *et al.*, 2004].

These results indicate that the inhibition of the NMDA subtype of glutamate receptors and its related pathways is the crucial point of the

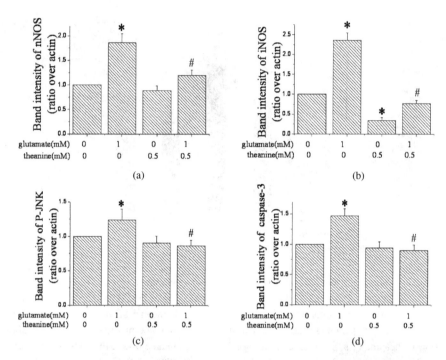

Figure 6-12. Effect of L-theanine and glutamate on protein expression of (a) nNOS, (b) iNOS, (c) p-JNK, and (d) Caspase-3 in APPsw cell. a) Statistical results of band integrated optical density of nNOS; b) Statistical results of band integrated optical density of iNOS; c) Statistical results of band integrated optical density of p-JNK; d) Statistical results of band integrated optical density of Caspase-3. The data are expressed as the integrated optical density of the bands and the internal parameters (Aβ-actin) and the mean ± SEM after comparison with the control group, n = 5. *, $p < 0.05$ means that compared with the control group, means that compared with glutamate injury group.

neuroprotective effect of L-theanine in this AD cell model, and supports the notion that L-theanine may provide effective prophylaxis and treatment for Alzheimer's disease.

6.5.2 *Reduction effect of L-theanine on Aβ levels*

It was shown that oral treatment of L-theanine dose-dependently reduced Abeta42 levels and the Abeta42-induced neuronal cell death in the cortex and hippocampus of the brain and significantly attenuated Abeta42-induced learning and memory impairment in an AD mouse model, in which mice were subjected to a single intracerebroventricular injection of Aβ peptides. Moreover, L-theanine caused elevation of glutathione levels and significantly reduced oxidative products of proteins and lipids in brains of these mice. It was also found that L-theanine inhibited Abeta42-induced extracellular signal-regulated kinase (ERK) and p38 MAPK as well as the activity of NF-kappa B [Zhang *et al.*, 2018b; Mandel *et al.*, 2008]. The result suggested that the positive effects of L-theanine on memory and learning may be contributed by the suppression of these signaling pathways as well as the reduction of macromolecular oxidative damages.

6.5.3 *Anti-inflammatory and oxidative stress of L-theanine*

L-theanine has anti-inflammatory, antioxidant, and possible antidiabetic activities, and it may synergistically protect against dementia. L-theanine improved memory function, as determined by water maze and passive avoidance tests; by potentiating the hippocampal insulin signaling and reducing inflammation, L-theanine primarily reduced tumor necrosis factor-α. In the metabolomics analysis of the hippocampus lysates, the concentration of proline, phenylpyruvic acid, and normetanephrine decreased in the L-theanine compared to AD-control. Norepinephrine contents were lower in the AD-dextrin than non-AD rats with a high fat diet with 0.2% dextrin, whereas L-theanine and L-theanine inhibited the decrease. L-theanine increased glucose infusion rates and decreased hepatic glucose output under basal and hyperinsulinemic conditions, indicating improved whole-body and hepatic insulin sensitivity. Disturbances in glucose-stimulated insulin secretion during hyperglycemic clamp were most effectively corrected by the L-theanine treatments. In conclusion, the hypothesis of the study was accepted. L-theanine prevented AD-like symptom, possibly by improving hippocampal insulin signaling, norepinephrine metabolisms, and decreasing neuro-inflammation [Park *et al.*, 2018]. L-Theanine prevented increased expression of NF-κB and down-regulated the pro-inflammatory (interleukin (IL)-1β and IL-6) and profibrotic cytokines transforming growth factor β (TGF-β) and connective tissue growth factor (CTGF). Furthermore, the levels of messenger RNA encoding these proteins decreased in agreement with the expression levels. L-Theanine promoted the expression of the anti-inflammatory cytokine IL-10 and the fibrolytic enzyme metalloproteinase-13 [Pérez-Vargas *et al.*, 2016].

L-theanine may be a useful therapeutic option for preventing and/or delaying the progression of memory dysfunction.

6.5.4 *Inhibiting effect of L-theanine on AGE formation d-galactose-induced brain damage*

As the irreversible products of the non-enzymatic reduction of sugars and the amino groups of proteins or peptides, advanced glycation end products (AGEs) are metabolized and excreted via the kidneys. However, if AGEs are not metabolized, they are deposited in the kidneys and bind to AGE receptors (RAGE), which can induce various pathological changes, including oxidative stress, apoptosis, and inflammation. The effects of L-theanine on Aβ and AGEs generation were investigated in this study. Decreased AGEs and Aβ 1-42 levels were reflected by increased acetyl-choline (ACh) concentration and acetylcholinesterase (AChE) activity inhibition compared to model rats. L-Theanine also inhibited nuclear factor-κB (p65) protein expression by activating sirtuin1 (SIRT1), reducing inflammatory factor expression, and downregulating the mRNA and protein expression of AGE receptors (RAGE). Superoxide dismutase 2 and catalase protein expressions were markedly upregulated by L-theanine, whereas oxidative stress-related injury was alleviated. The expression of peroxisome proliferator-activated receptor-γ coactivator 1α (PGC-1α) was also found to be increased. The apoptosis of hippocampal neurons was mitigated by decreased Bax and cleaved-caspase-3 protein expression and the increase of Bcl-2 protein expression. Moreover, L-theanine increased the gene and protein expression of brain-derived neurotrophic factor (BDNF) [Zeng *et al.*, 2021]. A study used the D-galactose (DG)-induced rat model to explore the potential role and mechanism of L-theanine in inhibiting AGEs/RAGE-related signaling pathways in renal tissues. L-theanine increased the activities of glutathione peroxidase (GSH-Px) and total antioxidant capacity (T-AOC) while downregulating the contents of malondialdehyde (MDA) and AGEs in renal tissues induced by DG ($p < 0.05$). By inhibiting the upregulation of RAGE protein expression attributed to AGEs accumulation ($p < 0.05$), L-theanine downregulated phosphorylated nuclear factor (p-NF-κB (p65)), Bax, and cleaved-caspase-3 expression and increased Bcl-2 protein expression ($p < 0.05$), thereby alleviating the oxidative stress damage and reducing the inflammation and cell injury induced by DG [Zeng *et al.*, 2021].

These findings suggest that the potential preventive effects of L-theanine against AD may be attributed to its regulation of SIRT1 and BDNF proteins and its mitigation of AGEs/RAGE signaling pathways in the brain tissue of AD model rats.

6.5.5 *Protection of L-theanine on mitochondrial function against mitochondrial damage*

We studied the effect of L-theanine on mitochondrial fusion protein of AD transgenic (APPsw) cells interact with mitochondria, damage mitochondrial function, and increase free radical production. Mitochondria continue to undergo two opposite processes: division and fusion. The destruction of this dynamic balance may indicate cell injury or death, and lead to neurodegenerative diseases. Mitochondrial morphology is controlled by proteins that regulate fusion and division events. Fusion is mediated by two outer membrane GTPases, namely mitochondrial dynamics related proteins, mitotic fusion proteins mfn1 and Mfn2. Our results showed that Mfn1 in the hippocampus of AD transgenic mice aged three and six months was significantly higher than that of control mice. Compared with control mice, Mfn2 in hippocampus of 3-month-old AD transgenic mice was significantly increased. In 6-month-old mice, the situation of Mfn2 was just the opposite. The difference in Mfn2 levels between the 6-month-old control group and the 3-month-old control group may be due to age-dependent expression. The significant increase of Mfn2 suggests that there may be great differences in mitochondrial motility in these ages. More importantly, we found that these mitochondrial abnormalities became obvious before the formation of senile plaques. Consistent with the mitochondrial morphology, the changes in the expression of Mfn1 and Mfn2 suggest that they are markers of abnormal mitochondrial morphology. These results suggest that early mitochondrial damage may be the cause of AD [Wu *et al.*, 2010]. Several mitochondrial morphology related proteins were evaluated under routine and hypoxic conditions. The three cell lines were treated under hypoxia and non-hypoxia for eight hours, then lysed, and the protein concentration was measured. Mfn2 and mitochondrial motility related protein in APPsw mutant cells were higher than those in carrier cells. Interestingly, the expression of Mfn1 and Mfn2 in APPsw mutant cells and vector cells under hypoxia is similar. Mitochondrial motility related proteins respond strongly to hypoxia. Our results showed that L-theanine significantly increased the expression of Mfn1 in control cells and significantly decreased the expression of Mfn2 in APP overexpression and mutant cells [Wu *et al.*, 2014].

These data suggest that L-theanine may protect mitochondria from damage by regulating mitochondrial fusion. L-theanine treatment may have neuroprotective effect by regulating the expression of mitochondrial fusion/division protein. However, the effect of L-theanine on AD patients need to be further studied in clinical trials.

6.5.6 *Inhibition effect of L-theanine on hippocampal long-term enhancement and memory damage*

Synaptic refinement improves synaptic efficiency, which provides a possibility to improve memory in AD. Current study aimed to investigate the role of L-theanine, a natural constituent in green tea, in hippocampal synaptic transmission and to assess its potential to improve memory in transgenic AD mice. It found that L-theanine bath application facilitated hippocampal synaptic transmission and reduced paired-pulse facilitation (PPF). Moreover, L-theanine enhanced PKA phosphorylation via dopamine D1/5 receptor activation. L-theanine did not influence hippocampal long-term potentiation (LTP) in the slices obtained from wild-type mice, but rescued the impairment of hippocampal LTP in AD mice. Importantly, systemic application of L-theanine also improved memory and hippocampal LTP in AD mice. The results demonstrate that L-theanine administration promotes hippocampal dopamine and noradrenaline release, and stimulates PKA phosphorylation. The results reveal that L-theanine ameliorates the impairment of memory and hippocampal LTP in AD mice, likely through dopamine D1/5 receptor-PKA pathway activation [Zhu *et al.*, 2018]

A study explored the protective effect of L-theanine on hippocampal long-term enhancement and memory damage in AD mice, and evaluated its potential to improve the memory of transgenic AD mice. Initially, L-theanine was found to promote hippocampal synaptic transmission; antagonists of N-methyl-D-aspartate receptor and dopamine D1/5 receptor, and selective protein kinase A (PKA) inhibitors can block hippocampal synaptic transmission. L-theanine can also enhance PKA phosphorylation through dopamine D1/5 receptor activation. L-theanine does not affect the long-term enhancement of hippocampus in wild-type mouse brain slices long-term potential (LTP), but it can save the damage of hippocampal LTP in AD mice. Importantly, the application of L-theanine also improves the memory and hippocampal LTP of AD mice. These results show that taking L-theanine can promote the release of dopamine and norepinephrine in hippocampus and stimulate PKA phosphorylation. L-theanine may improve the memory and hippocampus of AD mice by activating dopamine D1/5 receptor PKA pathway LTP damage. These data demonstrate that L-theanine is a candidate drug for the treatment of AD [Wu *et al.*, 2014].

6.5.7 *Reduction of L-theanine on aluminum brain toxicity by antioxidant*

Aluminum (Al) is one of the most extended metals in the Earth's crust. Its abundance, together with the widespread use by humans, makes Al-related toxicity particularly relevant for human health. It remains controversial whether low doses of this metal may contribute to developing AD, probably because of the multifactorial and highly variable presentation of the disease. Two key aspects related to Al neurotoxicity and AD are metabolic impairment and iron (Fe) alterations [Colomina & Peris-Sampedro, 2017]. The investigated the role of oxidative stress and the status of antioxidant system in the management of aluminum chloride (AlCl$_3$) induced brain toxicity in various rat brain regions and further elucidate the potential role of L-theanine in alleviating such negative effects. Pretreatment with L-theanine at a dose of 200 mg/kg b.w. significantly increased the antioxidant status and activities of membrane bound enzymes and decreased the level of LPO and the activities of marker enzymes, when compared with aluminum-induced rats. Aluminum induction also caused histopathological changes in the cerebral cortex, cerebellum, and hippocampus of rat brain which was reverted by pretreatment with L-theanine [Sumathi *et al.*, 2015].

6.6 Conclusion

Considering the multi-etiological character of AD, the current pharmacological approaches using drugs oriented towards a single molecular target possess limited ability to modify the course of the disease and thus, only offer a partial benefit to the patient, unfortunately posing some side effects. In line with this concept, novel strategies include the use of a cocktail of several drugs and/or the development of a single molecule, possessing two or more active neuroprotective-neurorescue moieties that simultaneously manipulate multiple targets involved in AD pathology. A consistent observation in AD is a steady state maladjustment of metal ions (Fe^{2+}, Cu^{2+}, and Zn^{2+}) homeostasis and consequential induction of oxidative stress, associated with beta-amyloid aggregation and neurite plaque formation. In particular, iron has been demonstrated to modulate the Alzheimer's amyloid precursor holo-protein expression by a pathway similar to that of ferritin L-and H-mRNA translation through iron-responsive elements in their 50UTRs (untranslated region). Two separate scenarios concerning multiple therapy targets in AD, share a common implementation of iron chelation activity and oxidative stress caused by ROS and RNS: 1) novel multimodal brain-permeable iron-chelating drugs, possessing neuroprotective-neurorescue and amyloid precursor protein-processing regulatory activities, and 2) natural tea polyphenols (flavonoids), such as green tea epigallocatechin gallate (EGCG), L-theanine, and curcumin, reported to have access to the brain and to possess multifunctional activities, such as metal chelation, free radical scavenging, anti-inflammation, and neuroprotection [Mandel *et al.*, 2007]. In summary, the pathway and the mechanism of tea polyphenols in prevention and treatment of AD are shown in Figure 6-13.

In the future, more efforts should be made to use clinical trials to evaluate the efficacy of tea polyphenols in the prevention and intervention

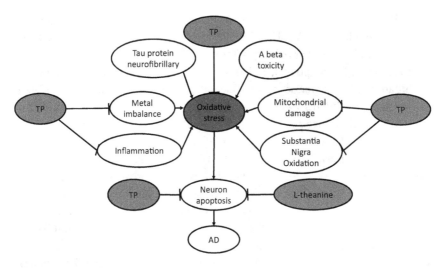

Figure 6-13. The pathway and mechanism of tea polyphenols in prevention and treatment of AD. TP: tea polyphenols.

of AD. Since most natural antioxidants have complex components, standardized formulations can be used for individual antioxidants. In order to avoid the differences between different trials, a research group with good characteristics is also needed. The absorption, metabolism, and transport of individual tea polyphenols to the brain need to be characterized to determine the effective dose and promote the evaluation of clinical trial efficacy. In addition, most studies have focused on Aβ and tau protein mediated pathology. Since tea polyphenols may have different effects on the pathology of Aβ and tau protein, it is important to test their effects in the model of simultaneous neuropathological lesions of Aβ and tau. Isolation and purification can further understand the protective mechanism of tea polyphenols against AD.

References

Alkon DL, Sun MK, Nelson TJ. (2007) PKC signaling deficits: A mechanistic hypothesis for the origins of Alzheimer's disease. *Trends Pharmacol Sci*, **28**, 51–60.

Ali B, Jamal QM, Shams S, *et al*. (2016) In silico analysis of green tea polyphenols as inhibitors of AChE and BChE enzymes in Alzheimer's disease treatment. *CNS Neurol Disord. Drug Targets*, **15**, 624–628.

Anantharaman M, Tangpong J, Keller JN, *et al*. (2006) Beta-amyloid mediated nitration of manganese superoxide dismutase: implication for oxidative stress in a APPNLH/NLH X PS-1P264L/P264L double knock-in mouse model of Alzheimer's disease. *Am J Pathol*, **168**(5), 1608–1618.

Ankner BAY, Lu T (2009). Amyloid beta-protein toxicity and the pathogenesis of Alzheimer disease. *J Biol Chem*, **284**(8), 4755–4759.

Ansari MA, Scheff SW. (2010) Oxidative stress in the progression of Alzheimer disease in the frontal cortex. *J Neuropathol Exp Neurol*, **69**(2), 155–167.

Arab L, Biggs ML, O'Meara ES, *et al*. (2011) Gender differences in tea, coffee, and cognitive decline in the elderly: the cardiovascular health study. *J Alzheimers Dis*, **27**, 553–566.

Arai M, Matsuura T, Kakuda T. (1989) Variation of the yields and the chemical composition in leaves on nutriculture of tea plant. *Nippon Dojo Hiryogaku Zasshi*, **60**, 157–159.

Atwood CS, Moir RD, Huang X, Scarpa RC, Bacarra NM, Romano DM, Hartshorn MA, Tanzi RE, Bush AI. (1998) Dramatic aggregation of Alzheimer abeta by Cu(II) is induced by conditions representing physiological acidosis. *J Biol Chem*, **273**, 12817–12826.

Beal MF. (2002) Oxidatively modified proteins in aging and disease. *Free Radic Biol Med*, **32**(9), 797–803.

Beckman JS. (1996) Oxidative damage and tyrosine nitration from peroxynitrite. *Chem Res Toxicol*, **9**, 836–844.

Behl C, Davis JB, Lesley R, Schubert D. (1994) Hydrogen peroxide mediates amyloid beta protein toxicity. *Cell*, **77**(6), 817–827.

Biasibetti R, Tramontina AC, Costa AP, Dutra MF, Quincozes-Santos A, Nardin P, Bernardi CL, Wartchow KM, Lunardi PS, Goncalves CA. (2013) Green tea (-)epigallocatechin-3-gallate reverses oxidative stress and reduces acetylcholinesterase activity in a streptozotocin-induced model of dementia. *Behav Brain Res*, **236**, 186–193.

Bryan J. (2008) Psychological effects of dietary components of tea: caffeine and L-theanine. *Nutr Rev*, **66**(2) 82–90.

Bush AI, Pettingell WH, Multhaup G, d Paradis M, Vonsattel JP, Gusella JF, Beyreuther K, Masters CL, Tanzi RE. (1994) Rapid induction of Alzheimer A beta amyloid formation by zinc. *Science*, **265**, 1464–1467.

Bush AI. (2002) Metal complexing agents as therapies for Alzheimer's disease. *Neurobiol Aging*, **23**, 1031–1038.

Butterfield DA, Kanski J. (2001) Brain protein oxidation in age-related neurode-generative disorders that are associated with aggregated proteins. *Mech Ageing Dev*, **122**(9), 945–962.

Casimir J, Jadot J. (1960) Separation and characterization of N-ethyl-gamma-glutamine from Xerocomus badius. *Biochimica et biophysica acta*, **39**, 462.

Casley CS, Canevari L, Land JM, Clark JB, Sharpe MA (2002). Beta-amyloid inhibits integrated mitochondrial respiration and key enzyme activities. *J Neurochem*, **80**(1), 91–100.

Caspersen C, Wang N, Yao J, Sosunov A, Chen X, Lustbader JW, Xu HW, Stern D, McKhann G, Yan SD. (2005) Mitochondrial Abeta: a potential focal point for neuronal metabolic dysfunction in Alzheimer's disease. *FASEB J*, **19**(14), 2040–2041.

Colomina MT, Peris-Sampedro F. (2017) Aluminum and Alzheimer's disease. *Adv Neurobiol*, **18**, 183–197.

De Felice FG, Velasco PT, Lambert MP, Viola K, Fernandez SJ, Ferreira ST, Klein WL. (2007) Abeta oligomers induce neuronal oxidative stress through an N-methyl-D-aspartate receptor-dependent mechanism that is blocked by the Alzheimer drug memantine. *J Biol Chem*, **282**(15), 11590–11601.

de la Monte SM, Wands JR. (2006) Molecular indices of oxidative stress and mitochondrial dysfunction occur early and often progress with severity of Alzheimer's disease. *J Alzheimers Dis*, **9**(2), 167–81.

Di X, Yan J, Zhao Y, Zhang J, Shi Z, Chang Y, Zhao B. (2010) L-theanine protects the APP (Swedish mutation) transgenic SH-SY5Y cell against glutamate-induced excitotoxicity via inhibition of the NMDA receptor pathway. *Neuroscience*, **168**(3), 778–786.

Dominguez LJ, Barbagallo M. (2018) Nutritional prevention of cognitive decline and dementia. *Acta Biomed*, **89**(2), 276–290.

Dumont M, Wille E, Stack C, Calingasan N, Beal M, Lin M. (2009) Reduction of oxidative stress, amyloid deposition, and memory deficit by manganese superoxide dismutase overexpression in a transgenic mouse model of Alzheimer's disease. *FASEB J*, **23**(8), 2459–2466.

Echtay KS. (2007) Mitochondrial uncoupling proteins--what is their physiological role? *Free Radic Biol Med*, **43**(10), 1351–1371.

Ekborg-Ott KH, Taylor A, Armstrong DW. (1997) Varietal differences in the total and enantiomeric composition of theanine in tea. *J Agric Food Chem*, **45**(2), 353–363.

Eskelinen MH, Ngandu T, Tuomilehto J, Soininen H, Kivipelto M. (2009) Midlife coffee and tea drinking and the risk of late-life dementia: a population-based CAIDE study. *J Alzheimers Dis*, **16**, 85–91.

Feng L, Gwee X, Kua EH, Ng TP. (2010) Cognitive function and tea consumption in community dwelling older Chinese in Singapore. *J Nutr Health Aging*, **14**, 433–438.

Feng L, Li J, Ng TP, Lee TS, Kua EH, Zeng Y. (2012) Tea drinking and cognitive function in oldest-old Chinese. *J Nutr Health Aging*, **16**, 754–758.

Fernando WMADB, Somaratne G, Goozee KG, *et al*, (2017) Diabetes and Alzheimer's disease: can tea phytochemicals play a role in prevention? *J Alzheimers Dis*, **59**(2), 481–501.

Finger A, Kuhr S, Engelhardt UH. (1992) Chromatography of tea constituents. *J Chromatogr*, **624**(1-2), 293–315.

Gabbita SP, Lovell MA, Markesbery WR. (1998) Increased nuclear DNA oxidation in the brain in Alzheimer's disease. *J Neurochem*, **71**, 2034–2040.

Ghosh D, Scheepens A. (2009) Vascular action of polyphenols. *Mol Nutr Food Res*, **53**(3), 322–331.

Gu YJ, He CH, Li S, Zhang SY, Duan SY, Sun HP, Shen YP, Xu Y, Yin JY, Pan CW. (2017) Tea consumption is associated with cognitive impairment in older Chinese adults. *Aging Ment Health*, 1–7.

Haque AM, Hashimoto M, Katakura M, Hara Y, Shido O. (2008) Green tea catechins prevent cognitive deficits caused by Aβ1–40 in rats. *J Nutr Biochem*, **19**, 619–626.

Haque AM, Hashimoto M, Katakura M, Tanabe Y, Hara Y, Shido O. (2006) Long-term administration of green tea catechins improves spatial cognition learning ability in rats. *J Nutr*, **136**, 1043–1047.

Huang CQ, Dong BR, Zhang YL, Wu HM, Liu QX. (2009) Association of cognitive impairment with smoking, alcohol consumption, tea consumption, and exercise among Chinese nonagenarians/centenarians. *Cogn Behav Neurol*, **22**, 190–196.

Huang X, Atwood CS, Hartshorn MA, *et al*. (1999) The amyloid-b-peptide of Alzheimer's disease directly produces hydrogen peroxide through metal ion reduction. *Biochemistry*, **38**, 7609–7616.

Hynd MR, Scott HL, Dodd PR. (2004) Glutamate-mediated excitotoxicity and neurodegeneration in Alzheimer's disease. *Neurochem Int*, **45**(5), 583–595.

Ide K, Matsuoka N, Yamada H, Furushima D, Kawakami K. (2018) Effects of tea catechins on Alzheimer's disease: recent updates and perspectives. *Molecules*, **23**(9), 2357.

Kakuda T, Nozawa A, Sugimoto A, Niino H. (2002) Inhibition by theanine of binding of [3H]AMPA, [3H]kainate, and [3H]MDL 105,519 to glutamate receptors. *Biosci Biotechnol Biochem*, **66**(12), 2683–2686.

Kakutani S, Watanabe H, Murayama N. (2019) Green tea intake and risks for dementia, Alzheimer's disease, mild cognitive impairment, and cognitive impairment: a systematic review. *Nutrients*, **11**(5), 1165.

Kamat PK, Kalani A, Rai S, *et al.* (2016) Mechanism of oxidative stress and synapse dysfunction in the pathogenesis of Alzheimer's disease: understanding the therapeutics strategies. *Mol Neurobiol*, **53**(1), 648–661.

Kaur T, Pathak CM, Pandhi P, Khanduja KL. (2008) Effects of green tea extract on learning, memory, behavior and acetylcholinesterase activity in young and old male rats. *Brain Cogn*, **67**, 25–30.

Kim HK, Kim M, Kim S, Kim M, Chung JH. (2004) Effects of green tea polyphenol on cognitive and acetylcholinesterase activities. *Biosci Biotechnol Biochem*, **68**, 1977–1979.

Kim GH, Kim JE, Rhie SJ, Yoon S. (2015) The role of oxidative stress in neurodegenerative diseases. *Exp Neurobiol*, **24**, 325–340.

Kito M, Kokura H, Izaki J, *et al.* (1968) Theanine, a precursor of the phloroglucinol nucleus of catechins in the tea plants. *Phytochemistry*, **7**(4), 599–603.

Kuriyama S, Hozawa A, Ohmori K, Shimazu T, Matsui T, Ebihara S, Awata S, Nagatomi R, Arai H, Tsuji I. (2006) Green tea consumption and cognitive function: a cross-sectional study from the Tsurugaya Project 1. *Am J Clin Nutr*, **83**, 355–361.

Lagalwar S, Guillozet-Bongaarts AL, Berry RW, Binder LI. (2006) Formation of phospho-SAPK/JNK granules in the hippocampus is an early event in Alzheimer disease. *J Neuropathol Exp Neurol*, **65**(5), 455–464.

Lee SY, Lee JW, Lee H, Yoo HS, Yun YP, Oh KW, Ha TY, Hong JT. (2005) Inhibitory effect of green tea extract on beta-amyloid-induced PC12 cell death by inhibition of the activation of NF-kappaB and ERK/p38 MAP kinase pathway through antioxidant mechanisms. *Brain Res Mol Brain Res*, **140**(1–2), 45–54.

Lee YJ, Choi DY, Yun YP, Han SB, Oh KW, Hong JT. (2013) Epigallocatechin-3-gallate prevents systemic inflammation-induced memory deficiency and amyloidogenesis via its anti-neuroinflammatory properties. *J Nutr Biochem*, **24**, 298–310.

Levites Y, Amit T, Mandel S, Youdim MB. (2003) Neuroprotection and neurorescue against Abeta toxicity and PKC-dependent release of nonamyloidogenic soluble precursor protein by green tea polyphenol (−)-epigallocatechin-3-gallate. *FASEB J*, **17**, 952–954.

Lim GP, Chu T, Yang F, Beech W, Frautschy SA, Cole GM. (2001) The curry spice curcumin reduces oxidative damage and amyloid pathology in an Alzheimer transgenic mouse. *J Neurosci*, **21**(21), 8370–8377.

Liu Q, Jie Zhang, Hua Zhu, Chuan Qin, Qi Chen, Baolu Zhao. (2007) Dissecting the signalling pathway of nicotine-mediated neuroprotection in a mouse Alzheimer disease model. *FASEB J*, **21**, 61–73.

Liu X, Du X, Han G, Gao W. (2017) Association between tea consumption and risk of cognitive disorders: a dose-response meta-analysis of observational studies. *Oncotarget*, **8**, 43306–43321.

Luo Y–F, Zhang J, Liu N-Q, Luo Y, Zhao B-L. (2011) Copper ions influence the toxicity of β-amyloid(1-42) in a concentration-dependent manner in a Caenorhabditis elegans model of Alzheimer's disease. *Sci China Life Sci*, **54**, 1–8.

Lynch T, Cherny RA, Bush AI. (2000) Oxidative process in Alzheimer's disease: the role of A_-metal interactions. *Exp Gerontol*, **35**, 445–451

Manczak M, Anekonda TS, Henson E, Park BS, Quinn J, Reddy PH. (2006) Mitochondria are a direct site of A beta accumulation in Alzheimer's disease neurons: implications for free radical generation and oxidative damage in disease progression. *Hum Mol Genet*, **15**(9), 1437–1449.

Mandel S, Amit T, Bar-Am O, Youdim MBH. (2007) Iron dysregulation in AD: Multimodal brain permeable iron chelating drugs, possessing neuroprotective-neurorescue and amyloid precursor protein-processing regulatory activities as therapeutic agents. *Prog Neurobiol*, **82**, 348–360.

Mandel SA, Amit T, Kalfon L, Reznichenko L, Youdim MB. (2008) Targeting multiple neurodegenerative diseases etiologies with multimodal-acting green tea catechins. *J Nutr*, **138**(8), 1578S–1583S.

Markesbery WR, Carney JM. (1999) Oxidative alterations in Alzheimer's disease. *Brain Pathol*, **9**, 133–146.

Mattson MP. (1997) Cellular actions of beta-amyloid precursor protein and its soluble and fibrillogenic derivatives. *Physiol Rev*, **77**(4) 1081–132.

Matsui T, Ingelsson M, Fukumoto H, Ramasamy K, Kowa H, Frosch MP, Irizarry MC, Hyman BT. (2007) Expression of APP pathway mRNAs and proteins in Alzheimer's disease. *Brain Res*, **1161**, 116–123.

Mecocci PL, MacGarvey U, Beal MF. (1994) Oxidative damage to mitochondrial DNA is increased in Alzheimer's disease. *Ann Neurol*, **36**, 747–750.

Morales I, Guzman-Martinez L, Cerda-Troncoso C, Farias GA, Maccioni RB. (2014) Neuroinflammation in the pathogenesis of Alzheimer's disease. A rational framework for the search of novel therapeutic approaches. *Front Cell Neurosci*, **8**, 112.

Multhaup G, Scheuermann S, Schlicksupp A, Simons A, Strauss M, Kemmling A, Oehler C, Cappai R, Pipkorn R, Bayer TA. (2002) Possible mechanisms of APP-mediated oxidative stress in Alzheimer's disease. *Free Radic Biol Med*, **33**, 45–51.

Ng TP, Feng L, Niti M, Kua EH, Yap KB. (2008) Tea consumption and cognitive impairment and decline in older Chinese adults. *Am J Clin Nutr*, **88**, 224–231.

Nishida Y, Yokota T, Takahashi T, Uchihara T, Jishage K, Mizusawa H. (2006) Deletion of vitamin E enhances phenotype of Alzheimer disease model mouse. *Biochem Biophys Res Commun*, **350**(3), 530–536.

Nurk E, Refsum H, Drevon CA, Tell GS, Nygaard HA, Engedal K, Smith AD. (2009) Intake of flavonoid-rich wine, tea, and chocolate by elderly men and women is associated with better cognitive test performance. *J Nutr*, **139**, 120–127.

Obregon DF, Rezai-Zadeh K, Bai Y, Sun N, Hou H, Ehrhart J, Zeng J, Mori T, Arendash GW, Shytle D, Town T, Tan J. (2006) ADAM10 activation is required for green tea (-)-epigallocatechin-3-gallate-induced alpha-secretase cleavage of amyloid precursor protein. *J Biol Chem*, **281**, 16419–16427.

Oda A, Tamaoka A, Araki W. (2009) Oxidative stress up-regulates presenilin 1 in lipid rafts in neuronal cells. *J Neurosci Res*, **88**(5), 1137–1145.

Omar RA, Chyan YJ, Andorn AC, Poeggeler B, Robakis NK, Pappolla MA. (1999) Increased expression but reduced activity of antioxidant enzymes in Alzheimer's disease. *J Alzheimers Dis*, **1**(3), 139–145.

Opazo C, Huang X, Cherny RA, Moir RD, Roher AE, White AR, Cappai R, Masters CL, Tanzi RE, Inestrosa NC, Bush AI. (2002) Metalloenzyme-like activity of Alzheimer's disease beta-amyloid. Cu-dependent catalytic conversion of dopamine, cholesterol, and biological reducing agents to neurotoxic H_2O_2. *J Biol Chem*, **277**, 40302–40308.

Paduraru M, Ciobica A, Hritcu L, Stoica B, Bild W, Stefanescu C. (2010) Changes of some oxidative stress markers in the serum of patients with mild cognitive impairment and Alzheimer's disease. *Neurosci Lett*, **469**(1), 6–10.

Parameshwaran K, Dhanasekaran M, Suppiramaniam V. (2008) Amyloid beta peptides and glutamatergic synaptic dysregulation. *Exp Neurol*, **210**(1), 7–13.

Park S, Kim DS, Kang S, Kim HJ. (2018) The combination of luteolin and l-theanine improved Alzheimer disease-like symptoms by potentiating hippocampal insulin signaling and decreasing neuroinflammation and norepinephrine degradation in amyloid-β-infused rats. *Nutr Res*, **60**, 116–131.

Pérez-Vargas JE, Zarco N, Vergara P, Shibayama M, Segovia J, Tsutsumi V, Muriel P. (2016) L-Theanine prevents carbon tetrachloride-induced liver fibrosis via inhibition of nuclear factor κB and down-regulation of transforming growth factor β and connective tissue growth factor. *Hum Exp Toxicol*, **35**(2), 135–146.

Pîrşcoveanu DFV, Pirici I, Tudorică V, Bălşeanu TA, Albu VC, Bondari S, Bumbea AM, Pîrşcoveanu M. (2017)Tau protein in neurodegenerative diseases — a review. *Rom J Morphol Embryol*, **58**(4), 1141–1150.

Pratico D, Sung S. (2004) Lipid peroxidation and oxidative imbalance: early functional events in Alzheimer's disease. *J Alzheimers Dis*, **6**(2), 171–175.

Praticò D. (2008a) Evidence of oxidative stress in Alzheimer's disease brain and antioxidant therapy. *Ann N Y Acad Sci*, **1147**, 70–78.

Praticò D. (2008b) Oxidative stress hypothesis in Alzheimer's disease: a reappraisal. *Trends Pharmacol Sci*, **29**(12), 609–615.

Querfurth H, LaFerla F. (2009) Alzheimer's disease. *N Engl J Med*, **362**(4), 329–344.

Rezai-Zadeh K, Shytle D, Sun N, Mori T, Hou H, Jeanniton D, Ehrhart J, Townsend K, Zeng J, Morgan D, Hardy J, Town T, Tan J. (2005) Green tea epigallocatechin-3-gallate (EGCG) modulates amyloid precursor protein cleavage and reduces cerebral amyloidosis in Alzheimer transgenic mice. *J Neurosci*, **25**(38), 8807–8814.

Roberts AM, Jagadapillai R, Vaishnav RA, *et al.* (2016) Increased pulmonary arteriolar tone associated with lung oxidative stress and nitric oxide in a mouse model of Alzheimer's disease. *Physiol Rep*, **4**(17), e12953.

Rodriguez-Callejas JD, Fuchs E, Perez-Cruz C. (2020) Increased oxidative stress, hyperphosphorylation of tau, and dystrophic microglia in the hippocampus of aged Tupaia belangeri. *Glia*, **68**(9), 1775–1793.

Rousset S, Alves-Guerra MC, Mozo J, Miroux B, Cassard-Doulcier AM, Bouillaud F, Ricquier D. (2004) The biology of mitochondrial uncoupling proteins. *Diabetes*, **53**(Suppl 1), S130–S135.

Sakato Y. (1949) Studies on the chemical constituents of tea. Part III. On a new amide theanine. *Nippon Nogeikagaku Kaishi*, **23**, 262–267.

Sambamurti K, Kinsey R, Maloney B, Ge YW, Lahiri DK. (2004) Gene structure and organization of the human beta-secretase (BACE) promoter. *FASEB J*, **18**(9), 1034–1036.

Sang S, Tian S, Wang H, Stark RE, Rosen RT, Yang CS, Ho CT. (2003) Chemical studies of the antioxidant mechanism of tea catechins: radical reaction products of epicatechin with peroxyl radicals. *Bioorg Med Chem*, **11**, 3371–3378.

Seeram NP, Henning SM, Niu Y, Lee R, Scheuller HS, Heber D. (2006) Catechin and caffeine content of green tea dietary supplements and correlation with antioxidant capacity. *J Agric Food Chem*, **54**, 1599–1603.

Shen W, Xiao YY, Ying XH, Li ST, Zhai YJ, Shang XP, Li FD, Wang XY, He F, Lin JF. (2015) Tea consumption and cognitive impairment: a cross-sectional study among Chinese elderly. *PLoS ONE*,

Shen CC, Chen Y, Liu H, Zhang K, Zhang T, Lin A, Jing N. (2008) Hydrogen peroxide promotes Abeta production through JNK-dependent activation of gamma-secretase. *J Biol Chem*, **283**(25), 17721–17730.

Smith M, Hirai K, Hsiao K, Pappolla M, Harris P, Siedlak S, Tabaton M, Perry G. (1998) Amyloid-beta deposition in Alzheimer transgenic mice is associated with oxidative stress. *J Neurochem*, **70**(5), 2212–2215.

Smith MA, Rottkamp CA, Nunomura A, Raina AK, Perry G. (2000) Oxidative stress in Alzheimer's disease. *Biochim Biophys Acta*, **1502**(1), 139–144.

Spiers JG, Chen HC, Bourgognon JM, Steinert JR. (2019) Dysregulation of stress systems and nitric oxide signaling underlies neuronal dysfunction in Alzheimer's disease. *Free Radic Biol Med*, **134**, 468–483.

Srividhya R, Gayathri R, Kalaiselvi P. (2012) Impact of epigallo catechin-3-gallate on acetylcholine-acetylcholine esterase cycle in aged rat brain. *Neurochem Int*, **60**, 517–522.

Stadtman ER. (1990) Metal ion-catalyzed oxidation of proteins: biochemical mechanism and biological consequences. *Free Radical Biol Med*, **9**, 315–325.

Strausak D, Mercer JF, Dieter HH, Stremmel W, Multhaup G. (2001) Copper in disorders with neurological symptoms: Alzheimer's, Menkes, and Wilson diseases. *Brain Res Bull*, **55**, 175–185.

Sumathi T, Shobana C, Thangarajeswari M, Usha R. (2015) Protective effect of L-Theanine against aluminium induced neurotoxicity in cerebral cortex, hippocampus and cerebellum of rat brain — histopathological, and biochemical approach. *Drug Chem Toxicol*, **38**(1), 22–31.

Tsushida T, Takeo T. (1984) Occurrence of theanine in Camellia japonica and Camellia sasanqua seedlings. *Agric Biol Chem*, **48**(11), 2861–2862.

Varadarajan S. (2000) Alzheimer's amyloidpeptide-associated free radical oxidative stress and neurotoxicity. *J Struct Biol*, **130**, 184–208.

Wang R, Reddy PH. (2017) Role of Glutamate and NMDA Receptors in Alzheimer's Disease. *J Alzheimers Dis*, **57**(4), 1041–1048.

Wan L, Nie G, Zhang J, Luo Y, Zhang P, Zhang Z, Zhao B. (2011) β-amyloid peptide increases levels of iron content and oxidative stress in human cell and C. elegans models of Alzheimer's disease. *Free Rad Biol Med*, **50**(1), 122–129.

Wan L, Nie G, Zhang J, Zhao B. (2012) Overexpression of human wild-type amyloid-β protein precursor decreases the iron content and increases the oxidative stress of neuroblastoma SH-SY5Y cells. *J Alzheimer's Dis*, **30**, 523–530.

Wang Z. (1980) Biochemistry of tea. Beijing: Agricultural Press, 48–56.

Ward R, Zucca FA, Duyn JH, Crichton RR, Zecca L. (2014) The role of iron in brain ageing and neurodegenerative disorders. *Lancet Neurol*, **13**, 1045–1060.

Weinreb O, Amit T, Mandel S, Youdim MB. (2009) Neuroprotective molecular mechanisms of (–)-epigallocatechin-3-gallate: a reflective outcome of its antioxidant, iron chelating and neuritogenic properties. *Genes Nutr*, **4**, 283–296.

Williams TI, Lynn BC, Markesbery WR, Lovell MA. (2006) Increased levels of 4-hydroxynonenal and acrolein, neurotoxic markers of lipid peroxidation, in the brain in Mild Cognitive Impairment and early Alzheimer's disease. *Neurobiol Aging*, **27**(8), 1094–1099.

Wu KJ, Hsieh MT, Wu CR, Wood WG, Chen YF. (2012) Green tea extract ameliorates learning and memory deficits in ischemic rats via its active component polyphenol epigallocatechin-3-gallate by modulation of oxidative stress and neuroinflammation. *Evid Based Complement Alternat Med*, **2012**, 163106.

Wu Z, Zhang J, Zhao B. (2009) Superoxide anion regulates the mitochondrial free Ca2+ through uncoupling proteins. *Antioxid Redox Signal*, **11**(8), 1805–1818.

Wu Z, Zhao Y, Zhao B. (2010) Superoxide anion, Uncoupling Proteins and Alzheimer's Disease. *J Clin Biochem Nutri*, **46**, 187–194.

Wu Z, Zhu Y, Cao X, Sun S, Zhao B. (2014) Mitochondrial toxic effects of Aβ through mitofusins in the early pathogenesis of Alzheimer's disease. *Mol Neurobiol*, **50**, 986–996.

Xie Y-X, Bezard E, Zhao Bo-L. (2005) Unraveling the receptor-independent neuroprotective mechanism in mitochondria. *J Biol Chem*, **37**, 32405–32412.

Xin W-J, Zhao B-L, Zhang J-Z. (1984) Studies on the property of sulfhydryl groups binding sites on the lung normal cell and cancer cell membrane of Chinese hamster with maleimide spin labels. *Sci Sin B*, **28**, 1008–1014.

Yao M, Nguyen TV, Pike CJ. (2005) Beta-amyloid-induced neuronal apoptosis involves c-Jun N-terminal kinase-dependent downregulation of Bcl-w. *J Neurosci*, **25**(5), 1149–1158.

Zeng L, Lin L, Chen L, Xiao W, Gong Z. (2021) L-Theanine ameliorates d-galactose-induced brain damage in rats via inhibiting AGE formation and regulating sirtuin1 and BDNF Signaling pathways. *Oxid Med Cell Longev*, **2021**, 8850112.

Zhang J, Liu Q, Chen Q, Liu NQ, Li FL, Lu ZB, Qin C, Zhu H, Huang YY, He W, Zhao BL. (2006a) Nicotine attenuates beta-amyloid-induced neurotoxicity by regulating metal homeostasis. *FASEB J*, **20**, 1212–1214.

Zhang J,Mori A,Chen Q, Zhao B. (2006b) Fermented papaya preparation attenuates β-amyloid precursor protein: β-amyloid–mediated copper neurotoxicity in β-amyloid precursor protein and β-amyloid precursor protein Swedish mutation overexpressing SH-SY5Y cells. *Neuroscience*, **143**, 63–72.

Zhang YH, Wang DW, Xu SF, *et al.* (2018) Alpha-Lipoic acid improves abnormal behavior by mitigation of oxidative stress, inflammation, ferroptosis, and tauopathy in P301S Tau transgenic mice. *Redox Biol*, **14**, 535–548.

Zhao B-L. (2005) Natural antioxidant for neurodegenerative diseases. *Mol Neurobiol*, **31**, 283–293.

Zhao B-L. (2009) Natural antioxidants protect neurons in Alzheimer's disease and Parkinson's Disease. *Neurochem Res*, **34**, 630–638.

Zhao B-L. (2012) Natural Antioxidant green tea polyphenols and health. *Acta Biophysica Sinica*, **28**, 26–36.

Zhao Y, Zhao B-L. (2012) Oxidative stress, natural antioxidants protect neurons against Alzheimer's Disease. *Front Biosci*, **15**, 454–461.

Zhao B-L, Wan L. (2012) Imbalance of metal ion metabolism and early pathogenesis of Alzheimer disease. *Prog Biochem Biophys*, **39**, 735–743.

Zheng W, Xin N, Chi ZH, Zhao BL, Zhang J, Li JY, Wang ZY. (2009). Divalent metal transporter 1 is involved in amyloid precursor protein processing and Abeta generation. *FASEB J*, **23**, 4207–4217.

Zhu G, Yang S, Xie Z, Wan X. (2018) Synaptic modification by L-theanine, a natural constituent in green tea, rescues the impairment of hippocampal long-term potentiation and memory in AD mice. *Neuropharmacology*, **138**, 331–340.

Chapter 7

Tea Polyphenols and Parkinson's Disease

Baolu Zhao

Institute of Biophysics, Chinese Academy of Sciences, Beijing, China

7.1 Introduction

Jame Parkinson first reported Parkinson's disease (PD) in 1817, thereafter, it was called Parkinson's disease, also known as paralytic excitability. It is an age-related disease characterized by pathological degeneration of substantia nigra and striatum pathway. PD is a common motor neurodegenerative disease. Its etiology is multifactorial, which has caused an increasing burden on the aging society. PD is characterized by nigrostriatal degeneration and may involve oxidative stress, α-Synuclein (α-Syn) agreggation, imbalance of redox metal homeostasis, and neurotoxicity. Its incidence rate is 1% among the people over 55 years old, second only to Alzheimer's disease (AD), the second largest neurodegenerative disease of mankind. When the degeneration and deletion of selective dopaminergic neurons in substantia nigra and striatum reach about 20% of the normal value, patients with PD will have clinical symptoms, which are characterized by tremor, stiffness, abnormal posture, and motor retardation [Parkinson, 1817].

More and more evidence show that oxidative stress is involved in the pathological process of PD. The excessive production of oxidative stress leads to the depletion of dopaminergic neurons, which is also the main reason for the further damage of residual neurons. Dopamine is a neurotransmitter that can be oxidized by chemical and enzymatic reactions to produce metabolites, reactive oxygen species (ROS), and reactive nitrogen (RNS). It was also found that metal ions such as iron ions in substantia nigra of PD patients were overloaded. Excessive ROS, such as hydrogen peroxide, can react with transition metal iron to produce more active species and hydroxyl radicals, increase oxidative stress, deplete cellular antioxidants, and destroy antioxidant defense system. Redox imbalance can block mitochondrial respiratory chain, leading to respiratory failure and energy crisis. In addition, there is feedback between mitochondrial damage and oxidative stress. Finally, dopaminergic neurons died, and the typical clinical characteristics of patients with PD were observed [Trist *et al.*, 2019].

Tea, one of the drinks with the largest consumption in the world, contains specific polyphenols. Green tea and black tea are rich in polyphenols, among which epigallocatechin gallate (EGCG) and theaflavin are the most abundant. It plays an important role in delaying the onset or preventing the progression of PD. Cumulative evidence suggests that tea drinking is associated with a lower risk of PD. Studies have shown that green tea can prevent the occurrence of nervous system diseases and promote health. Oxidative stress, inflammation, and ROS play a key role in neurodegenerative diseases, supporting the clinical application of free radical scavengers, transition metal chelators, and natural antioxidant tea polyphenols for prevention and treatment of PD. Therefore, green tea polyphenols are currently considered to prevent brain aging and may be used as a neuroprotective agent for neurodegenerative diseases such as PD. Experiments have shown that drinking three cups of green tea a day can reduce the incidence rate of PD. Much data show that it is related to polyphenols, the main component of tea. Green tea polyphenols can prevent hippocampal neuron apoptosis and improve recognition function [Dutta & Mohanakumar, 2015; Camilleri *et al.*, 2013]. The neuroprotective effect of tea polyphenols has been verified in various PD models [Caruana & Vassallo, 2015; Zhao, 2020]. At present, consistent mechanism data showed the effects of tea polyphenols on the neuroprotective and nerve regeneration, indicating that tea polyphenols may interfere directly *in vitro* and animal models α-Syn protein aggregates and regulates intracellular signaling pathways. Nevertheless, despite the important data on its potential neuroprotective effect, clinical research is still very limited. So far, only EGCG has reached the phase II trial.

This chapter discusses the damaging effect of oxidative stress on AD and the inhibitory effect of tea polyphenols on oxidative stress and summarizes the existing knowledge and research results of tea polyphenols for PD, analyzes the effect and mechanism of tea polyphenols in the prevention and treatment of PD, and looks forward to the potential of tea polyphenols as nutritional drugs for various pathologies of PD.

7.2 Parkinson's disease

Parkinson's disease, also known as tremor paralysis, is named after the first detailed description of the disease by British doctor James Parkinson in 1817. Clinically, it is often manifested as tremor, muscle rigidity, movement retardation, and balance disorder [Parkinson, 1817]. Patients of PD may be accompanied by non-motor symptoms such as depression, constipation, and sleep disorders. Levodopa preparation is still the most effective drug. Now there is surgical treatment, and the outcome is also good as it serves as an effective supplement to drug treatment. Rehabilitation treatment, psychotherapy, and good nursing can also improve symptoms to a certain extent. However, the current treatment methods can only improve the symptoms, cannot prevent the progress of the disease, and cannot cure the disease. Although, effective treatment can significantly improve the quality of life of patients. At present, the exact cause of this pathological change is still unclear. Many factors, such as genetic factors, environmental factors, aging, oxidative stress, and so on, may participate in the degeneration and death process of PD dopaminergic neurons. Only 5–10% of PD have family history, which is the first pathogenic gene of PD α. Since the discovery of synuclein (Park1), about 5–10% of patients have inherited mono-genotype PD. So far, at least 23 PD genes and 19 pathogenic genes have been found, but more genetic risk genes and variants of sporadic PD phenotype have been found in various association studies. The study of mutant protein products reveals the potential pathogenic pathway and provides insights into the neurodegenerative mechanism of familial and sporadic PD. At present, environmental factors are considered to be the main factors. For example, synthetic heroin taken by drug users contains a neurotoxic substance of 1-methyl-4-phenyl-1,2,3,6-tetrahydropyridine (MPTP). The substance is converted into highly toxic 1-methyl-4-phenyl-pyridine ion MPP^+ in the brain, and selectively enters substantia nigra dopaminergic neurons to inhibit the activity of mitochondrial respiratory chain complex I and promote oxidative stress response, resulting in the degeneration and death of dopaminergic neurons. In addition, the chemical structures of some herbicides and pesticides are similar to MPTP. Mitochondrial dysfunction may be one of the pathogenic factors of PD. The activity of mitochondrial respiratory chain complex I decreased selectively in substantia nigra in patients with primary PD [Wu, 2015].

7.2.1 *Pathogenic factors of PD*

PD is caused by many factors, such as age, heritage, environments, and so on. The main pathological change of PD is the degeneration and death of dopaminergic neurons in the substantia nigra of the midbrain, which leads to the significant reduction of dopamine content in the striatum. There are many exact causes leading to this pathological change, such as genetic factors, environmental factors, aging, oxidative stress, and so on, which may be involved in the degeneration and death process of PD dopaminergic neurons [Chinta & Andersen, 2005].

The incidence rate and prevalence of PD increased with age. PD mostly occurs in people over the age of 60, which suggests that aging is related to the onset of PD. Data show that dopaminergic neurons in substantia nigra of normal adults gradually decrease with age. However, the prevalence of PD in the elderly over the age of 65 is not high. Therefore, aging is only one of the risk factors causing PD.

The role of genetic factors in the pathogenesis of PD has attracted more and more attention. In the late 1990s, the first pathogenic gene of PD α-Synuclein (α-Syn) was discovered, and since then, at least six pathogenic genes have been associated with familial PD. However, only 5–10% of PD have a family history, and most of them are sporadic cases. Genetic factors are only one of the factors caused PD [Rocha *et al.*, 2018; Belin & Westerlund, 2008].

In the 1980s, American scholars Langston and others found that some drug users would quickly develop typical PD like symptoms and were effective for L-dopamine preparations. The study found that synthetic heroin used by drug addicts contained a neurotoxic substance of MPTP. The substance is converted into highly toxic MPP^+ in the brain, and selectively enters substantia nigra dopaminergic neurons to inhibit the activity of mitochondrial respiratory chain complex I and promote oxidative stress response, resulting in the degeneration and death of dopaminergic neurons. Therefore, the scholars suggest that mitochondrial dysfunction may be one of the pathogenic factors of PD. In subsequent studies, it was also confirmed that the activity of mitochondrial respiratory chain complex I decreased selectively in substantia nigra in patients with primary PD. The chemical structures of some herbicides and insecticides are similar to MPTP. With the discovery of MPTP, people realize that some chemicals, similar to MPTP in the environment, may be one of the pathogenic factors of PD. However, only a small number of drug users exposed to MPTP have the disease, suggesting that PD may be the result of multiple factors [Langston, 1987].

In addition to aging and genetic factors, factors such as brain injury, smoking, and drinking coffee may also increase or reduce the risk of PD. Smoking is negatively correlated with the occurrence of PD, which has been consistent in many studies. Caffeine has a similar protective effect. Severe brain trauma may increase the risk of PD [Ascherio & Schwarzschild, 2016].

In conclusion, PD may be the result of the interaction of multiple genes and environmental factors.

7.2.2 *Pathophysiology of PD*

The prominent pathological changes of PD are the degeneration and death of dopaminergic neurons in the substantia nigra of the midbrain, the significant decrease of dopamine content in the striatum, and the presence of eosinophilic inclusion bodies in the cytoplasm of residual neurons in the substantia nigra, namely Lewy body. When clinical symptoms appear, the death of dopaminergic neurons in substantia nigra is at least more than 50%, and the content of dopamine in striatum is reduced by more than 80%. In addition to the dopaminergic system, the non-dopaminergic system of patients with PD is also significantly damaged. For example, cholinergic neurons in the basal nucleus, noradrenergic neurons in the locus coeruleus, serotonergic neurons in the raphe nucleus of the brainstem, and neurons in the cerebral cortex, brainstem, spinal cord, and peripheral autonomic nervous system. The significant decrease of dopamine content in striatum is closely related to the occurrence of motor symptoms of PD. The significant decrease of dopamine concentration in midbrain limbic system and midbrain cortical system is closely related to the decline of intelligence and affective disorder in patients with PD [Birtwistle & Baldwin, 1998; Sezgin *et al.*, 2019].

7.2.3 *Clinical manifestations*

Parkinson's disease starts covertly and progresses slowly. The first symptom of limb tremor, usually involves one side or the other side of the limb. The main clinical manifestations are static tremor, bradykinesia, myotonia, and postural and gait disorders. In recent years, more and more people have noticed that non-motor symptoms such as depression, constipation, and sleep disorders are also common complaints of patients with PD, and their impact on the quality of life of patients even exceeds that of motor symptoms.

About 70% of patients take tremor as the first symptom, which mostly starts at the distal end of one upper limb, appears obvious at rest, reduces or stops at random movement, intensifies when nervous, and disappears after falling asleep. Hand static tremor is aggravated during walking. The typical manifestation is "pill like" tremor with a frequency of 4–6 Hz. Some patients may be complicated with postural tremor. The typical complaint of the patient is, "My hand often shakes. The more I keep it still, the more I shake. I don't shake when I work and take things. I also shake badly when I meet a stranger or get excited, and I don't shake when I fall asleep" [Jiménez & Vingerhoets, 2012].

When the examiner moves the patient's limbs, neck, or trunk, it can detect obvious resistance. The increase of resistance is uniform in all directions, which is similar to the feeling of bending a soft lead pipe. When the patient is combined with limb tremor, there may be intermittent pauses in the uniform resistance, such as rotating the gear. The patient's typical complaint is, "My limbs are stiff and stiff". In the early stage of the disease, sometimes myotonia is not easy to detect. At this time, the patient can actively move one limb, and the muscle tension of the affected limb will increase [Ferreira-Sánchez *et al.*, 2020].

Slow motion refers to difficulty in starting and loss of active motion. The patient's range of motion will be reduced, especially when repeated exercise. According to the different involved parts, bradykinesia can be manifested in many aspects. The reduction of facial expression, movement, and blink. The voice is monotonous and low, and the enunciation is not clear. Writing becomes slower and smaller. Clumsiness and inflexibility in washing, dressing, and other fine movements. The walking speed becomes slow and often drags, and the swing range of the arm will gradually decrease or even disappear. The step becomes smaller. Salivation occurs due to inability to swallow saliva actively. Difficulty turning over can occur at night. In the early stage of the disease, patients often mistook bradykinesia for weakness, and often misdiagnosed as cerebrovascular disease or cervical spondylosis due to the acid swelling and weakness of one limb. Therefore, when patients slowly develop weakness of one limb and accompanied by increased muscle tension, they should be alerted to the possibility of Parkinson's disease. [Bologna *et al.*, 2013].

The disappearance of postural reflex often occurs in the middle and late stage of the disease. It is difficult for patients to maintain their body balance and it is possible to fall on a slightly uneven road. Postural reflex can be detected by a pull-back test. The examiner stands behind the patient and asks the patient to pull his shoulders when he is ready. Normal people can return to normal standing within one step backward. Patients with postural reflex disappearance often have to step back more than three steps or need help to stand upright. PD patients often walk faster and faster and are not easy to stop, which is called flustered gait. Patients with advanced PD can have freezing phenomenon, which is characterized by the sudden inability to walk for a short time when walking, and their feet seem to stick to the ground. They must pause for a few seconds before they can move on or else, they cannot start again. Freezing is common at the beginning of walking, when turning around, when approaching the target, or when worried about not being able to cross a known obstacle [Debû *et al.*, 2018].

In addition to motor symptoms such as tremor and bradykinesia, patients with PD can also have non-motor symptoms such as depression, anxiety, sleep disorder, and cognitive impairment. Fatigue is also a common non-motor symptom of PD. The typical symptoms of the patient are: feeling very tired and powerless, poor sleep and often inability to sleep, trouble defecating once every few days, bad mood and always unhappy, poor memory, and slow brain response.

7.2.4 *Treatments of PD*

The main treatment methods of PD include drug treatment, symptomatic treatment, prevention and treatment of complications, surgical treatment, and traditional Chinese medicine treatment, but none of these treatments can stop the progression of the disease [Olanow *et al.*, 2009; Zhou and Chen, 2004].

Drug treatment is the most comprehensive treatment for PD. L-dopamine preparation is still the most effective drug. Surgical treatment is an effective supplement to drug treatment. Rehabilitation treatment, psychotherapy, and good nursing can also improve symptoms to a certain extent. At present, the main treatment is to improve symptoms, but it cannot stop disease progression. Medication principle: The dosage should be gradually increased from a small dose. Medication should also emphasize individualization. The best treatment scheme was adopted according to the patient's condition, age, occupation, and economic conditions. During drug treatment, it should not only control the symptoms, but also try to avoid the occurrence of drug side effects and try to control the clinical symptoms of patients for a long time from a long-term perspective [Olanow *et al.*, 2009].

Protective treatment: In principle, once PD is diagnosed, it should be treated as soon as possible. At present, B monoamine oxidase inhibitor

(MAO-B) is the main protective drug in clinic. Recent studies have shown that MAO-B inhibitors may delay the progress of the disease, but there is no conclusion at present [Zhou & Chen, 2004].

When to start medication: If the disease affects the patient's daily life or work ability, or the patient requests to control the symptoms as soon as possible, symptomatic treatment should be started. When the disease is mild in the early stage and has no obvious impact on daily life or work, the medication can be suspended. Patients first choose dopamine (DA) agonists, MAO-B inhibitors, or amantadine/anticholinergic drugs in the early stage. When the original drugs cannot control the symptoms well, compound levodopa should be added for treatment. When the symptom control is not ideal in the middle stage, the dose should be appropriately increased or DA agonist, MAO-B inhibitor, amantadine inhibitor should be added. Due to the disease progression and the emergence of sports complications, the treatment of advanced patients is relatively complex and difficult. Therefore, at the beginning of treatment, we should formulate a reasonable treatment plan in combination with the actual situation of patients, in order to delay the occurrence of sports complications as far as possible.

Motor complications can occur in patients with advanced PD. Patients with PD may have mental symptoms in the late stage of the disease, such as hallucination, euphoria, illusion, etc. Anti-PD drugs can also cause mental symptoms. The most common are phenylhexol hydrochloride and amantadine. Patients with PD may have sleep disorders such as difficulty falling asleep, dreaminess, easy to wake up, and early awakening. If the sleep disorder of PD is caused by the aggravation of the disease at night, you can take levodopa controlled-release agent before going to bed at night.

There are two main surgical methods, neuro-nuclear destruction and deep brain stimulation. Due to its low cost and certain curative effect, neuronucleotomy is still used in some places. Deep brain electrical stimulation has been the first choice for surgical treatment because it is minimally invasive, safe, and effective. Patients with PD who have obvious decreased curative effect or dyskinesia and cannot improve their symptoms well after drug adjustment, can consider surgical treatment. The outcome of operation on limb tremor and myotonia is better, but there will be no significant improvement on axial symptoms such as abnormal posture and gait and dysphagia. Surgery, like drug treatment, can only improve symptoms, not cure the disease, nor prevent disease progression. Surgical treatment was ineffective in patients with secondary Parkinson's syndrome and Parkinson's superposition syndrome. Patients with early PD and patients with good drug treatment effect are not suitable for early operation.

Traditional Chinese medicine, acupuncture, and other treatment methods can play a positive role in improving symptoms, but they need to be treated in regular medical institutions under the guidance of professional doctors [He, 2005].

7.3 Oxidative stress and Parkinson's disease

The characteristics of neurodegenerative disease are that various pathological conditions have similar key processes, such as oxidative stress, free radical activity, protein aggregation, mitochondrial dysfunction, and energy failure. Oxidative stress has been widely believed to be an important pathogenetic mechanism of neuronal apoptosis in PD [Halliwall, 1992]. Overproduction of ROS species can lead to oxidative damage in the brain of PD, as shown by increased lipid peroxidation and DNA damage in the substantia nigra. Increased protein oxidation is also apparent in many areas of the brain, whereas substantia nigra is particularly vulnerable [Jenner & Olanow, 1998]. Under physiological conditions, 6-OHDA is rapidly and non-enzymatically oxidized by molecular oxygen to form hydrogen peroxide (H_2O_2) and the corresponding quinone [Soto-Otero *et al.*, 2000]. H_2O_2 can react with iron (II) to form the reactive and damaging hydroxyl free radical. Quinone then undergoes an intramolecular cyclization, followed by a cascade of oxidative reactions, resulting in the formation of an insoluble polymeric pigment related to neuro-melanin [Graham *et al.*, 1978]. PD is a progressive neurodegenerative disorder, and the hallmark of this disease is selective loss of dopaminergic neurons in the substantia nigra pars compacta [Hirsch *et al.*, 1997]. Recently, the death of dopaminergic neurons has been reported to occur by apoptosis [Tompkins *et al.*, 1997].

It has been found that no matter how many factors and changes lead to the pathogenesis of PD, oxidative stress is the most important point. Whether oxidative stress is the cause or result of PD, it is still closely related to the pathogenesis and course of PD. There is increasing evidence that brain aging and degenerative nervous system diseases are associated with abnormal and excessive neuronal apoptosis [Yuan & Yankner, 2000]. Recent studies have shown that apoptosis is related to the injury and loss of dopaminergic neurons. Apoptosis or programmed cell death is a kind of death regulated by genes, that is, senile death. Apoptosis plays an important role in the development of nervous system and is a normal developmental regulation mechanism. However, various stresses *in vivo* and *in vitro*, especially apoptosis induced by oxidative stress, are closely related to the occurrence of degenerative nervous system diseases such as PD. Experiments show that the apoptosis pathway activated by oxidative stress can lead to the degeneration and damage of dopamine (DA) neurons, and the inhibition of apoptosis pathway can prevent the occurrence of DA neuron degeneration. It is considered that excessive oxidative stress is an important reason for the degeneration of dopamine neurons and the further damage of residual neurons. The levels of α-synuclein or dopamine in endoplasmic reticulum may be related to cellular oxidative stress and PD symptoms.

7.3.1 *Dopamine metabolism produces ROS through chemical reaction or enzymatic reaction*

PD syndrome is caused by excessive loss of dopamine in the forebrain striatum due to the death of dopaminergic neurons in the dense part of the substantia nigra and midbrain. Dopaminergic neurons in patients with PD degenerate and dopaminergic transmitters decrease, resulting in patients with PD who cannot move or move slowly. Dopamine is also involved in mood changes, and the decrease in learning and memory. Cells secrete serotonin, which acts on the receptor and promotes the receptor cells to secrete enkephalin. Enkephalin continues to stimulate the target cells to produce aminobutyric acid, and then the dopamine neurons receiving the signal secrete dopamine, which stimulates the reward center, ultimately affects our mood. Therefore, many PD patients have some mental symptoms such as anxiety and depression. Dopamine damage and reduction affect the motor system and produce symptoms of PD. Dopamine autoxidation can lead to the production of quinones, semiquinones, and (other) ROS, and finally form neuro-melanin. More importantly, dopamine forms hydrogen peroxide and its degradation products under the action of monoamine oxidase (MAO). Excess hydrogen peroxide produces more active hydroxyl radicals ($\cdot OH$) in the reaction with transition metal iron, which increases cellular oxidative stress, consumes antioxidants, and destroys cellular antioxidant defense system.

L-dopa induces dopamine synthesis and oxidative stress in serotonergic cells. L-dopa is the precursor of dopamine synthesis and the main drug for the treatment of PD. However, L-dopa treatment has side effects that may be attributed to non-dopaminergic mechanisms. Synthetic dopamine produces reactive by-products, hydrogen peroxide, and hydroxyl radicals through enzymatic degradation of monoamine oxidase (MAO), or form highly active quinones that bind to proteins and lose their function through automatic oxidation, resulting in neurotoxicity. As the aromatic amino acid decarboxylase (AADC) in dopamine and serotonin neurons can decarboxylate L-dopa, it is assumed that serotonin neurons convert L-dopa into dopamine, resulting in excessive ROS and quinolone proteins, leading to the death of serotonin neurons. A study showed that RN46A-B14 cells contained AADC and could synthesize dopamine after incubation with L-dopa. Furthermore, L-dopa dose-dependently increased intracellular ROS and cell death. Dopamine, ROS production, and cell death were attenuated by co-incubation with the AADC inhibitor, NSD-1015, the MAO inhibitor, pargyline, also attenuated cell death and ROS after L-dopa treatment. Also, quino-protein formation was enhanced significantly by incubation with L-dopa [Stansley & Yamamoto, 2013]. These data illustrate that serotonergic cells can produce dopamine and that the accumulation of dopamine after L-dopa and its subsequent degradation can lead to ROS production and death of serotonergic cells.

7.3.2 *Metal ion unbalance causes oxidative stress*

It was found that when the contents of copper, zinc, aluminum, and manganese are increased, the contents of antioxidants and enzymes, such as

GSH, GPX, and catalase in substantia nigra of PD patients, decreased. The balance of metal ions was destroyed, such as the content of total iron and transferrin increased, and the ratio of iron (II)/iron (III) decreased. Moreover, oxidative damage to DNA and polyunsaturated fatty acids was also found. These evidence suggest that the overall redox balance of neurons is destroyed, the defense ability is reduced, and finally lead to apoptosis and degradation. They are mediated or triggered by the imbalance of metal ions, leading to the changes of key biological systems and a series of events, eventually leading to neurodegeneration and cell death. The reason is multifactorial. Although the source of oxidation steady-state change is not clear, the current evidence shows that the balance of redox transition metals has changed, especially iron, copper, and other trace metals [Jellinger, 2013]. PD is a characterized by abnormalities in the brain α-Syn deposition. Changes in homeostasis of metal-induced oxidative stress and α-Syn amyloid assembly may play a key role in the progress and pathogenesis of PD. Contrary to α-Syn, β-synuclein (βS) is not involved in the PD etiology. α-Syn and βS share similar abilities to coordinate Cu(II). A study showed the importance of M10K mutation, which induces different Cu(I) chemical environments [De Ricco *et al.*, 2015]. It has been suggested that the βS/α-Syn ratio is altered in PD, which caused oxidative stress and brain damage, indicating that a correct balance of these two proteins is implicated in the inhibition of α-Syn aggregation.

Glutathione disulfide, the main cellular disulfide, releases zinc from metallothionein (MT) and metal oxide from MT through zinc mercaptan/disulfide exchange. The interaction of rabbit liver MT-II with other selected biological disulfides (coenzymeA/glutathione mixed disulfide, coenzyme a disulfide and cystine) was studied. The interaction between rabbit liver MT-II and other selected biological disulfides (coenzyme A/glutathione mixed disulfide, coenzyme A disulfide, and cystamine) was investigated by measuring the concomitant release of radioactive 65-zinc from MT. These disulfides react more rapidly than glutathione disulfide, thus, underscoring the reactivity of zinc sulfur bonds in the clusters of metallothionein and the importance of the MT/disulfide interaction as a chemical mechanism for mobilizing zinc from a thermodynamically stable zinc complex. Two implications of these *in vitro* findings are discussed. In the case of zinc, which is redox inert, nature has availed itself of the redox activity of the cysteine ligand to mobilize the metal, and presumably, to permit redox-control of cellular zinc distribution. The mobilization of zinc from MT suggests a possible function of MT as a physiological zinc donor. A shift of the glutathione redox balance under conditions of oxidative stress will accelerate metal release from metallothionein [Maret, 1995].

These new insights into the bioinorganic chemistry of copper and synuclein proteins are a basis to understand the molecular mechanism by which βS might inhibit α-Syn aggregation. Such a disturbance of metal metabolism has important consequences for the progression of diseases such as PD where oxidative stress occurs in affected brain tissue. Metals play an important role in the pathogenesis and pathophysiology of major neurodegenerative diseases such as PD, and chelating agents are potential therapeutic methods.

7.3.3 *Mitochondrial electron transport chain obstruction, respiratory failure, increased oxidative stress, and energy crisis*

Mitochondrial dysfunction in patients with PD is closely related to oxidative stress. The unique genetic characteristics of mitochondria also make its pathogenic role in PD and other late-onset and sporadic neurodegenerative diseases worthy of consideration and play a key role. It was found in autopsy that the enzyme activity and immunostaining of mitochondrial enzyme I (complex I) decreased in 30–40% of PD patients. The loss of mitochondrial enzyme I function leads to the obstruction of mitochondrial electron transport chain and electron leakage, the production of excess ROS, the decoupling of oxidative phosphoric acid, the production of ATP, and the reduction of cells. The final manifestation is the degeneration and deletion of dopamine (DA) neurons. The general mitochondrial abnormalities associated with the disease include mitochondrial electron transport chain damage, mitochondrial morphological and kinetic changes, mitochondrial DNA mutation, and abnormal calcium homeostasis. Mitochondria are important organelles with multiple functions. Their dysfunction can lead to decreased energy production, ROS production, and stress-induced apoptosis [Subramaniam & Chesselet, 2013]. We found that antioxidant nicotine inhibited MPP$^+$ and calcium-induced mitochondria high amplitude swelling and cytochrome c release from intact mitochondria. Intra-mitochondria redox state was also maintained by nicotine, which could be attributed to an attenuation of mitochondria permeability transition. Further investigation revealed that nicotine did not prevent MPP$^+$ or calcium-induced mitochondria membrane potential loss, but instead decreased the electron leak at the site of respiratory chain complex I. In the presence of mecamylamine hydrochloride, a non-selective nicotinic acetylcholine receptor inhibitor, nicotine significantly postponed mitochondria swelling and cytochrome c release induced by a mixture of neurotoxins (MPP$^+$ and 6-hydroxydopamine) in SH-SY5Y cells, suggesting that there is a receptor-independent nicotine-mediated neuroprotective effect of nicotine [Xie *et al.*, 2005]. These results showed that the interaction of antioxidant nicotine with mitochondria respiratory chain, together with its antioxidant effects, should be considered in the neuroprotective effects of nicotine.

Mitochondrial dysfunction is caused by bioenergy defects, mitochondrial DNA mutations, mitochondrial-related nuclear DNA gene mutations, mitochondrial dynamic changes (such as fusion or division), size and morphology changes, transport or transport changes, and mitochondrial movement changes. PD is related to transcriptional damage and the existence of mitochondrial-related mutant proteins. PD involves bioenergy defects, mitochondrial DNA mutations, nuclear DNA gene mutations, changes in mitochondrial dynamics, changes in transport/transport and mitochondrial movement, abnormal size and morphology, transcriptional damage, and the existence of mitochondrial related mutant proteins [Parker & Swerdlow, 1998; Bose & Beal, 2016].

In these studies, we see the mechanism of mitochondrial dysfunction in PD and push new signaling pathways that may participate in PD to the forefront. Some new signaling pathways, such as reverse transcriptase transport pathway and its role in PD, provide strategies for improving the treatment of mitochondrial defects in PD, the mechanism of mitochondrial dysfunction in PD, and the promotion of new signaling pathways that may be involved in PD. For the mitochondrial functions affected in the pathogenesis of sporadic and familial PD, it is valuable to focus on the research and development of mitochondrial targets that may be used for PD neuroprotective interventions in the future.

7.3.4 *Reactive oxygen species (ROS) and reactive nitrogen species (RNS) cause oxidative stress*

Oxidative stress is an intracellular or extracellular state that can lead to the production of ROS, RNS, and their metabolites from chemical or metabolic sources. On the one hand, ROS such as $\bullet O_2^-$ and H_2O_2 and RNS such as NO and NO_2, play an important role in cell signal regulation as intracellular second messengers. The role of excessive oxidative stress induction in the pathogenesis of brain aging and aging-related degenerative nervous system diseases has also been confirmed. The oxygen metabolism rate of brain tissue is very high, the content of metal ions such as iron is also very high, and the protective mechanism of antioxidants is relatively lacking. This leads to the imbalance of ROS and RNS metabolism in brain tissue. In addition, many kinds of cells, such as cortical neurons, cerebellar granule cells, and astrocytes, can produce excessive NO and ROS under pathological conditions. The reaction between NO and $\bullet O_2^-$ produces $ONOO^-$ and its metabolite $\bullet OH$ with stronger reactivity, resulting in extensive oxidative damage in cells. Many environmental factors and metabolic toxins can induce excessive ROS and RNS, resulting in varying degrees of cytotoxicity. Excessive ROS and RNS will not only affect the permeability of cell membrane, cause the increase of intracellular Ca^{2+} influx and the content of second messenger, and then activate a variety of redox pathways, but can also directly start the apoptotic pathway through mutual feedback regulation with mitochondria. Mitochondria are one of the organelles sensitive to ROS and RNS. Excessive ROS and RNS affect the normal function of mitochondrial "energy factory", increase the production of ROS and RNS, release apoptotic factor cytochrome C, activate caspase protein family, and finally lead to apoptosis [Yuan & Yankner, 2000; Hengartner, 2000; Nie *et al.*, 2000a, 2000b; Guo *et al.*, 2005, 2007]. A large number of studies have shown that excessive nitric oxide can induce neuronal apoptosis. Under normal physiological conditions, the concentration of NO and superoxide anion in brain cells is very low, so it is difficult to react. However, under the pathological conditions of PD, the production of ROS and NOS increases greatly, and they easily react to produce peroxynitrite. Much evidence show that NO is involved in the degradation of PD dopaminergic neurons. In the process of MPTP-induced apoptosis of dopaminergic neurons, MPP^+ is concentrated in mitochondria, inhibiting the activity of complex I, and leading to the production of superoxide anion. The latter reacts with NO produced by nNOS and iNOS to produce peroxynitrite, which promotes the oxidation of lipids and proteins, causes DNA strand breaks, and finally leads to cell death, showing the action mode of MPTP in the substantia nigra striatum system [Nie *et al.*, 2000a, 2000b; Smeyne & Jackson-Lewis, 2005]. In SH-SY5Y cells, tea polyphenols inhibited 6-OHDA-induced NF-κB transport and binding [Levites, 2002c], suggesting that tea polyphenols may play a protective role by inhibiting the activity of iNOS [Guo *et al.*, 2005, 2007]. We studied the mechanism of PD induced by 6-OHDA through nitric oxide and peroxynitrite in PC12 and SH-SY5Y cell systems [Nie *et al.*, 2002b; Guo *et al.*, 2005]. PC12 cells treated with 6-OHDA were used as the oxidative damage model of PD, and 6-OHDA caused PC12 cell viability to significantly reduce (Fig. 7-1). The relationship

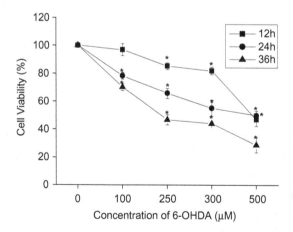

Figure 7-1. Effects of 6-OHDA on PC12 cell viability. PC12 cells were incubated in drug free medium or medium containing different concentrations of 6-OHDA (0 μM, 100 μM, 250 μM, 300 μM, and 500 μM). Cell viability was estimated by MTT assay after the cells treated with 6-OHDA for 12 h, 24 h, and 36 h. Data are mean ± SD, n = 7. *, $p < 0.01$ compared with normal cells (6-OHDA = 0).

between the morphology of PC12 cells treated with 6-OHDA, the temporal dynamics of intracellular calcium, and apoptosis was observed. The results showed that the mode of action of 6-OHDA was obviously concentration-dependent and time-dependent, resulting in cytotoxicity. It was accompanied by the increase of ROS and RNS, the decrease of mitochondrial membrane potential, the increase of intracellular calcium and nitric oxide, the expression of nNOS and iNOS, and the level of protein bound nitrotyrosine. We also studied the mechanism of nitric oxide peroxynitrite-induced PD in rats. It showed that MPTP induced oxidative damage, increased monoamine oxidase located in the outer membrane of mitochondria in the mouse brain, produced ROS and RNS during the catalytic reaction, and led to apoptosis [Nie *et al.*, 2002b; Guo *et al.*, 2005].

In another animal experiment of injecting 6-OHDA. It was found that after 6-OHDA injection, the rotation behavior of rats increased with time and reached a stable level at three weeks, indicating that the apoptosis of dopaminergic neurons in the dense part of substantia nigra caused by unilateral injection of 6-OHDA into the medial forebrain tract was time-dependent. This rat model of PD is the result of the slow action of 6-OHDA. After 6-OHDA injection, ESR experiment showed that compared with the negative control group, 6-OHDA injection increased the contents of ROS and NO free radicals in the midbrain and substantia nigra, and 6-OHDA exacerbated the damage to neurons through diffusion. The level of lipid peroxidation TBARS in the first week was higher than that in the negative control group and increased with time in the midbrain and hippocampus. In midbrain and striatum, tea polyphenol pretreatment significantly reduced the level of TBARS and had a time-dependent effect. Further detection of the production of nitrate/nitrite, the end product of NO showed that compared with the negative control group, nitrate/nitrite in the midbrain and striatum on the injection side of 6-OHDA increased significantly in a time-dependent manner [Guo *et al.*, 2007]. After 6-OHDA injection, the protein expression levels of nNOS and iNOS proteins in midbrain and striatum increased and had a time-dependent effect. Compared with the negative control, 6-OHDA can significantly increase the content of protein bound nitrotyrosine in the middle brain and striatum. It also showed that the number of neurons in the normal group was higher than that in the injured side of 6-OHDA three weeks after

6-OHDA injection, indicating that 6-OHDA had a significant damaging effect on dopaminergic neurons in midbrain. Three weeks after 6-OHDA injection, a large amount of apoptosis of dopaminergic neurons were found in substantia nigra. At the same time, some nerve fibers showed broken DNA fragments in 6-OHDA-induced apoptotic neurons, which further showed that at least a small part of neuronal DNA breakage after 6-OHDA injection was caused by apoptosis. The 6-OHDA injection operation significantly increased ROS and NO in midbrain and striatum, significantly increased the level of lipid peroxidation, and significantly reduced the antioxidant capacity of brain homogenate, indicating that 6-OHDA injury selectively increased ROS in substantia nigra and striatum, oxidized lipid components in cells, and reduced the content of antioxidants in brain tissue while antagonizing ROS [Guo *et al.*, 2007].

In addition, NO molecules are 1000 times smaller than Cu and Zn SOD, and the diffusion speed is fast. The reaction rate of NO with superoxide anion is 10 times faster than that of SOD. Therefore, most of the superoxide anion produced in the cells is captured by NO. Peroxynitrite is a strong oxidant, which can react with protein sulfhydryl or nitroaromatic amino acids and affect its signal transduction function. Peroxynitrite can oxidize lipids, proteins, and DNA, thereby destroying their functions [Cohen *et al.*, 1985; Groves & Wang, 2000; Ghafourifar & Richter, 1999; Torreilles *et al.*, 1999]. NO can also react with iron ligands to reduce Fe(III) by binding to NO and the resulting $Fe^{2+}-NO^+$ can further react with intracellular sulfhydryl groups to nitrose them.

NO plays an important role in the process of apoptosis and participates in the degradation of dopaminergic neurons in PD. 6-OHDA injection increased the concentration of NO and its end product nitrate/nitrite, the expression of nNOS and iNOS, and the content of protein-bound nitrotyrosine. There are many ways to produce NO *in vivo*, among which 6-OHDA induced increased iNOS activity and produced a large amount of NO; MtNOS may also produce a large amount of NO under oxidative stress. NO can react rapidly with superoxide anion to produce more harmful peroxynitrite. Peroxynitrite will further decompose to produce hydroxyl radical, which are the most toxic ROS *in vivo*. They can react with many biological target molecules by oxidizing or nitrifying proteins, lipids, and DNA, so as to destroy their functions and cause lipid peroxidation, oxidize protein and non-protein sulfhydryl groups, cause hydroxylation and nitration of aromatic substances, induce DNA damage, and activate p38 MAPK signal pathway.

7.3.5 *Oxidative stress of aging substantia nigra and the etiology of Parkinson's disease*

Midbrain dopamine neurons of aging non-human primates demonstrates that markers of correlates of dopamine neuron degeneration in PD, including impaired proteasome/lysosome function, oxidative/nitrative damage, and inflammation, all increase with advancing age and are exaggerated in the ventral tier substantia nigra dopamine neurons most vulnerable to degeneration in PD. PD is characterized by dopaminergic neuronal loss in the substantia nigra pars compacta and intracellular inclusions called Lewy bodies (LB). The prevalence of PD rises exponentially with age, and aging is associated with impairment of cellular pathways, which increases susceptibility of dopaminergic neurons to cell death. The causes of motor dysfunction are loss of dopaminergic neurons in the pars compacta of substantia nigra and depletion of dopamine in the striatal pathway of substantia nigra. Although the specific biochemical mechanism is still unclear, oxidative stress plays an undeniable role in the complex progressive neurodegenerative cascade. Molecular factors lead to the high homeostasis of oxidative stress in healthy substantia nigra during aging, and create a chemical environment that makes neurons vulnerable to oxidative damage in PD [Trist *et al.*, 2019]. ROS are considered as key regulators in the development of PD. Despite in-depth studies, the antioxidant-dependent molecular mechanism of PD occurrence and development is still controversial. When the antioxidant capacity of cells decreases, free radicals will cause serious damage and death to dopamine producing cells. Many intracellular reactions produce ROS, including NADPH oxidase activation, mitochondrial dysfunction, and hydrogen peroxide (H_2O_2) decomposition. On the contrary, natural antioxidants, vitamins, proteins, and antioxidant signaling pathways are the main factors that neutralize ROS and their destructive effects. The functional roles of nuclear factor

E2 related factor 2, heme oxygenase-1, and selenium in the initiation and progression of anti-ROS dependent PD were elucidated [Subramaniam & Chesselet, 2013].

Oxidative stress of during aging caused redox imbalance. Especially, the oxidative stress of substantia nigra during aging puts forward a contemporary view on the central biochemical role of the etiology of PD, which can improve the ability to accurately measure oxidative stress. Dopaminergic neurotransmission and cell death pathway *in vivo* are very important for the development of new therapies and the identification of biomarkers of PD. Oxidative stress in the process of aging and as a contributing factor of the pathogenesis of PD are therapeutic methods targeting cellular redox activity.

7.3.6 *Inflammation leads to oxidative stress in Parkinson's disease*

It has been speculated that brain inflammation plays an important role in the pathogenesis of nervous system disease. Treatment of primary rat midbrain mixed neuron glial cultures resulted in the activation of microglia, resident immune cells in the brain, and the subsequent death of dopaminergic neurons. Injection of Lipopolysaccharide (LPS) into the rat substantia nigra (SN) led to the activation of microglia and degeneration of dopaminergic neurons. Microglial activation was observed as early as six hours and loss of dopaminergic neurons was detected three days after the LPS injection. Furthermore, the LPS-induced loss of dopaminergic neurons in the substantia nigra was time- and LPS concentration-dependent [Liu *et al.*, 2000]. The inflammation-mediated degeneration of dopaminergic neurons in the rat substantia nigra, resulting from the targeted injection of LPS, may be a useful model to gain further insights into the pathogenesis of PD. The existing data support the importance of non-cellular autonomous pathological mechanisms in PD, which are mainly mediated by activated glial cells and peripheral immune cells. The response of such cells to neurodegeneration can trigger harmful events, such as oxidative stress and cytokine receptor-mediated apoptosis, which may eventually lead to dopaminergic cell death, leading to disease progression [Hirsch & Hunot, 2009]. Different animal and human studies on PD have focused on oxidative stress, which is due to the induction of DA degeneration pathway in substantia nigra striatum due to increased inflammation and cytokine-dependent neurotoxicity. Chronic inflammation is the main feature of PD and the basis of neurodegeneration. The aging related to glial cell activation caused by nerve injury may be due to immune changes and genetic susceptibility, leading to the deregulation of inflammatory pathway before the onset of PD. A family of inducible transcription factors, nuclear factor-κB (NF-κB), is found to show expression in various cells and tissues, such as microglia, neurons, and astrocytes which play an important role in activation and regulation of inflammatory intermediates during inflammation. Standard and non-standard NF-κB pathway is involved in the regulation of stimulated cells. In the prodromal/ asymptomatic stage of age-related PD, chronic neuro-inflammation may be a driver of neurological dysfunction. Polyphenols are a class of compounds that naturally exist in medicinal plants. They have attracted attention because of their antioxidant and anti-neuro-inflammatory properties in neurodegenerative diseases. In this regard, NF-κB and the role of polyphenols in NF-κB have possible therapeutic effects in B-mediated neuro-inflammation [Singh *et al.*, 2020]. Although the exact mechanism leading to neuronal degeneration and related pathological changes is unclear, the key role of oxidative stress in the pathogenesis of PD is related to several proteins (e.g., α-Synuclein, DJ-1, amyloid β, and tau protein) and some signal pathways (such as extracellular regulated protein kinase, phosphatidylinositol 3 kinase/protein kinase B pathway, and extracellular signal regulated kinase 1/2). Based on the evidence collected in the past decade on a variety of pathogenic proteins and their important signal pathways in PD and the pathogenesis related to oxidative stress, the function of proteins and related signal pathways may be a promising treatment [Jiang *et al.*, 2016].

Oxidative stress reflects the imbalance between the excessive production and incorporation of free radicals and the dynamic ability of biological systems to detoxify active intermediates. Free radicals produced by oxidative stress are one of the common characteristics of several disease experimental models. Free radicals affect the structure and function of nerve cells and lead to a wide range of neurodegenerative diseases, including PD. Possible treatment strategies should include immunomodulatory drugs and therapeutic immunity, aimed at downregulating these inflammatory processes, which may be important to slow down the progression of PD. Antioxidants can protect nerves, effectively eliminate excess ROS, and have a certain therapeutic effect. They have attracted much attention because of their antioxidant and anti-inflammatory properties in neurodegenerative diseases, including PD. However, some of them are controversial in terms of anti-oxidation, and these need to be further studied in patients with PD through clinical application. Tea polyphenol is a natural compound in medicinal plants. The evidence collected in the past decade on the functions of various pathogenic proteins and their important signal pathways in PD and the pathogenesis related to oxidative stress as well as the related signal pathways, may be a promising therapeutic method. Due to their antioxidant and anti-inflammatory properties in neurodegenerative diseases, including PD, they have received extensive attention.

7.4 Preventive and therapeutic effects of tea polyphenols on Parkinson's disease

Parkinson's disease syndrome is caused by excessive loss of dopamine in the forebrain striatum due to the death of dopaminergic neurons in the dense part of substantia nigra. Parkinson's disease with multifactorial etiology is considered to ideally require one or more drugs acting on multiple sites to alleviate symptoms. Accumulating new data suggests that green tea polyphenols may well fulfill the requirement for a putative neuroprotective drug because of their diverse pharmacological activities. Not only have there been a large number of cell system studies, but there has also been a large number of animal system and human epidemiological studies proving that tea polyphenols can prevent and treat PD. We have two kinds of cell and two kinds of animal experiments that have proved the role of tea polyphenols in preventing PD. In PC12 and SH-5Y cells, tea polyphenols can scavenge ROS free radicals and prevent apoptosis of 6-hydroxydopamine cells. In mouse experiments, tea polyphenols have obvious protective effects on tyrosine hydroxylase damage induced by drug MPTP and 6-hydroxydopamine (6-OHDA), and can also regulate antioxidant enzymes. In the left brain injury experiment of rats injected with 6-OHDA, feeding tea polyphenols for 1–3 weeks can significantly reduce the behavior of PD in mice, significantly reduce ROS and RNS free radicals, and significantly reduce the damage to striatum, and have a protective effect on nerve cells [Nie *et al.*, 2002a, 2002b; Guo *et al.*, 2005; Zhao, 2020].

Following, we will discuss the protective effects of tea polyphenols on oxidative stress injury and PD in detail.

7.4.1 *Preventive and therapeutic effects of tea polyphenols on PD in human epidemiological data*

In recent years, green tea polyphenols have attracted extensive attention in the prevention of oxidative stress-related diseases such as cancer, cardio-vascular diseases, and degenerative diseases. A study undertook a case-control study to examine the relationship between tea drinking, factors, and risk of PD among ethnic Chinese in population. The 300 PD and 500 population controls were initially screened. The results demonstrated a dose-dependent protective effect of PD in tea drinkers in a Chinese population [Tan *et al.*, 2003]. A cross-sectional study investigated green tea consumption and cognitive function. The objective was to examine the association between green tea consumption and cognitive function in humans. It analyzed cross-sectional data from a community-based Comprehensive Geriatric Assessment (CGA) conducted in 2002. The subjects were 1,003 Japanese subjects aged 70 years old and above. They completed a self-administered questionnaire that included questions about the frequency of green tea consumption. Results showed that higher consumption of green tea was associated with a lower prevalence of cognitive impairment [Kuriyama *et al.*, 2006]. In elderly Japanese subjects, it was found that higher consumption of green tea was associated with a lower prevalence of cognitive impairment and in the United States, people that consumed two or more cups of tea per day presented a decreased risk of PD [Checkoway *et al.*, 2002]. A recent prospective 13-year study of nearly 30,000 Finnish adults, demonstrated that drinking three or more cups of tea is associated with a reduced risk of PD. A study examined the association of tea consumption with the risk of incident PD among 29,335 Finnish subjects aged 25–74 years without a history of PD at baseline. During a mean follow-up of 12.9 years, 102 men and 98 women developed an incident PD. In both sexes combined, the multivariate-adjusted HRs of PD for subjects drinking three or more cups of tea daily compared with tea non-drinkers was 0.41 [Hu *et al.*, 2007]. These epidemiological results suggest that tea drinking is associated with a lower risk of PD. Increased tea drinking is associated with a lower risk of PD. These findings indicate that the risk of PD is reduced in consumers of green and black tea.

Green tea polyphenols and their major constituents, such as EGCG, have diverse pharmacological activities, such as anti-mutagenic, anti-carcinogenic effects, and anti-neurodegenerative diseases. It is believed that these beneficial effects of green tea polyphenols are the result of their potent antioxidative properties. In fact, it was demonstrated that green tea polyphenols serve as powerful antioxidants against free radicals such as DPPH radicals [Zhao *et al.*, 1989], superoxide anion [Guo *et al.*, 1996], lipid free radicals, and hydroxyl radicals [Zhao *et al.*, 2001]. In the central nervous system (CNS), there is also evidence to show that oral administration of green tea polyphenols and flavonoid-related compounds has preventive effects on iron-induced lipid peroxide accumulation and age-related accumulation of neurotoxic lipid peroxides in the rat brain [Inanami *et al.*, 1998].

7.4.2 *Preventive and therapeutic effects of tea polyphenols on PD in cell system*

Neurodegenerative diseases are related to the dysfunction of cellular redox state. Tea catechins (TCs) are usually expected to be scavengers of free radicals, but different components have different functions. A large number of cell system studies show that tea polyphenols can prevent and treat PD. Our two kinds of cell and two kinds of animal experiments, have proved the role of tea polyphenols in the prevention of PD. In PC12 and SH-SY5Y cells, it can eliminate ROS free radicals and prevent 6-hydroxy-dopamine induced cell apoptosis. [Nie *et al.*, 2002a, 2002b; Guo *et al.*, 2005; Zhao, 2020].

We investigated the different effects of five main components of green tea polyphenols (ECG, (–)-epicatechin gallate (EGCG), (–)-epicatechin

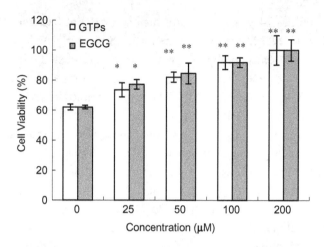

Figure 7-2. Effects of green tea polyphenols and EGCG on 6-OHDA-induced apoptosis decreased in the PC12 cell viability. Cells viability was determined by MTT assay after 24 hours. Different concentrations of green tea polyphenols and EGCG were added 30 minutes before 250 μM 6-OHDA treatments. Date were analyzed by the student *t* test and presented as mean ± SD, n = 7, $p < 0.05$ was considered significant. *, $p < 0.05$ in comparison with the control cells (treatment with 250 μM 6-OHDA); **, $p < 0.01$ in comparison with the control cells.

(EC), (+)-catechin(C), and (−)-epicatechin gallate (ECG)) on 6-hydroxy-dopamine (6-OHDA)-induced apoptosis of PC12 cells, an *in vitro* model of PD. When the cells were treated with five catechins for 30 minutes before exposure to 6-OHDA or MPTP, respectively, EGCG and ECG were in the range of 50–200 μM, while EC, (+)-C, and EGC have obvious protective effects on cell viability (Fig. 7-2) and reduced the ROS concentration (Fig. 7-3). Through the analysis of cell viability, fluorescence microscopy, flow cytometry, and DNA fragment electrophoresis, the five catechins also showed the same and different action modes on 6-OHDA-induced apoptosis. The current results show that 200 μM EGCG or ECG can significantly inhibit the typical apoptotic characteristics of PC12 cells, while other catechins have little protective effect on 6-OHDA-induced cell death. Therefore, the order of the classified protective effects of the five catechins is ECG > or = EGCG >> EC > or = (+)-C >> EGC. The anti-apoptotic activity seems to be structurally related to the 3-gallic acid group of green tea polyphenols. The current data suggest that EGCG and

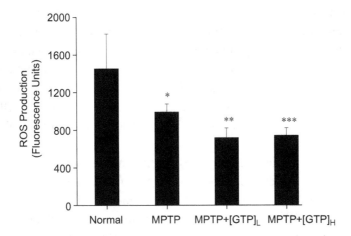

Figure 7-3. ROS production using of DCFH-DA as a probe in crude homogenate from MPTP- and GTP-treated mice cortex. Normal group: without MPTP and GTP treatment; MPTP group: treated with MPTP as a dose of 24 mg/kg/day; MPTP+[GTP]$_L$: treated with both the same amount of MPTP and GTP as a dose of 0.5 mg/kg/day; MPTP+[GTP]$_H$: treated with MPTP and GTP as a dose of 2.5 mg/kg/day. *, $p < 0.1$ comparing with normal group; **, $p < 0.05$ comparing with normal group; ***, $p < 0.01$ comparing with normal group.

ECG may be more effective neuroprotective agents for PD [Nie *et al.*, 2002a].

Our studies demonstrated that TC could protect PC12 cells against apoptosis caused by 6-OHDP. We investigated the effects of the exposure of PC12 cells to 6-OHDA alone or associated with pretreatment with TC. TCs displayed significantly inhibitory effects against PC12 cell death. EGCG and ECG were more effective than TC, but EGC, EC, and (+)-C were less effective. 6-OHDA-induced apoptosis was greatly inhibited by green tea polyphenols at 200–400 μM. From 50–400 μM, the protective effects increased with the concentrations and EGCG was better than green tea polyphenols at the same concentrations. The apoptotic cells were inhibited substantially by 200–400 μM of green tea polyphenols and EGCG. The nuclear changes characteristic of apoptosis disappeared, especially in the EGCG-protected PC12 cells. The DNA ladder also disappeared in the channels of 200–400 μM of green tea polyphenols and EGCG.

In another study, we evaluated the neuroprotective effect of green tea polyphenols on the pathological cell model of PD. The results showed that natural antioxidants could significantly inhibit apoptosis induced by oxidative stress. In this study, 6-OHDA-induced catecholaminergic PC12 cell apoptosis was selected as the *in vitro* model of PD. The concentration is 200 μM green tea polyphenols and its main component, EGCG, has a significant protective effect on PC12 cell apoptosis induced by 6-OHDA. EGCG is more effective than the mixture of green tea polyphenols [Nie *et al.*, 2002b].

The above studies have proved that the main green tea polyphenol, EGCG, plays a powerful neuroprotective role in the mouse model of PD. In a study, EGCG (0.1–1 μM) pretreatment can reduce the death of human neuroblastoma SH-SY5Y cells, which are composed of 6-OHDA (50 μM) 24-hour exposure induction [Levites *et al.*, 2002b]. This study demonstrated highly potent antioxidant-radical scavenging activities of green tea and black tea extracts on brain mitochondrial membrane fraction, against iron (2.5 μM)-induced lipid peroxidation. Both extracts (0.6–3 μM total polyphenols) were shown to attenuate the neurotoxic action of 6-OHDA induced neuronal death. The 6-OHDA (350 μM and 50 μM) activated the iron-dependent inflammatory redox-sensitive nuclear factor (NF)-κB in PC12 and SH-SY5Y cells, respectively. Immunofluorescence and electromobility shift assays showed increased nuclear translocation and binding activity of NF-κB after exposure to 6-OHDA in SH-SY5Y cells, with a concomitant disappearance from the cytoplasm. Introduction of green tea extract (0.6 μM and 3 μM total polyphenols) before 6-OHDA inhibited both NF-κB nuclear translocation and binding activity induced by this toxin in SH-SY5Y cells [Levites *et al.*, 2002c].

Neuroprotection was attributed to the potent antioxidant and iron-chelating actions of the polyphenolic constituents of tea extracts, preventing nuclear translocation and activation of cell death promoting NF-κB. Mandel's group also demonstrated that EGCG restored the reduced protein kinase C (PKC) and extracellular signal-regulated kinases (ERK1/2) activities caused by 6-OHDA toxicity. However, the neuroprotective effect of EGCG on cell survival was abolished by pretreatment with PKC inhibitor GF 109203X (1 μM). EGCG increased phosphorylated PKC, which suggests that PKC isoenzymes are involved in the neuroprotective action of EGCG against 6-OHDA. In addition, gene expression analysis revealed that EGCG prevented both the 6-OHDA-induced expression of several mRNAs, such as Bax, Bad, and Mdm2, and the decrease in Bcl-2, Bclw, and Bcl-xL. These results suggest that the neuroprotective mechanism of EGCG against oxidative-stress-induced cell death includes stimulation of PKC and modulation of cell survival/cell cycle genes [Mandel *et al.*, 2004]. Green tea polyphenols (GTP) are considered to help prevent oxidative stress-related diseases, such as neurodegenerative diseases and aging. We studied the protective mechanism of GTP on SH-SY5Y cells against apoptosis induced by 6-hydroxydopamine (6-OHDA). GTP saved the changes of concentrated nuclei and apoptotic bodies, weakened the early apoptosis induced by 6-OHDA, prevented the decrease of mitochondrial membrane potential, and inhibited the accumulation of ROS and intracellular free Ca^{2+}. GTP can also resist the increase of nitric oxide and the overexpression of nNOS and iNOS induced by 6-OHDA, and reduce the concentration of ROS (Fig. 7-4) and the level of protein bound

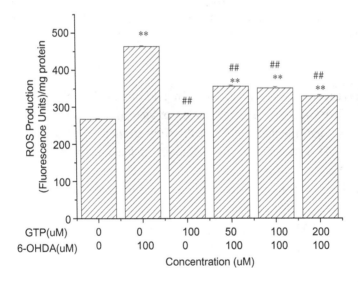

Figure 7-4. Green tea polyphenols attenuates 6-OHDA-induced accumulation of ROS. Cells were exposed to 6-OHDA with different concentration of GTP for 24 hours. Then 2′,7′-dichlorofluorescin diacetate was added to a final concentration of 100 μM and DCF fluorescence was determined 30 minutes later. Data are the mean ± SEM of a ratio between fluorescence intensity and protein content, n = 5. **, $p < 0.01$ compared with control cells; ##, $p < 0.01$ compared with 6-OHDA treated cells.

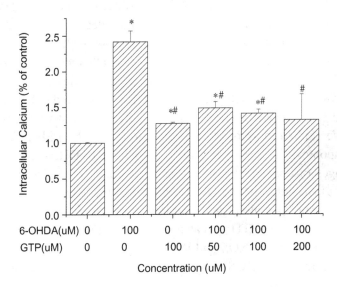

Figure 7-5. Green tea polyphenols attenuates 6-OHDA-induced elevation of intracellular $[Ca^{2+}]_i$ in SY5Y. Cells were exposed to 6-OHDA and/or green tea polyphenols, and $[Ca^{2+}]_i$ was measured 24 hours later. Data are the mean ± SEM of a ratio between fluorescence intensity and protein content, n = 5. *, $p < 0.01$ compared with control cells; #, $p < 0.01$ compared with 6-OHDA treated cells.

3-nitrotyrosine (3-NT) (Fig. 7-5). In addition, GTP inhibited the self-oxidation of 6-OHDA and scavenged oxygen free radicals in a dose and time-dependent manner. Their results showed that the protective effect of GTP on SH-SY5Y cells was mediated at least in part by controlling ROS-NO pathway [Guo *et al.*, 2005].

Above studies have shown that the main green tea polyphenol, EGCG, plays a strong neuroprotective role in the mouse model of PD. The antioxidant function of green tea polyphenols may be the reason for this neuroprotective effect. These studies support the potential efficacy of green tea polyphenols as a neuroprotective agent in the treatment of neurodegenerative diseases PD. It provides a new basis for the basic theory and clinical research of PD. Green tea polyphenols are usually expected to be a potent chemo-preventive agent because of their scavenging free radicals and chelating metal ions ability. However, not all the actions of

GTPs are necessarily beneficial, which need suitable concentration, in a suitable time, and by a particular treatment method.

7.4.3 *Neuroprotective effect of green tea polyphenols on PD animal model induced by MPTP*

The growth of biochemical knowledge of the metabolic process of dopaminergic neurons has opened up a new way to design and screen new therapeutic drugs, which can improve the vitality of dopaminergic neurons by destroying one or more targets in the process of apoptosis. In this regard, some dietary components in fruits, tea, and vegetables may play an important role in alleviating some diseases, such as PD. On the basis of PD model, we established MPTP-induced PD mouse model and 6-OHDA-induced PD rat model, and evaluated the neuroprotective effect of green tea polyphenols, and it was found that tea polyphenols can significantly reduce the behavior of PD, significantly reduce ROS and RNS free radicals, significantly reduce the damage to striatum, and protect nerve cells [Nie *et al.*, 2002a, 2002b; Guo *et al.*, 2005; Zhao, 2020].

Green tea extract and its main polyphenol EGCG have strong neuroprotective activity in PD mouse model. On the basis of PD model, the established MPTP induced PD mouse model and evaluated the neuroprotective effect of green tea polyphenols. It was found that green tea polyphenols had a concentration-dependent inhibitory effect on MAO (monoamine oxidase) activity. The antibacterial constants and IC_{50} values of different compounds were different. The inhibition mechanism may be non-competitive. It is believed that the possible neuroprotective molecular target of green tea polyphenols is apoptosis-related proteins. Green tea polyphenols play a neuroprotective role on dopaminergic neurons by regulating the ratio of Bcl-2 and Bax protein [Nie *et al.*, 2002b]. An animal study demonstrated the neuroprotective property of green tea extract and ECG in the MPTP-treated mice model of PD. MPTP neurotoxin caused dopamine neuron loss in substantia nigra concomitant with a depletion in striatal dopamine and tyrosine hydroxylase protein levels. Pretreatment with either green tea extract or ECG prevented these effects.

In addition, the neurotoxin caused an elevation in striatal antioxidant enzymes superoxide dismutase (SOD) and catalase activities, both effects being prevented by ECG. ECG also increased the activities of both enzymes in the brain. The brain-penetrating property of polyphenols, as well as their antioxidant and iron-chelating properties, might make such compounds an important class of drugs to be developed for treatment of neurodegenerative diseases where oxidative stress has been implicated [Levites *et al.*, 2001]. The brain-penetrating property of polyphenols, as well as their antioxidant properties, might make such compounds an important class of drugs to be developed for treatment of neurodegenerative diseases where oxidative stress has been implicated. In this regard, some dietary components in fruits, tea, and vegetables may play an important role in alleviating some diseases, such as PD.

Monoamine oxidase (MAOB) plays a key role in the pathogenesis of Parkinson's syndrome induced by MPTP. MPTP selectively causes degeneration of nigrostriatal dopaminergic neurons and activates MAOB. In brain and platelets, 3-hydrogen MPTP has consistent binding sites with MAOB. By irreversibly inhibiting MAOB in the brain, blocking the degradation of dopamine, and relatively increasing the content of dopamine, it can achieve the goal of treating PD. We studied the effect of tea polyphenols on ROS and MAOB produced in MPTP treated in mice. In order to establish a subacute PD model with apoptosis as the mechanism of cell damage, we established a PD mouse model: MPTP hydrochloride was dissolved in sterile normal saline with a final concentration of 2 mg/ml and was injected subcutaneously 24 mg/kg/day (free form: 20 mg/kg/day) according to the method in the literature [Levites, 2001], which was administered continuously for five days. The animals were divided into four groups: normal control group, MPTP injury group, GTP low dose protection group (0.5 mg/kg/day), and GTP high dose protection group (2.5 mg/kg/day). The mice were given GTP one day in advance, once every six hours, twice in a row. The results showed that tea polyphenols could significantly reduce the MAOB and ROS in MPTP treated mice (Figs. 7-6, 7-7, Table 7-1). The MAO activities of GTP-treated mice was significantly reduced in cerebellum, cortex, hippocampus, striatum, and midbrain (Table 7-2).

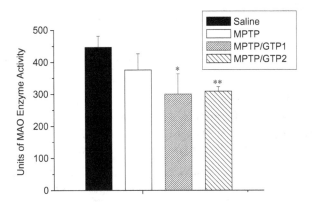

Figure 7-6. Green tea polyphenols inhibition on MAO enzyme activity in the midbrain of MPTP-treated mice. Kyruamine was used as the MAO substrate. The unit of MAO enzyme activity was denoted as transforming 1 pM of 4-hydroxyquinoline per hour per mg tissue. *, $p < 0.05$ comparing to saline group; **, $p < 0.01$ comparing to saline group, one-way ANOVA. GTP1: 0.5 mg/kg; GTP2: 2.5 mg/kg (n ≥ 3).

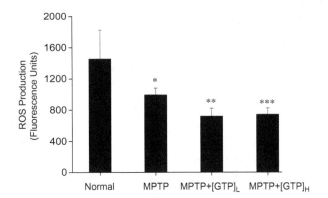

Figure 7-7. ROS production using DCFH-DA as a probe in crude homogenate from MPTP- and GTP-treated mice cortex. Normal group: without MPTP and GTP treatment; MPTP group: treated with MPTP as a dose of 24 mg/kg/day; MPTP+ $[GTP]_L$: treated with both the same amount of MPTP and GTP as a dose of 0.5 mg/kg/day; MPTP+$[GTP]_H$: treated with MPTP and GTP as a dose of 2.5 mg/kg/day. *, $p < 0.1$ comparing with normal group; **, $p < 0.05$ comparing with normal group; ***, $p < 0.01$ comparing with normal group. GTP_L: 0.5 mg/kg; GTP_H: 2.5 mg/kg.

Table 7-1.   Inhibition of green tea polyphenols on MAO (mM).

	EGCG	ECG	EC	(+)-C	EGC	Ferulic acid
Ki	0.33	0.10	0.21	0.39	1.80	0.27
IC_{50}	0.51	0.40	0.48	0.71	0.75	0.25

Table 7-2.   MAO activity of normal and MPTP- and GTP-treated mice. Kyruamine was used as the MAO substrate. The unit of MAO enzyme activity was denoted as transforming 1 pM of 4-hydroxyquinoline per hour per mg tissue. *, $p < 0.05$ comparing to saline group; **, $p < 0.01$ comparing to saline group, one-way ANOVA. GTP1: 0.5 mg/kg; GTP2: 2.5 mg/kg (n ≥ 3).

	Cerebellum	Cortex	Hippocampus	Striatum	Midbrain
Normal mice	164 ± 24	126 ± 10	117 ± 15	236 ± 18	446 ± 35
MTPT mice	147 ± 15	86 ± 10**	81 ± 25	166 ± 46	376 ± 50
MPTP/GTP1	135 ± 16	78 ± 13**	117 ± 49	160 ± 32*	300 ± 63*
MPTP/GTP2	146 ± 22	90 ± 5**	107 ± 26	181 ± 23**	309 ± 14**

7.4.4 *Neuroprotective effect of green tea polyphenols on PD animal model induced by 6-OHDA*

Nitric oxide and its related pathways are considered to play an important role in the pathogenesis of PD. Our *in vitro* experiments show that green tea polyphenols (GTP) may protect dopamine neurons by inhibiting NO and ROS. We established a semi-PD rat model with 6-OHDA to explore the protective mechanism of tea polyphenols. The results showed that the rotation behavior induced by 6-OHDA was time-dependent. Tea polyphenols could reduce the rotation behavior induced by 6-OHDA in a concentration- and time-dependent manner, and reduce the contents of ROS and NO, antioxidant level, degree of lipid peroxidation, nitrate/nitrite content, and protein-bound nitrotyrosine in midbrain and striatum. At the same time, the expression levels of nNOS and iNOS were reduced. Tea polyphenol pretreatment could increase the survival neurons and reduce the apoptotic cells in the dense part of substantia nigra. The results show that GTP treatment protects dopaminergic neurons in a dose-dependent manner by preventing the increase of ROS and NO levels, lipid peroxidation,

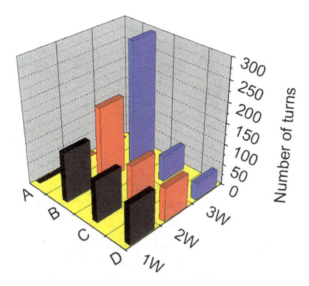

Figure 7-8. Effects of GTP on the mean number of asymmetrical turning executed in the 30 minutes following apomorphine administration (0.25 mg/kg, i.p.) at different time after an unilateral 6-OHDA-induced lesion. (A) sham group; (B) 6-OHDA injection group; (C) low dose; (D) high dose group of GTP. A, B, C, D: one week after injection.

nitrite/nitrate content, inducible nitric oxide synthase, and protein bound 3-nitrotyrosine induced by 6-OHDA in the midbrain and striatum. In addition, GTP treatment retained the free radical scavenging ability of midbrain and striatum in a dose-dependent manner. Oral tea polyphenols can effectively protect brain tissue from nerve cell death caused by 6-OHDA injury, and its protective effect may be realized through the pathway of ROS and NO. Figure 7-8 shows the average number of asymmetric rotations within 30 minutes at different times after unilateral 6-oxda-induced injury within one week after apomorphine administration (0.25 mg/kg, i.p.: intraperitoneal injection.), which significantly increased two weeks and three weeks after injection, but decreased two weeks and three weeks after tea polyphenol injection. It showed that tea polyphenols significantly inhibited the PD behavior induced by 6-oxda-induced injury in rats. Figure 7-9 shows the ESR spectra of ROS and NO captured in midbrain homogenate. ROS and NO captured after unilateral 6-oxda-induced injury significantly increased, while ROS and NO captured two and three weeks

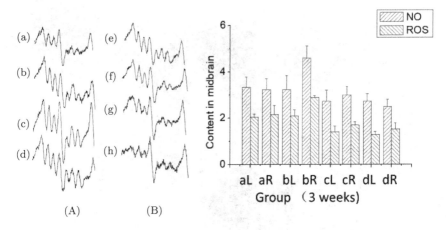

Figure 7-9. The generation of ROS and NO in the midbrain detected by ESR. (A) ESR spectra of ROS trapped by PBN and NO trapped by DETC-ferrous complex in brain homogenates. Spectra are from (a) left side of A group at one week; (b) right side of B group at one week; (c) right side of B group at two weeks; (d) right side of B group at three weeks; (e) right side of C group at three weeks; (f) right side of D group at three weeks; (g) left side of C group at three weeks; (h) left side of D group at three weeks. (B) The contents of NO and ROS in brain homogenates at three weeks, L: left side; R: right side.

after tea polyphenol injection significantly decreased (Fig. 7-10). GTP can also resist 6-OHDA-induced increase in nitric oxide and overexpression of nNOS and iNOS, and reduce protein-bound 3-nitrotyrosine (3-NT) level (Fig. 7-11) [Guo *et al.*, 2007].

These results support the protective effect of GTP on 6-OHDA-induced neuron injury, and suggest that the neuroprotective effect of GTP can be used in the prevention and treatment of PD. The experimental evidence provides experimental evidence for the neuroprotective theory of tea polyphenols and a new idea and strategy for the prevention and treatment of PD.

A significant pathological feature of PD is the abnormal accumulation of iron in reactive microglia of substantia nigra compacta and the abnormal accumulation of iron related to neuromelanin in dopamine (DA) neurons containing melanin. The ability of free iron to enhance and promote the formation of toxic ROS has been discussed countless times. Recent observations have found that iron induces inertia α-Syn aggregates

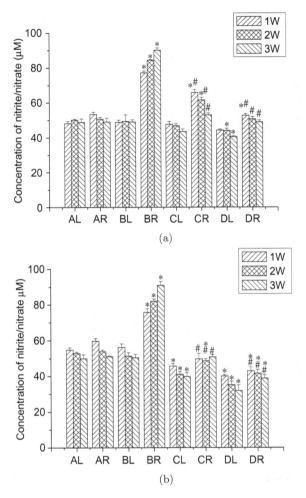

Figure 7-10. Nitrite/nitrate content in brain homogenate of (a) midbrain and (b) striatum. Every date come from six animals. *, $p < 0.01$ indicated significant difference comparing with sham group; #, $p < 0.01$ indicated significant difference comparing with 6-OHDA lesion group.

into toxic aggregates, which strengthens the key role of iron in the pathogenesis of oxidative stress-induced DA neuron degeneration and protein degradation through ubiquitination. Mandel *et al.* [2004] studied the effects of tea polyphenols on iron and iron in substantia nigra of mice treated with MPTP α-Syn. The results showed the role of neuroprotective

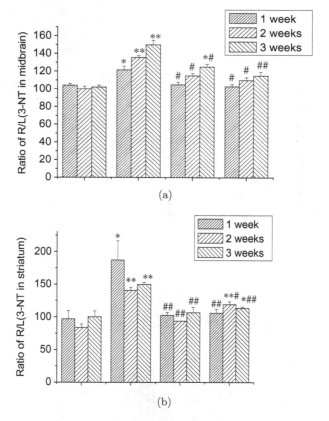

Figure 7-11. Protein bound 3-NT content in brain homogenate of (a) midbrain and (b) striatum. Every date comes from six animals, expressed as percentage of the ratio of right side to left side ± SEM. *, $p < 0.01$ indicated significant difference comparing with sham group; #, $p < 0.01$ indicated significant difference comparing with 6-OHDA lesion group.

drugs r-apomorphine and green tea polyphenol EGCG. Neurodegeneration induced by MPTP and 6-hydroxydopamine in rodents and non-human primates is associated with substantia nigra pars compacta. It is related to the increase of α-Syn. The accumulation of iron in MPTP-induced neurodegeneration is related to the nitric oxide-dependent mechanism, leading to significant degradation of iron regulatory protein through ubiquitination. Free radical scavengers, green tea polyphenols EGCG, and the recently developed brain permeable vk-28 series-derived iron chelators, have neuroprotective effects on these neurotoxins in mice and rats and can prevent substantia nigra pars compacta iron and α-Accumulation of α-Syn [Mandel *et al.*, 2004].

These studies show that tea polyphenols can prevent and treat PD in animal models. The antioxidant effect of tea polyphenols may be an important method for neuroprotection of PD and other neurodegenerative diseases. In addition, the brain permeability and antioxidant and iron chelating properties of polyphenols may make these compounds an important class of drugs for the treatment of neurodegenerative diseases involving oxidative stress.

7.4.5 *EGCG protects melanogaster against PD by reshaping the intestinal microbiota*

Background: Parkinson's disease (PD) is a neurodegenerative disorder with no absolute cure. The evidence of the involvement of gut microbiota in PD pathogenesis suggests the need to identify certain molecule(s) derived from the gut microbiota, which has the potential to manage PD. Recent evidence provides support for the involvement of microbiota-gut-brain axis in PD pathogenesis. We propose that a pro-inflammatory intestinal milieu, due to intestinal hyper-permeability and/or microbial dysbiosis, initiates or exacerbates PD pathogenesis. One factor that can cause intestinal hyper-permeability and dysbiosis is chronic stress which has been shown to accelerate neuronal degeneration and motor deficits in Parkinsonism rodent models. In addition, RS promoted a number of rotenone-induced effects in the brain, including reduced number of resting microglia and a higher number of dystrophic/phagocytic microglia as well as (FJ-C+) dying cells in the substantia nigra (SN), increased lipopolysaccharide (LPS) reactivity in the SN, and reduced dopamine (DA) and DA metabolites in the striatum compared to control mice [Dodiya *et al.*, 2020]. It is well known that the main polyphenol EGCG in green tea has beneficial effects on patients with PD. Alteration of the gut microbiota may contribute to the development of inflammatory bowel disease (IBD). The results of a study revealed that anti-inflammatory effect and colonic barrier integrity were enhanced by oral, but not rectal, EGCG.

Epigallocatechin-3-gallate (EGCG), a major bioactive constituent of green tea, is known to be beneficial in IBD alleviation. Prophylactic EGCG attenuated colitis and significantly enriched short-chain fatty acids (SCFAs)-producing bacteria such as Akkermansia and SCFAs production in DSS-induced mice. Microbiota from EGCG-dosed mice alleviated the colitis over microbiota from control mice and SFF shown by superiorly anti-inflammatory effect and colonic barrier integrity, and also enriched bacteria such as Akkermansia and SCFAs [Wu *et al.*, 2021]. Some mechanisms considered that EGCG mediated protection is achieved by remodeling intestinal microbiota. In this study, EGCG was given to *Drosophila melanogaster* together with PINK1 (PTEN induced hypothetical kinase 1) mutation (PD model), and its behavioral performance and neuronal/mitochondrial morphology were measured. According to the results, mutant pink1b9 *Drosophila melanogaster* showed dopaminergic, survival, and behavior defects, which were saved by supplementing EGCG. At the same time, EGCG caused profound changes in the intestinal microbial components of pink1b9 *Drosophila melanogaster* and restored the abundance of a group of bacteria. However, noting that the protective effect of EGCG is weakened when the intestinal microbiota is destroyed by antibiotics. This study showed that EGCG can improve neuronal and behavioral defects by reshaping intestinal microbiota and TotM expression of the *Drosophila* model of PD. Transcriptomic analysis identified TotM as the central gene responding to EGCG or microbial manipulations. Genetic ablation of TotM blocked the recovery activity of EGCG, suggesting that EGCG-mediated protection warrants TotM. Apart from familial form, EGCG was also potent in improving sporadic PD symptoms induced by rotenone treatment, wherein gut microbiota shared regulatory roles [Xu *et al.*, 2020].

These results suggest the relevance of the gut microbiota-TotM pathway in EGCG-mediated neuroprotection, providing insight into indirect mechanisms underlying nutritional intervention of PD.

7.4.6 *Theaflavin protects neuron against PD*

Theaflavin is a golden yellow pigment existing in black tea and the product of tea fermentation. In biochemistry, theaflavin is a kind of polyphenol hydroxyl substance with benzophenone structure. It is the first compound with exact pharmacological effect found in black tea. Tea theaflavin accounts for 0.5–2% of the weight of dry black tea, depending on the method of black tea processing. Theaflavin plays a certain role in the bright color and strong taste of tea soup. It is an important quality index of black tea. Theaflavin has a variety of pharmacological properties, including effective antioxidant, anti-apoptotic, and anti-inflammatory effects.

Evidence from clinical and experimental studies show that the degeneration of dopaminergic neurons in substantia nigra and striatum is the pathological feature of PD. A study aimed to investigate the neuroprotective effects of theaflavin on oxidative stress, monoamine transporters, and behavioral abnormalities in MPTP-induced neurodegeneration. Theaflavin is known to have neuroprotective effects on ischemia, AD, and other neurodegenerative diseases, but its beneficial mechanism on MPTP-induced dopaminergic neurodegeneration is not clear. Administration of MPTP increased oxidative stress and decreased the expression of behavior patterns, nigrostriatal dopamine transporter (DAT), and vesicular monoamine transporter 2 (VMAT2). Theaflavin pretreatment can reduce oxidative stress, improve exercise behavior and the expression of DAT and VMAT2 in striatum and substantia nigra [Anandhan *et al.*, 2012a]. Theaflavin improves behavioral defects, biochemical indexes, and monoamine transporter expression in subacute MPTP-induced PD mouse model. Long-term administration of MPTP/probenecid (MPTP/P) can lead to oxidative stress, apoptosis, and loss of dopaminergic neurons, resulting in motor disorders. Another study evaluated the effect of theaflavin on MPTP/P-induced neurasthenia in C57BL/6 mice. It was found that theaxanthin could attenuate MPTP/P-induced apoptosis and neurodegeneration, which was characterized by increased expression of tyrosine hydroxylase and dopamine transporter in substantia nigra, decreased apoptosis markers, and normalized behavioral characteristics. This may be due to antioxidant and anti-apoptotic activities [Anandhan *et al.*, 2012b]. Theaflavin protects substantia nigra dopaminergic neurons against chronic MPTP/probenecid-induced PD. Another study focused on the possible anti-inflammatory and anti-apoptotic effects of theaflavin, a black tea polyphenol against MPTP-induced neurotoxicity in mice. C57BL/6 male mice were treated with 10 doses of MPTP and probenecid for 3.5 days interval. MPTP/p treatment upregulates the release of interleukin-1beta, IL-6, tumor necrosis factor-alpha, IL-10, glial fibrillary acidic protein, and Bax, and downregulates anti-apoptotic marker Bcl-2. Oral treatment of black tea polyphenol theaflavin significantly attenuates MPTP-induced neuro-inflammation as well as apoptosis. Behavioral studies (catalepsy and akinesia) were carried out to confirm these molecular studies [Anandhan *et al.*, 2013].

These results suggest that theaflavin may reduce MPTP-induced dopaminergic neuron injury, possibly through its neuroprotective and antioxidant capacity. These results demonstrate the protective effect of theaflavin on neuro-inflammation and apoptosis in chronic MPTP/probenecid model of PD, and theaflavin mediated its neuroprotection against chronic MPTP-induced toxicity through the involvement of multiple molecular events. It was concluded that theaflavin may provide a precious therapeutic strategy for the treatment of progressive neurodegenerative disease such as PD in future. These data suggest that theaflavin may provide a valuable therapeutic strategy for the treatment of progressive neurodegenerative diseases such as PD.

7.5 Mechanisms of the protective effects of tea polyphenols on neurons against Parkinson's disease

It can be found that the protective mechanism of tea polyphenols on PD is multifaceted from the above discussion. These include anti-oxidation, free radical scavenging, iron chelating properties, mitochondria, apoptosis, cellular signal regulated kinase, protein kinase C signal pathway activation, and cell survival/cell cycle gene regulation, etc.

7.5.1 *Protective effects of tea polyphenols on neurons against PD by antioxidant mechanism*

Accumulation studied have shown that green tea and black tea have strong antioxidant effect, free radical scavenging activity, and resist lipid peroxidation induced by iron in brain mitochondrial membrane

[Guo *et al.*, 1996, 2005; Zhao *et al.*, 1989, 2001, 2020]. Oxidative stress induced by glutamate excitotoxicity is related to neurodegenerative diseases such as PD. For example, a study showed that green tea polyphenols protected the cortical neurons against glutamate-induced neurotoxicity. Green tea polyphenols were then showed to inhibit the glutamate-induced ROS release and SOD activity reduction in the neurons [Cong *et al.*, 2016]. Another study demonstrated the administration of EGCG-attenuated anesthesia-induced memory deficit in young mice, potentially via the modulation of nitric oxide expression and oxidative stress [Ding *et al.*, 2018]. In 6-OHDA-induced rat and SH-SY5Y cell and animal PD models, tea polyphenols can regulate oxidative stress, protect neurons by regulating NO, and inhibit PD behavior [Guo *et al.*, 2005, 2007].

Green tea and black tea could attenuate the neurotoxic effect of 6-OHDA-induced neuronal death. 6-OHDA activates iron-dependent inflammatory redox sensitive nuclear factor NF-κB in rat PC12 and human neuroblastoma SH-SY5Y cells. Immunofluorescence and electromigration analysis showed that NF-κB decreased after exposure to 6-OHDA in SH-SY5Y cells. The translocation of nuclear NF-κB and binding activity increased and disappeared from the cytoplasm. The introduction of green tea extract before 6-OHDA inhibited the NF-κB induced by the toxin in neuroblastoma SH-SY5Y cells nuclear translocation and binding activity. The neuroprotective effect is attributed to the effective antioxidant and iron chelation of polyphenols in tea extract, preventing nuclear NF-κB translocation and promoting cell death activation. The brain permeability of polyphenols may make them an important drug for the treatment of neurodegenerative diseases [Levites *et al.*, 2002c]. The biological properties of green tea polyphenols reported in the literature include antioxidant effect, free radical scavenging, iron chelating property, activation of protein kinase C or extracellular signal regulated kinase signaling pathway, and cell survival/cell cycle gene regulation. Tea polyphenols may play a protective role by inhibiting the activity of iNOS. It can protect MPP induced neurotoxicity. Depreny is a special inhibitor of MAO-B, which can reduce the production of hydrogen peroxide and thus, reduce oxidative stress. In MPTP-induced PD mice and monkeys, 7-nitroindazole (7-NI) can effectively prevent the decrease of striatal dopamine and the degradation of substantia nigra cells. The possible mechanism is that 7-NI is an inhibitor of NO, which regulates oxidative stress and protects neurons by regulating no signal.

Pathologically accumulated metal ions (iron species and Mn^{3+}) and abnormally up-regulated monoamine oxidase B (MAOB) activity-induce endogenous dopamine (DA) oxidation, which leads to mito-chondrial damage, lysosomal dysfunction, proteasome inhibition, and selective dopamine neuron vulnerability, which is related to the patho-genesis of PD. The dopamine oxidation can generate deleterious ROS and highly reactive dopamine quinones (DAQ) to induce dopamine-related toxicity, which can be alleviated by dopamine oxidation sup-pressors, ROS scavengers, dopamine quinones quenchers, and MAOB inhibitors. However, the nuclear factor erythroid 2-related factor 2 (Nrf2)-Keap1 and peroxisome proliferator-activated receptor gamma coactivator 1-alpha (PGC-1α) anti-oxidative and proliferative signaling pathways, play roles in anti-oxidative cell defense and mitochondria biogenesis, which is implicated in dopamine neuron protections. Therefore, agents with capabilities to suppress dopamine-related toxic-ity, including inhibition of dopamine oxidation, scavenge of ROS, detoxification of dopamine quinones, inhibition of MAOB, and modu-lations of anti-oxidative signaling pathways, can be protective to dopa-mine neurons. Tea polyphenols with more phenolic hydroxyl groups and ring structures have stronger protective functions. The protective capabilities of tea polyphenols is further strengthened by evidence that phenolic hydroxyl groups can directly conjugate with dopamine qui-nones. However, GSH and other sulfhydyl groups containing agents have weaker capabilities to abrogate DA oxidation, detoxify ROS and dopamine quinones, and inhibit MAOB. The tea polyphenols are identi-fied to protect against overexpression of mutant A30P α-Syn-induced dopamine neuron degeneration and dopamine-like symptoms in trans-genic *Drosophila* [Zhou *et al.*, 2019].

Oxidative stress is associated with neuronal loss and age-related cog-nitive decline in neurodegenerative diseases. Tea polyphenols are consid-ered to play a beneficial role in protecting the central nervous system from oxidative and excitotoxic stress, although its mechanism is unclear. Using oxidized low-density lipoprotein (oxLDL) as oxidative damage, the mechanism of neurotoxicity was studied to identify the possible action sites of the effective protective flavonoids epicatechin in cultured primary neurons. The potential roles of extracellular signal regulated kinase 1/2 (ERK1/2) and c-Jun N-terminal kinase (JNK) were explored. OxLDL stimulated Ca^{2+}-dependent activation of ERK1/2 and JNK, which was strongly inhibited by low micro molar concentration epicatechin pretreatment. However, inhibition of ERK1/2 activation by mitogen activated protein kinase kinase (MEK) inhibitors neither reduced nor enhanced oxLDL induced neurotoxicity, suggesting that this cascade is unlikely to be involved in oxLDL toxicity or the protective effect of flavonoids. Oxidized LDL caused sustained activation of JNK, resulting in phosphorylation of transcription factor c-Jun, which was abolished in neurons pretreated with flavonoids. In addition, oxidized low density lipoprotein induced the cleavage of pre-cysteine protease-3 and increased cysteine protease-3-like protease activity in neurons, which was strongly inhibited by pre-exposure to epicatechin. In addition, caspase-3 inhibitor reduced oxLDL-induced neuronal death, suggesting an apoptotic mechanism [Schroeter *et al.*, 2001].

Above evidence show that tea consumptions are related to deceased PD prevalence. The protective capabilities of tea polyphenols and other PD relevant agents inhibit DA-related toxicity and protect against environmental or genetic factors-induced dopamine neuron degeneration *in vitro* and *in vivo*. Tea polyphenols significantly suppressed dopamine-related toxicity to protect dopamine neurons. The tea polyphenols can protect dopamine neurons via inhibition of dopamine oxidation, conjugation with dopamine quinones, scavenge of ROS, inhibition of MAOB, and modulations of Nrf2-Keap1 and PGC-1α anti-oxidative signaling pathways. Therefore, tea polyphenols may be a protective agent for inhibiting oxidative stress and preventing oxidative damage of neurons through cell responses activated in many ways. Based on achievements from these results, the excellent and versatile protective capabilities of tea polyphenols are highlighted, which will contribute and benefit future anti-PD therapy.

7.5.2 *Protective effects of tea polyphenols against PD by anti-apoptotic mechanism*

Apoptosis and autophagy are important intracellular processes to maintain environmental stability and promote cell survival. Autophagy selectively degrades damaged organelles and protein aggregates, while apoptosis clears damaged or aging cells. Maintaining the balance between autophagy and apoptosis is very important for cell fate, especially for long-lived cells such as neurons. On the contrary, their imbalance is related to neurodegenerative diseases such as PD, which is characterized by the progressive loss of dopaminergic neurons in the dense part of substantia nigra. Restoring the balance between autophagy and apoptosis is a promising strategy for the treatment of PD. Some core proteins are involved in the interaction between apoptosis and autophagy, including members of the B-cell lymphoma-2 (bcl-2) family. Bcl-2 members play an important role in apoptosis and autophagy regulation [Liu *et al.*, 2019]. When ERK1/2 lost phosphorylation and was partially inactivated, JNK phosphorylation and activity increased. They jointly regulate apoptosis. 6-OHDA activates nuclear transcription factor NF-κB in PC12 cells and SH-SY5Y cells, increasing nuclear transport ability and binding ability to DNA. Tea polyphenols can increase the mRNA expression of bad, Bax, caspase10, caspase7, caspase1, and MDM2, and reduce the mRNA levels of anti-apoptotic genes Bcl-2 and Bcl XL. The anti-apoptotic response of tea polyphenols in low concentration (1–10 μM) was significantly different compared with high concentration (50–500 μM). The expression pattern of pro-apoptotic genes was observed. The results demonstrated that DA, EGCG, and melatonin showed similar gene expression and protein profiles. The imbalance of internal environment stability is related to PD, which is characterized by the progressive loss of dopaminergic neurons in the dense part of substantia nigra. In addition, the results showed that green tea polyphenols could restore the dysfunction of mitochondrial pro- or anti-apoptotic proteins Bax, Bcl-2, and caspase-3 caused by glutamate [Weinreb *et al.*, 2003].

These results show that apoptosis is induced not only by oxidative stress, but also by regulating intracellular signal transduction and gene expression. Dietary tea polyphenols may act as a protective agent against neuronal apoptosis through selective effects in stress-activated cellular responses, including protein kinase signal cascade.

7.5.3 *Protective effects of tea polyphenols against PD by cellular signal regulation*

The above studies reveal that many effects of EGCG on PD are independent of antioxidant mechanisms. In particular, EGCG directly interacts with proteins and phospholipids in plasma membrane, DNA methylation, second messenger PKC, α-Syn, transcription factors, mitochondrial function and autophagy, regulates signal transduction pathways, and plays many useful biological roles.

EGCG is usually classified as an antioxidant according to its chemical structure. However, in the presence of iron(III), cells treated with EGCG produce hydrogen peroxide and hydroxyl radicals. These novel molecular mechanisms of action for EGCG, particular, EGCG directly interacts with proteins and phospholipids in the plasma membrane and regulates signal transduction pathways, transcription factors, DNA methylation, mitochondrial function, activation of second messengers, and autophagy to exert many of its beneficial biological actions [Kim *et al.*, 2014]. EGCG stimulated signal transduction pathway in vascular smooth muscle cells and stimulated restenosis [Ahn *et al.*, 1999]. Moreover, EGCG regulates activities of cell surface growth factor receptors, especially receptor tyrosine kinases (RTK), including epidermal growth factor receptor (EGFR), vascular endothelial growth factor receptor (VEGFR), insulin-like growth factor receptor (IGFR), and the insulin receptor (InsR) [Ahn *et al.*, 1999; Kondo *et al.*, 2002]. Tea polyphenols can protect nerve cells and prevent PD through this pathway.

Synuclein is a potential protein that may lead to PD α-Syn constitutes, the main protein component of Lewy body. PD is a characterized by abnormalities in the brain α-Syn deposition. Changes in homeostasis and metal-induced oxidative stress may be contributed by α-Syn amyloid assembly, which plays a key role in the progress and pathogenesis of PD. The emergence of transgenic *Drosophila* models with similar terms meets the requirements of most PD models, which can open up a new signal way of PD research. In particular, the *Drosophila* model has advantages over other existing models, such as progressive, age-dependent, selective DA neuron deletion, and the formation of fibrous α-Syn containing cell connotation. These models show that α-Syn is closely related to the pathogenesis of PD, α-Syn selectively damages DA neurons, the relationship between other factors such as α-Syn, Lewy body, ubiquitin, and oxidative stress, and that α-Syn plays a role in the pathogenesis and progression of PD. Theaflavin is an effective inhibitor of Aβ and α-Syn and antioxidants of fibril formation. Similar to EGCG, theaflavin stimulates Aβ and α-Syn is assembled into non-toxic spherical aggregates, which do not play a role in the formation of seed amyloid protein β, and the fibrils were remolded into non-toxic aggregates. Compared with EGCG, theaflavin is less sensitive to air oxidation and has higher efficacy under oxidation conditions. These findings suggest that theaflavin may be used to remove toxic amyloid deposits [Grelle *et al.*, 2011]. There are consistent mechanism data on the neuroprotective and nerve regeneration effects of tea polyphenols, indicating that tea polyphenols may interfere directly *in vitro* and animal models α-Syn protein aggregates and regulates intracellular signaling pathways.

Protein kinase C is a cytoplasmic enzyme. Once there is a second messenger, PKC will become a membrane-bound enzyme. It can activate enzymes in the cytoplasm and participate in the regulation of biochemical reactions. At the same time, it can also act on transcription factors in the nucleus and participate in the regulation of gene expression. Research found that EGCG restored the decreased activity of protein kinase C (PKC) and extracellular signal regulated kinase (ERK1/2) caused by 6-OHDA toxicity. However, after pretreatment with PKC inhibitor, the neuroprotective effect of EGCG on cell survival was eliminated. Since EGCG increases phosphorylated PKC, PKC isozymes are involved in the neuroprotective effect of EGCG on 6-OHDA. In addition, gene expression analysis showed that EGCG prevented the expression of several mRNA induced by 6-OHDA, such as Bax, bad, and MDM2, as well as the reduction of Bcl-2, Bcl-w, and Bcl-X (L). These results suggest that the neuroprotective mechanisms of EGCG against oxidative stress-induced cell death include stimulating PKC and regulating cell survival/cell cycle genes [Levites *et al.*, 2002a]. PKC with the inhibitor or by downregulation of PKC, blocked the EGCG-induced sAPPα secretion, suggesting the involvement of PKC. Indeed, EGCG induced the phosphorylation of PKC, thus identifying a novel PKC-dependent mechanism of EGCG action by activation of the non-amyloidogenic pathway. EGCG markedly increased PKCα and PKC in the membrane and the cytosolic fractions of mice hippocampus [Levites *et al.*, 2003]. Iron chelating, PKC, or extracellular signal regulated kinase signal pathway activation and cell survival/cell cycle gene regulation. Tea polyphenols may play a protective role by inhibiting the activity of iNOS. It can protect MPTP induced neurotoxicity. Thus, EGCG has protective effects against Aβ-induced neurotoxicity and regulates secretory processing of non-amyloidogenic APP via PKC pathway.

Studies indicated the mechanism of concentration-dependent neuroprotective and pro-apoptotic effects of EGCG by gene expression and protein assay. It demonstrated a concentration- and time-dependent correlation between EGCG, and melatonin in modulation of cell survival/cell death-related gene pathways. Unlike the effects of low concentrations, where an anti-apoptotic response was manifest, a pro-apoptotic pattern of gene expression was observed at high toxic concentrations of EGCG (e.g., increase in caspases, fas, and gadd45) [Weinreb *et al.*, 2003]. Another result demonstrated that green tea polyphenols restored the dysfunction of mitochondrial pro- or anti-apoptotic proteins Bax, Bcl-2, and caspase-3 caused by glutamate. Interestingly, the neuroprotective effect of green tea polyphenols was abrogated when the neurons were incubated with siBcl-2 [Cong *et al.*, 2016]. These results have provided novel insights into the gene mechanisms involved in both the neuroprotective and pro-apoptotic activities of neuroprotective drugs. It showed that EGCG and melatonin exhibit similar gene expression and protein profiles.

Therefore, these results suggest that many effects of EGCG are independent of antioxidant mechanisms. In particular, EGCG directly interacts with proteins and phospholipids in the plasma membrane and regulates signal transduction pathways, transcription factors, DNA methylation, and gene expression to play many useful biological roles. The pathway and the mechanism of tea polyphenols in prevention and treatment of PD are summarized in Figure 7-12.

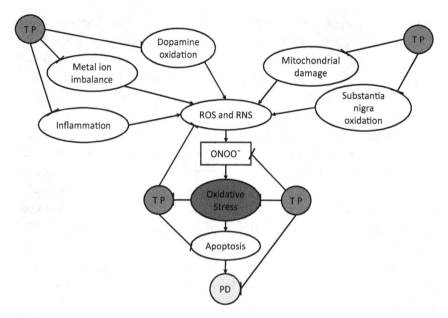

Figure 7-12.   The pathways and mechanisms of tea polyphenols in prevention and treatment of PD. TP: tea polyphenols.

7.6 Conclusion

From the above discussion, we can see that oxidative stress plays an important role in leading to nerve injury and PD, and the randomized epidemiological study of the population supports habitual drinking tea to reduce the risk of PD and effectively control the disease syndrome. A large number of studies at the cellular and animal levels have shown that tea polyphenols can regulate signal transduction pathways, transcription factors, DNA methylation, mitochondrial function, and autophagy through anti-oxidation, anti-apoptosis, inhibition of inflammation, and play many useful biological roles to effectively control PD syndrome. Therefore, it is suggested that tea cooperates with established drugs (such as levodopa) in clinical treatments to maximize its role at a specific level of the typical pathway of disease phenotype-induced PD.

References

Ahn HY, Hadizadeh KR, Seul C, Yun YP, Vetter H, Sachinidis A. (1999) Epigallocathechin-3 gallate selectively inhibits the PDGF-BB-induced intracellular signaling transduction pathway in vascular smooth muscle cells and inhibits transformation of sis-transfected NIH 3T3 fibroblasts and human glioblastoma cells (A172). *Mol Biol Cell*, **10**, 1093–1104.

Anandhan A, Tamilselvam K, Radhiga T, Rao S, Essa MM, Manivasagam T. (2012a) Theaflavin, a black tea polyphenol, protects nigral dopaminergic neurons against chronic MPTP/probenecid induced Parkinson's disease. *Brain Res*, **1433**, 104–113.

Anandhan A, Janakiraman U, Manivasagam T. (2012b) Theaflavin ameliorates behavioral deficits, biochemical indices and monoamine transporters expression against subacute 1-methyl-4-phenyl-1,2,3,6-tetrahydropyridine (MPTP)-induced mouse model of Parkinson's disease. *Neuroscience*, **218**, 257–267.

Anandhan A, Essa MM, Manivasagam T. (2013) Therapeutic attenuation of neuroinflammation and apoptosis by black tea theaflavin in chronic MPTP/probenecid model of Parkinson's disease. *Neurotox Res*, **23**(2), 166–173.

Ascherio A, Schwarzschild MA. (2016) The epidemiology of Parkinson's disease: risk factors and prevention. *Lancet Neurol*, **15**(12), 1257–1272.

Belin AC, Westerlund M. (2008) Parkinson's disease: a genetic perspective. FEB J, **275**(7), 1377–1383.

Bose A, Beal MF. (2016) Mitochondrial dysfunction in Parkinson's disease. *J Neurochem*, **139**(Suppl 1), 216–231.

Birtwistle J, Baldwin D. (1998) Role of dopamine in schizophrenia and Parkinson's disease. *Br J Nurs*, **7**(14), 832–834, 836, 838–841.

Bologna M, Fabbrini G, Marsili L, Defazio G, Thompson PD, Berardelli A. (2013) Facial bradykinesia. *J Neurol Neurosurg Psychiatry*, **84**(6), 681–685.

Camilleri A, Zarb C, Caruana M, Ostermeier U, Ghio S, Hogen T, Schmidt F, Giese A, Vassallo N. (2013) Mitochondrial membrane permeabilisation by amyloid aggregates and protection by polyphenols. *Biochim Biophys Acta*, **1828**, 2532–2543.

Caruana M, Vassallo N. (2015) Tea polyphenols in Parkinson's Disease. *Adv Exp Med Biol*, **863**, 117–137.

Checkoway H, Powers K, Smith-Weller T, Franklin GM, Longstreth WT Jr, Swanson PD. (2002) Parkinson's disease risks associated with cigarette smoking, alcohol consumption, and caffeine intake. *Am J Epidemiol*, **155**(8), 732–738.

Chinta SJ, Andersen JK. (2005) Dopaminergic neurons. *Int J Biochem Cell Biol*, **37**(5), 942–946.

Cohen G, Pasik P, Cohen B, *et al.* (1985) Pargyline and deprenyl prevent the neurotoxicity of1-methyl-4-phenyl-1,2,3,6-tetrahydropyridine (MPTP) in monkeys. *Eur J Pharmacol*, **106**, 209–210.

Cong L, Cao C, Cheng Y, Qin XY. (2016) Green tea polyphenols attenuated glutamate excitotoxicity via antioxidative and antiapoptotic pathway in the primary cultured cortical neurons. *Oxid Med Cell Longev*, **2016**, 2050435

Debû B, De Oliveira Godeiro C, Lino JC, Moro E. (2018) Managing gait, balance, and posture in Parkinson's disease. *Curr Neurol Neurosci Rep*, **18**(5), 23.

De Ricco R, Valensin D, Dell'Acqua S, *et al*. (2015) Differences in the binding of copper(I) to α- and β-synuclein. *Inorg Chem*, **54**(1), 265–272.

Ding L, Gao X, Hu J, Yu S. (2018) (-)Epigallocatechin-3-gallate attenuates anesthesia-induced memory deficit in young mice via modulation of nitric oxide expression. *Mol Med Rep*, **18**(6), 4813–4820.

Dodiya HB, Forsyth CB, Voigt RM, Engen PA, Patel J, Shaikh M, Green SJ, Naqib A, Roy A, Kordower JH, Pahan K, Shannon KM, Keshavarzian A. (2020) Chronic stress-induced gut dysfunction exacerbates Parkinson's disease phenotype and pathology in a rotenone-induced mouse model of Parkinson's disease. *Neurobiol Dis*, **135**, 104352.

Dutta D, Mohanakumar KP. (2015) Tea and Parkinson's disease: constituents of tea synergize with antiparkinsonian drugs to provide better therapeutic benefits. *Neurochem Int*, **89**, 181–190.

Ferreira-Sánchez MDR, Moreno-Verdú M, Cano-de-la-Cuerda R. (2020) Quantitative measurement of rigidity in Parkinson's disease: a systematic review. *Sensors (Basel)*, **20**(3), 880.

Ghafourifar P, Richter C. (1999) Mitochondrial nitric oxide synthase regulates mitochondrial matrix pH. *Biol Chem*, **380**, 1025–1028.

Graham D, Tiffany SM, Bell WR Jr, Gutknecht WF. (1978) Autoxidation versus covalent binding of quinines as the mechanism of toxicity of dopamine, 6-hydroxydopamine, and related compounds toward C1300 neuroblastoma cells in vitro. *Mol Pharmacol*, **14**, 644–653.

Grelle G, Otto A, Lorenz M, Frank RF, Wanker EE, Bieschke J. (2011) Black tea theaflavins inhibit formation of toxic amyloid-beta and alpha-synuclein fibrils. *Biochemistry*, **50**(49), 10624–10636.

Groves JT, Wang CC. (2000) Nitric oxide synthase: models and mechanisms. *Curr Opin in Chem Biol*, **4**, 687–695.

Guo Q, Zhao B-L, Li M-F, Shen S-R, Xin W-J. (1996) Studies on protective mechanisms of four components of green tea polyphenols (GTP) against lipid peroxidation in synaptosomes. *Biochem Biophys Acta*, **1304**, 210–222.

Guo SH, Bezard E, Zhao BL. (2005) Protective effect of green tea polyphels on the SH-SY5Y cells against 6-OHDA induced apoptosis through ROS-pathway. *Free Rad Biol Med*, **39**, 682–695.

Guo SH, Bezard E, Zhao BL. (2007) Protective effects of green tea polyphels in the 6-OHDA rat model of Parkinson's disease through inhibition of ROS-pathway. *Biol Psychiatry*, **62**(12), 1353–1362.

Halliwall B. (1992) Reactive oxygen species and the central nervous system. *J Neurochem*, **59**, 1609–1623.

He J. (2005) Rethinking on the treatment of Parkinson's disease with traditional Chinese medicine — Also on "nourishing the kidney and calming the liver, resolving phlegm and activating blood circulation, detoxifying and dispersing knots" is the basic treatment principle of Parkinson's disease. *Bull Tradit Chin Med*, **1**, 12–14.

Hengartner MO. (2000) The biochemistry of apoptosis. *Nature*, **407**(6805), 770–776.

Hirsch EC, Faucheux B, Damier P, Prigent A, Agid Y. (1997) Neuronal vulnerability in Parkinson's disease. *J Neural Transm*, **50**, 79–88.

Hirsch EC, Hunot S. (2009) Neuroinflammation in Parkinson's disease: a target for neuroprotection? *Lancet Neurol*, **8**(4), 382–397.

Hu G, Bidel S, Jousilahti P, Antikainen R, Tuomilehto J. (2007) Coffee and tea consumption and the risk of Parkinson's disease. *Mov Disord*, **22**, 2242–2248.

Inanami O, Watanabe Y, Syuto B, Nakano M, Tsuji M, Kuwabara M. (1998) Oral administration of (–) catechin protects against ischemia–reperfusion-induced neuronal death in the gerbil. *Free Radical Res*, **29**, 359–365.

Jellinger KA. (2013) The relevance of metals in the pathophysiology of neurodegeneration, pathological considerations. *Int Rev Neurobiol*, **110**, 1–47.

Jiang T, Sun Q, Chen S. (2016) Oxidative stress: a major pathogenesis and potential therapeutic target of antioxidative agents in Parkinson's disease and Alzheimer's disease. *Prog Neurobiol*, **147**, 1–19.

Jenner P, Olanow CW. (1998) Understanding cell death in Parkinson's disease. *Ann Neurol*, **44**(Suppl 1), S72–S84.

Jiménez MC, Vingerhoets FJ. (2012) Tremor revisited: treatment of PD tremor. *Parkinsonism Relat Disord*, **18**(Suppl 1), S93–S95.

Kim HS, Quon MJ, Kim JA. (2014) New insights into the mechanisms of polyphenols beyond antioxidant properties; lessons from the green tea polyphenol, epigallocatechin 3-gallate. *Redox Biol*, **2**, 187–195.

Kondo T, Ohta T, Igura K, Hara Y, Kaji K. (2002) Tea catechins inhibit angiogenesis *in vitro*, measured by human endothelial cell growth, migration and tube formation, through inhibition of VEGF receptor binding. *Cancer Lett*, **180**, 139–144.

Kuriyama S, Hozawa A, Ohmori K, Shimazu T, Matsui T, Ebihara S, Awata S, Nagatomi R, Arai H, Tsuji I. (2006) Green tea consumption and cognitive function: a cross-sectional study from the Tsurugaya Project 1. *Am J Clin Nutr*, **83**, 355–361.

Langston JW. (1987) MPTP: insights into the etiology of Parkinson's disease. *Eur Neurol*, **26**(Suppl 1), 2–10.

Levites Y, Weinreb O, Maor G, Youdim MBH, Mandel S. (2001) Green tea polyphenol epigallocatechin-3-gallate prevents MPTP induced dopaminergic neurodegeneration. *J Neurochem*, **78**, 1073–1082.

Levites Y, Amit T, Mandel S, Youdim MBH. (2003) Neuroprotection and neuro-rescue against amyloid beta toxicity and PKC-dependent release of non-amyloidogenic soluble precursor protein by green tea polyphenol (–)-epigallocatechin-a-gallate. *FASEB J*, **17**, 952–954.

Levites Y, Weinreb O, Maor G, Youdim MB, Mandel S. (2002a) Green tea poly-phenol (–)-epigallocatechin-3-gallate prevents N-methyl-4-phenyl-1,2,3,6-tetrahydropyridine-induced dopaminergic neurodegeneration. *J Neurochem*, **78**, 1073–1082.

Levites Y, Youdim MB, Maor G, Mandel S. (2002b) Attenuation of 6-hydroxydo-pamine (6-OHDA)-induced nuclear factor-kappaB (NF-κB) activation and cell death by tea extracts in neuronal cultures. *Biochem Pharmacol*, **63**, 21–29.

Levites Y, Amit T, Youdim MBH, Mandel S. (2002c) Involvement of protein kinase C activation and cell survival/cell cycle genes in green tea polyphe-nol–epigallocatechin-3-gallate neuroprotective action. *J Biol Chem*, **277**(34), 30574–30580.

Liu B, Jiang JW, Wilson BC, Du L, Yang SN, Wang JY, Wu GC, Cao XD, Hong JS. (2000) Systemic infusion of naloxone reduces degeneration of rat sub-stantia nigral dopaminergic neurons induced by intranigral injection of lipopolysaccharide. *J Pharmacol Exp Ther*, **295**, 125–132.

Liu J, Liu W, Yang H. (2019) Balancing apoptosis and autophagy for Parkinson's disease therapy: targeting BCL-2. *ACS Chem Neurosci*, **10**(2), 792–802.

Mandel S, Maor G, Youdim MB. (2004) Iron and alpha-synuclein in the substan-tia nigra of MPTP-treated mice: effect of neuroprotective drugs R-apomorphine and green tea polyphenol (–)-epigallocatechin-3-gallate. *J Mol Neurosci*, **24**(3), 401–416.

Maret W. (1995) Metallothionein/disulfide interactions, oxidative stress, and the mobilization of cellular zinc. *Neurochem Int*, **27**(1), 111–117.

Nie GJ, Jin C-F, Cao Y-L, Shen S-R, Zhao B-L. (2002a) Distinct effects of tea catechins on 6-hydroxydopamine-induced apoptosis in PC12 cells. *Arch Biochem Biophys*, **397**, 84–90.

Nie GJ, Cao YL, Zhao B-L. (2002b) Protective effects of green tea polyphenols and their major component, (–)-epigallocatechin-3-gallate (EGCG), on 6-hydroxyldopamine-induced apoptosis in PC12 cells. *Redox Report*, **7**, 170–177.

Olanow CW, Stern MB, Sethi K. (2009) The scientific and clinical basis for the treatment of Parkinson disease. *Neurology*, **72**(21 Suppl 4), S1–S136.

Parker Jr WD, Swerdlow RH. (1998) Mitochondrial dysfunction in idiopathic Parkinson disease. *Am J Hum Genet*, **62**, 758–762.

Parkinson J. (1817) An essay on the Shaking Palsy. London: Sherwood, Nelly and Jones, 1817.

Rocha EM, De Miranda B, Sanders LH. (2018) Alpha-synuclein: Pathology, mitochondrial dysfunction and neuroinflammation in Parkinson's disease. *Neurobiol Dis*, **109**(Pt B), 249–257.

Schroeter H, Spencer JP, Rice-Evans C, Williams RJ. (2001) Flavonoids protect neurons from oxidized low-density-lipoprotein-induced apoptosis involving c-Jun N-terminal kinase (JNK), c-Jun and caspase-3. *Biochem J*, **358**, 547–557.

Sezgin M, Bilgic B, Tinaz S, Emre M. (2019) Parkinson's disease dementia and lewy body disease. *Semin Neurol*, **39**(2), 274–282.

Singh SS, Rai SN, Birla H, *et al.* (2020) NF-kappaB-mediated neuroinflammation in Parkinson's disease and potential therapeutic effect of polyphenols. *Neurotox Res*, **37**(3), 491–507.

Smeyne RJ, Jackson-Lewis V. (2005) The MPTP model of Parkinson's disease. *Mol Brain Res*, **134**, 57–66.

Soto-Otero R, Méndez-Álvarez E, Hermida-Ameijeiras Á, Muñoz-Patiño AM, Labandeira-Garcia JL. (2000) Autoxidation and neurotoxicity of 6-hydrodopamine in the presence of some antioxidants: potential implication in relation to the pathogenesis of Parkinson's disease. *J Neurochem*, **74**, 1605–1612.

Stansley BJ, Yamamoto BK. (2013) L-dopa-induced dopamine synthesis and oxidative stress in serotonergic cells. *Neuropharmacology*, **67**, 243–251.

Subramaniam SR, Chesselet MF. (2013) Mitochondrial dysfunction and oxidative stress in Parkinson's disease. *Prog Neurobiol*, **106–107**, 17–32.

Tan EK, Tan C, Fook-Chong SM, Lum SY, Chai A, Chung H, Shen H, Zhao Y, Teoh ML, Yih Y, Pavanni R, Chandran VR, Wong MC. (2003) Dose-dependent protective effect of coffee, tea, and smoking in Parkinson's disease: a study in ethnic Chinese. *J Neurol Sci*, **216**(1), 163–167.

Tompkins MM, Basgall EJ, Zamrini E, Hill WD. (1997) Apoptotic-like changes in Lewy-body-associated disorders and normal aging in substantia nigra neurons. *Am J Pathol*, **150**, 119–131.

Torreilles FO, Salman-Tabcheh S, Guerin MC, Torreilles J. (1999) Neurodegenerative disorders: the role of peroxynitrite. *Brain Res Rev*, **30**, 153–163.

Trist BG, Hare DJ, Double KL. (2019) Oxidative stress in the aging substantia nigra and the etiology of Parkinson's disease. *Aging Cell*, **18**(6), e13031.

Weinreb O, Mandel S, Youdim MB. (2003) Gene and protein expression profiles of anti- and pro-apoptotic actions of dopamine, R-apomorphine, green tea polyphenol (–)-epigallocatechine-3-gallate, and melatonin. *Ann N Y Acad Sci*, **993**, 351–361.

Wu J. (2015) Neurology (Third Edition) People's Health Publishing House, Beijing, 291–299.

Wu Z, Huang S, Li T, Li N, Han D, Zhang B, Xu ZZ, Zhang S, Pang J, Wang S, Zhang G, Zhao J, Wang J. (2021) Gut microbiota from green tea polyphenol-dosed mice improves intestinal epithelial homeostasis and ameliorates experimental colitis. *Microbiome*, **9**(1), 184.

Xie Y-X, Bezard E, Zhao Bo-L. (2005) Unraveling the receptor-independent neuroprotective mechanism in mitochondria. *J Biol Chem*, **37**, 32405–32412.

Xu Y, Xie M, Xue J, Xiang L, Li Y, Xiao J, Xiao G, Wang HL. (2020) EGCG ameliorates neuronal and behavioral defects by remodeling gut microbiota and TotM expression in Drosophila models of Parkinson's disease. *FASEB J*, **34**(4), 5931–5950.

Yuan J, Yankner B. (2000) Apoptosis in the nervous system. *Nature*, **407**, 802–809.

Zhao B-L, Li X-J, He R-G, Cheng S-J, Xin W-J. (1989) Scavenging effect of extracts of green tea and natural antioxidants on active oxygen radicals. *Cell Biophys*, **14**, 175–181.

Zhao B-L, Guo Q, Xin W-J. (2001) Free radical scavenging by green tea polyphenols. *Methods Enzymol*, **335**, 217–231.

Zhao B-L. (2020) The pros and cons of drinking tea. *Tradit Med Mod Med*, **3**(3), 1–12.

Zhou H, Chen S. (2004) Progress in drug treatment of Parkinson's disease. *World Clinical Drugs*, **25**, 518.

Zhou ZD, Xie SP, Saw WT, Ho PGH, Wang H, Lei Z, Yi Z, Tan EK. (2019) Implications of tea polyphenols against dopamine (DA) neuron degeneration in Parkinson's disease (PD). *Cells*, **8**(8), 911.

Chapter 8

Tea Polyphenols and Cardiovascular and Cerebrovascular Health

Baolu Zhao

Institute of Biophysics, Chinese Academy of Sciences, Beijing, China

8.1 Introduction

The incidence and mortality of cardiovascular and cerebrovascular disease are the highest among all disease in China and all over the world, and are still increasing with the rising living standards. Cardiovascular and cerebrovascular diseases are generally referred to ischemic or hemorrhagic diseases of the heart, brain, and systemic tissues caused by hyperlipidemia, blood viscosity, atherosclerosis, and hypertension. According to the WHO, cerebrovascular diseases are the second leading cause of death worldwide and the major cause of disability in adults. Cardiovascular and cerebrovascular disease is a common disease that seriously threatens the health of human beings, especially the middle-aged and elderly over 50 years old. It has the characteristics of high prevalence, high disability rate, and high mortality. Even with the application of the most advanced and perfect treatment methods, more than 50% of the survivors of cerebrovascular accidents cannot take care of themselves. The number of people dying from cardiovascular and cerebrovascular diseases in the world is as high as 15 million every year, ranking first among various causes of death. Cardiovascular and cerebrovascular disease is a kind of common circulatory disease. The circulatory system is composed of heart, brain, blood vessels, and neuro-humoral tissues regulating blood circulation. Circulatory system diseases are also known as the diseases of all the above tissues and organs. These common diseases can significantly affect the labor force of patients. There are two kinds of heart diseases and one brain disease — congenital heart disease and acquired heart disease. Congenital heart disease is caused by the abnormal development of the heart in the fetal period, and the later heart disease is caused by external or internal factors. There are many variations, such as coronary atherosclerotic heart disease, rheumatic heart disease, hypertensive heart disease, pulmonary heart disease, infectious heart disease, endocrine heart disease, hematological heart disease, nutritional metabolic heart disease, etc. The brain disease mainly is stroke. Stroke represents 3–4% of the healthcare spending in developed countries [Evers *et al.*, 2004]. In recent years, the incidence rate of hyperlipidemia, hyperglycemia, hypertension, cardiovascular, and cerebrovascular disease has increased dramatically in China and world. If the treatment is not timely, it will leave sequela and even lead to death, which will bring enormous pressure and burden to the family and society. This will be a severe medical and social problem for people. Therefore, there is an urgent need to explore the preventive and therapeutic effects and mechanisms of effective methods, avoid and reduce the risk, and prevent and treat cardiovascular and cerebrovascular disease as well as develop safe and effective health products and drugs for the general population and promote cardiovascular and cerebrovascular health [Zhao, 1999, 2007, 2008, 2016].

Many studies have shown that cardiovascular and cerebrovascular diseases are closely related to oxidative stress. A large number of reactive oxygen species (ROS) and reactive nitrogen (RNS) free radicals are produced in the pathogenesis of cardiovascular and cerebrovascular diseases, resulting in cardiovascular and cerebrovascular damage. Epidemiology and many experimental studies show that antioxidants in tea polyphenols can scavenge ROS and RNS free radicals, inhibit oxidative stress, relieve cardiovascular and cerebrovascular injury, and prevent cardiovascular and cerebrovascular diseases. All these will be discussed in this chapter.

8.2 Cardiovascular and cerebrovascular health

The main causes of heart and brain diseases are atherosclerosis, arteriosclerosis, arteritis, vascular factors, hyperlipidemia, diabetes, abnormal hemorheology, leukemia, anemia, thrombocytosis, and other factors. The most dangerous factor is hypertension, which can thicken or harden the arterial wall and narrow the lumen, thus affecting the blood supply to the heart and brain. Hypertension can aggravate the heart load, increased risk of left ventricular hypertrophy and blood viscosity, further leading to hypertensive heart disease and heart failure. The pace of modern life is tensed, the pressure of family and career is increasing, and people's mood are becoming more and more unstable; at the same time, excessive drinking, eating too much fat, lack of necessary exercise, coupled with the pollution of the living environment, the content of negative ions in the air decreases sharply, and the intake of negative ions in the body is insufficient. These factors directly lead to the slowdown of human metabolism, the slowdown of blood flow rate and the rapid increase of blood viscosity, resulting in insufficient blood supply to the heart and brain. If not prevented and conditioned in time, it will cause cardiovascular and cerebrovascular diseases such as coronary heart disease, hypertension, and cerebral thrombosis [Zhao, 2002, 2007, 2008, 2016].

Stroke is a major clinical manifestation of cerebral ischemia and hemorrhagic injury. It has high incidence rate, mortality, and disability rate. It can be divided into hemorrhagic stroke (intracerebral hemorrhage or subarachnoid hemorrhage) and ischemic stroke (cerebral infarction and cerebral thrombosis). Cerebral apoplexy is one of the most fatal diseases in the world. The mortality of stroke also tends to increase with age. Its mortality is second only to heart disease and cancer, ranking the third cause of death. At the same time, stroke is also the most important disabling disease in adults, and the second cause of dementia after Alzheimer's disease (AD), which is the main cause of long-term disability in adults. Epidemiological survey shows that the incidence rate of stroke in China is higher than that in developed countries such as the United States. The annual incidence of new stroke is about 2 million higher than the average level in the world, and in recent years, with the improvement of industrialization and living standards, there is an upward trend. With the advent of an aging society, stroke will cause a heavy economic burden to society and families, and significantly reduce the quality of life for the elderly. Therefore, the research on the pathogenesis, preventive means, and therapeutic drugs for stroke is an important topic faced by the scientific and medical circles [Zhao, 1999, 2007, 2008, 2016].

Hypertension is a clinical syndrome characterized by increased systemic arterial blood pressure (systolic and/or diastolic blood pressure) (systolic blood pressure $\geq$ 140 mmHg, diastolic blood pressure $\geq$ 90 mmHg), which can be accompanied by functional or organic damage to heart, brain, kidney, and other organs. Hypertension is the most common chronic disease and the main risk factor of cardiovascular and cerebrovascular diseases. The blood pressure of normal people fluctuates within a certain range with the changes of internal and external environment. In the whole population, the blood pressure level gradually increases with age, especially the systolic blood pressure, but the diastolic blood pressure shows a downward trend after the age of 50, and the pulse pressure difference also increases. In recent years, people have deepened their understanding on the role of multiple risk factors of cardiovascular and cerebrovascular disease and the protection of target organs of heart, brain, and kidney, and the diagnostic criteria of hypertension are constantly adjusted. At present, it is considered that patients with the same blood pressure level have different risks of cardiovascular and cerebrovascular disease. Therefore, there is the concept of blood pressure stratification, that is, patients with different risks of cardiovascular and cerebrovascular disease. Appropriate blood pressure levels should be different. With the improvement of living standards, the extension of human life, and the arrival of an aging society, the proportion of people with hypertension is becoming higher and higher. Hypertension is the most common chronic disease, which seriously endangers human health. Although hypertension is a chronic disease, it is indeed a healthy "invisible killer"! Hypertension is the culprit of common cardiovascular and cerebrovascular diseases. It can cause stroke (cerebral hemorrhage, cerebral infarction), transient ischemic attack, etc. Some data show that 70% of stroke is related to hypertension, and stroke is an important cause of vascular dementia. Moreover, long-term hypertension will cause serious damage to organs such as heart, kidney, and eyes, and seriously endanger human health. It is found that hypertension is closely related to nitric oxide [Zhao, 2002, 2007, 2008, 2016].

Cigarette smoking and alcohol abuse are very dangerous factors leading to heart disease. The study found that cigarette smokers had a much higher incidence rate than non-smokers. The incidence rate of coronary heart disease was 3.5 times higher than that of non-smokers, and the mortality of coronary heart disease and cerebrovascular disease was 6 times that of non-smokers, and that of subarachnoid hemorrhage was 3–5.7 times higher than that of non-smokers. Cigarette smoking is the first risk factor for cerebral infarction. Nicotine can increase the content of adrenaline in plasma, promote platelet aggregation and endothelial cell contraction, and cause the increase of blood viscosity factors. Alcohol intake has a direct dose correlation with hemorrhagic stroke. The risk of heart and brain infarction increased when the daily alcohol intake was more than 50 g. Long-term heavy drinking can increase platelets in the blood, resulting in poor blood flow regulation, arrhythmia, hypertension, and hyperlipidemia, leading to high risk of cardiovascular and cerebrovascular diseases. Drinking a small amount of alcohol is beneficial and drinking a large amount of alcohol is harmful [Zhao, 2002, 2007, 2008, 2016].

Diabetes is an independent risk factor for heart disease or ischemic stroke. With the progression of diabetes, various cardiovascular and cerebrovascular complications will gradually emerge, such as coronary atherosclerosis, cerebral infarction, and the formation of atherosclerotic plaques in the lower extremities. For example, obesity, insulin resistance, age, gender (male incidence is higher than female), race, and heredity are all risk factors related to cardiovascular disease. The heart, brain, and blood vessels of the human body are an interconnected and interactive whole. If there is a problem with any, there will be some problems in the other two, and even the overall health. A key problem affecting the health of heart, brain, and blood vessels is the blockage of blood vessels by atherosclerosis caused by hyperlipidemia. In this way, the blood flow will be partly blocked or even completely blocked, and fatal diseases such as hypertension, heart disease, encephalopathy, myocardial infarction, and cerebral infarction will occur (Fig. 8-1).

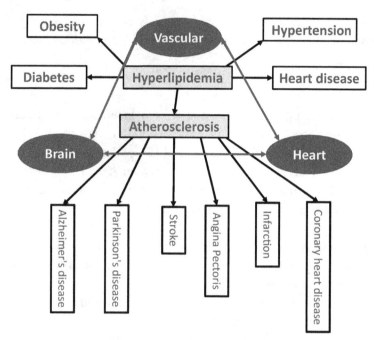

Figure 8-1.   Risk factors and consequences of cardiovascular and cerebrovascular diseases.

8.3 Cardiovascular and cerebrovascular disease and oxidative stress

To find effective treatment and prevention of the diseases, it is important to thoroughly understand the pathobiology of cardiovascular and cerebrovascular disease. Increasing evidence have shown that ROS and RNS play important roles in the development of the diseases [Ferrari *et al.*, 2004]. Generation of ROS and RNS normally need to be tightly regulated. At low and moderate concentrations, ROS/RNS mediate signal transduction cascades involved in a variety of cellular and physiological functions and are important for defense against infectious agents. In contrast, overproduction of ROS/RNS results in oxidative stress, a deleterious process that causes damages to cell structures including lipids, proteins, and DNA, which can eventually lead to cellular senescence, apoptosis, and cardiovascular and cerebrovascular diseases [Afanas'ev, 2005; Zhao, 1999, 2002; Zhao *et al.*, 1989, 1996a, 2000].

8.3.1 *Oxidative stress during ischemic-reperfusion of cardiovascular and cerebrovascular*

Oxidative stress is a condition in which oxidant metabolites exert their toxic effect because of an increased production or an altered cellular mechanism of protection. The heart needs oxygen avidly and although it has powerful defense mechanisms, it is susceptible to oxidative stress, which occurs, for instance, during post-ischemic reperfusion. Ischemia causes alterations in the defense mechanisms against oxygen free radicals, mainly a reduction in the activity of mitochondrial superoxide dismutase (SOD) and a reduction content of reduced glutathione in tissue. At the same time, production of oxygen-free radicals increases in the mitochondria and leukocytes and toxic oxygen metabolite production is exacerbated by the re-admission of oxygen during reperfusion. Oxidative stress, in turn, causes oxidation of thiol groups and lipid peroxidation leading first to reversible damage, and eventually to necrosis of cardiovascular and cerebrovascular cells. In man, there is evidence of oxidative stress during surgical reperfusion of the whole heart, or after thrombolysis, and it is related to transient left ventricular dysfunction or stunning. Data on oxidative stress in the failing heart are scant. It is not clear whether the defense mechanisms of the myocyte are altered or whether the production of oxygen free radicals is increased, or both. Recent data have shown a close link between oxidative stress and apoptosis. Relevant to heart failure is the finding that tumor necrosis factor, which is found increased in endothelial cells, induces a rapid rise in intracellular ROS and apoptosis. This series of events is not only confined to the myocytes, but occurs also at the level of endothelium, where tumor necrosis factor causes expression of inducible nitric oxide synthase (iNOS), production of the reactive radical nitric oxide, oxidative stress, and apoptosis. Therefore, it is possible that the immunological response to heart failure results in endothelial and myocyte dysfunction through oxidative stress mediated apoptosis [Ferrari *et al.*, 2004].

1. ROS generated during myocardial ischemia-reperfusion

ROS include hydroxyl radical ($\cdot$OH), superoxide free radical ($\cdot O_2^-$), hydrogen peroxide (H_2O_2), singlet oxygen, and other reactive molecules containing oxygen atoms. Due to the high rate of oxidative metabolic activity, the heart and brain are extremely susceptible to the damage caused by oxygen radicals. During ischemia/reperfusion (IR), ROS can be generated from mitochondrial respiratory chain, ischemia-activated xanthine/hypoxanthine oxidase, and lipid metabolism. Electron paramagnetic resonance (EPR) or electron spin resonance (ESR) spectroscopy can be applied to measure free radical directly in this process. To determine the free radical generation during ischemia-reperfusion, Zweier *et al.* examined perfused rabbit hearts and Zhao *et al.* examined the ischemia/reperfusion myocardial damage in the rabbit and rats using EPR. The results demonstrated that a burst of ROS and RNS free radicals were produced in hearts during ischemia and that damage of hearts occurred within moments of reperfusion [Zweier *et al.*, 1987, 1989; Cheng *et al.*, 1990; Huang *et al.*, 1990; Zhao *et al.*, 1989a, 1996a, 2000; Zhao, 1999, 2002].

Using EPR spectroscopy and the spin trap 5,5-dimethyl1-pyrroline-N-oxide (DMPO), Sanders *et al.* determined the species and sources of the free radicals generated in a hyperoxic endothelial cell model. They found that sheep pulmonary microvascular endothelial cells exposed to 100% O_2 for 30 minutes exhibited a prominent signal of trapped hydroxyl radical, DMPO-OH, and occasionally, additional smaller DMPO-R signals thought to arise from the trapping of superoxide anion ($\cdot O_2^-$), hydroxyl ($\cdot OH$), and alkyl ($\cdot R$) radicals. SOD quenched both signals, suggesting that these observed radicals were derived from $\cdot O_2^-$. The generation of $\cdot R$ occurred secondary to the formation of $\cdot OH$ from $\cdot O_2^-$ via an iron-mediated Fenton reaction since removing iron with deferoxamine decreased the $\cdot R$ signal. Blocking mitochondrial electron transport with rotenone (20 mM) markedly decreased radical generation, suggesting that endothelial cells exposed to hyperoxia produced free radicals via mitochondrial electron transport. Cell mortality increased slightly in these oxygen-exposed cells, which was not altered by SOD or deferoxamine, or different from the mortality observed in air-exposed cells [Sanders *et al.*, 1993].

Similar results were obtained by our laboratory when we measured the free radical generation in ischemia and reperfused rabbit hearts using EPR spectroscopy. The hearts were freeze-clamped at 100K during control perfusion after 150 minutes of global ischemia or following post-ischemic reperfusion with oxygenated perfusate for 15 seconds. The spectra of the ischemia hearts exhibited two different signals with different power saturation and temperature stability. Signal 1 was identical to oxygen radicals, which disappeared when superoxide dismutase (SOD) and catalase were added to the perfusion solution, indicating that they conformed to be the oxygen free radicals generated from ischemia-reperfusion heart [Zhao *et al.*, 1989]. Furthermore, in the isolated rat heart, EPR imaging instrumentation demonstrated a transmural gradient in the rate of myocardial radical clearance with a slower clearance in the endocardium, suggesting that endocardium is more susceptible to ischemia injury [Kuppusamy *et al.*, 1994; Zhao *et al.*, 1996a]. Therefore, under the experimental conditions of ischemia/reperfusion brain injury, a large number of ROS free radicals are produced, and lead to heart brain injury and cell death.

Using ESR technology, we found that the EPR spectrum of normal myocardium showed semiquinone-free radical and transition metal cation in the myocardium. The signal at g = 2.04 of ESR could be decreased by the addition of NG-nitro-L-arginine methyl ester (NAME, the inhibitor of NO synthase) in the reperfusion solution, while increased by L-arginine (substrate of NOS), indicating that it might be associated with NO generation. Both the signals at g = 2.04 and g = 2.03 of ESR were decreased by SOD/catalase but increased by the addition of Fe/H_2O_2 or xanthine/xanthine oxidase in the reperfusion solution, indicating that they were both dependent on ROS production (Table 8-1). At the same time, it can be seen that with the addition of L-arginine, the signal can be increased, and the metabolites of myocardial injury, L-arginine hydroxylase (LDH), and creatine kinase (CK), also increase significantly (Table 8-1) [Zhao *et al.*, 1996b].

Ischemia-reperfusion causes damages to myocardium, which is demonstrated by increased activities of lactate dehydrogenase (LDH) and creatine kinase (CK). It was also shown that ischemia caused a significant increase in serum thiobarbituric acid reaction substance (TBARS) concentration, which was further increased during the reperfusion. The increased lipid peroxidation and enzyme leakage from the ischemia-reperfusion myocardium were significantly decreased by SOD/catalase treatment [Zhao *et al.*, 1996b; Shen *et al.*, 1998, 2000b]. In contrast, the addition of Fe^{2+}/H_2O_2 or xanthine/xanthine oxidase into the reperfusion buffer significantly increased the elevated activities of LDH and CK in the coronary artery effluent [Zhao *et al.*, 1996a]. Meanwhile, SOD treatment reduced the elevation of heart rate and the incidence of ventricular arrhythmias

Table 8-1. Free radical generation and injury (LDH and CK) of myocardium during ischemia-reperfusion.

	ESR signal at g = 2.04 relative signal high	ESR signal at g = 2.03 relative signal high	LDH U/L	CK U/L
N (n = 5)	0	0	4.0 ± 1.1	2.7 ± 1.2
N + LA1 (n = 5)	0	0	4.2 ± 1.2	2.5 ± 1.0
N + LA10 (n = 5)	0	0	4.0 ± 1.0	2.6 ± 1.2
N + LA100 (n = 5)	0	0	3.9 ± 1.4	2.7 ± 1.2
I (n = 10)	$5.7 \pm 2.8*$	$4.1 \pm 2.7*$	—	—
IR (n = 10)	$7.5 \pm 3.2* +$	$5.8 \pm 1.0* +$	$25.5 \pm 5.8*$	$16.0 \pm 4.6*$
IR + NAME (n = 10)	$5.9 \pm 1.1*\#$	$4.5 \pm 1.0*\#$	$7.8 \pm 1.9*\#$	$6.3 \pm 2.3*\#$
IR + LA1 (n = 10)	$6.0 \pm 1.1*$	$4.6 \pm 1.0*$	$14.3 \pm 2.8*\#$	$9.2 \pm 2.9*$
IR + LA10 (n = 10)	$8.1 \pm 2.6*$	$5.8 \pm 1.4*$	$31.6 \pm 8.0*$	$18.0 \pm 2.6*$
IR + LA100 (n = 10)	$10.7 \pm 2.5*\#$	$6.7 \pm 2.7*$	$46.7 \pm 7.8*\#$	$25.9 \pm 7.4*$
IR + SOD/CAT (n = 10)	$2.9 \pm 1.0*\#$	$2.9 \pm 0.8*\#$	$6.0 \pm 2.3*\#$	$5.6 \pm 2.5*\#$
IR + X/XO (n = 5)	$20.2 \pm 4.3*\#$	$16.0 \pm 2.8*\#$	$85.0 \pm 5.6*\#$	$71.0 \pm 12.6*\#$
IR + Fe^{2+}/H_2O_2 (n = 5)	$20.5 \pm 3.5*\#$	$16.5 \pm 9.1*\#$	$113.0 \pm 18.3*\#$	$85.5 \pm 10.5*\#$
IR + Fe^{2+} (n = 5)	$14.0 \pm 1.7*\#$	$16.2 \pm 2.3*\#$	$50.5 \pm 3.5*\#$	$45.5 \pm 5.5*\#$
IR + H_2O_2 (n = 5)	$16.5 \pm 1.8*\#$	$16.6 \pm 2.5*\#$	$47.5 \pm 4.5*\#$	$41.5 \pm 6.5*\#$

Abbreviations: N: normal; IR: ischemia-reperfusion; LA1: 1 mmol/L L-arginine; LA10: 10 mmol/L L-arginine; LA100: 100 mmol/L L-arginine; NAME: 100 mmol/L N^G-nitro-L-arginine methyl ester; X: xanthine; XO: xanthine oxidase; CAT: catalase. The errors are the standard deviation. $*, p < 0.05$ versus N; #, $p < 0.05$ versus IR; +, $p < 0.05$, versus I.

caused by ischemia-reperfusion [Shen *et al.*, 1998, 2000a]. These results indicated that ROS could contribute to the myocardium injury induced by ischemia-reperfusion and the reduction of ROS and lipid peroxidation might prevent myocardium damage and cardiac injury.

2. NO free radicals generated during myocardium ischemia-reperfusion

Nitric oxide (NO) is a simple gas with free radical properties. It appears as a major signaling molecule in cardiovascular, immune, and nervous systems. NO is generated by three isoforms of NO synthase (NOS) in the body, endothelial NO synthase (eNOS), neuronal NO synthase (nNOS), and inducible NO synthase (iNOS), and is important for regulating many physiological processes such as blood pressure and vascular tone [Bredt *et al.*, 1990]. However, excessive NO produced by iNOS results in inhibition of cardiac contractility, impairment of mitochondrial respiration, and apoptosis, thus contributing to ischemia-reperfusion injury [Ghafourifar *et al.*, 1999]. The level of NO in the media was increased by hypoxia, but it was decreased by re-oxygenation. In addition, the levels of TBARS and LDH in the media were slightly increased by hypoxia and significantly augmented by re-oxygenation. The results suggest that hypoxia improved NO production in the cardiomyocytes, while re-oxygenation caused oxygen burst and led to lipid peroxidation injury.

On the above experimental basis, we used ESR spin trapping agents DETC/ Fe^{2+} and studied the NO free radical generation in rat heart during ischemia-reperfusion. It was shown that there was a baseline of the EPR signals with characteristic of $NOFe^{2+}$ DETC complex in normal heart, perhaps represented the physiological level of NO. The EPR signal intensity of $NOFe^{2+}$ DETC complex in rat myocardium was increased remarkably after 30 minutes of ischemia. L-arginine increased while NG-nitro-L-arginine (NNA, inhibitor of NOS) decreased the signal intensity, indicating that the signal originated from NO. After 10 minutes of reperfusion followed 30 minutes of ischemia, the signal intensity of $NOFe^{2+}$ DETC in rat hearts was significantly lower than that in ischemia-only myocardium. It appears that the NO production increases during ischemia and decreases during reperfusion. Interestingly, the signal intensity of $NOFe^{2+}$ DETC complex was increased significantly by SOD treatment, suggesting that superoxide anions might contribute to the decrease of NO level in the ischemia-reperfusion myocardium, because SOD scavenged the superoxide free radicals and saved NO (Figs. 8-2, 8-3) [Shen *et al.*, 2000a, 2000b].

We measured the level of nitric oxide metabolite, NO_2^-/NO_3^-, produced by myocardium during myocardial ischemia and reperfusion. The results showed that the level of NO_2^-/NO_3^- produced by myocardium during myocardial ischemia (HO) was very high, and the level of NO_2^-/NO_3^- produced during reperfusion (IR) was slightly decreased. When nitric

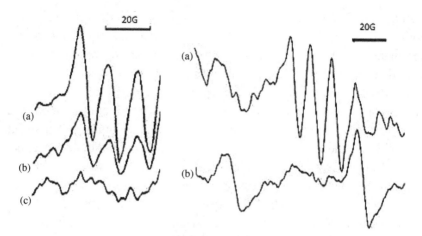

Figure 8-2.   Left: Spectra of nitric oxide generated by rat heart homogenate. The spectrum generated from homogenate containing (a) L-arginine; (b) L-arginine +NMMA; (c) -L-Arginine. Right: ESR spectra of NO complex produced from normal cardiomyocytes in culture media, trapped by $DETC_2$-Fe^{2+} and extracted by ethyl acetate, (a) in the presence of cardiomyocytes, and (b) in the absence of cardiomyocytes.

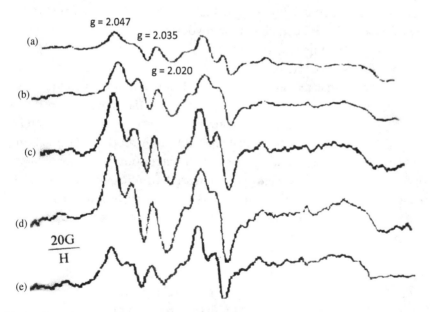

Figure 8-3.   ESR spectra of NO free radical generated from ischemia-reperfusion heart *in vivo*. (a) Normal heart; (b) Ischemia heart; (c) Ischemia and reperfusion heart; (d) Ischemia and reperfusion + SOD heart; (e): Ischemia and reperfusion + antioxidant heart.

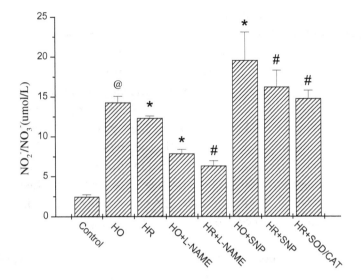

Figure 8-4. Levels of NO_2^-/NO_3^- (µmol/L) in the cultured media of cardiomyocytes. HO, 24 hours of hypoxia; HR, 24 hours of hypoxia plus 4 hours of re-oxygenation; L-NAME, N^G-nitro-L-arginine methyl ester 100 µmol/L; SNP, sodium nitroprusside 5 µmol/L; SOD/CAT, superoxide di mutase and catalase 100 U/ml, each. @, $p < 0.05$ *vs* control; *, $p < 0.05$ *vs* HO; #, $p < 0.05$ *vs* HR, n = 3–5.

oxide synthase inhibitor L-NAME was added during myocardial ischemia and reperfusion, the level of NO_2^-/NO_3^- further decreased. The addition of sodium nitroprusside (SNP) that produces nitric oxide during myocardial ischemia/reperfusion will greatly increase the level of NO_2^-/NO_3^-, while the addition of SOD during myocardial ischemia/reperfusion will reduce the level of NO_2^-/NO_3^- (Fig. 8-4).

It has also been shown that iNOS expression plays a role in myocardial chamber dilation and hypertrophy induced by pressure overload. In wild-type mice, chronic transverse aortic constriction resulted in myocardial iNOS expression, cardiac hypertrophy, ventricular dilation and dysfunction, and fibrosis, while mice deficient in iNOS displayed much less cardiac hypertrophy, dilation, fibrosis, and dysfunction. Consistent with these findings, transverse aortic constriction resulted in marked increases of myocardial atrial natriuretic peptide (ANP), 4-hydroxy-2-nonenal a marker of lipid peroxidation, and nitrotyrosine a marker for peroxynitrite in wild-type but not in iNOS-deficient mice. In response to transverse aortic constriction, there was an increase in ROS production in myocardium, which was accompanied by the expression of iNOS, suggesting that iNOS was a source for the increased oxidative stress. Furthermore, selective iNOS inhibition with 1,400 W significantly attenuated transverse aortic constriction-induced myocardial hypertrophy and pulmonary congestion. These data implicated iNOS in the maladaptive response to systolic overload and suggested that selective inhibition on iNOS activity might be effective for treatment of systolic overload-induced cardiac dysfunction [Zhang *et al.*, 2007]. In the process of ischemia/reperfusion brain damage, the content of nitrite/nitrate (the end product of NO) increased, and of NO detected by ESR decreased [Zhang *et al.*, 2004].

3. Nitric oxide is a double-edged sword in the process of myocardial ischemia-reperfusion injury

The role of nitric oxide in cardiovascular and cerebrovascular diseases and health and its mechanism need further research and discussion. Studies show that nitric oxide is a double-edged sword in the process of myocardial ischemia-reperfusion injury. We used ESR technique to study the free radicals produced by myocardial ischemia reperfusion injury *in vivo*. In addition, if the production of NO free radicals is excessive, or there are oxygen free radicals, it will lead to cell damage and even increase the occurrence of cardiovascular and cerebrovascular diseases. In the myocardial ischemia reperfusion experiment of rats, we added L-arginine (LA), the substrate of nitric oxide synthase. At a low concentration (5–100 μM), the effect of L-arginine on arrhythmia of myocardial ischemia reperfusion injury was reduced in a concentration-dependent manner. At 100 μM, there was no arrhythmia at all with normal hearts, but with the increase of concentration, arrhythmia appeared at 100 μM. If nitro arginine (NNA), an inhibitor of nitric oxide synthase, is added, the effect on arrhythmia of myocardial ischemia-reperfusion injury will increase; if L-arginine is added at the same time (NNA + LA), the effect on arrhythmia of myocardial ischemia-reperfusion injury will decrease (Table 8-2).

Why is nitric oxide a double-edged sword in myocardial ischemia-reperfusion injury? Since nitric oxide reacts with superoxide anion radical to produce peroxynitrite, its reactivity and toxicity are far greater than that

Table 8-2. Effect of L-arginine-NO pathway on arrhythmias (TAR) induced by myocardial ischemia-reperfusion injury (mean ± SE).

Group	BG	TG	VP	VF	VT	TAR(%)
N	0/9(0)	0/9(0)	0/9(0)	0/9(0)	0/9(0)	0
I	0/10(0)	0/10(0)	0/10(0)	0/10(0)	0/10(0)	10.0
IR	0/16(0)	0/16(0)	3/16(18.75)	3/16(18.75)	1/16(6.25)	43.75
IR + LA						
5	0/6(0)	0/6(0)	1/6(16.67)	0/6(0)	0/6(0)	16.67
50	0/8(0)	0/8(0)	1/8(12.5)	1/8(12.5)	0/8(0)	25.0
100	0/6(0)	0/6(0)	0/6(0)	0/6(0)	0/6(0)	0
500	0/9(0)	1/9(11.11)	1/9(11.11)	1/9(11.11)	0/9(0)	33.33
IR + NNA	2/10(20)	0/10(0)	2/10(20)	1/10(10)	0/10(0)	50
IR + NNA + LA	1/6(16.67)	0/6(0)	1/6(16.67)	0/6(0)	0/6(0)	33.33

Results are shown as "arrhythmias case/total case (%)". Abbreviations: BG: bigeminy; TG: trigeminy; VP: ventricular premature beat; VF: ventricular fibrillation; VT: ventricular tachycardia; TAR: total arrhythmias rate; N: normal control heart; I: ischemia heart; IR: ischemia-reperfusion heart; IR + LA: L-arginine (5, 50, 100, 500 mg/kg, i.p); NNA50: N^G-nitro-L-arginine (50 mg/kg, i.p.); NNA + LA: NNA(50 mg/kg, i.p.) + L-arginine (500 mg/kg, i.p.).

of nitric oxide and hydroxyl radical. The reaction activity and toxicity of peroxynitrite are far greater than that of nitric oxide and hydroxyl free radicals, which can cause cell and tissue damage, further causing lipid peroxidation damage of cardio cerebral cell membrane, DNA and mitochondrial damage, calcium dys-homeostasis, ATP depletion that leads to cardio cerebral vascular damage through the signal pathway tumor necrosis factor (TNF-α), and nuclear factor-κB (NF-κB), P^{53} activation that leads to apoptosis of heart and brain cells (Fig. 8-5).

4. Oxidative stress-induced myocardium damage and apoptosis in ischemia-reperfusion of the heart

The role of NO in ischemia-reperfusion-induced myocardium damage is more complicated. In isolated heart, high concentration of L-arginine increased the activities of LDH and CK in the coronary artery effluent while NAME decreased the activities of LDH and CK in the coronary artery effluent, suggesting that high concentration of NO was associated with cardiac injury [Zhao *et al.*, 1996a; Shen *et al.*, 1998]. Administration

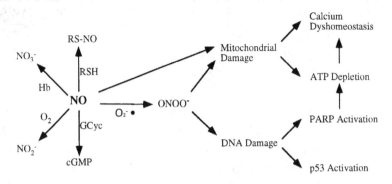

Figure 8-5.   NO will react with superoxide to form ONOO⁻, with more reactivity than •OH.

of L-arginine (50 mg/kg/wt) to rat hearts subjected to ischemia-reperfusion *in vivo* caused a reduction in lipid peroxidation and CK activity as well as a decrease in the elevated heart rate and the incidence of ventricular arrhythmias caused by ischemia-reperfusion [Shen *et al.*, 1998, 2000]. Thus, moderated production of NO may be beneficial for protecting the heart from ischemia-reperfusion, while excess NO causes injury to mycardium [Shen *et al.*, 2000b]. In fact, the cardio-protective function of NO in myocardial ischemia-reperfusion has been well documented by Bolli in a review paper, *Cardio-protective function of inducible nitric oxide synthase and role of nitric oxide in myocardial ischemia and preconditioning: an overview of a decade of research* [Bolli, 2001].

Annexin V-FLOUS was simultaneously applied with PI to differentiate necrotic cells from apoptotic cells in the study. As shown in Figure 8-6, the statistical data was showed that exposure to 24 hours of hypoxia alone induced mainly apoptotic cell death while 24 hours of hypoxia plus 4 hours of re-oxygenation caused mainly necrotic cell death. The rates of apoptosis and necrosis were increased by sodium nitroprusside (SNP) and inhibited by L-NAME respectively in both hypoxic and hypoxia-re-oxygenated cardiomyocytes, suggesting that extraneous and endogenous NO might be contributed to apoptotic and necrotic cell death. SOD/catalase significantly inhibited both apoptotic and necrotic cell death in

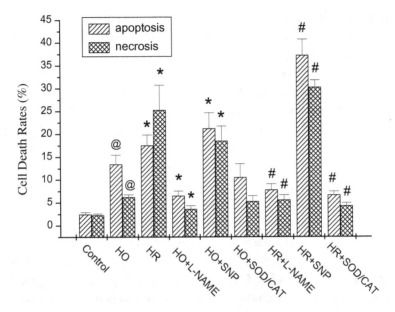

Figure 8-6.   Apoptotic and necrotic cell death rates (%) in cultured cardiomyocytes based on Annexin V-FLOUS / Propidium iodide (PI) double staining FACS measurement. HO, 24 hours of hypoxia; HR, 24 hours of hypoxia plus 4 hours of reoxygenation; L-NAME, N^G-nitro-L-arginine methyl ester 100 µmol/L; D-NAME, N^G-nitro-D-arginine methyl ester 100 µmol/L; SNP, sodium nitroprusside 5 µmol/L; SOD/CAT, superoxide dismutase plus catalase 100 U/ml, each. @, $p < 0.05$ *vs* control; *, $p < 0.05$ *vs* HO; #, $p < 0.05$ *vs* HR, n = 3–5.

hypoxia-re-oxygenated cardiomyocytes but without protective effects on hypoxic cardiomyocytes. The results provide the evidence that superoxide anions contribute to cell death in hypoxia-re-oxygenation, but not hypoxia alone [Shen *et al.*, 2000a].

Oxidative stress leads cells to the exposure of ROS, such as superoxide, hydroxyl radicals, and NO, and triggers apoptosis [Beckman & Koppeno, 1996]. In a cardiomyocyte model, we found that hypoxia caused a slight increase in lipid peroxidation and cell damage as manifested by increased LDH activity, while re-oxygenation of the hypoxic cardiomyocytes induced oxygen burst and a much severe lipid peroxidation injury [Shen *et al.*, 2003]. This is consistent with the observation that hypoxia induced apoptosis which was further increased by re-oxygenation

[Shen *et al.*, 2000b, 2003]. SOD/catalase inhibited apoptosis in hypoxia-re-oxygenated cardiomyocytes, suggesting that superoxide may be one of the important mediators in apoptotic cell death during re-oxygenation injury. In this cell model, NO production was increased by hypoxia, but decreased by re-oxygenation [Shen *et al.*, 2000b]. The augmentation of NO production during hypoxia may be caused by an up-regulation of NOS activity in response to hypoxia [Shen *et al.*, 2003]. It is also possible that a direct disproportion or reduction of nitrite to NO under the acidic and highly reduced conditions, such as ischemia or hypoxia, leads to the accumulation of NO during hypoxia. In fact, the enzyme-independent mechanism has been demonstrated to not only contribute to the post-ischemic myocardial injury, but also reverse the protective effects of NOS inhibitors.

To further elucidate the apoptotic mechanisms of hypoxia-re-oxygenation injury, we examined the alteration in the expression levels of bcl-2 and p53 proteins associated with hypoxia and hypoxia-re-oxygenation-induced cell death. The changes in the protein levels of p53 and bcl-2 in response to hypoxia and hypoxia-re-oxygenation showed that hypoxia induced the up-regulation of bcl-2 and p53 proteins in cardiomyocytes, while re-oxygenation further upregulated p53 protein, but downregulated bcl-2 protein [Shen *et al.*, 2003]. These results suggested that nitric oxide and oxygen radicals induced apoptosis via bcl-2 and p53 pathway in hypoxia-re-oxygenated cardiomyocytes. The reaction of superoxide with NO results in the formation of $ONOO^-$. Therefore, the decrease of detectable NO may be attributed to the generation of oxygen radicals in re-oxygenated cardiomyocytes. Inhibition of NO production by L-NAME inhibited the apoptosis induced by hypoxia and re-oxygenation, suggesting the involvement of endogenous NO in cardiomyocyte apoptosis during hypoxia and re-oxygenation [Shen *et al.*, 2003]. The cytotoxicity of NO has been shown to be associated with the formation of $ONOO^-$ [Chen *et al.*, 2007]. Low concentration of $ONOO^-$ promotes apoptosis whereas large amounts of $ONOO^-$ rapidly lead to necrotic cell death, a primary mechanism responsible for cell death from ischemia. In cultured cardiomyocytes, hypoxia induced mainly cell apoptosis while re-oxygenation led to both apoptotic and necrotic cell death, which may be contributed by the formation of oxygen radicals and $ONOO^-$. Increased O_2 and NO production is a key mechanism of mitochondrial dysfunction in myocardial ischemia/reperfusion injury. Succinate ubiquinone reductase (SQR or Complex II) is a crucial component of the mitochondrial electron transport chain. The intrinsic protein S-glutathionylation at cysteine residue (Cys) of the 70-kDa flavin adenine dinucleotide (FAD)-binding subunit of SQR makes the protein susceptible to redox change induced by oxidative stress. Ischemia-reperfusion of rat hearts *in vivo* or *in vitro* caused deglutathionylation of the 70-kDa FAD-binding subunit and significantly decreased the electron transfer activity of SQR. IR also caused enhancement of tyrosine nitration of SQR. Site-specific nitration at the 70-kDa FAD-binding subunit with peroxynitrite impaired the interaction of SQR with Complex III and the electron transfer activity. It was found that S-glutathionylation protected the protein from oxidative modification and impairment was mediated by peroxynitrite [Chen *et al.*, 2008]. These studies on SQR give a hint on how ROS generated during ischemia-reperfusion may lead to mitochondrial dysfunction and IR injury. Adenosine monophosphate (AMP)-activated protein kinase (AMPK) is a key regulatory enzyme regulating myocardial metabolism and protein synthesis, which can be activated by cellular stresses that increase the AMP-to-ATP ratio, such as hypoxia/anoxia, glucose deprivation, and pressure overload-induced hypertrophy. The activation of AMPK has been shown to have protective effect against IR injury by promoting ischemic pre-conditioning [Young, 2008]. A number of downstream signaling pathways have been implicated in the cardio-protective mechanisms of AMPK such as the activation of eNOS, p38 mitogen-activated protein kinase, and the inactivation of eEF2 [Li *et al.*, 2004, 2005; Terai *et al.*, 2005]. Activation of AMPK also attenuates hypertrophy in cultured cardiac myocytes. In mice deficient in AMPKα2 gene, transverse aortic constriction (TAC)-induced ventricular hypertrophy and dysfunction were significantly exacerbated, suggesting that regulation of AMPKα2 may be a potential therapeutic approach to attenuate pressure overload-induced ventricular hypertrophy [Zhang *et al.*, 2008].

8.3.2 *Oxidative stress of stroke leading to brain nerve injury*

Stroke is one of the most common causes of death and disability in the world. The ischemic injury behind stroke is complex, involving complex interactions between many biological functions, including energy metabolism, vascular regulation, hemodynamics, oxidative stress, inflammation, platelet activation, and tissue repair. At present, the first choice of drug treatment is to supplement blood to ischemic tissue in time. However, reperfusion may cause additional damage to tissues through a process called ischemia/reperfusion injury. Therefore, new drugs that supplement reperfusion by providing neural and cardiovascular protection and targeting a variety of antioxidants in ischemia are receiving more and more attention.

Similar to cardiac ischemia-reperfusion, ROS and RNS are closely-related to cerebral ischemia and hemorrhagic stroke. Under physiological conditions the steady-state concentrations of ROS and RNS are finely regulated for proper cellular functions. Reduced surveillance of endogenous antioxidant defenses and/or increased ROS/RNS production leads to oxidative stress with consequent alteration of physiological processes. Nitric oxide is an inhibitor of strong vasodilator, platelet aggregation, and leukocyte adhesion. By inhibiting the obstruction of leukocytes and platelets to capillaries, NO can improve the blood supply of brain tissue after ischemia-reperfusion and play a protective role. After 24 hours of ischemia, cerebral infarction and lipid peroxidation increased, and anti-oxidant treatment restored these parameters. Oxidative stress markers were affected by ischemic injury, which showed that the level of ROS/RNS increased, the expression of SOD increased, the activity of SOD decreased, and the antioxidant defense ability of non-enzymes (GSH and vitamin C) decreased [Hansel *et al.*, 2014]. Oxidative stress was induced by upregulated inducible nitric oxide synthase (iNOS) and heat shock proteins (HSPs) in SH-SY5Y cells were induced by oxygen glucose deprivation (OGD) [Lee *et al.*, 2010].

We used ESR spin trapping to detect ROS and nitric oxide free radicals produced during ischemia-reperfusion in Mongolian gerbil stroke model. One hour after the ischemia-reperfusion operation, ESR test showed that the ROS captured in the ischemia-reperfusion group increased by about 36.89% compared with the negative control group (sham). Compared with the ischemia-reperfusion group, pretreatment with anti-oxidant hawthorn extract (CE) for 15 days significantly reduced ROS by 17.37% and 31.14% in the low-dose group and the high-dose group, respectively. The production of ROS in the high-dose group was even lower than that in the negative-control group, but the difference was not significant. Ischemia reperfusion injury increased the TBARS level in brain homogenate by 74.04% compared with the negative-control group. Oral administration of hawthorn extract for 15 days reduced the TBARS level in a dose-dependent manner. Compared with the simple ischemia-reperfusion group (IR), TBARS levels in the low-dose and high-dose treatment groups decreased by 24.25% and 47.39%, respectively.

Compared with the increase of ROS during ischemia-reperfusion, the production of nitric oxide decreased by 19.17% in the ischemia-reperfusion group compared with the negative-control group ($p > 0.05$) (Fig. 8-7). Hawthorn extract pretreatment for 15 days significantly increased the production of nitric oxide in the brain after ischemia-reperfusion. Compared with the ischemia-reperfusion group, the production of nitric oxide in the low-dose and high-dose groups increased by 44.67% and 77.56%, respectively. The production of nitric oxide in these two groups was even higher than that in the negative-control group. There is evidence

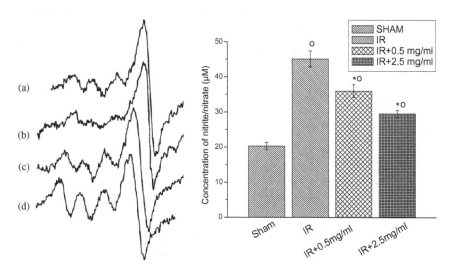

Figure 8-7. Left: ESR spectra of NO trapped by DETC-ferrous complex in brain homogenates one hour after five minutes ischemic damage. Spectra from (a) sham group; (b) ischemia/reperfusion group; (c) low dose; (d) high dose of antioxidant hawthorn extract group. Right: Statistical results from lift ESR spectra.

that ROS produced immediately after acute ischemic stroke increase rapidly, rapidly destroy the antioxidant defense system and causing further tissue damage in the brain ischemic insults of Mongolian gerbil stroke model. These ROS can damage cell macromolecules, leading to autophagy, apoptosis, and necrosis. In addition, the rapid recovery of blood flow increases the level of tissue oxygenation and leads to the generation of second ROS, resulting in reperfusion injury. iNOS was implied in delayed neuron death after brain ischemic damage could increase the protein level of tumor necrosis factor (TNF)-α and nuclear factor-kappa B (NF-κB), and decrease the mRNA level of NOS estimated by western blotting and RT-PCR (Fig. 8-8) [Zhang *et al.*, 2004].

The oxidative stress caused by ischemic stroke overwhelms the neutralization capacity of the body's endogenous antioxidant system, which leads to an overproduction of ROS and RNS and eventually results in cell death. The overproduction of ROS compromises the functional and structural integrity of brain tissue. Cerebral ischemia causes over-activation of membrane receptors and accumulation of extracellar glutamate and intracellular calcium, which activates neuronal nitric oxide synthase, causing damage to lipids, proteins, and nucleic acids, and reduces energy sources with consequent functional deterioration, leading to cell death. Restoration processes normally repair genes with few errors. However, ischemia

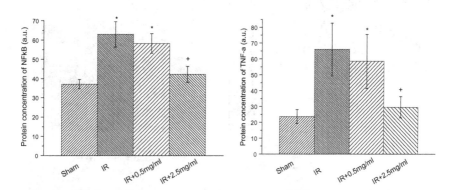

Figure 8-8.   Protein level of NFκB (left) and TNF-a (right) in the hippocampus of Gerbils 48 hours after five minutes ischemia/reperfusion damage (every result was obtained from 5–6 animals and repeated at least three times).

elevates oxidative DNA lesions despite these repair mechanisms. These episodes concurrently occur with the induction of immediate-early genes that critically activate other late genes in the signal transduction pathway. The results of the studies show that treatments with 7-nitroindazole, a specific inhibitor of nitric oxide synthase known to attenuate nitric oxide, oxidative DNA lesions, and necrosis, increase intact c-fos mRNA levels after stroke [Liu, 2003; Zhang *et al.*, 2004].

These studies indicate that ROS and RNS are produced during ischemia/reperfusion brain injury during stroke, which leads to oxidative stress, signal pathway and gene changes, and apoptosis. For the brain nerve cell injury, the accurate repair of gene expression may restore the cell function after brain injury.

8.3.3 *Inflammation and oxidative stress leading to heart and brain injury*

Inflammation caused by oxidative stress is the cause of most chronic human diseases, including human aging and heart disease. The inflammatory response after cerebral ischemia/reperfusion is an important cause of neurological damage and repair. After cerebral ischemia/reperfusion, microglia are activated and a large number of circulating inflammatory cells infiltrate the affected area. This leads to the secretion of inflammatory mediators and an inflammatory cascade that eventually causes secondary brain damage, including neuron necrosis, blood-brain barrier destruction, cerebral edema, and an oxidative stress response. Activation of inflammatory signaling pathways plays a key role in the pathological process of ischemic stroke. Inflammatory response after cerebral ischemia/reperfusion, activation, and infiltration of inflammatory cells, increases the expression of inflammatory-related cytokines and pro-inflammatory and anti-inflammatory factors [Cao *et al.*, 2021]. We found two peak kinetic curves of NO in phorbol-stimulated macrophage by chemiluninencence [Li *et al.*, 1996]. Compared with wild-type mice, atherosclerotic hypertensive (LDLr-/-:hApoB+/+, ATX) mice prematurely developed cognitive decline associated with cerebral micro-hemorrhages, loss of microvessel density and brain atrophy, cerebral endothelial cell senescence and dysfunction, brain inflammation, and oxidative stress associated with blood-brain barrier leakage and brain hypoperfusion [de Montgolfier *et al.*, 2019]. Despite the fact that reactive oxygen and nitrogen species (ROS and RNS) are the by-products of normal metabolic processes and mediate important physiological processes, they can inflict damage to the cell if produced in excess due to oxidative stress. A paper studied the cellular and molecular aspects of ROS and RNS generation and its role in the pathogenesis of stroke produced by hypoxia-reperfusion (H-R) phenomena that elicit oxidative stress and the reasons for the vulnerability of the brain to ischaemic insult, chronic infection, and inflammation as well as the natural defense mechanisms against radical mediated injury. It also found that ROS and RNS had effect on intracellular signaling pathways together with the phenomena of apoptosis, mitochondrial injury, and survival associated with these pathways [Saeed *et al.*, 2007].

Oxidative stress mainly comes from mitochondria that produce ROS/RNs. Most key steps can be identified in the pathophysiology of atherosclerosis and the clinical manifestations of brain vascular disease. The intracellular signaling mechanisms influenced by reactive species can have significant effects on the outcome of the hypoxia-reperfusion condition. Future studies should focus on understanding the molecular mechanisms involved in the action of anti-radicals' agents and their mode of action. This is the reason why the brain is vulnerable to ischemic injury, chronic infection, and inflammation, and the influence of ROS and RNS on intracellular signal pathways lead to mitochondrial damage and neuronal apoptosis.

In conclusion, oxidative stress plays an important role in the pathogenesis of cardio-cerebrovascular disease. Its pathogenesis is related to dyslipidemia, insulin resistance, and metabolic syndrome. It increases oxidative stress and hyperlipidemia, thereby increasing the incidence rate of atherosclerosis, stroke, and coronary heart and brain diseases in diabetes patients. Oxidative stress is an important factor in the formation of atherosclerosis. Cardio-cerebrovascular diseases caused by ischemia-reperfusion injury play an important role. Inflammation and ischemia-reperfusion leads to a large number of ROS and RNS free radicals, forming peroxynitrite ions, leading to cardiovascular and cerebrovascular injury, further causing lipid peroxidation injury of cardiac and cerebral

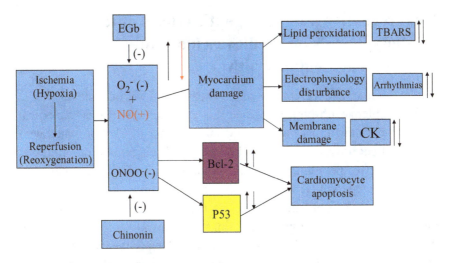

Figure 8-9. Mechanism of oxidative stress-induced myocardial and cerebrovascular injury and apoptosis during ischemia-reperfusion.

cell membrane, and leading to apoptosis of cardiac and cerebral cells through signal pathway tumor necrosis factor (TNF)-α and nuclear factor-kappa B (NF-κB), and c-FOS gene (Fig. 8-9).

Of course, there are many other factors related to oxidative stress of cardiovascular and cerebrovascular diseases, such as the pathophysiology of atherosclerosis and the clinical manifestations of cardiovascular diseases, as well as traumatic brain injury traumatic brain injury (TBI) results in cerebral circulation dysfunction, microvascular impairment, and moderate hypoxia [Toth *et al.*, 2016; Veenith *et al.*, 2016], lipid metabolism, plaque rupture, thrombosis, and blood-brain barrier (BBB) breakdown, edema formation, excitotoxicity, etc. Revealing the mechanisms related to oxidative stress in these cardiovascular and cerebrovascular diseases is of great significance for the discovery and development of antioxidant therapy in the future. Some antioxidants may be designated as recent drugs for the treatment of cardiovascular diseases and cerebrovascular [Pashkow, 2011]. Further understanding of redox signaling molecules and cell biology paves the way for more effective antioxidant drugs to prevent cardiovascular disease and cerebrovascular and prolong healthy life.

8.4 Preventive and therapeutic effects of tea polyphenols on cardiovascular and cerebrovascular health

Cardiovascular and cerebrovascular disease is a serious global public health problem with high incidence rate and mortality. Epidemiological studies have shown that regular tea drinking is negatively correlated with the risk of cardiovascular and cerebrovascular diseases. In cross-sectional and prospective population studies, tea intake and polyphenols intake in tea were related to the reduction of the risk of cardiovascular and cerebrovascular diseases. Epidemiological observation in Asian countries shows that habitual drinking of green tea drinks is negatively correlated with the incidence of cardiovascular events. Clinical and experimental studies have confirmed that there is a positive correlation between green tea consumption and cardiovascular and cerebrovascular health. In addition, a large number of *in vitro* and *in vivo* experimental studies have shown that tea and its bioactive compounds are effective in preventing cardiovascular and cerebrovascular diseases. The two main teas are green tea and black tea, both rich in polyphenols and flavonoids. Existing evidence shows that the benefits of drinking tea on cardiovascular and cerebrovascular health mainly come from flavonoids and tea polyphenols. Many possible pathways and mechanisms have been studied. Related mechanisms include reducing blood lipid, reducing ischemia/reperfusion injury, inhibiting oxidative stress, enhancing endothelial function, reducing inflammation, and protecting cardiomyocyte and cerebrovascular function. In animal models, flavonoids isolated from tea have also been shown to inhibit the development of atherosclerosis. Tea and tea flavonoids can enhance nitric oxide status and improve endothelial function, which may be at least partially beneficial to cardiovascular and cerebrovascular health. In addition, some experiments have proven that the protective effect of tea on cardiovascular and cerebrovascular diseases is through affecting the activities of receptors and signal transduction kinases. Tea polyphenols in green tea and theaflavins in black tea play a variety of beneficial effects to cardiovascular and cerebrovascular functions.

8.4.1 *Epidemiological and clinical results*

Several epidemiological and clinical studies have shown that regular consumption of foods and drinks rich in flavonoids such as tea can reduce the risk of many diseases such as hypertension, coronary heart disease, stroke, and dementia.

1. Drinking tea can reduce the risk of cardiovascular and cerebrovascular diseases

Some statistical studies on the relationship between green tea consumption and cardiovascular disease risk show that drinking five or more cups of green tea a day is negatively correlated with cardiovascular and cerebrovascular disease and mortality from various causes. Among cardiovascular and cerebrovascular mortality, the strongest negative correlation of stroke mortality with drinking tea was observed. The aim of a meta-analysis was to evaluate the association between green tea consumption, cardiovascular and cerebrovascular diseases, and ischemic-related diseases. All observational studies and randomized trials that were published through October 2014 and that examined the association between green tea consumption and risk of cardiovascular and cerebrovascular diseases and ischemic-related diseases as the primary outcome were included in this meta-analysis. A total of nine studies including 259,267 individuals were included in the meta-analysis. The results showed that those who did not consume green tea had higher risks of cardiovascular and cerebrovascular diseases, intracerebral hemorrhage, and cerebral infarction compared to <1 cup of green tea per day. Those who drank 1–3 cups of green tea per day had a reduced risk of myocardial infarction and stroke compared to those who drank <1 cup/day. Similarly, those who drank ≥4 cups/day had a reduced risk of myocardial infarction compared to those who drank <1 cup/day. Those who consumed ≥10 cups/day of green tea were also shown to have lower LDL compared to the <3 cups/day group [Pang *et al.*, 2016]. A study found that green tea polyphenols are incorporated into LDL particles in non-conjugated forms after the incubation of green tea extract and reduced the oxidizability of LDL [Suzuki-Sugihara *et al.*,

2015]. In July 2014, Zhang *et al.* conducted a meta-analysis of prospective observational studies in order to summarize the evidence regarding the association between tea consumption and major cardiovascular outcomes or total mortality. Of the 736 citations identified from database searches, it included 22 prospective studies from 24 articles reporting data on 856,206 individuals, and including 8,459 cases of coronary heart disease, 10,572 of stroke, 5,798 cardiac deaths, 2,350 stroke deaths, and 13,722 total deaths. Overall, an increase in tea consumption by three cups per day was associated with a reduced risk of coronary heart disease ($p = 0.045$), cardiac death ($p < 0.001$), stroke ($p = 0.001$), total mortality ($p = 0.003$), cerebral infarction ($p = 0.023$), and intracerebral hemorrhage ($p < 0.001$), but had little or no effect on stroke mortality ($p = 0.260$). The findings from this meta-analysis indicated that increased tea consumption is associated with a reduced risk of coronary heart disease, cardiac death [Zhang *et al.*, 2015]. Saito *et al.* studied 90,914 Japanese recruited between 1990 and 1994. After 18.7 years of follow-up, 12,874 deaths were reported. Hazard ratios for all-cause mortality among men who consumed green tea compared with those who drank less than 1 cup/day were 0.96 for 1–2 cups/day, 0.88 for 3–4 cups/day, and 0.87 for more than 5 cups/day ($p < 0.001$). Corresponding hazard ratios for women were 0.90, 0.87, and 0.83 ($p < 0.001$) [Saito *et al.*, 2015]. The Ohsaki National Health Insurance Cohort Study, a population-based, prospective cohort study was initiated in 1994 among 40,530 Japanese adults aged 40–79 years without history of stroke, coronary heart disease, or cancer at baseline. Participants were followed up for up to 11 years (1995–2005) for all-cause mortality and for up to seven years (1995–2001) for cause-specific mortality. Green tea consumption was inversely associated with mortality due to all causes and due to cardiovascular disease. The inverse association with all-cause mortality was stronger in women. In men, the multivariate hazard ratios of mortality due to all causes associated with different green tea consumption frequencies were 1.00 for less than 1 cup/day, 0.93 (95% confidence interval for 1–2 cups/day, 0.95 for 3–4 cups/day, and 0.88 for ≥5 cups/day, respectively ($p = 0.03$). The corresponding data for women were 1.00, 0.98, 0.82, and 0.77 respectively ($p < 0.001$). The inverse association with cardiovascular disease mortality was stronger than that with all-cause mortality. This inverse association was also stronger in women. Among the types of cardiovascular disease mortality, the strongest inverse association was observed for stroke mortality. More than half of the randomized controlled trials have demonstrated the beneficial effects of green tea on cerebrovascular disease (CVD) risk profiles. These results from randomized controlled trials suggest a plausible mechanism for the beneficial effects of green tea and provide substantial support for the observations [Kuriyama *et al.*, 2006]. A study found a 37% relative reduction in all-cause mortality comparing the highest versus the lowest quintiles of total polyphenol intake (hazard ratio (HR) = 0.63; P = 0.12). These results may be useful to determine optimal polyphenol intake that may reduce the risk of all-cause mortality [Tresserra-Rimbau *et al.*, 2014].

These meta-analysis provides evidence that consumption of green tea is associated with favorable outcomes with respect to risk of cardiovascular and cerebrovascular diseases and ischemic related diseases. Whether male or female, the death rate of cardiovascular and cerebrovascular diseases is negatively related to drinking enough green tea.

2. Drinking tea reduces blood pressure

Hypertension is one of the most dangerous factors causing cardiovascular and cerebrovascular diseases. Blood pressure changes is associated with cardiovascular and cerebrovascular disease and related outcomes. There is evidence that some polyphenol-rich foods, including berry fruits rich in anthocyanins, cocoa and green tea rich in flavan-3-ols, almonds and pistachios rich in hydroxycinnamic acids, and soy products rich in isoflavones, are able to improve blood pressure levels. An epidemiological study showed that drinking tea can reduce blood pressure. Evidence from animal and observational studies supports the beneficial effect of green tea on reducing blood pressure. In a statistical analysis of a randomized controlled trial, the effect of green tea supplementation on blood pressure control measures was evaluated. The results showed that there were 24 trials and 1,697 subjects were included in meta-analysis. The comprehensive results showed that green tea could significantly reduce systolic blood pressure (−1.17 mm Hg; 95% confidence interval: −2.18 to −0.16 mm Hg; $p = 0.02$) and diastolic blood pressure (−1.24 mm Hg; 95% confidence interval: −2.07 to −0.40 mm Hg; $p = 0.004$) [Bogdanski *et al.*, 2012]. A study investigated the effect of drinking black tea on the rate of ambulatory blood pressure change. In a randomized, controlled, double-blind, 6-month parallel design trial, men and women with systolic blood pressure between 115–150 mmHg at the time of screening (n = 111) were recruited to mainly evaluate the impact on blood pressure changes. Participants drank three cups of black tea solid powder or flavonoid-free caffeine matching beverage every day (control group). The 24-hour ambulatory blood pressure levels and blood pressure change rates were assessed at baseline, day 1, month 3, and month 6. The results showed that at three time points, compared with the control group, the change rates of systolic blood pressure (p = 0.0045) and diastolic blood pressure (p = 0.016) of black tea decreased by about 10% at night (22:00–06:00). These effects occurred immediately on day 1 and lasted for more than six months, independent of blood pressure and heart rate levels. The change rate of blood pressure did not change significantly during the day (08:00–20:00) [Hodgson *et al.*, 2013]. A randomized study analyzed the effects of black tea on fasting and postprandial blood pressure and fluctuating reflex in patients with hypertension. Hypertension and arteriosclerosis are independent predictors of cardiovascular mortality. Flavonoids may have some vascular protective effects. The effects of black tea on blood pressure and fluctuation reflex in patients with hypertension before and after diet were studied. According to a randomized, double-blind, controlled, crossover design, 19 patients were assigned to drink black tea (129 mg flavonoids) or placebo twice a day for eight days. The blood pressure at one, two, three, and four hours before and after drinking black tea was measured. Compared with placebo, the systolic and diastolic blood pressure decreased (-3.2mmhg, p < 0.005 and -2.6mmhg, p < 0.0001) respectively after drinking black tea; it can also prevent the increase of blood pressure after diet (p < 0.0001) [Grassi *et al.*, 2015].

These findings suggest that regular consumption of green tea or black tea may be related to the protection of cardiovascular and cerebrovascular health. Moreover, a component in green tea and black tea solid affects the rate of change of nocturnal blood pressure and drinking green tea. Black tea also reduces fasting blood pressure and may significantly affect the rate of blood pressure change. In order to further study the effects of green tea and black tea supplementation on blood pressure control and clinical hypertension, larger and longer-term trials are needed.

There are two studies to determine whether black tea and green tea can reduce blood pressure in stroke-prone spontaneously hypertensive rats. Male rats (n = 15) recovered for two weeks after intraperitoneal implantation of blood pressure transmitter. The rats were divided into three groups: the control group drank tap water 30 ml/d; the black tea polyphenol group drank water containing thearubin 3.5 g/L, theaflavin 0.6 g/L, flavone 0.5 g/L, and catechin 0.4 g/L; the green tea polyphenol group drank water containing 3.5 g/L catechin, 0.5 g/L flavonoids, and 1 g/L polyflavonoids. It was found that during the day, the systolic and diastolic blood pressure of black tea polyphenol group and green tea polyphenol group were significantly lower than those of the control group. Green tea polyphenols significantly increased the expression of catalase in aorta, while black tea polyphenols and green tea polyphenols significantly decreased the expression of phosphorylated myosin light chain. The results show that black tea and green tea polyphenols reduce the increase of blood pressure through their antioxidant properties in stroke prone spontaneously hypertensive rats. In addition, because the amount of polyphenols used in the experiment is equivalent to about 1 L of polyphenols contained in tea, regular drinking of black tea and green tea may also play a certain protective effect on human hypertension [Negishi *et al.*, 2004].

The above population epidemiological survey and animal experimental research show that regular drinking of green tea and black tea may be related to the protection of cardiovascular and cerebrovascular health. Therefore, it is concluded that in general, the polyphenols in green tea and black tea significantly reduced systolic and diastolic blood pressure during the short-term test. Regular drinking of black tea can also reduce blood pressure but its effect on blood pressure needs to be further studied.

3. Drinking tea reduces atherosclerosis

Atherosclerosis is one of the most dangerous factors causing cardiovascular and cerebrovascular diseases. Atherosclerosis changes is associated with cardiovascular and cerebrovascular disease and related outcomes. The risk factors for atherosclerosis includes hypertension, diabetes mellitus, hyperlipidemia, obesity, and cigarette smoking. The anti-oxidative activity of polyphenols can prevent the pathology of atherosclerosis, including endothelial dysfunction, low-density lipoprotein oxidation, vascular smooth muscle cell proliferation, inflammatory process by monocytes, macrophages or T lymphocytes, and platelet aggregation. Polyphenols in green tea have been shown to reduce a risk of coronary heart disease in epidemiological studies. Also, it has been reported that polyphenols have hypolipidemic and antioxidant effects. A study investigated the effects of ground green tea drinking on the susceptibility of plasma and low-density lipoprotein (LDL) to the oxidation by CuSO4 *ex vivo*, and evaluated daily food consumption using semi-quantitative questionnaire. Polyphenols in green tea were shown to reduce a risk of coronary heart disease in epidemiological studies. The lag time of conjugated dienes formation was increased in all subjects after ground green tea consumption from 67 ± 19 minutes to 118 ± 42 minutes in plasma and from 47 ± 6 minutes to 66 ± 10 minutes in LDL. The cholesterol contents in plasma and LDL decreased 10 mg/dl after ground green tea consumption. The beta-carotene, alpha-tocopherol, vitamin C, and uric acid contents in plasma did not change after ground green tea consumption. The superoxide dismutase (SOD) activity in plasma also remained unchanged during this study periods [Gomikawa et al., 2008]. Another study elucidated whether or not green tea consumption may have these effects *in vivo*, which may be protective against atherosclerotic disease. It was found that MDA-LDL concentrations (84 ± 45 IU/L vs. 76 ± 40 IU/L, $p < 0.05$) and the ratio of MDA-LDL/LDL-cholesterol (0.74 ± -0.21 vs. 0.65 ± 0.20, $p < 0.02$) significantly decreased at the end of the green tea period. These results suggest that green tea consumption may inhibit LDL oxidation *in vivo* and daily consumption of green tea decreased serum MDA-LDL concentrations. These findings also indicated that ground green tea consumption decreased susceptibility of plasma and LDL to oxidation and modulated cholesterol metabolism and might prevent initiation and progression of atherosclerosis [Hirano-Ohmori et al., 2005]. The strong radical-scavenging properties of polyphenols would exhibit anti-inflammation and antioxidative effects. Polyphenols reduce ROS production by inhibiting oxidases, reducing the production of superoxide, inhibiting OxLDL formation, suppressing vascular smooth muscle cell proliferation and migration, reducing platelet aggregation, and improving mitochondrial oxidative stress. Polyphenol consumption also inhibits the development of hypertension, diabetes mellitus, hyperlipidemia, and obesity [Cheng *et al.*, 2017].

These findings suggest that regular consumption of green and black tea may be related to the protection of cardiovascular and cerebrovascular health and tea polyphenols, a component in green and black tea can affect the rate of changes in blood lipids and drinking green and black tea can also reduce atherosclerosis. In order to further study the effects of green tea and black tea supplementation on blood lipid and atherosclerosis, larger and longer-term trials are needed.

4. Drinking tea increases blood flow

A number of large-scale epidemiological studies have demonstrated an association between the consumption of tea and cognitive function, as well as improved blood flow. The effect of epicatechin on cognition is due to the effect of increased cerebral blood flow [Haskell-Ramsay *et al.*, 2018]. This randomized, double-blind, placebo-controlled, parallel groups trial randomized (N = 155), 50–70 year old male and female participants who were assessed for the cognitive (N = 140), mood (N = 142), BP (N = 133), and CBF (N = 57) effects of two doses of Greek mountain tea (475 mg and 950 mg) as well as an active control of 240 mg *Ginkgo biloba*, and a placebo control, following acute consumption (Day 1) and following a month-long consumption period (Day 28). Results found that relative to the placebo control, 950 mg Greek mountain tea evinced significantly fewer false alarms on the Rapid Visual Information Processing (RVIP) task on Day 28 and significantly reduced state anxiety following 28 days consumption (relative also to the active, Ginkgo control). This higher dose of Greek mountain tea also attenuated a reduction in accuracy on the picture recognition task, on Day 1 and Day 28, relative to Ginkgo and both doses of Greek mountain tea trended towards significantly faster speed of attention on both days, relative to Ginkgo. Both doses of Greek mountain tea, relative to placebo, increased oxygenated haemoglobin (HbO) and oxygen saturation (Ox %) in the prefrontal cortex during completion of cognitively demanding tasks on Day 1. The higher dose also evinced greater levels of total (THb) and deoxygenated (Hb) haemoglobin on Day 1 but no additional effects were seen on CBF on Day 28 following either dose of Greek mountain tea [Wightman *et al.*, 2018].

Modern Chinese medicine believes that the meridians are defined as "channels for the movement of qi and blood". The team of Fudan University released for the first time a systematic image that conforms to the traditional description of human meridians, allowing people to "see the meridians". The related paper, *Tea stimulated human body infrared image showing meridian system*, was published in the journal Quantitative Biology in March. After drinking 68 kinds of tea, the research team made a statistical result, and the consistency of all volunteer reports reached 96%. The meridian tropism of different tea leaves has an extremely obvious law. Green tea corresponds to the sun pulse, green tea corresponds to the yang ming pulse, black tea corresponds to the shaoyang pulse, white tea corresponds to the taiyin pulse, and black tea corresponds to the jueyin pulse. Only yellow tea is not clear. The research team collected 512 kinds of tea from various provinces in southern China, India, Sri Lanka, Japan, the United States, New Zealand, and other countries. After drinking, they took infrared images of their bodies to observe the time and way of infrared radiation. Some of these teas have a strong sense of meridian flow after drinking. The instrument detects that infrared radiation will cause a temperature difference of 5–8°C on the body surface. NMR also sees a large amount of liquid flowing in the intercellular substance [Jin *et al.*, 2020].

The above research shows that tea polyphenols have the function of dredging blood vessels, activating blood circulation and removing blood stasis, and can reduce cardiovascular and cerebrovascular damage by increasing blood flow.

5. Drinking tea can reduce the risk factors of cardiovascular and cerebrovascular diseases

Some studies have evaluated the effects of long-term tea drinking on cardiovascular and cerebrovascular health. In addition to hypertension and atherosclerosis, studies have found that drinking tea can reduce a variety of other risk factors leading to cardiovascular and cerebrovascular diseases, including, diabetes, hyperlipidemia, vascular endothelial function, elasticity of large arteries, anti-inflammatory and antioxidant, obesity, hyperglycemia, as well as the antioxidant activity of polyphenols [Cheng *et al.*, 2017]. A study assessed the health effects of long-term drinking of tea. Drinking tea, especially green tea, can significantly reduce the risk of stroke, diabetes, and depression, and improve blood sugar, cholesterol, abdominal obesity, and blood pressure. In large epidemiological studies, habitual drinking of green tea can reduce the mortality of all-cause death and cardiovascular death. Moreover, green tea intake was negatively correlated with heart failure, stroke, diabetes, and certain cancers. Surprisingly, tea can also reduce the risk of atrial and ventricular arrhythmias [Bhatti *et al.*, 2013]. Increasing blood flow is a potential way to improve brain function. The prospect of using dietary tea polyphenols to increase cerebral blood flow is very broad. These therapeutic effects are mainly associated with inhibition of NF-κB that is anti-cardiac hypertrophy, inhibition of myeloperoxidase (MPO) activity that is anti-myocardial infarction, reduction in plasma glucose and glycated haemoglobin level that is anti-diabetes, reduction of inflammatory markers that is anti-inflammatory, and the inhibition of ROS generation that is antioxidant. Polyphenols may lower the risk of cerebrovascular disease (CVD) and other chronic diseases due to their antioxidant and anti-inflammatory properties, as well as their beneficial effects on blood pressure, lipids, and insulin resistance [Eng *et al.*, 2018]. EGCG is a promising therapeutic to combat cardiovascular complications associated with the metabolic diseases characterized by reciprocal relationships between insulin resistance and endothelial dysfunction that include obesity, metabolic syndrome, and type 2 diabetes [Kuppusamy *et al.*, 1994].

Through epidemiological and clinical studies in the above countries and regions, it can see that drinking tea can protect cardiovascular and cerebrovascular health and prevent cardiovascular and cerebrovascular diseases. Despite a large number of *in vivo* and *in vitro* studies, more advanced clinical trials are still needed to confirm the efficacy of polyphenols in the treatment of atherosclerosis related vascular diseases. In addition, although drinking tea does not affect osteoporosis, people who drink tea should pay attention that the concentration of tea should not be too high, and adequate calcium intake may be particularly important.

8.4.2   *The mechanism of tea polyphenols on reducing cardiovascular diseases*

Not only does epidemiological investigation show that drinking tea can protect cardiovascular and cerebrovascular health against the development and progression of cardiovascular and cerebrovascular disease and reduce cardiovascular and cerebrovascular diseases, but also a large number of animal experimental studies support the results of epidemiological investigation and clinical research [Zhao & Zhao, 2010; Zhao, 2020]. Additionally, the mechanism of tea polyphenols in preventing and treating cardiovascular and cerebrovascular diseases has been proposed.

1. Tea polyphenols reduce oxidative damage of brain synaptosomal membrane

We compared the protective effects of four components of green tea polyphenols (GTP) — (-)-epigallocatechin gallate, EGCG; (-)-epicatechin gallate, ECG; (-)epigallocatechin, EGC; and (-)epicatechin, EC, against iron-induced lipid peroxidation in synaptosomes. Results showed that 1) the inhibitory effects of those compounds on TBA reactive materials from lipid peroxidation decreased in the order of EGCG > ECG > EGC > EC; (2) the scavenging effects of those compounds on lipid free radicals produced by lipid peroxidation could be classified as follows: ECG > EGCG > EC > EGC. Moreover, we found that the ability of those compounds to protect synaptosomes from the damage of lipid peroxidation initiated by $Fe^{2+}/^{3+}$ was dependent not only on their iron-chelating activity and free-radical scavenging activity, but also on the stability of their semiquinone free radicals formed after reaction with ROS [Guo *et al.*, 1996, 1999]. These results indicate that tea polyphenols can protect synaptosomes of brain cells from oxidative damage through oxidative damage of brain synaptic membrane.

2. Tea polyphenols improve ischemia-reperfusion myocardium injury

We studied the scavenging effect of tea polyphenols on oxygen free radicals during ischemia-reperfusion in isolated rat hearts. The addition of tea polyphenols to the reperfusion solution significantly reduced the oxygen free radical, and the study on the relationship between efficacy and dose showed that the scavenging effect of tea polyphenols on the free radical produced by myocardial ischemia-reperfusion injury had an obvious concentration dependence [Zou *et al.*, 1995]. It was found that cardiomyocytes produced superoxide anion free radicals as well as NO during ischemia-reperfusion. In Langendorff-perfused rat hearts, submicromolar concentrations of ECG administered either before or after ischemia reduced infarct size by more than 40%, decreased lactate dehydrogenase release, and improved the recovery of cardiac function. ECG protection was blocked by PKCε inhibition and attenuated by mitochondrial KATP channel inhibition. In a unique mammalian cell system with depleted Na/K-ATPase α1 expression, ECG-induced PKCε activation persisted but protection against ischemia/reperfusion was blunted [Qi *et al.*, 2019]. NO reacts with superoxide anion radical to form peroxynitrite (ONOO⁻). Tea polyphenols can significantly scavenge ONOO⁻ and inhibit its oxidation activity [Zhao *et al.*, 1996c]. The protective effect of EGCG is due to its ability to decrease lipid peroxidation, oxidative stress, and the production of NO radicals by inhibiting the expression of iNOS. EGCG also ameliorates the overproduction of pro-inflammatory cytokines and mediators, and reduces the activity of NF-κB and AP-1 and the subsequent formation of peroxynitrite with NO and reactive oxygen species [Tipoe *et al.*, 2007].

Thus, EGCG effectively mitigates cellular damage by lowering the inflammatory reaction and reducing the lipid peroxidation and NO-generated radicals leading to the oxidative stress. Green tea is proposed to be a dietary supplement in the prevention of cardiovascular diseases, in which oxidative stress and pro-inflammation are the principal causes.

3. Tea polyphenols improve injury from stroke

Tea polyphenols can cross the blood-brain barrier, inhibit apoptosis, and play a neuroprotective role against cerebral ischemia. Tea polyphenols can also decrease DNA damage caused by free radicals. A study demonstrated that intraperitoneal injection of tea polyphenols immediately after reperfusion significantly reduced apoptosis in the hippocampal CA1 region; this effect started six hours following reperfusion. Tea polyphenols could reverse the ischemia/reperfusion-induced reduction in the expression of DNA repair proteins [Wang *et al.*, 2012]. A study showed that long-term consumption fermented herbal tea by Wistar rats significantly reduced brain edema and neuronal apoptosis but did not attenuate BBB damage following cerebral ischemia. Analysis of whole-brain homogenates showed significantly reduced lipid peroxidation levels, increased total antioxidant capacity, and resulted in improved neuro-behavioral outcomes in long-term consumption of fermented rooibos herbal tea-treated rats when compared with untreated animals [Akinrinmade *et al.*, 2017].

Generation of AOS is suggested to be concerned with various senile disorders. Tea catechins, (+) catechin (CA), EC, and EGCG, are polyhydroxy-fravan derivatives from tea leaves and have been proposed to possess active oxygen scavenging effect. Tea polyphenols protected the cultured newborn-mouse cerebral nerve cells from death induced by glucose oxidase. The protective potency of EGCG was weaker than those of EC and CA. Learning ability of mice was assessed by a step-down-type passive avoidance test, and memory impairment of mice was achieved by intracerebral injection of glucose oxidase or cerebral ischemia induced by 10 minutes occlusion of the common carotid arteries. Intracerebral injection of EC improved the memory impairment induced by intracerebral glucose oxidase, and intravenous (IV) injection of CA or EC improved that induced by the cerebral ischemia. CA and EC depressed IV-induced edema in rat hind paw, but EGCG did not [Matsuoka *et al.*, 1995]. Study of protective effect of green tea polyphenol EGCG against neuronal damage and brain edema after unilateral cerebral ischemia in gerbils found that EGCG reduced excitotoxin-induced MDA production and neuronal damage in the culture system. In the *in vivo* study, treatment of gerbils with the lower EGCG dose failed to show neuroprotective effects, however, the higher EGCG dose attenuated the increase in MDA level caused by cerebral ischemia. EGCG also reduced the formation of post-ischemic brain edema and infarct volume [Lee *et al.*, 2004].

These results suggest that tea polyphenols can reduce DNA damage caused by free radicals. Tea polyphenols repair DNA damage and inhibit neuronal apoptosis during global cerebral ischemia/reperfusion. Studies have shown that tea polyphenols can reverse DNA repair proteins induced by ischemia/reperfusion, and intraperitoneal injection of tea polyphenols immediately after reperfusion can significantly reduce apoptosis in hippocampal CA1 region. Tea polyphenols ameliorate the injuries or impairments induced by scavenging intracellular active oxygens.

4. Tea polyphenols prevent atherosclerosis

A study showed that tea can help reduce the occurrence of aortic atherosclerosis in rabbits, which indicates the potential beneficial effect of tea on cardiovascular and cerebrovascular health. Many studies have demonstrated that polyphenols also have good effects on the vascular system by lowering blood pressure, improving endothelial function, increasing anti-oxidant defenses, inhibiting platelet aggregation and low-density lipopro-tein oxidation, and reducing inflammatory responses. Several polyphenols have been found to have effects on several cardiovascular risk factors such as hypertension, oxidative stress, atherosclerosis formation, endothelial dysfunction, carotid artery intima-media thickness, diabetes, and lipid disorders. It has been proven that these compounds have many cardio protective functions: they alter hepatic cholesterol absorption, triglyceride biosynthesis and lipoprotein secretion, the processing of lipoproteins in plasma, and inflammation [Giglio *et al.*, 2018]. A study found that EGCG could ameliorate *Porphyromonas gingivalis*-induced atherosclerosis. It found that atherosclerotic lesion areas of the aortic sinus caused by *P. gingivalis* infection decreased in EGCG-treated groups, wherein EGCG reduced the production of C-reactive protein, monocyte chemoattractant protein-1, and oxidized low-density lipoprotein (LDL), and slightly low-ered LDL/very LDL cholesterol in *P. gingivalis*-challenged mice serum [Cai *et al.*, 2013]. In this study, the anti-atherosclerosis formation effects of tea catechins were examined in atherosclerosis-susceptible C57BL/6J, apoprotein (apo) E-deficient in mice. Tea ingestion did not influence plasma cholesterol or triglyceride concentrations. Plasma lipid peroxides were reduced in the tea group at Week 8, suggesting that the *in vivo* oxida-tive state is improved by tea ingestion. Atheromatous areas in the aorta from the arch to the femoral bifurcation and aortic weights were both significantly attenuated by 23% in the tea group compared with the con-trol group. Aortic cholesterol and triglyceride contents were 27% and 50% lower, respectively, in the tea group than in the control group [Miura *et al.*, 2001]. A study was undertaken to evaluate the cardio-protective role of EGCG against fluoride-induced oxidative stress-mediated cardiotoxicity in rats. The animals exposed to NaF for four weeks exhibited a significant increase in the levels of cardiac troponins. The cardiac serum markers, lipid peroxidative markers, and plasma total cholesterol, triglycerides, phospholipids free fatty acids, low density lipoprotein cholesterol, very low density lipoprotein cholesterol as well as cardiac lipids profile, with the significant decrease of high density lipoprotein cholesterol and cardiac phospholipids. The mitochondrial Ca^{2+} ion level was also significantly reduced along with the significant decrease in the levels of enzymatic and non-enzymatic antioxidants. Furthermore, NaF treatment significantly increased the DNA fragmentation, upregulate cardiac pro-apoptotic markers and inflammatory markers, and downregulate the anti-apoptotic markers in the cardiac tissue. Pre-administration of EGCG in fluoride intoxicated rats remarkably recovered all these altered parameters to near normalcy through its antioxidant nature [Miltonprabu and Thangapandiyan, 2015]. EGCG protects cellular damage by inhibiting DNA damage and oxidation of LDL [Tipoe *et al.*, 2007].

These results indicate that long-term intake of tea extract may improve the oxidation state *in vivo*, reduce the production of oxidized low-density lipoprotein (LDL), and prevent the development of atherosclerosis through the effective antioxidant activity of tea. EGCG protects cellular damage by inhibiting DNA damage and oxidation of LDL. The protective effect of EGCG is due to its ability to decrease lipid peroxidation, oxidative stress, and the production of nitric oxide (NO) radicals by inhibiting the expression of iNOS.

5. Tea polyphenols can enhance NO and improve cardiovascular and cerebrovascular health

Endothelial function, the elastic properties of large arteries and the magnitude and timing of wave reflections, are important determinants of cardiovascular performance. Several epidemiological studies suggest that the regular consumption of foods and beverages rich in flavonoids is associated with a reduction in the risk of several pathological conditions ranging from hypertension to coronary heart disease, stroke, and dementia. The impairment of endothelial function is directly related to aging and an association between decreased cerebral perfusion and dementia has been shown to exist. Cerebral blood flow (CBF) must be maintained to ensure a constant delivery of oxygen and glucose as well as the removal of waste products. Increasing blood flow is one potential way of improving brain function, and the prospect for increasing CBF with dietary polyphenols is extremely promising. The major polyphenols shown to have some of these effects in humans are primarily from cocoa, wine, grape seed, berries, tea, tomatoes (polyphenolics and non-polyphenolics), soy, and pomegranate [Ghosh & Scheepens, 2009]. Nitric oxide is an endothelial cell relaxation factor, which can increase blood flow, reduce blood pressure, and prevent cardiovascular and cerebrovascular diseases. We found that appropriately increasing nitric oxide can reduce myocardial ischemia-reperfusion injury, arrhythmia and cardiomyocyte injury, and improve myocardial redox state [Zhao *et al.*, 1996a; Shen *et al.*, 1998, 2000a]. There is now consistent data indicating that tea and tea flavonoids can enhance nitric oxide status and improve endothelial function, which may be at least partly responsible for benefits of cardiovascular health. There is also evidence to suggest benefits of green tea on body weight and body fatness. Dietary flavonoids and tea consumption have been described to improve endothelial function and flow-mediated dilation. A proposed mechanism by which dietary flavonoids could affect flow-mediated dilation is that they improve the bioactivity of the endothelium-derived vasodilator NO by enhancing NO synthesis or by decreasing superoxide-mediated NO breakdown. The endothelium plays a pivotal role in arterial homeostasis. Reduced NO bioavailability with endothelial dysfunction is considered the earliest step in the pathogenesis of atherosclerosis [Grassi *et al.*, 2013].

A study found that the phosphatidylinositol 3-kinase-dependent transcription factor, FOXO1, may mediate effects of EGCG to regulate the expression of ET-1 in endothelial cells. EGCG decreases ET-1 expression and secretion from endothelial cells, in part, via Akt- and AMPK-stimulated FOXO1 regulation of the ET-1 promoter. These findings may be relevant to beneficial cardiovascular actions of green tea [Reiter *et al.*, 2010]. However, theaxanthin and thearubigins has a protective effect on the heart by blocking the expression of nitric oxide gene. Theaflavin may influence activation of transcription factors such as NFnB or AP-1 that ultimately hinder the formation of nitric oxide expression gene. Likewise, black tea contains a unique amino acid, theanine, that acts as a neurotransmitter owing to its ability to cross the blood-brain barrier. Moreover, it boasts immunity by enhancing the disease-fighting ability of gamma delta T cells. Theaflavin and thearubigins act to safeguard against oxidative stress, thereby being effective in cardiac functioning. The mechanistic approach of these antioxidants is likely to be associated with the inhibition of redox sensitive transcription factors and pro-oxidant enzymes such as xanthine oxidase or nitric oxide synthase. However, their involvement in anti-oxidative enzyme induction as in glutathione-S-transferases is also well documented. They act as curative agent against numerous pathological disorders by disrupting the electron chain, thus inhibiting the progression of certain ailments [Butt *et al.*, 2014].

Endothelial dysfunction has been considered an important and independent predictor of future development of cardiovascular risk and events. The association between brachial NO-dependent flow-mediated dilation and cardiovascular disease risk suggesting NO-mediated blood flow expansion can reduce myocardial ischemia-reperfusion injury.

6. Preventive effect of tea polyphenols on cardiovascular and cerebrovascular diseases induced by inflammation

Tea polyphenols have a significant effect on cells, inhibiting neutrophil migration and inflammatory response at low concentrations. EGCG has been found to have a wide range of therapeutic properties, including anti atherosclerosis, anti-myocardial hypertrophy, anti-myocardial infarction, and anti-diabetes. One study investigated the cellular and molecular mechanisms of cardiovascular protection of green tea polyphenols (especially EGCG), with emphasis on antioxidant and anti-inflammatory effects. EGCG is the main and most active ingredient in green tea. It has been demonstrated that peripheral inflammation can disrupt the blood brain barrier (BBB) by various pathways, resulting in different central nervous system (CNS) diseases. Recently, clinical research also showed CNS complications following SARS-CoV-2 infection and chimeric antigen receptor (CAR)-T cell therapy, which both lead to a cytokine storm in the circulation. Therefore, elucidation of the mechanisms underlying the BBB disruption induced by peripheral inflammation will provide an important basis for protecting the CNS in the context of exacerbated peripheral inflammatory diseases [Huang *et al.*, 2021]. A study found multiple sclerosis (MS) to be the most associated with the presence of Th17 cells and IL-17 among the different inflammatory conditions of the CNS. In particular, many studies using the murine model for MS, experimental autoimmune encephalomyelitis, found a clear association of Th17 and IL-17 with disease severity and progression of stroke. There is also evidence that IL-17 plays a pathogenic role in the post-ischemic phase of stroke as well as its experimental model [Waisman *et al.*, 2015]. Consumption of green tea has been inversely associated with the development and progression of cardiovascular diseases and cardiovascular risk factors. The protective effects of green tea and green tea constituents, particularly tea polyphenols on the cardiovascular system, include multiple ways. Mechanisms that have been suggested as being involved in the anti-atherosclerotic effects of green tea consumption and tea polyphenols primarily entail anti-oxidative, anti-inflammatory, anti-proliferative, and antithrombotic properties, as well as beneficial effects on endothelial function. Emerging evidence has shown that tea polyphenols and their metabolites have many additional mechanisms of action by affecting numerous sites, potentiating endogenous antioxidants, and eliciting dual actions during oxidative stress, ischemia, and inflammation [Tipoe *et al.*, 2007]. Tea polyphenols have proven to modulate apoptosis at various points in the sequence, including altering expression of anti- and pro-apoptotic genes. Their anti-inflammatory effects are activated through a variety of different mechanisms, including modulation of nitric oxide synthase isoforms. Tea polyphenols' actions of attenuating oxidative stress and the inflammatory response may, in part, account for their confirmed neuroprotective capabilities following cerebral ischemia [Sutherland *et al.*, 2006]. Tea polyphenols also exhibit anti-inflammatory and antioxidant effects, which have the potential to impact various biological mechanisms for reducing the initiation and progression of periodontitis [Basu *et al.*, 2018]. Green tea polyphenols can inhibit H_2O_2-induced oxidative stress through the Akt/GSK-3β/caveolae pathways in cardiac cells. They could prevent the activation of NF-κb and the inhibition on PI3K/Akt signaling for the acute myocardial infarction (MI stress). Moreover, green tea polyphenols also could improve mitochondria dysfunction associated with alterations of lipid metabolism, the adaptor 14-3-3 ε protein signaling, and chaperone-induced stress response during post-MI remodeling [Hsieh *et al.*, 2014; Ding *et al.*, 2017]. When (-)catechin solution instead of drinking water was orally administered ad libitum for two weeks, dose-dependent protection against neuronal death following by transient ischemia and reperfusion was observed. These results suggested that orally-administered (-)catechin was absorbed, passed through the blood-brain barrier, and that delayed neuronal death of hippocampal CA1 after ischemia-reperfusion was prevented due to its antioxidant activities [Inanami *et al.*, 1998].

The above results indicate that tea polyphenols and their metabolites have many mechanisms of action on inflammation of cardiovascular and cerebrovascular ischemia reperfusion caused oxidative stress. The effects include many parts, for example, anti-inflammatory effect is activated by a variety of different mechanisms, including regulation of nitric oxide synthase subtypes, enhancement of endogenous antioxidants and initiation of dual effects, anti-hypertensive, varying the values of various blood pressure regulatory factors, such as vascular compliance, peripheral vascular resistance, and total blood volume via anti-inflammatory and anti-oxidant actions, and these effects are caused by the inhibition of NF-κ B, IL-17, and PI3K/Akt signals caused by a variety of inflammatory factors, including nuclear factors.

8.5 The mechanism and conclusion regarding the protective effects of tea polyphenols on cardiovascular and cerebrovascular health

In conclusion, cardiovascular and cerebrovascular ischemia reperfusion is caused by an interruption of blood flow to the brain which generally leads to irreversible brain and heart ischemia reperfusion damage. Ischemic injury is associated with vascular leakage, inflammation, tissue injury, and cell death. Cellular changes associated with ischemia include impairment of metabolism, energy failure, free radical production, excitotoxicity, altered calcium homeostasis, and activation of proteases all of which affect brain and heart functions, also contribute to long-term disabilities including cognitive decline and heart failure. Inflammation, mitochondrial dysfunction, increased oxidative/nitrosative stress, and intracellular calcium overload contribute to brain injury including cell death, brain edema, and cardiac hypertrophy. Tea polyphenols have antioxidant, anti-inflammatory, and anti-apoptotic properties and their protective effects on mitochondrial functioning, glutamate uptake, and regulating intracellular calcium levels in ischemic injury *in vitro* have been demonstrated. The heart and brain protective effects of polyphenols involve a variety of mechanisms, including anti-oxidation, vasodilation, anti-hypertension, anti-inflammation, anti-atherosclerosis and lipid-lowering effects, and anti-apoptosis. Tea polyphenols can reduce oxidative stress by inhibiting oxidase, reducing the production of superoxide, inhibiting the formation of OxLDL, inhibiting the proliferation and migration of vascular smooth muscle cells, reducing platelet aggregation, reducing the production of ROS and RNS, and improving oxidative stress. The intake of tea polyphenols will also promote the prevention and treatment of cardiovascular and cerebrovascular health and diseases, as follows:

1. Tea polyphenols protect cardiovascular and cerebral vessels by reducing blood lipid and blood pressure.
2. Tea polyphenols protect cardiovascular and cerebral vessels by scavenging peroxynitrite caused by ROS and RNS.
3. Tea polyphenols present antioxidant activity by scavenging free radicals, chelating redox active transition-metal ions, inhibiting redox active transcription factors, inhibiting pro-oxidant enzymes, and inducing antioxidant enzymes.
4. Tea polyphenols inhibit the key enzymes involved in lipid biosynthesis and reduce intestinal lipid absorption, thereby improving blood lipid profile.
5. Tea polyphenols regulate vascular tone by activating endothelial nitric oxide.
6. Tea polyphenols prevent vascular inflammation that plays a critical role in the progression of atherosclerotic lesions. The anti-inflammatory activities of tea polyphenols may be due to their suppression of leukocyte adhesion to endothelium and subsequent transmigration through inhibition of transcriptional factor NF-κB-mediated production of cytokines and adhesion molecules both in endothelial cells and inflammatory cells.
7. Tea polyphenols inhibit proliferation of vascular smooth muscle cells by interfering with vascular cell growth factors involved in atherosclerosis.

Tea polyphenols have attracted considerable attention in the prevention of cardiovascular and cerebrovascular diseases. In comparison to tumor cells, the elucidation of their molecular targets in cardiovascular and cerebrovascular relevant cells is still at the beginning. Although promising experimental and clinical data demonstrate protective effects for the cardiovascular and cerebrovascular system, little information is actually available on how these beneficial effects of tea polyphenols are mediated at the cellular level. By affecting the activity of receptor and signal transduction kinases, both catechins and theaflavins — the major ingredients of green and black tea, respectively, exert a variety of cardio-vascular and cerebrovascular beneficial effects. In general, the number and positions of galloyl groups have major influence on the potency of polyphenols. Compared to their broad impact on cellular signal transduction, tea polyphenols reveal little transcriptional effects (Fig. 8-10).

However, more detailed and profound analysis of molecular actions in different cells of the relevant system is necessary before safe clinical use of tea polyphenols for treatment of cardiovascular and cerebrovascular diseases will become possible. Taken together, tea polyphenols may provide a treatment and protection against cardiovascular and cerebrovascular diseases in a variety of ways. A rigorous assessment of the effects of green

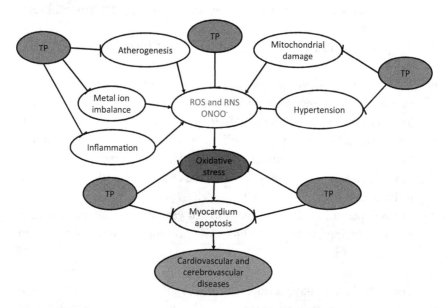

Figure 8-10. The mechanisms involved in the cardio-protective effects of tea polyphenols. TP: tea polyphenols.

tea polyphenols in well-controlled human trials will be required for better understanding of the effects of green tea in cardiovascular and cerebrovascular health.

References

Afanas'ev IB. (2005) On mechanism of superoxide signaling under physiological and pathophysiological conditions. *Med Hypotheses*, **64**, 127–129.

Akinrinmade O, Omoruyi S, Dietrich D, Ekpo O. (2017) Long-term consumption of fermented rooibos herbal tea offers neuroprotection against ischemic brain injury in rats. *Acta Neurobiol Exp (Wars)*, **77**(1), 94–105.

Basu A, Masek E, Ebersole JL. (2018) Dietary polyphenols and periodontitis — a mini-review of literature. *Molecules*, **23**(7), 1786.

Beckman JS, Koppenol WH. (1996) Nitric oxide, superoxide and peroxynitrite: the good, the bad, and the ugly. *Am J Physiol*, **271**, C1424–C1437.

Bhatti SK, O'Keefe JH, Lavie CJ. (2013) Coffee and tea: perks for health and longevity? *Curr Opin Clin Nutr Metab Care*, **16**(6), 688–697.

Bogdanski P, Suliburska J, Szulinska M, Stepien M, Pupek-Musialik D, Jablecka A. (2012) Green tea extract reduces blood pressure, inflammatory biomarkers, and oxidative stress and improves parameters associated with insulin resistance in obese, hypertensive patients. *Nutr Res*, **32**(6), 421–427.

Bolli R. (2001) Cardioprotective function of inducible nitric oxide synthase and role of nitric oxide in myocardial ischemia and preconditioning: an overview of a decade of research. *J Mol Cell Cardiol*, **33**, 1897–1918.

Bredt DS, Hwang PM, Snyder SH. (1990) Localization of nitric oxide synthase indicating a neural role for nitric oxide. *Nature*, **347**(6295), 768–770.

Butt MS, Imran A, Sharif MK, Ahmad RS, Xiao H, Imran M, Rsool HA. (2014) Black tea polyphenols: a mechanistic treatise. *Crit Rev Food Sci Nutr*, **54**(8), 1002–1011.

Cai Y, Kurita-Ochiai T, Hashizume T, Yamamoto M. (2013) Green tea epigallo-catechin-3-gallate attenuates Porphyromonas gingivalis-induced atherosclerosis. *Pathog Dis*, **67**(1), 76–83.

Cao BQ, Tan F, Zhan J, Lai PH. (2021) Mechanism underlying treatment of ischemic stroke using acupuncture: transmission and regulation. *Neural Regen Res*, **16**(5), 944–954.

Chen YR, Chen CL, Pfeiffer DR, Zweier JL. (2007) Mitochondrial complex II in the post-ischemic heart: oxidative injury and the role of protein S-glutathionylation. *J Biol Chem*, **282**, 32640–32654.

Chen C-L, Chen J, Rawale S, Varadharaj S, Kaumaya PPT, Zweier JL, Chen Y-R. (2008) Protein tyrosine nitration of the flavin subunit is associated with oxidative modification of mitochondrial complex ii in the post-ischemic myocardium. *J Biol Chem*, **283**, 27991–28003.

Cheng YC, Sheen JM, Hu WL, Hung YC. (2017) Polyphenols and oxidative stress in atherosclerosis-related ischemic heart disease and stroke. *Oxid Med Cell Longev*, **2017**, 8526438.

Cheng S, Zhao B-L, Xin WJ, Tang Z-S. (1990) Myocadium damage during ischemia-reperfusion of rat heart. *Chin Circ*, **5**, 222–226.

de Montgolfier O, Pouliot P, Gillis MA, Ferland G, Lesage F, Thorin-Trescases N, Thorin É. (2019) Systolic hypertension-induced neurovascular unit disruption magnifies vascular cognitive impairment in middle-age atherosclerotic LDLr -/-:hApoB + / + mice. *Geroscience*, **41**(5), 511–532.

Ding ML, Ma H, Man YG, Lv HY. (2017) Protective effects of a green tea polyphenol, epigallocatechin-3-gallate, against sevoflurane-induced neuronal apoptosis involve regulation of CREB/BDNF/TrkB and PI3K/Akt/mTOR signalling pathways in neonatal mice. *Can J Physiol Pharmacol*, **95**(12), 1396–1405.

Eng QY, Thanikachalam PV, Ramamurthy S. (2018) Molecular understanding of Epigallocatechin gallate (EGCG) in cardiovascular and metabolic diseases. *J Ethnopharmacol*, **210**, 296–310.

Evers SMAA, Struijs JN, Ament AJHA, van Genugten MLL, Jager JC, van den Bos GAM. (2004) International comparison of stroke cost studies. *Stroke*, **35**(5), 1209–1215.

Ferrari R, Guardigli G, Mele D, Percoco GF, Ceconi C, Curello S. (2004) Oxidative stress during myocardial ischaemia and heart failure. *Curr Pharm Des*, **10**, 1699–1711.

Ghafourifar P, Schenk U, Klein SD, Richter C. (1999) Mitochondrial nitric-oxide synthase stimulation causes cytochrome c release from isolated mitochondria. Evidence for intramitochondrial peroxynitrite formation. *J Biol Chem*, **274**, 31185–31188.

Giglio RV, Patti AM, Cicero AFG, Lippi G, Rizzo M, Toth PP, Banach M. (2018) Polyphenols: potential use in the prevention and treatment of cardiovascular diseases. *Curr Pharm Des*, **24**(2), 239–258.

Grassi D, Draijer R, Desideri G, Mulder T, Ferri C. (2015) Black tea lowers blood pressure and wave in fasted and postprandial conditions in hypertensive patients: a randomised study. *Nutrients*, **7**(2), 1037–1051.

Ghosh D, Scheepens A. (2009) Vascular action of polyphenols. *Mol Nutr Food Res*, **53**(3), 322–331.

Gomikawa S, Ishikawa Y, Hayase W, Haratake Y, Hirano N, Matuura H, Mizowaki A, Murakami A, Yamamoto M. (2008) Effect of ground green tea drinking for 2 weeks on the susceptibility of plasma and LDL to the oxidation ex vivo in healthy volunteers. *Kobe J Med Sci*, **54**(1), E62–E72.

Grassi D, Desideri G, Di Giosia P, De Feo M, Fellini E, Cheli P, Ferri L, Ferri C. (2013) Tea, flavonoids, and cardiovascular health: endothelial protection. *Am J Clin Nutr*, **98**(6 Suppl), 1660S–1666S.

Guo Q, Zhao B-L, Li M-F, Shen S-R, Xin W-J. (1996) Studies on protective mechanisms of four components of green tea polyphenols (GTP) against lipid peroxidation in synaptosomes. *Biochem Biopphys Acta*, **1304**, 210–222.

Guo Q, Zhao B-L, Hou J-W, Xin W-J. (1999) ESR study on the structure-antioxidant activity relationship of tea catechins and their epimers. *Bichim Biophys Acta*, **1427**, 13–23.

Hansel G, Ramos DB, Delgado CA, *et al.* (2014) The potential therapeutic effect of guanosine after cortical focal ischemia in rats. *PLoS One*, **9**(2), e90693.

Haskell-Ramsay CF, Schmitt J, Actis-Goretta L. (2018) The impact of epicatechin on human cognition: the role of cerebral blood flow. *Nutrients*, **10**(8), 986.

Hirano-Ohmori R, Takahashi R, Momiyama Y, Taniguchi H, Yonemura A, Tamai S, Umegaki K, Nakamura H, Kondo K, Ohsuzu F. (2005) Green tea consumption and serum malondialdehyde-modified LDL concentrations in healthy subjects. *J Am Coll Nutr*, **24**(5), 342–346.

Hodgson JM, Croft KD, Woodman RJ, Puddey IB, Fuchs D, Draijer R, Lukoshkova E, Head GA. (2013) Black tea lowers the rate of blood pressure variation: a randomized controlled trial. *Am J Clin Nutr*, **97**(5), 943–950.

Hsieh SR, Cheng WC, Su YM, Chiu CH, Liou YM. (2014) Molecular targets for anti-oxidative protection of green tea polyphenols against myocardial ischemic injury. *BioMedicine*. **4**, 23.

Huang N, Chen Y, Zhao B-L, Xin W-J. (1990) Studies on free radicals generated during ischemia-reperfusion of rat heat. *J Chin Med*, **70**, 691–694.

Huang X, Hussain B, Chang J. (2021) Peripheral inflammation and blood-brain barrier disruption: effects and mechanisms. *CNS Neurosci Ther*, **27**(1), 36–47.

Inanami O, Watanabe Y, Syuto B, Nakano M, Tsuji M, Kuwabara M. (1998) Oral administration of (-)catechin protects against ischemia-reperfusion-induced neuronal death in the gerbil. *Free Rad Res*, **29**, 359–365.

Jin WL, *et al.* (2020) Infrared imageries of human body activated by teas indicate the existence of meridians system. *Research Square*, PPR: PPR150771.

Kuppusamy P, Chzhan M, Vij K, Shteynbuk M, Lefer DJ, Giannella E, Zweier JL. (1994) Three-dimensional spectral-spatial EPR imaging of free radicals in the heart: a technique for imaging tissue metabolism and oxygenation. *Proc Natl Acad Sci USA*, **91**(8), 3388–3392.

Kuriyama S, Shimazu T, Ohmori K, Kikuchi N, Nakaya N, Nishino Y, Tsubono Y, Tsuji I. (2006) Green tea consumption and mortality due to cardiovascular disease, cancer, and all causes in Japan: the Ohsaki study. *JAMA*, **296**(10), 1255–1265.

Lee H, Bae JH, Lee SR. (2004) Protective effect of green tea polyphenol EGCG against neuronal damage and brain edema after unilateral cerebral ischemia in gerbils. *J Neurosci Res*, **77**(6), 892–900.

Lee DH, Lee YJ, Kwon KH. (2010) Neuroprotective effects of astaxanthin in oxygen-glucose deprivation in SH-SY5Y cells and global cerebral ischemia in rat. *J Clin Biochem Nutr*, **47**(2), 121–129.

Li J, Hu X, Selvakumar P, Russell RR, Cushman SW, Holman GD, Young LH. (2004) Role of the nitric oxide pathway in AMPK-mediated glucose uptake and GLUT4 translocation in heart muscle. *Am J Physiol*, **287**, E834–E841.

Li J, Miller EJ, Ninomiya-Tsuji J, Russell RR, Young LH. (2005) AMP-activated protein kinase activates p38 mitogen-activated protein kinase by increasing recruitment of p38 MAPK to TAB1 in the ischemic heart. *Circ Res*, **97**, 872–879.

Li H-T, Zhao B-L, Hou J-W, Xin W-J. (1996) Two peak kinetic curve of chemil-uninencence in phorbol stimulated macrophage. *Biochem Biophys Res Commn*, **223**, 311–314.

Liu PK. (2003) Ischemia-reperfusion-related repair deficit after oxidative stress: implications of faulty transcripts in neuronal sensitivity after brain injury. *J Biomed Sci*, 10(1), 4–13.

Matsuoka Y, Hasegawa H, Okuda S, Muraki T, Uruno T, Kubota K. (1995) Ameliorative effects of tea catechins on active oxygen-related nerve cell injuries. *J. Pharmacol Exp Ther*, **274**, 602–608.

Miltonprabu S, Thangapandiyan S. (2015) Epigallocatechin gallate potentially attenuates fluoride induced oxidative stress mediated cardiotoxicity and dyslipidemia in rats. *J Trace Elem Med Biol*, **29**, 321–335.

Miura Y, Chiba T, Tomita I, Koizumi H, Miura S, Umegaki K, Hara Y, Ikeda M, Tomita T. (2001) Tea catechins prevent the development of atherosclerosis in apoprotein E-deficient mice. *J Nutr*, **131**(1), 27–32.

Negishi H, Xu JW, Ikeda K, Njelekela M, Nara Y, Yamori Y. (2004) Black and green tea polyphenols attenuate blood pressure increases in stroke-prone spontaneously hypertensive rats. *J Nutr*, **134**(1), 38–42.

Pang J, Zhang Z, Zheng TZ, Bassig BA, Mao C, Liu X, *et al.* (2016) Green tea consumption and risk of cardiovascular and ischemic related diseases: a meta-analysis. *Int J Cardiol*, **202**, 967–974.

Pashkow FJ. (2011) Oxidative stress and inflammation in heart disease: do anti-oxidants have a qi. Y, Yang C, Jiang Z, Wang Y, Zhu F, Li T, Wan X, Xu Y, Xie Z, Li D, Pierre SV. (2019) Epicatechin-3-gallate signaling and protection against cardiac ischemia/reperfusion injury. *J Pharmacol Exp Ther*, **371**(3), 663–674.

Qi Y, Yang C, Jiang Z, Wang Y, Zhu F, Li T, Wan X, Xu Y, Xie Z, Li D, Pierre SV. (2019) Epicatechin-3-gallate signaling and protection against cardiac ischemia/reperfusion injury. *J Pharmacol Exp Ther*, **371**(3), 663–674.

Reiter CE, Kim JA, Quon MJ. (2010) Green tea polyphenol epigallocatechin gal-late reduces endothelin-1 expression and secretion in vascular endothelial cells: roles for AMP-activated protein kinase, Akt, and FOXO1. *Endocrinology*, **151**(1), 103–114.

Saeed SA, Shad KF, Saleem T, Javed F, Khan MU. (2007) Some new prospects in the understanding of the molecular basis of the pathogenesis of stroke. *Exp Brain Res*, **182**(1), 1–10.

Saito E, Inoue M, Sawada N, Shimazu T, Yamaji T, Iwasaki M, *et al.* (2015) Association of green tea consumption with mortality due to all causes and major causes of death in a Japanese population: the Japan public health center-based prospective study (JPHC Study). *Ann Epidemiol*, **25**, 512–518.

Sanders SP, Zweier JL, Kuppusamy P, Harrison SJ, Bassett DJ, Gabrielson EW, Sylvester JT. (1993) Hyperoxic sheep pulmonary microvascular endothelial cells generate free radicals via mitochondrial electron transport. *J Clin Invest*, **91**, 46–52.

Shen J-G, Wang J, Zhao B-L, Hou J-W, Gao T-L, Xin W-J. (1998) Effects of EGb-761 on nitric oxide, oxygen free radicals, myocardial damage and arrhythmias in ischemia-reperfusion injury in vivo. *Biochim Biophys Acta*, **1406**, 228–236.

Shen J-G, Guo X-S, Jiang B, Li M, Xin W-J, Zhao B-L. (2000a) Chinonin, a novel drug against cardiomyocyte apoptosis induced by hypoxia and reoxy-genation. *Biochim Biophyscs Acta*, **1500**, 217–226.

Shen J-G, Li M, Xin W-J, Zhao B-L. (2000b) Effects of chinonin on nitric oxide free radical, myocardial damage and arrhythmia in ischemia-reperfusion injury in vivo. *Appl Magn Reson*, **19**, 9–19.

Shen J-G, Qiu X-S, Jiang B, Zhang D-L, Xin W-J, Fung PCW, Zhao B-L. (2003) Nitric oxide contributes to redistribution of phosphatidylserine and triggers apoptosis via peroxynitrite and p53 pathway in hypoxia-reoxygenated cardiomyocytes. *Sci China*, **46**, 27–39.

Sutherland BA, Rahman RM, Appleton I. (2006) Mechanisms of action of green tea catechins, with a focus on ischemia-induced neurodegeneration. *J Nutr Biochem*, **17**(5), 291–306.

Suzuki-Sugihara N, Kishimoto Y, Saita E, *et al.* (2015) Green tea catechins prevent low-density lipoprotein oxidation via their accumulation in low-density lipoprotein particles in humans. *Nutr Res*, **36**(1), 16–23.

Terai K, Hiramoto Y, Masaki M, Sugiyama S, Kuroda T, Hori M, Kawase I, Hirota H. (2005) AMP-activated protein kinase protects cardiomyocytes against hypoxic injury through attenuation of endoplasmic reticulum stress. *Mol Cell Biol*, **25**, 9554–9575.

Tipoe GL, Leung T-M, Hung M-W, Fung M-Lg. (2007) Green tea polyphenols as an anti-oxidant and anti-inflammatory agent for cardiovascular protection. *Cardiovasc Hematol Disord Drug Targets*, **7**(2), 135–144.

Toth P, Szarka N, Farkas E, Ezer E, Czeiter E, Amrein K, Ungvari ZI, Hartings JA, Buki A, Koller A. (2016) Traumatic brain injury-induced autoregulatory dysfunction and spreading depression-related neurovascular uncoupling: pathomechanism and therapeutic implications. *Am J Physiol Heart Circ Physiol*, **311**(5), H1118–H1131.

Tresserra-Rimbau A, Rimm EB, Medina-Remón A, *et al.* (2014) Polyphenol intake and mortality risk: a re-analysis of the PREDIMED trial. *BMC Med*, **12**, 77.

Veenith TV, Carter EL, Geeraerts T, Grossac J, Newcombe VF, Outtrim J, Gee GS, Lupson V, Smith R, Aigbirhio FI, *et al.* (2016) Pathophysiologic mechanisms of cerebral ischemia and diffusion hypoxia in traumatic brain injury. *JAMA Neurol*, **73**, 542–550.

Waisman A, Hauptmann J, Regen T. (2015) The role of IL-17 in CNS diseases. *Acta Neuropathol*, **129**(5), 625–637.

Wang Z, Xue R, Lei X, Lv J, Wu G, Li W, Xue L, Lei X, Zhao H, Gao H, Wei X. (2012) Tea polyphenols increase X-ray repair cross-complementing protein 1 and apurinic/apyrimidinic endonuclease/redox factor-1 expression in the hippocampus of rats during cerebral ischemia/reperfusion injury. *Neural Regen Res*, **7**(30), 2355–2361.

Wightman EL, Jackson PA, Khan J, Forster J, Heiner F, Feistel B, Suarez CG, Pischel I, Kennedy DO. (2018) The acute and chronic cognitive and cerebral blood flow effects of a Sideritis scardica (Greek mountain tea) extract: a double blind, randomized, placebo controlled, parallel groups study in healthy humans. *Nutrients*, **10**(8), 955.

Young LH. (2008) AMP-activated protein kinase conducts the ischemic stress response orchestra. *Circulation*, **117**, 832–840.

Zhang C, Qin YY, Wei X, Yu FF, Zhou YH, He J. (2015) Tea consumption and risk of cardiovascular outcomes and total mortality: a systematic review and meta-analysis of prospective observational studies. *Eur J Epidemiol*, **30**, 103–113.

Zhang D-L, Yin J-J, Zhao B-L. (2004) Oral administration of Crataegus extraction protects against ischemia/reperfusion brain damage in the Mongolian gerbils. *J Neur Chem*, **90**, 211–219.

Zhang P, Xu X, Hu X, Deel E, Zhu G, Chen Y. (2007) iNOS deficiency protects the heart from systolic overload induced ventricular hypertrophy and congestive heart failure. *Circ Res*, **100**, 1089–1098.

Zhang P, Hu X, Fassett JE, Zhu G, Viollet B, Xu W, Wiczer B, Bernlohr DA, Bache RJ, Chen Y. (2008) AMPKα2 deficiency exacerbates pressure-overload induced left ventricular hypertrophy and dysfunction in mice. *Hypertension*, **52**, 918–924.

Zhao B-L, Zhou W-A, Ni Y-C, Hou J-W, Gao T-L, Xin W-J. (2000) Kinetic scavenging effects of chinonin on NO and oxygen free radicals generated from ischemia reperfusion myocardium and its protection effects on the myocardium. *Res Chem Intermed*, 747–762.

Zhao B-L. (1999) Oxygen free radicals and natural antioxidants (First Edition). Science Press, Beijing.

Zhao B-L. (2002) Oxygen free radicals and natural antioxidants (Revised Edition). Science Press, Beijing.

Zhao B-L. (2007) Free radicals and natural antioxidants and health. China Science and Culture Press, Hong Kong.

Zhao B-L. (2008) Nitric oxide free radical. Science Press, Beijing.

Zhao B-L. (2020) The pros and cons of drinking tea. *Tradit Med Mod Med*, 1–12.

Zhao Y, Zhao B. (2010) Protective effect of natural antioxidant on heart against ischemis-reperfusion damage. *Curr Pharm Biotechnol*, **11**(8), 868–874.

Zhao B-L. (2016) Nitric oxide free radical biology and medicine. Science Press, Beijing.

Zhao B-L, Shen J-G, Li M, Xin W-J. (1996a) Synergic effect of NO and oxygen free radicals in ischemia-reperfusion rabbit myocardium. *Sci China*, **26**, 331–338.

Zhao B-L, Shen J-G, Li M, Xin W-J. (1996b) Scavenging effect of Chinonin on NO and oxygen free radicals generated from ischemia reperfusion myocadium. *Biachem Biophys Acta*, **1317**, 131–137.

Zhao B-L, Xin W-J, Yang W-D, Zhu H-L. (1989a) Direct measurement of active oxygen free radicals from ischemia-reperfusion rabbit myocardium. *Chin Sci Bull*, **34**, 780–787.

Zhao B-L, Li X-J, Xin W-J. (1989b) ESR study on oxygen consumption during the respiratory burst of human polymophonuclear leukocytes. *Cell Biol Intern Rep*, **13**, 317.

Zhao B-L, Wang J, Hou J, Xin W. (1996c) Scavenging effect of tea polyphenols on methyl radical produced by peroxynitrite oxidation dimethyl sulfoxide. *Sci Bull*, **41**, 925–927.

Zou X-L, Wan Q, Li M-F, Zhao B-L, Xin W-J. (1995) Scavenging effect of green tea polyphenols on the oxygen free radicals generated from isolated ischemic reperfused rat myocardium. *Chin J Magn. Res*, **12**, 237–244.

Zweier JL, Flaherty JT, Weisfeldt ML. (1987) Direct measurement of free radical generation following reperfusion of ischemic myocardium. *Proc Natl Acad Sci USA*, **84**, 1404–1407.

Zweier JL, Kuppusamy P, Williams R, Rayburn BK, Smith D, Weisfeld ML, Flahery JT. (1989) Measurement and characterization of postischemic free radical generation in the isolated perfused heart. *J Biol Chem*, **261**, 18890–18895.

Chapter 9

Tea Polyphenols and Delayed Aging

Yushun Gong, Ying Chen

Key Laboratory of Tea Science of Ministry of Education, Hunan Agricultural University, Changsha, China

9.1 Introduction

Aging is a global problem that is becoming increasingly common in today's society. The field of aging-related population trends, organismal heterogeneity, and prolonged life via using external factors, has emerged as an important area of modern life science [Harman, 2006]. Aging is characterized by a progressive loss of physiological integrity and function and increase of vulnerability to death, which is widely accepted with nine major bio-hallmarks: 1) genomic instability, 2) epigenetic alterations, 3) loss of proteostasis, 4) deregulated nutrient sensing, 5) mitochondrial dysfunction, 6) cellular senescence, 7) stem cell exhaustion, 8) altered intercellular communication, and 9) telomere attrition [Harman, 1981].

Oxidative stress is one of the common mechanisms contributing to aging and age-related diseases [Buford, 2016]. In response to this progress, antioxidants removing excess oxygen free radicals come into play. The result is a significant extent to improve disease progression and quality of life. Further work in pro-oxidant for healthy aging, in the recent year, supplements the importance function of reactive oxygen species (ROS) related regulatory mechanisms, especially in endogenous oxidative stress [Viña *et al.*, 2020]. Although the effect for roles of pro/anti-oxidant in delaying aging is limited, changes in oxidation level in pathological states alter the risk of age-related disease. As mechanisms leading to aging and oxidative stress continue to be intensively investigated, it is identified reasonable and effective to extend healthy life via regulation of the endogenous oxidative stress system.

Numerous studies show that active ingredients of food sources have health benefits and life-extending effects [Okoro *et al.*, 2021]. Tea polyphenol attracts attention because of outstanding biological activity. Tea polyphenol is a class of polyphenol compounds in tea, including catechins, flavonoids, and phenolic acids. Among them, catechins (mainly including epigallocatechin gallate (EGCG), epicatechin gallate (ECG), epigallocatechin (EGC), and epicatechin (EC), are the main components of tea polyphenols and the main substances that exert their biological activities. As a commended antioxidant, tea polyphenols show remarkable effect of intervention in targeting various metabolic and neurological diseases, due to B and C rings in its molecular structure increasing the number of hydroxyl groups and showing a strong hydrogen supply capacity. Tea polyphenols have a wide range of regulatory effects on intercellular signaling molecules and cellular energy metabolism. Furthermore, tea polyphenols have been shown to positively influence, including anti-cancer, anti-inflammation, anti-aging, anti-radiation, regulation of immune response, closely related to the length and quality of life of organisms [Yan *et al.*, 2020]. In this review, we discuss in detail the effects of oxidative stress on aging and the potential of tea polyphenols to delay lifespan and describe the mechanism of action of tea polyphenols in the treatment of age-related diseases (Fig. 9-1). Finally, we present an outlook of the application prospects of tea polyphenols in improving the quality of survival and extending healthy lifespan.

9.2 Genome theory and free radical theory of aging

Aging is an irreversible and inevitable process in organism, closely related with the characteristic change of individuals morphologic and function [Barth *et al.*, 2020]. The focus on molecular mechanisms, cellular processes, and biomarkers associated with aging in mammals has great

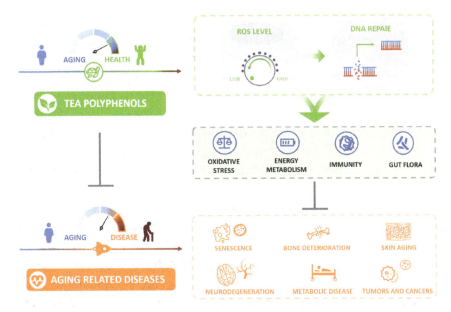

Figure 9-1. Mechanisms of tea polyphenols action of biological aging. The biological activity of tea polyphenols is carried out through the modulation of ROS levels. After improving the oxidative stress state of the body and reducing DNA damage, tea polyphenols work via four ways: 1) mitochondrial function, 2) metabolism, 3) immunity, and 4) intestinal flora. These four ways can extend an organism's healthy lifespan by mitigating aging-related diseases that exacerbate aging and accelerate death.

reference significance for understanding the basic biological mechanism in humans [Cebe *et al.*, 2014]. Therefore, many theories and hypotheses about aging have been put forward, among which there are two kinds that have the greatest impact and can explain aging, namely, genome theory, and free radical theory.

9.2.1 *Genome theory*

Genomic instability is one of the main theories of aging. First, DNA damage increases with age [Gems, 2022], including bulky DNA adduct production, base sites, DNA single strand breaks, DNA double strand breaks, base misfits, insertions, and deletions. These genomic lesions have been associated with age-related diseases [Schumacher *et al.*, 2021]. Second, DNA damage is a driver of aging. It does not only lead to genomic instability, but also initiate several signaling cascades and penetrate the entire cell, that if unresolved, contribute to accelerate aging and inflammation, and therefore exacerbate aging-related diseases. Notably, DNA repair relies primarily on non-homologous end-joining, an error-prone pathway, another process that repairs DNA double-strand breaks, rather than homologous recombination. High levels of DNA damage cause cells to age and stop proliferating. Another study showed that aging is not only coupled with less efficiency DNA repair, but that its process also initiates more complex DNA repair mechanisms, leading to more mutations being introduced. These inefficient DNA repair processes have been linked to aging and cancer. In addition, telomeres are regions at the end of linear chromosomes that consist of DNA and proteins. They become shorter as cells divide, which cause cellular senescence. Telomere defects have also been shown to accelerate aging in mice and humans. In addition, epigenetic modifications such as DNA and histone methylation, polyadenosine diphosphate ribosylation modifications, and acetylation, combine to affect the tertiary structure of chromatin, thereby strongly influencing chromatin activity and function, such as transcription and replication. Although the genome is relatively genetically stable, epigenetic modifications provide opportunities for organisms to adapt to the external environment and to cope with different stress conditions, whether DNA replication errors or spontaneous hydrolytic and oxidative reactions.

The occurrence of genomic instability is caused by both endogenous (mitochondrial ROS, DNA replication errors, etc.) and exogenous (environmental and iatrogenic) factors. Among them, endogenous DNA damage as a potential driver of senescence and aging is well understood. Research has shown that ROS produced during mitochondrial-based aerobic metabolism are a major source of endogenous DNA damage [Robinson *et al.*, 2018]. This accelerates aging and increases susceptibility to cancer and neurodegenerative diseases. Dysfunctional mitochondria and ROS can also cause chromatin aneuploidy-driven senescence. Most studies suggest that the aging process is mostly caused by mitochondrial dysfunction, which increases ROS generation and leads to chromatin aneuploidy ageing [Lefkimmiatis *et al.*, 2021]. In addition, telomeres are thought to be particularly susceptible to oxidative damage. ROS and mitochondria also play a key role in regulating the relationship between physiological stress and telomere maintenance [Lin & Epel, 2022]. In addition to directly inducing single-strand breaks and the collapse or loss of replication forks on telomeres, ROS can damage telomeric sequences, leading to DNA damage. In view of these, the role of oxygen radicals in aging has been widely explored and intensively studied in recent years.

9.2.2 *Free radical theory*

Free radical theory, also called oxidative stress theory (OST), was introduced to the world in 1955, as a senescence marker by the American Dr. Harman. Today, it has become one of the most important and disputed theories of aging. Free radical production usually occurs during cellular metabolism and is usually thought to be mainly generated by mitochondrial and nicotinamide adenine dinucleotide phosphate oxidase (NADPH oxidase/NOX) pathways. Mitochondria are the primary sites of ROS production in cells, possessing a small genome composed of high copy number circular DNA molecules [Vidoni *et al.*, 2013], capable of maintaining optimal function through proteasomal degradation, export of damaged proteins, mitochondrial fusion and division, and mitochondrial autophagy, a physiological function that progressively decreases with age. Oxygen radical damage is manifested in two major ways: 1) ROS target organelles such as mitochondria, peroxisomes, and endoplasmic reticulum, and cause cumulative oxidative damage to macromolecules such as lipids, DNA, and proteins, as well as induce nucleotide oxidation and base damage, leading to an endogenous oxidant/antioxidant imbalance and consequent age-related functional loss [Golubev *et al.*, 2018]; 2) Continued exposure to ROS will trigger inappropriate signaling and conversion to more damaging secondary ROS resulting in cumulative effects [Knaus, 2021].

Today, significant advances have been made in physiological regulation and disease treatment through the targeted control of specific ROS-mediated signaling pathways, including enzymatic defense systems (e.g., those controlled by the stress response transcription factors Nrf2 and nuclear factor-κB), the action of trace elements such as selenium, the use of redox drugs, and the regulation of environmental factors (e.g., nutrition, lifestyle, and irradiation), including neurological, cardiovascular and immune systems, skeletal muscle and metabolic regulation, as well as ageing and cancer. The effectiveness of natural antioxidants in delaying lifespan and activating genes involved in longevity, in particular, supports the validity of the free radical theory of aging. However, the failure of clinical trials of antioxidants to delay lifespan and the possible involvement of ROS as signaling molecules in delaying aging present a clear challenge to this theory. Notably, oxygen radicals can only trigger damage and aging when they accumulate in disproportionate amounts. Since ROS are not only toxic byproducts of oxidative metabolism but also serve as important mediators of signaling during growth and development [Back *et al.*, 2012; Bodega *et al.*, 2019; Hansen *et al.*, 2020], the hormesis phenomenon suggests that oxidative stress that has not reached a state of damage is instead beneficial to health [Banerjee *et al.*, 2017; Calabrese *et al.*, 2012]. This suggests that in the context of free radical theory, changes in oxidative stress levels that depend on dosage effects are an effective means of intervening in aging signaling [Valko *et al.*, 2007; Yoon *et al.*, 2002].

9.3 Redox homeostasis and aging

Studies have shown that maintaining the balance between the production and elimination of free radicals in the body and the balance between oxidation and antioxidation can ensure the health and longevity of the body. Once the balance is broken, the imbalance of free radical production and elimination in the body and the imbalance of oxidation and antioxidation may lead to body damage. Oxidative stress will lead to disease and aging. Therefore, we should try our best to maintain this balance in the body.

9.3.1 *Oxidative stress levels during aging*

The accumulation of ROS or markers of oxidative stress have very strong correlativity with actual age. Its products disrupt protein homeostasis, alter genomic stability, and lead to cell death [Jones, 2015]. Numerous studies have shown that oxidative stress in humans and model organisms (fruit flies, mice, fish, nematodes) increases with age [Martínez de Toda *et al.*, 2020; Pérez de la Cruz *et al.*, 2014; Xiao *et al.*, 2020]. In addition, researchers have found, in both mice and humans, that older subjects have fewer antioxidants and higher levels of oxidation, while long-lived individuals have oxidation levels closer to those of adults [Martínez de Toda *et al.*, 2020]. Furthermore, high expression of the ROS-generating enzyme Nox4 accelerates disease aggravation and aging in lung-injured mice, and targeted inhibition of NOX4-Nrf2 is effective in attenuating aging and reversing lung injury progression [Hecker *et al.*, 2014]. During photoaging-induced aging, it is the continued accumulation of ROS and RNS that causes degradation of extracellular matrix (e.g., collagen) [Kammeyer & Luiten, 2015].

A significant body of evidence shows that changes in oxidative stress in the tissues and organs of the elderly and young people are characterized by a reduction of antioxidants, and a decrease in mechanisms that maintain oxidative homeostasis in the body. Bodega *et al.* found that the microvesicles in umbilical vein endothelial cells have an integrated and well-developed antioxidant system during youth, whereas they became larger and lost the capacity of the antioxidant after aging [Bodega *et al.*, 2017]. Young donors are found to have better transplant outcomes than older donors in allogeneic hematopoietic stem cell transplantation because of significantly lower levels of ROS in their six progenitor cell lines and bone marrow [Yao *et al.*, 2021]. In clinical studies, muscle mitochondrial protein abundance is found to be lower in older than in adults and muscle cysteine oxidation increased after exercise, showing significant disruption of redox homeostasis [Pugh *et al.*, 2021]. This supports findings from past studies regarding variation in the skeletal muscle with age. A recent study revealed that the decline in muscle function is likely to be dependent on reduced levels of NAD [Janssens *et al.*, 2022]. In humans, levels of NADH in brain tissue also gradually increase with ageing of the body, with a subsequent decrease in NAD levels [Massudi *et al.*, 2012; Zhu *et al.*, 2015]. This explains the relationship between abnormal redox status and the development of age-related metabolic and neurodegenerative disorders.

Intriguingly, advanced oxidative protein products (AOPPs), a novel marker of oxidative stress, induce trophoblast senescence by increasing senescence-associated β-galactosidase, senescence-associated heterochromatin foci, mitochondrial membrane potential levels, and reducing the cell cycle [Li *et al.*, 2022]. Some studies indicated that AOPPs not only accumulate at high levels in diseased organisms [Ammar *et al.*, 2022; Vinereanu *et al.*, 2021], but also in patients with juvenile metabolic syndrome [Celec *et al.*, 2021], as well as promote disease progression and increase the risk of death [Zeng *et al.*, 2021; Zhou *et al.*, 2021]. These affirms the change of ROS levels with age and provides useful evidence for enriching our knowledge on human aging and its relationship with redox state.

9.3.2 *Alteration of redox state during aging*

Levels of oxidative stress are difficult to use to predict aging because they change extensively throughout life. Indeed, effective dietary, environmental, and pharmacological interventions based on changes in the body's oxidation levels hold great promise for maintaining health and improving age-related diseases [Wen *et al.*, 2020].

1. Antioxidants

Antioxidants balance the biological activity of ROS, modulate ROS signaling, and minimize toxicity [Kumar *et al.*, 2018]. Their action depends on three pathways: 1) scavenging or quenching of ROS, 2) decreasing of ROS-producing enzymes (e.g., NADPH oxidase, xanthine oxidase), and 3) upregulation of endogenous antioxidant enzymes (e.g., SOD, catalase) and the antioxidant system (e.g., Keap1/ Nrf2). The common antioxidants glutathione, vitamin C, coenzyme Q10, selenium, and plant polyphenols are all clinically effective in intervening in the pathogenesis of aging and ageing-related diseases [Csaba, 2019; Gusarov *et al.*, 2021; Liguori *et al.*, 2018; Sadowska-Bartosz & Bartosz, 2014], and improving indicators of aging (e.g., mobility, independence in activities of daily living, and cognitive function) [Justice *et al.*, 2018].

A major role for mitochondria-targeted antioxidants in preventing and treating age-related diseases has been demonstrated [Madreiter-Sokolowski *et al.*, 2018]. In fact, mitochondrial dysfunction plays a key role in the development of disease and acceleration of ageing during old age [Amorim *et al.*, 2022], especially in neurodegenerative diseases [Stanga *et al.*, 2020]. Young and Franklin [2019] found that the mitochondria-targeted antioxidant mito Q in mice with Alzheimer's disease reduced oxidative stress, synaptic loss, astrocyte proliferation, microglia proliferation, Aβ accumulation, cystein, activation, and tau hyperphosphorylation, significantly extending lifespan while rescuing memory deficits. However, the disruption of energy metabolism caused by mitochondrial dysfunction is accompanied by a variety of high-risk chronic diseases and inflammatory conditions. It has been demonstrated that some antioxidants also have caloric restriction (CR) effects, such as metformin, rapamycin, and resveratrol, which can play a dual role intervening with nicotinamide adenine dinucleotide precursors, small molecule activators that reduce sirtuins and senolytics, among other ageing-related markers [Palliyaguru *et al.*, 2019].

NAD/NADP and NADH/NADPH maintain cellular redox homeostasis and energy metabolism [Koju *et al.*, 2022]. NADPH oxidases (NOXs) are transmembrane proteins that produce ROS as their sole function and are currently the main drug targets for inhibiting ROS production [Altenhöfer *et al.*, 2015]. NOX-generated ROS act as intra/interneuronal redox signaling molecules and are widely present in various neurological diseases [Villegas *et al.*, 2021]. Targeted inhibition of NOX1 by antioxidants can prevent aging-related vascular aging and mitigate vascular system aging [Li *et al.*, 2021a]. Targeted inhibition of NOX5 improves oxidative stress-mediated blood-brain barrier breakdown and alleviates cognitive dysfunction during aging [Cortés *et al.*, 2021].

In recent years, supplementation with antioxidants to extend lifespan has been widely debated. Since the relationship between ROS and lifespan is complex, ROS exert beneficial or detrimental effects on lifespan depending on the species and conditions [Shields *et al.*, 2021]. Numerous studies have also confirmed that antioxidants do not significantly extend lifespan as expected, especially in mammals. The introduction of antioxidants into the diet may also affect the endogenous antioxidant system and even have detrimental effects [Deepashree *et al.*, 2022; Le Bourg, 2001]. Izabela *et al.* suggest that 1) the effect of antioxidants may not be due to their direct antioxidant effect, but to their indirect antioxidant effect (induction of endogenous antioxidant mechanisms); 2) the "antioxidant" class of compounds may have other excessive effects in the body, completely unrelated to the antioxidant effect [Sadowska-Bartosz & Bartosz, 2014].

2. Pro-oxidants

Although antioxidants effectively reduce ROS-induced oxidative damage, insufficient activation of the endogenous antioxidant system still causes homeostatic disruption and organismal damage. In the context of hormesis effect, pro-oxidants may play a positive role in longevity and disease prevention by inducing physiological concentrations of ROS to activate endogenous antioxidant systems and scavenging excess intracellular ROS to avoid toxic damage. In particular, the use of small molecules for indirect pro-oxidant or induced antioxidant defense is expected to improve quality of life and longevity [Schiffers *et al.*, 2021], which suggests an important potential for some degree of lifespan extension by pro-oxidants. Plant components with antioxidant activity have been found to improve health and longevity by activating anti-stress pathways and defensive cellular responses through pro-oxidant responses [Martel *et al.*, 2019]. In addition, pro-oxidant-activated ROS acting as secondary messengers in many immune and non-lymphoid cells, also have a positive effect on damage repair [Dunnill *et al.*, 2017]. Short-term intracellular pro-oxidant responses induced by regular exercise also exhibit a wide range of health benefits [Louzada *et al.*, 2020].

Altogether, this evidence supports the ameliorating effect by antioxidant/pro-oxidant between changes in the body's oxidation levels and in the risk of age-related diseases. Since there is a causal relationship between these changes and organismal decline, it is the extension of healthy lifespan that is the ultimate goal of intervention in the redox state of organism [Campisi *et al.*, 2019].

9.4 Tea polyphenols delay aging

A large number of animal and cell experiments have shown that tea polyphenols can remove excess free radicals produced in the body, reduce oxidative stress damage, help the body to maintain redox balance, and have a direct anti-aging effect. In addition, tea polyphenols can also prevent aging caused by some diseases and play an indirect anti-aging role. However, tea polyphenols need a certain dose and action time to delay aging, and the mechanism is relative complex.

9.4.1 *Direct effects of tea polyphenols in delaying aging*

1. Direct effects of tea polyphenols on longevity

Like other health foods, tea is believed to have life-extending abilities due to its richness in polyphenols [Augustyniak *et al.*, 2011; Uysal *et al.*, 2013]. Studies have shown that long-term consumption of green tea can effectively alleviate the symptoms of the aging process, such as a decrease in total phospholipids in liver cell membranes, an increase in lipid peroxidation products, and an increase in charge density on the surface of cell membranes [Dobrzynska *et al.*, 2008]. In nematodes, the lifespan-promoting and healthy lifespan effects of oolong tea extract and oolong flavonoids were positively correlated with resilience [Duangjan & Curran, 2022]. Black tea extract significantly enhances the survival of nematodes under heat stress and UV stress [Xiong *et al.*, 2014]. A 2.5 µM EGCG concentration and ECG increased nematode health, lifespan and stress resistance [Tian *et al.*, 2021]. Daily application of 220 µM EGCG not only increased nematode longevity but also survival under lethal oxidative stress conditions [Abbas & Wink, 2009]. The addition of green tea polyphenols to indirect feeding significantly increased lifespan and reduced brain tissue inflammation in mice [Aires *et al.*, 2012]. It has also been proposed that EGCG exerts a protective effect against heat stress and oxidative stress in *Hidradenia* nematodes, extending nematode lifespan in a dose-dependent manner [Zhang *et al.*, 2009].

2. Direct effects of tea polyphenols on skin aging

Skin aging is one of the features of aging [Victorelli *et al.*, 2019]. Tea aqueous extracts have shown to promote restoration of integrity in photoaging-induced wrinkles, sores caused by radiation damage, and in the treatment of wounds or certain skin conditions characterized by changes in cellular activity or metabolism [Hsu *et al.*, 2003a; Lee *et al.*, 2014a; Pajonk *et al.*, 2006]. A hound experiment found that continuous administration of 50 mg of tea polyphenols for 42 days was effective in protecting against radiation-induced organ and tissue damage [Dong *et al.*, 2019]. In addition, the four major components of tea polyphenols, EGCG, ECG, EGC, and EC, exhibited anti-aging effects on B16F10 melanoma cells and human skin fibroblasts [Chaikul *et al.*, 2020]. Among them, EGCG not only prevented skin connective tissue changes caused by ultraviolet (UV) radiation through collagen degradation [Bae *et al.*, 2008; Katiyar *et al.*, 2001], but also by modifying UVB-induced photoaging, as well as by inhibiting age-related decreases in muscle activity and the accumulation of age pigments (lipofuscin) [Duangjan & Curran, 2022]. The anti-damaging and anti-natural aging effects of theaflavin-3′-gallate (TF3′G) are demonstrated in human epidermal keratinized cells as evidenced by dual role in photoprotection and maintenance of cell homeostasis [Zheng *et al.*, 2021]. In addition, myketin is also shown to attenuate UVB-induced keratinocyte death in a concentration-dependent manner [Huang *et al.*, 2010].

3. Direct effects of tea polyphenols on bone aging

Clinical studies have shown that long-term consumption of oolong tea can increase bone density in women's heel bones after menopause, prevent bone loss, and maintain bone health [Duan *et al.*, 2020; Shen *et al.*, 2010; Wang *et al.*, 2014]. In particular, catechins can counteract the harmful effects of the imbalance between osteoblast and osteoclast production caused by osteoporosis [Chen *et al.*, 2019; Hsuan-Ti *et al.*, 2020]. In mice, feeding 400 mg/kg catechins reduced the typical aging and osteoporosis phenotype, delayed aging, and prevented aging-related osteoporosis [Huang & Liu, 2017]. An *in vitro* study of human alveolar osteogenic cells revealed that EGCG below 10 μM increased the osteogenic differentiation of cells, an effect that was attenuated at high concentrations of EGCG (25 μM), with a concomitant decrease in cell proliferation and migration capacity [Mah *et al.*, 2014].

9.4.2 *Indirect effects of tea polyphenols in delaying aging*

In addition to aging, another important factor in human lifespan is the death caused by some diseases. As senescent cells can stimulate a variety of aging phenotypes and disease markers through involuntary cellular effects, combating and delaying aging-induced diseases is currently the primary means and goals of maintaining healthy aging [Campisi *et al.*, 2019; Shetty *et al.*, 2018]. Epidemiological investigation have reported the statistical analysis results related to green tea and cardiovascular disease. Two prospective studies show that the intake of green tea was negatively correlated with all-cause mortality and cardiovascular mortality in 90,914 Japanese aged between 40–69 from 1990–1994 [Saito *et al.*, 2015a; Zhang *et al.*, 2010]. Three statistical analyses on green tea consumption and cardiovascular disease mortality support that the risk reduction is between 18–33% [Abe *et al.*, 2019; Tang *et al.*, 2015].

1. Effects of tea polyphenols on neurodegenerative diseases

Early studies suggested that improvement of cognitive decline, degenerative brain lesions, and aging-induced neuronal damage with regular doses of tea [Guoyuan *et al.*, 2017; Lee & Lee, 2007], in which tea polyphenols play a good neuroprotective role [Yang *et al.*, 2020]. Among them, EGCG has the potential to promote neuronal survival and regeneration and modulate brain plasticity [Ortiz-Lopez *et al.*, 2016]. EGCG and ECG are also effective in reducing β-amyloid, effectively reducing the risk of neurodegenerative diseases [Bastianetto *et al.*, 2006]. In rats, EGCG is hydrolyzed to epigallocatechin (EGC) and gallic acid after oral administration. It also enters the brain parenchyma as the small molecule metabolite 5-(3′,5′-dihydroxyphenyl)-gamma-valerolactone (EGC-m5) and its conjugated form to increase the number of neuro-synapses and exert the effect of reducing the risk of dementia [Unno *et al.*, 2017]. Studies have shown that continuous oral administration of EGCG (2 mg/kg or 6 mg/kg) for four weeks significantly improved cognitive dysfunction in mice, exhibiting antioxidant and anti-apoptotic neuroprotective effects in aging mice [He *et al.*, 2009]. EGCG was also associated with the rescue of aging and neurodegenerative diseases by chelating with iron and antioxidant, slowing or even reversing the rate of neuronal degeneration [Weinreb *et al.*, 2007].

2. Effects of tea polyphenols on metabolic diseases

A study of 4,579 adults aged 60 years and above found that habitual tea consumption was significantly associated with lower blood pressure component values and lower risk of hypertension [Yin *et al.*, 2017].

People who drank at least four cups of tea a day had a 16% lower risk of developing type 2 diabetes than non-tea drinkers [van Woudenbergh *et al.*, 2012]. Taking 15 mg of green tea extract continuously for 1–3 months after weight loss also helped maintain weight [Gilardini *et al.*, 2016]. Of these, catechins play a major role. For example, drinking 3–4 cups of tea (600–900 mg of catechins) per day is thought to be effective in reducing metabolic syndrome, diabetes, and preventing cardiovascular disease [Bahorun *et al.*, 2010; Sang *et al.*, 2007; Yang & Zhang, 2019]. Drinking green tea containing 583 mg of catechins or 96 mg of catechins for 12 consecutive weeks reduces body fat, systolic blood pressure, and LDL cholesterol, contributing to the effects of reducing obesity and cardiovascular disease risk [Nagao *et al.*, 2007]. Continuous consumption of tea containing 690 mg of catechins for 12 weeks reduced body fat, prevented and improved metabolic diseases, especially obesity [Nagao *et al.*, 2005] and improve the characteristics of metabolic syndrome, reduce the risk of diabetes, and other complications [Toolsee *et al.*, 2013]. In rats, the benefits of consuming green tea catechins for 45 consecutive days were more significant in reducing body weight and liver fat than for 30 days, suggesting that continued intake of tea polyphenols would be more beneficial for maintaining a healthy body weight [Yan *et al.*, 2013]. In addition, EGCG treatment significantly reversed changes in biochemical and histological parameters in young and old hypercholesterolemic rats and effectively prevented atherosclerosis in young rats [Krishnan *et al.*, 2014]. EGCG has also been shown to reduce early biomarkers of autoimmune disease-related exocrine gland dysfunction in mice, alleviating complications of type 1 diabetes and other autoimmune diseases [Ohno *et al.*, 2012].

3. Effects of tea polyphenols on tumors and cancer

Numerous studies have shown a significant correlation between tea consumption and reduced risk of cancer, including lung cancer [Laurie *et al.*, 2005; Li *et al.*, 2008; Lin *et al.*, 2012], breast cancer [Crew *et al.*, 2012; Nakachi *et al.*, 1998; Wu *et al.*, 2003], ovarian cancer [Lee *et al.*, 2013], oral and esophageal cancers [Chatterjee *et al.*, 2008; Yang *et al.*, 1999; Yang *et al.*, 2007]. EGCG is the main component of the anti-cancer effect of tea polyphenols, which not only exerts anti-proliferative and anti-cancer effects on different types of cancers [Didiano *et al.*, 2004; Li *et al.*,

2021b], but also can play a preventive and protective role by reducing the risk of cancer [Lee *et al.*, 2013]. EGCG treatment can effectively induce antitumor effects such as reducing cell proliferation, cell migration, telomerase activity, and increasing SA-β-galactosidase activity [Didiano *et al.*, 2004; Lee *et al.*, 2018]. EGCG treatment can also target and inhibit the migration and growth of highly chemo-resistant triple-negative breast cancer cells [Lee *et al.*, 2021]. Short-term (four weeks) and long-term (28 weeks) administration of a green tea catechin mixture and catechin gallate (0.5%) was found in mice to exhibited protective effects on both gastritis and the development of precancerous lesions [Ohno *et al.*, 2016]. Green tea extract (containing 45% EGCG) inhibited the growth of uterine smooth muscle tumor cells without side effects [Ahmed *et al.*, 2016]. Tea polyphenols, theaflavins, and especially EGCG also prevent the development of prostate cancer at an early stage [Harper *et al.*, 2007; Henning *et al.*, 2006; Henning *et al.*, 2015; McLarty *et al.*, 2009; Nguyen *et al.*, 2012]. In addition, EGCG inhibits photo-carcinogenesis in mice with angiogenic factors thus preventing UV-induced skin cancer [Mantena *et al.*, 2005]. It also protects normal epithelial cells from ROS and induces apoptosis in tumor cells to reduce the incidence of oral cancer [Yamamoto *et al.*, 2004]. Oral green tea extract (1000 mg/day) alone or in combination with low-dose cytarabine chemotherapy for at least six months was effective in improving the immune status of patients, and the improved status was maintained for 180 days [Calgarotto *et al.*, 2021]. This may be related to the inhibition of leukemic tumor cell activity by EGCG [Li *et al.*, 2000]. In addition, EGCG combined with radiotherapy not only significantly reduce the growth of human colorectal adenocarcinoma cells, but also improve the sensitivity of colorectal cancer cells to radiation and enhanced the therapeutic effect [Enkhbat *et al.*, 2018], confirming the potential of tea polyphenols (including green tea and black tea polyphenols) to reduce colon carcinogenesis [Henning *et al.*, 2013].

4. Others

EGCG reduces choroidal neovascularization and age-related macular degeneration (AMD)-induced vision loss in the elderly [Xu *et al.*, 2020]. It is also known to alter ocular angiogenesis inhibitors and their vascular permeability [Lee *et al.*, 2014b]. Both black and green tea aqueous extracts exhibit excellent antioxidant and anti-glycosylation effects [Ramlagan *et al.*, 2017]. Among them, catechins such as: EGCG, ECG, EGC, proanthocyanidin B-2, dihydromyricetin dimer, and quercetin show significant anti-glycosylation effects [Sun *et al.*, 2019]. In addition, EGCG can effectively inhibit the synthesis of lipofuscin (LF), showing the effect of blocking lipid peroxidation and glycosylation process [Cai *et al.*, 2016].

9.4.3 *Dose-dependent effects of tea polyphenols in delaying aging*

The results of numerous clinical and model biological studies have shown that the intake of high doses of tea polyphenols increases the risk of liver injury and kidney injury [Emoto *et al.*, 2014; Gupta *et al.*, 2001; Rasheed *et al.*, 2018]. Studies have shown that tea polyphenols, especially EGCG, cause discomfort at high concentrations (>800 mg/day) in humans, and even shorten the lifespan of *C.elegans* at 1000 μM [Chow *et al.*, 2003; Peng *et al.*, 2021]. The use of tea polyphenols in breast cancer treatment suggests the requirement for testing the maximum tolerable levels. Based on pharmacological levels of tea polyphenols in urine, 400 mg caused rectal bleeding in individual subjects, 600 mg caused weight gain, indigestion, and insomnia in individual subjects, and 800 mg caused abnormal liver function in subjects [Crew *et al.*, 2012]. In the treatment of prostate cancer, 69% of patients experienced varying degrees of negative effects (including nausea, vomiting, insomnia, fatigue, diarrhea, abdominal pain, and confusion) [Jatoi *et al.*, 2003]. Repeated intake of high doses of tea polyphenols (750 mg/kg) increased liver damage. Experiments in mice found that the ability to add high doses of green tea polyphenols to the diet was not only ineffective in the treatment of colitis, but also exacerbated antioxidant enzyme and heat shock protein expression causing kidney injury [Inoue *et al.*, 2011]. A single high dose of EGCG (1500 mg/kg) significantly increases plasma alanine aminotransferase (ALT) and decreases survival in male CF-1 mice [Lambert *et al.*, 2010]. In addition, high doses of EGCG (100 μM) exerts no antioxidant effect in diabetes treatment, but instead exacerbates oxidative damage to pancreatic β-cells [Suh *et al.*, 2010].

Actually, tea polyphenols act depending on their dose as well as on the adaptive system of the organism, and their effects show a strong dose-dependent relationship [Elbling *et al.*, 2005]. First, EGCG prolongs lifespan in an inverse u-shaped dose-dependent manner [Mähler *et al.*, 2013]. The intake of low doses of tea polyphenols is more beneficial for prolonging healthy lifespan and that their effects are more pronounced in young nematode than in the elderly [Xiong *et al.*, 2018]. Studies have shown that tea polyphenols activate the endogenous antioxidant system at sufficiently low concentrations and exert anti-inflammatory and chemopreventive effects. Low doses of catechins, especially EGCG, induce the expression of a number of representative antioxidant enzymes that enhance endogenous ROS defense and healthy lifespan of organisms by causing a transient surge of ROS [Rasheed *et al.*, 2018], which has been repeatedly demonstrated in model organisms [Tian *et al.*, 2021].

Second, repeated exposure to tea polyphenols may not necessarily exacerbate damage. Pre-exposure to physiological doses of EGCG instead attenuates physiological damage (e.g., smaller body size, shorter lifespan, reduced pharyngeal pumping rate, impaired locomotion) in nematodes exposed to high doses, suggesting that low doses of EGCG are effective in improving the stress feedback loop of the organism and possess the potential to effectively block post-stress adaptive responses [Lu *et al.*, 2021].

Further, in the act of tea drinking, damage caused by such high doses of tea polyphenols is extremely difficult to achieve and only very low doses of tea polyphenols can be ingested in the vast majority of cases [Murakami, 2014]. Notably, another major component of tea, theanine, can act synergistically with tea polyphenols, as demonstrated in nematodes, which restores other physiological indicators (e.g., locomotion) and effectively mitigates the damage caused by high doses of tea polyphenols [Peng *et al.*, 2021].

9.5 Anti-aging mechanism of tea polyphenols

The anti-aging mechanism of tea polyphenols is complex, including the need for a certain dose and action time. Tea polyphenols can remove excess free radicals ROS, inhibit the level of oxidative stress with appropriate dosage and connect signal pathways, regulate the immune system, and improve the intestinal microbiota, so as to maintain the inherent homeostasis of the antioxidant system in body and delay aging.

9.5.1 *The pharmacokinetics of tea polyphenols in delaying aging*

Oral bioavailability of tea polyphenols is low (0.1–0.2% on an empty stomach) [Chow *et al.*, 2005]. There are significant individual and behavioral differences [Scholl *et al.*, 2018; Van Amelsvoort *et al.*, 2001]. Metabolites of tea polyphenols are transported in the circulatory system and widely distributed in organs and tissues, and eventually excreted in the urine [Del Rio *et al.*, 2010]. Blood concentrations of tea polyphenols peak at 0.4–3.1 ng/mL after 60 minutes of oral administration [Narumi *et al.*, 2014]. The maximum plasma concentration of catechins (15–112 μg/mL) is reached after two hours of oral administration and reaches its half-life after 451–479 minutes [Abdelkawy *et al.*, 2020]. EGCG was the dominant component of free tea polyphenols in plasma, accounting for 85% of the total concentration and reaching peak plasma concentration at 90 minutes. The majority of tea polyphenols are rapidly absorbed via the small and large intestine, and then metabolized by intestinal microorganisms and excreted from the body. The remaining low concentrations of tea polyphenols and their metabolites can exist in internal organs for up to 24–48 hours, where organs such as the liver metabolize the remaining EGCG, which is converted by hepatocytes and finally excreted in the urine as a sulfate and glucuronide conjugate of EGC and EC [Chow *et al.*, 2001; Wang *et al.*, 2008]. The *in vivo* metabolic distribution of EGCG labeled by 4″-[11C] using positron emission tomography (PET) in rats confirms that orally-administered EGCG accumulates mainly in the stomach and small intestine and to a lesser extent in the liver, blood, and brain [Shimizu *et al.*, 2014]. These roles have been very well verified in humans [Andreu *et al.*, 2020; Manach *et al.*, 2005; Scholl *et al.*, 2018].

Being a hydrophilic compound, taken up orally through organism-specific transporters [Yang *et al.*, 2016], tea polyphenols enter and leave the cells mainly via drug metabolism enzymes (DMEs), efflux transporters (ETs), nuclear receptor superfamily (NRS), and other signal receptors [Yang *et al.*, 2016]. For example, after entering human intestinal epithelial cells, tea polyphenols are coupled by transport to the liver by catechol-O-methyltransferase (COMT), glucuronide synthase (UDP-glucuronosyl transferase (UGTs)), and sulfate transferase (SULTs), that undergo methylation, glucuronidation and sulfation reactions [Chow & Hakim, 2011; Miller *et al.*, 2012]. The enzymes are then dispersed to other sites of action by phase II metabolic enzymes interacting with efflux transfer proteins [Cialdella-Kam *et al.*, 2017; Feng, 2006; Lewandowska *et al.*, 2013; Pandit *et al.*, 2019; Roth *et al.*, 2011; Wang *et al.*, 2012].

The interaction of tea polyphenols with proteins is an important step in their involvement in the regulation of cellular physiology [Joyner, 2021]. Studies have shown that salivary proteins, adhesins, fibronectins, glycoproteins, glucagons, and insulin-like growth factor receptor 1 all interact with tea polyphenols and play a crucial role in several key signal transduction [Lorenz, 2013; Sur & Panda, 2017; Yang *et al.*, 2009]. Covalent modification of proteins is thought to be a key pathway for the biological activity of tea polyphenols, such as increase in histone acetylation and methylation [Ciesielski *et al.*, 2020]. The oxidation of specific amino acid residues such as cysteine, methionine, tyrosine, and proline has been shown to be strongly related to the regulation and signaling of various cellular processes. Among these, cysteine is the most oxidizable amino acid in proteins, a property commonly used as a regulatory feature of many protein and cellular signaling pathways. Ishii *et al.* found that EGCG forms covalent adducts with cysteine thiol residues in proteins via autoxidation by assaying metabolites in mouse urine and by simulating the binding of it to proteins [Ishii *et al.*, 2008]. Since covalent modification of thiol groups may impair protein function and indirectly affect cell signaling, this may lead to inhibition of tumor cell growth. The interaction with cysteine residues has also been demonstrated in other pathological mechanisms [Kato *et al.*, 2021]. This interaction relies on the tea polyphenol A-ring, C-ring, and ester group structures, and compounds that mimic these structures exhibit EGCG-like protein inhibitory effects [Kazi *et al.*, 2004]. However, when tea polyphenols are methylated, their ability to bind to the β5 subunit of the proteasome is reduced [Daniel *et al.*, 2006] .

9.5.2 *Tea polyphenols delay aging by inhibiting the level of oxidative stress*

As mentioned earlier, the redox state of mitochondria is directly responsive and consequential to regulating cellular senescence and age-related diseases. The current study confirms that the antioxidant effects of tea polyphenols [Fujiki *et al.*, 2002; Hsu *et al.*, 2003b] and pro-oxidant effects are both related to mitochondrial targeting [Bahadorani & Hilliker, 2008; Elbling *et al.*, 2005], including inhibition of mitochondrial oxidative phosphorylation.

1. Mitochondrial complexes

Changes in the stability of mitochondrial complex I as well as mutations in activity are responsible for many pathological phenotypes [Janssen *et al.*, 2006]. Tian *et al.* found that EGCG and ECG inhibited mitochondrial respiratory chain complex I, which resulted in a decrease in ATP production by mitochondrial OXPHOS production and a concomitant surge in ROS. This suggests that, in the context of aging and age-related diseases, tea polyphenols can extend healthy lifespan by directly interacting with mitochondria [Tian *et al.*, 2021]. Under conditions of mitochondrial energy deficit and oxidative stress, EGCG rescued mitochondrial complex I and ATP synthase catalytic activity, restoring OXPHOS efficiency and counteracting oxidative stress [Valenti *et al.*, 2013a]. The selective inhibition of mitochondrial complex I by EGCG was also demonstrated in a variety of biological models, including cancer cells, aging cells, and necrotic neuronal cells [Castellano-González *et al.*, 2016]. Notably, physiological concentrations of EGCG bind to mitochondrial complexes in damaged hepatocytes and thus exhibit hepatotoxicity, a phenomenon that does not occur in healthy hepatocytes [Weng *et al.*, 2014]. Conversely, high doses of EGCG are cytotoxic due to enhanced complex I-driven respiration-induced mitochondrial disorders [James *et al.*, 2018; Mezera *et al.*, 2016] .

Mitochondrial complex III is an enzyme complex oxidizing coenzyme Q (QH2) with cytochrome C as the electron acceptor, capable of producing superoxide in large quantities and subsequently rapidly mutates to form H_2O_2 [Andreyev *et al.*, 2005]. It has been shown that EGCG causes changes in the levels of intracellular mitochondrial complex III both at physiological and excessive concentrations [Tofolean *et al.*, 2016; Valenti *et al.*, 2013b]. Together with the level of mitochondrial complex I, it is considered to be a specific indicator that tea polyphenols alter the activity of the mitochondrial respiratory chain complex [James *et al.*, 2018; Mustata *et al.*, 2005; Santamarina *et al.*, 2015].

2. Mitochondrial membrane

The mitochondrial membrane is an important origin of intracellular ROS, and a decrease in membrane potential can lead to changes in its permeability, which consequently induces mitochondrial damage. Tea polyphenol pretreatment significantly improves this damage process and protects the functional integrity of mitochondria [Guoyuan *et al.*, 2017]. For example, in neurodegenerative diseases, protein mis-aggregation leads to increase mitochondrial membrane permeability, and black tea polyphenols can effectively inhibit it to avoid mitochondrial damage and functional disruption [Camilleri *et al.*, 2013]. Qi *et al.* demonstrated that the protective effect of tea polyphenol pro-oxidant activity was significantly associated with the regulation of mitochondrial dynamics and mitochondrial membrane potential [Guoyuan *et al.*, 2017]. EGCG improved the membrane potential and activity of respiratory chain complexes in liver mitochondria and reduced other injury-induced mitochondrial dysfunction [Lin *et al.*, 2021]. Like EGCG, theaflavin TF3′G inhibits the formation of senescence-associated heterochromatin foci (SAHF) during photoaging and natural aging of the skin by increasing the mitochondrial membrane potential [Srividhya & Kalaiselvi, 2013; Zheng *et al.*, 2021]. This effect is proposed to be associated with a substantial reduction in mitochondrial NADH, reduced proton penetration of the inner mitochondrial membrane, and lowered cellular oxygen consumption rates, demonstrating the ability of tea polyphenols to promote restoration of mitochondrial function [Jia *et al.*, 2021]. In addition, tea polyphenols also alter the OXPHOS protein NADH dehydrogenase-3 (ND3), which is associated with its ability to increase mitochondrial DNA copy number and inhibit superoxide production [Li *et al.*, 2012; Rehman *et al.*, 2014].

3. Antioxidant systems

The antioxidant systems maintain the body's intrinsic homeostatic and is widely present in organisms and provide antioxidant protection, providing antioxidant protection in response to overreactions of body. When ROS levels increase, superoxide dismutase (SOD) converts super-oxide ions to peroxides H_2O_2, through the peroxidase (PRX)/thioredoxin (TRX) system and glutathione peroxidase (GPX)/glutathione (GSH), resulting in lower H_2O_2 levels. Nrf2, one of the major regulators of the antioxidant system, is involved in GSH regeneration by increasing glutathione reductase (GR) expression as well as regulating antioxidant response element (ARE)-mediated expression of many antioxidants and detoxifying enzymes.

Numerous studies have shown that Keap1/Nrf2/ARE is the main pathway through which tea polyphenols modulate endogenous antioxidant capacity [Song *et al.*, 2021] and is consistent with a dose-effect [Na & Surh, 2008; Wang *et al.*, 2015]. These include effects on nuclear accumulation, ARE binding and transcriptional activity of Nrf2, and activation of downstream targets, including the antioxidant proteins Slc7A11, HO-1, and GPX4 [Patel & Maru, 2008; Xie *et al.*, 2020]. These mechanisms were verified in chondrocyte, hepatocyte of senescent mouse models, as well as in worm lifespan experiments [Tian *et al.*, 2021; Wang *et al.*, 2022; Xu *et al.*, 2021]. In addition, the activity of tea polyphenols on Nrf2 provides preventive and protective effects before the onset of disease [Zhang *et al.*, 2021a]. As the activity of tea polyphenols on Nrf2 was restored under pathological conditions, other signaling and anti-inflammatory pathways (e.g., p-Akt, p-JNK, p-ERK1/2 and p-P38, PPARγ, SIRT1) that exhibited effective long-term health benefits also improved [Wang *et al.*, 2020; Ye *et al.*, 2015].

9.5.3 *Signal pathway of tea polyphenols in delaying aging*

Tea polyphenols also increase the activity of key enzymes in glycolysis and tricarboxylic acid cycle, which are essential in energy metabolism [Luping *et al.*, 2021]. Wei *et al.* [2018] found that EGCG significantly inhibited the activities of the glycolytic enzymes hexokinase (HK), phosphofructokinase (PFK), and lactate dehydrogenase (LDH), and partially inhibited pyruvate kinase (PK) activity, as well as decreased the expression of hypoxia-inducible factor 1α (HIF1α) and glucose transporter protein 1 (GLUT1), thereby reducing glucose, lactate, and ATP levels in cancer cells.

1. Tea polyphenols activates AMPK

As an essential regulator of energy homeostasis, AMPK is a cellular energy sensor whose viability decreases with age, continually compromising cellular equilibrium, and aggravating the aging process [Salminen *et al.*, 2016]. Huang *et al.* found that EGCG and its analogs regulate AMPK activation in cells [Chen *et al.*, 2012; Hwang *et al.*, 2007], and their activation is dependent on gallate [Murase *et al.*, 2009]. The activation of EGCG and its analogs is dependent on the gallbladder group. Moreover, EGCG showed stronger activation of AMPK than the diabetic therapeutic drug metformin [Chen *et al.*, 2012]. EGCG induces changes in cellular energy charge through inhibition of action on mitochondrial respiratory chain complex I [Tian *et al.*, 2021]. During the delayed aging process, EGCG induced a short-term burst in ROS levels leading to a transient decrease in ATP levels [Bocchi *et al.*, 2018], in which ROS activates Calcium/calmodulin-dependent protein kinase (CaMKK), thereby altering downstream signaling pathways including AMPK [Collins *et al.*, 2007]. The mechanism of action of a large number of tea polyphenols on the AMPK pathway has been validated in chronic disease models such as obesity, although improvement of metabolic dysfunction and insulin resistance by tea polyphenols is often accompanied by AMPK activation [Rocha *et al.*, 2016]. However, a mouse experiment found that inhibition of hepatic gluconeogenesis by low concentrations of EGCG was not related to insulin signaling pathway, but only depended on AMPK activation. In contrast, when CAMKIIβ was knocked down, EGCG promoted a significant reduction in autophagic flux [Kim *et al.*, 2013].

AMPK activation induced by tea polyphenols is closely related to AMPKα levels [Serrano *et al.*, 2013], and increased AMPKα levels lead to reduced energy availability and thus lower energy uptake levels in the body [Serrano *et al.*, 2013]. Influenced by ROS homeostasis, AMPK and AKT directly or indirectly regulate phosphorylation, mTOR/FOXO signaling, and glucose supply through antagonism, affecting cell survival [Zhao *et al.*, 2017]. Reiter *et al.* [2010] proposed that tea polyphenols dose-dependently activate Akt (S(473)), FOXO1 (at the Akt phosphorylation site T(24)), and AMP-activated protein kinase (AMPKα) T(172) phosphorylation.

The G protein-coupled receptor PKA system is located downstream of AMPK and is also a key pathway for tea polyphenol bioactivity. It was shown that EGCG rescues mitochondrial complex I and ATP synthase catalytic activity, restores oxidative phosphorylation efficiency, and counteracts oxidative stress associated with enhanced PKA activity [Valenti *et al.*, 2013a].

2. Tea polyphenols upregulation mTOR

Like AMPK, mTOR is an important regulator that senses intracellular nutrient levels, regulates metabolic homeostasis, modulates cell growth, and plays a key role in mammalian aging. Holczer *et al.* found that EGCG was able to alleviate autophagy-dependent survival under endoplasmic reticulum stress conditions through upregulation of the mTOR/PKA-AMPK-dependent pathway death [Holczer *et al.*, 2018]. PI3K/AKT/mTOR signaling is the main pathway through which tea polyphenols interfere with autophagy [Yang *et al.*, 2021]. The important role of this pathway was verified by Jiao *et al.* using human vascular endothelial cells with knockdown mTOR [Meng *et al.*, 2020]. In addition, mTOR has two complex structures, mTORC1 and mTORC2. Among them, mTORC1 plays a central role as an AMPK substrate and regulatory node. It was shown that EGCG-induced activation of the PI3K/Akt pathway is precisely dependent on elevated levels of mTORc1 [Ding *et al.*, 2017].

3. Tea polyphenols upregulation of aging-associated genes such as *daf-16*, *sod-3*, and *skn-1*

Studies demonstrated that the significant longevity-extending effects of EGCG on *C. elegans* could be attributed to its *in vitro* and *in vivo* free radical-scavenging effects and its upregulating effects on stress-resistance-related proteins, including superoxide dismutase-3 (SOD-3) and heat shock protein-16.2 (HSP-16.2), in transgenic *C. elegans* with SOD-3::green fluorescent protein (GFP) and HSP-16::GFP expression. Quantitative real-time PCR results showed that the upregulation of aging-associated genes such as *daf-16*, *sod-3*, and *skn-1* could also contribute to the stress resistance attributed to EGCG [Zhang *et al.*, 2010].

9.5.4 *Tea polyphenols delay aging by regulating the immune system*

During aging, chronic inflammation, as well as the production of ROS, increases the occurrence of diseases associated with lipid peroxidation and tissue damage, and raises the risk of age-related diseases [Levy *et al.*, 2019]. The immune system is recognized to play a key role in regulating metabolic homeostasis throughout the body, acting through a variety of intrinsic and adaptive immune systems present in various organs, particularly adipose tissue and liver.

1. Modulation of immune cell function

Toll-like receptors (TRLs) play an important role in the innate immune system and are expressed by a variety of immune cells dependent on the MyD88-dependent and TRIF-dependent pathways. The TRL family contains 12 isoforms, of which TLR2 is a key target for immunotherapy, particularly in malignant diseases. When TLR2 is activated, NF-κB immediately increases and induces cytokine expression to enhance immunity. Studies have shown that tea polyphenols inhibit signal transducer and activator of transcription (STAT) 1/3 and NF-κB transcription factors by suppressing the activity of these transcription factors, which are critical in a variety of downstream pro-inflammatory signaling pathways [Menegazzi *et al.*, 2020; Saito *et al.*, 2015b]. In humans, TLR4 specifically recognizes bacterial LPS, endogenous molecules produced during tissue injury and other pathogenic components. EGCG3″Me significantly attenuates TLR4 expression in adipose tissue [Kumazoe *et al.*, 2017] and suppresses the activation of TLR4/NFκB p65 signaling pathway to regulate immune regulation of human neutrophils [Marinovic *et al.*, 2015].

Th1 and Th2 are key members of cellular and humoral mediated immunity, respectively. Pro-inflammatory cytokines produced by TH1 cells (IFNγ) and TH17 cells (IL-17) contribute to the inhibition of insulin signaling, particularly through activation of the NF-κB and JNK pathways. The TH2 cytokines IL-4 and IL-13 have been shown to promote insulin sensitivity and glucose homeostasis, as well as triggering the JAK-STAT-PPAR axis in metabolic organs. In organisms with a dynamic Th1/Th2 balance, tea polyphenols are able to reduce T cell proliferation and populations [Wu, 2016; Wu *et al.*, 2012], as well as affect signal transduction and transcription factors to inhibit the differentiation of initial T cell $CD4^+$ to Th1 and Th17 effectors. EGCG in mouse experiments inhibit mTOR and downstream HIF-1α activation due to HIF-1α affecting T cell subpopulation Th17/Treg differentiation and function. This exhibited inflammatory regulatory activity beyond the regulation of energy sensing by EGCG [Syed *et al.*, 2007].

2. Attenuation of heat shock proteins

Heat shock proteins (HSPs) act as molecular chaperones for other proteins that can be transferred between two cells and participate in the maintenance of proteostasis, have immunomodulatory functions, and are accompanied by decreasing levels of senescence, exacerbating oxidative stress in organisms, thereby accelerating cellular aging [Gomez, 2021]. In nematodes, EGCG is effective in alleviating heat stress-induced upregulation of heat shock proteins and the concomitant accumulation of ROS [Zhang *et al.*, 2009] and prolonged lifespan while enhancing nematode stress resistance [Abbas & Wink, 2009]. In cancer cells, black tea polyphenols attenuate the Akt signaling pathway by inhibiting upstream Hsp90 and consequently lead to elevate levels of downstream targets Wnt/β-catenin signaling, cyclin D1 and FOXO1, p27 [Halder *et al.*, 2012]. Inoue *et al.* verified that the cytoprotective effect of tea polyphenols was dose-dependent with HSP levels at low and medium doses [Inoue *et al.*, 2013].

HSPs are regulated by Keap1-Nrf2-ARE and heat shock transcription factor 1 (HSF1)-heat shock elements (HSE). Among them, in the Keap1-Nrf2-ARE pathway, Nrf2 normally binds to Keap1 and segregates into stable complexes in the cytoplasm. Transfer of Nrf2 to the nucleus to bind to the antioxidant response element (ARE) drives the expression of antioxidant genes. Most data support that EGCG exerts anti-inflammatory activity via Keap1-Nrf2-ARE [Shanmugam *et al.*, 2016; Zhu *et al.*, 2022] and activate the Nrf2/ARE signaling pathway in response to Nrf2 dysfunction [Mohan *et al.*, 2020], and that the effect of EGCG on KEAP-1 is completely lost in the absence of Nrf2 [Sun *et al.*, 2017]. This mechanism remains effective in the pro-oxidant effect of EGCG and partially restores the endogenous antioxidant activity of cells under stressful conditions [Sarkar & Sinha, 2018].

9.5.5 *Tea polyphenols delay aging by improving intestinal microbiota*

Although the bioavailability of tea polyphenols is limited and only briefly in organs such as the intestine and liver, it is extremely difficult to reach physiological concentrations in other tissues. Recent research has found that gut-derived peptides can reach the brain and "communicate" directly with it [Chandra *et al.*, 2020; Teichman *et al.*, 2020].

The systemic action of tea polyphenols through the brain-gut-microbe axis is a key step in the expression of biological activity [Liu *et al.*, 2021; Wei *et al.*, 2022; Zhang *et al.*, 2021b]. The intestinal flora plays a major role in this. Studies have shown that black tea polyphenols are associated with colonic microbial diversity for the treatment of intestinal and cardiovascular diseases [van Duynhoven *et al.*, 2013]. Ma *et al.* identified *Lachnospiraceae*, *Bacteroides*, *Alistipes*, and *Faecalibaculum* as possible key microbiota for the regulation of intestinal redox system by tea polyphenols [Ma *et al.*, 2019]. Zhou *et al.* found that tea polyphenols enhanced energy conversion by promoting the mitochondrial TCA cycle and urea cycle in the rat gut-microbiota [Zhou *et al.*, 2020], suggesting that the level of energy utilization in the host can be improved by the effect of green tea polyphenols on the intestinal flora. By collecting bile and intestinal contents of mice with long-term intake of different concentrations of tea polyphenols for compositional analysis, the assay results were found to be consistent with changes in the gut microbiome, which confirms the essential position of microorganisms in the improvement of metabolic diseases by tea polyphenols [Zhou *et al.*, 2018].

In addition to alterations in metabolic levels, interactions between the gut microbiota and the host are partially regulated through microbiota-derived neurological factors that affect the brain in a direct manner. Tea polyphenols have shown great potential to improve psychiatric disorders and neuroprotective effects by effectively altering host circadian rhythm disorders through intestinal microbes [Annunziata *et al.*, 2021; Guo *et al.*, 2019], as well as inducing intestinal microbes to secrete derived peptides that modulate vagal, immune system, and neuroendocrine pathways [Zhang *et al.*, 2021b].

Tea polyphenols can both depend on intestinal microorganisms for their biological activity and help the host maintain microecological homeostasis and intestinal health, creating a two-way effect [Hong *et al.*, 2022; Yan *et al.*, 2020]. While chronic intake of antibiotics can disrupt intestinal microecology, tea polyphenols have been shown to significantly increase the relative abundance of beneficial microorganisms such as *Lactobacillus*, *Akkermansia*, *Blautia*, *Roseburia*, and *Eubacterium* in the host, maintaining host gut homeostasis [Li *et al.*, 2021c]. In addition, long-term intake of tea polyphenols can improve the intestinal microbial structure of rats with rectal cancer [Wang *et al.*, 2018]. The metabolites of tea polyphenols also regulate intestinal microecology more effectively, further promoting human health [Sun *et al.*, 2021].

9.6 Conclusion

The efficacy of tea polyphenols in altering intracellular and inter-tissue oxidation levels of biological senescence is reflective of its strength and potential as a plant-derived therapeutic that can target crucial mechanistic pathways involved in aging. Tea polyphenols act mainly through the regulation of cellular redox homeostasis by dose-dependent, specifically in patients with disease and aging individuals. The interaction with complexes in the mitochondrial electron transport chain to modulate cellular ATP levels and maintain energy homeostasis via AMPK/mTOR. In patients, tea polyphenols exert anti-inflammatory/pro-inflammatory effects by modulating immune cell function and heat shock protein activity in patients. The systemic effects of the mechanisms may be dependent on gut microbial metabolism and pheromone secretion.

Aging is an irreversible physiological process. Laboratory data as well as clinical data on tea polyphenols in ameliorating age-related diseases support their potential to intervene in the aging process. In fact, some effect are not direct targets of the drug, rather impacting many of the similar hallmarks of aging. There is a need for more specific studies on how tea polyphenols affect each marker of aging. Thereby hopefully, the excavation of additional mechanisms will provide new pharmacotherapeutic avenues and interventions.

As the definition of aging has been updated, the goal has shifted from the pursuit of delaying death to reducing the incidence of disease or the risk of specific diseases and increasing healthy longevity, and has raised questions about personalized nutrition and aging body management related [Green & Hillersdal, 2021]. The modulatory mechanisms of tea polyphenols show clear therapeutic potential and have been partially validated. However, clear dose recommendations and dosing regimens for the elderly population as well as for patients with diseases are still lacking. The brain-gut-microbe axis may be an important breakthrough point.

References

Abbas S, Wink M. (2009) Epigallocatechin gallate from green tea (Camellia sinensis) increases lifespan and stress resistance in Caenorhabditis elegans. *Planta Med*, **75**, 216–221.

Abdelkawy KS, Abdelaziz RM, *et al.* (2020) Effects of green tea extract on Atorvastatin pharmacokinetics in healthy volunteers. *Eur J Drug Metab Pharmacokinet*, **45**, 351–360.

Abe SK, Saito E, *et al.* (2019) Green tea consumption and mortality in Japanese men and women: a pooled analysis of eight population-based cohort studies in Japan. *Eur J Epidemiol*, **34**, 917–926.

Ahmed RSI, Liu G, *et al.* (2016) Biological and mechanistic characterization of novel prodrugs of green tea polyphenol epigallocatechin gallate analogs in human leiomyoma cell lines. *J Cell Biochem*, **117**, 2357–2369.

Aires DJ, Rockwell G, *et al.* (2012) Potentiation of dietary restriction-induced lifespan extension by polyphenols. *Biochim Biophys Acta Mol Basis Dis*, **1822**, 522–526.

Altenhöfer S, Radermacher KA, *et al.* (2015) Evolution of NADPH oxidase inhibitors: selectivity and mechanisms for target engagement. *Antioxid Redox Signal*, **23**, 406–427.

Ammar M, Bahloul N, *et al.* (2022) Oxidative stress in patients with asthma and its relation to uncontrolled asthma. *J Clin Lab Anal*, e24345.

Amorim JA, Coppotelli G, *et al.* (2022) Mitochondrial and metabolic dysfunction in ageing and age-related diseases. *Nat Rev Endocrinol*, **18**, 243–258.

Andreu Fernandez V, Almeida Toledano L, *et al.* (2020). Bioavailability of epigallocatechin gallate administered with different nutritional strategies in healthy volunteers. *Antioxidants*, **9**.

Andreyev AY, Kushnareva YE, *et al.* (2005) Mitochondrial metabolism of reactive oxygen species. *Biochemistry (Mosc)*, **70**, 200–214.

Annunziata G, Sureda A, *et al.* (2021) The neuroprotective effects of polyphenols, their role in innate immunity and the interplay with the microbiota. *Neurosci Biobehav Rev*, **128**, 437–453.

Augustyniak A, Bylinska A, *et al.* (2011) Age-dependent changes in the proteolytic — antiproteolytic balance after alcohol and black tea consumption. *Toxicol Mech Methods*, 21, 209–215.

Back P, De Vos WH, *et al.* (2012) Exploring real-time in vivo redox biology of developing and aging Caenorhabditis elegans. *Free Radic Biol Med*, **52**, 850–859.

Bae J-Y, Choi J-S, *et al.* (2008) (-)Epigallocatechin gallate hampers collagen destruction and collagenase activation in ultraviolet-B-irradiated human dermal fibroblasts: Involvement of mitogen-activated protein kinase. *Food Chem Toxicol*, **46**, 1298–1307.

Bahadorani S, Hilliker AJ. (2008) Cocoa confers life span extension in Drosophila melanogaster. *Nutr Res (New York, N.Y.)*, **28**, 377–382.

Bahorun T, Luximon-Ramma A, *et al.* (2010) Black tea reduces uric acid and C-reactive protein levels in humans susceptible to cardiovascular diseases. *Toxicology*, **278**, 68–74.

Banerjee J, Khanna S, *et al.* (2017) MicroRNA regulation of oxidative stress. *Oxid Med Cell Longev*, **2017**, 2872156.

Barth E, Sieber P, *et al.* (2020) Robustness during aging-molecular biological and physiological aspects. *Cells*, **9**.

Bastianetto S, Yao Z-X, *et al.* (2006) Neuroprotective effects of green and black teas and their catechin gallate esters against beta-amyloid-induced toxicity. *Eur J Neurosci*, **23**, 55–64.

Bocchi L, Savi M, *et al.* (2018) Long-term oral administration of theaphenon-e improves cardiomyocyte mechanics and calcium dynamics by affecting phospholamban phosphorylation and atp production. *Cell Physiol Biochem*, **47**, 1230–1243.

Bodega G, Alique M, *et al.* (2017) The antioxidant machinery of young and senescent human umbilical vein endothelial cells and their microvesicles. *Oxid Med Cell Longev*, **2017**, 7094781–7094781.

Bodega G, Alique M, *et al.* (2019) Microvesicles: ROS scavengers and ROS producers. *J Extracell Vesicles*, **8**, 1626654.

Buford TW. (2016) Hypertension and aging. *Ageing Res Rev*, **26**, 96–111.

Cai S, Yang H, *et al.* (2016) EGCG inhibited lipofuscin formation based on intercepting amyloidogenic beta-sheet-rich structure conversion. *PLoS One*, **11**.

Calabrese EJ, Iavicoli I, *et al.* (2012) Hormesis: why it is important to biogerontologists. *Biogerontology*, **13**, 215–235.

Calgarotto AK, Longhini AL, *et al.* (2021) Immunomodulatory effect of green tea treatment in combination with low-dose chemotherapy in elderly acute myeloid leukemia patients with myelodysplasia-related changes. *Integr Cancer Ther*, **20**.

Camilleri A, Zarb C, *et al.* (2013) Mitochondrial membrane permeabilisation by amyloid aggregates and protection by polyphenols. *Biochimica et biophysica acta*, **1828**, 2532–2543.

Campisi J, Kapahi P, *et al.* (2019) From discoveries in ageing research to therapeutics for healthy ageing. *Nature*, **571**, 183–192.

Castellano-González G, Pichaud N, *et al.* (2016) Epigallocatechin-3-gallate induces oxidative phosphorylation by activating cytochrome c oxidase in human cultured neurons and astrocytes. *Oncotarget*, **7**, 7426–7440.

Cebe T, Atukeren P, *et al.* (2014) Oxidation scrutiny in persuaded aging and chronological aging at systemic redox homeostasis level. *Exp Gerontol*, **57**, 132–140.

Celec P, Janovičová Ĺ, *et al.* (2021) Circulating extracellular DNA is in association with continuous metabolic syndrome score in healthy adolescents. *Physiol Genomics*, 53, 309–318.

Chaikul P, Sripisut T, *et al.* (2020) Anti-skin aging activities of green tea (Camelliasinensis (L) Kuntze) in B16F10 melanoma cells and human skin fibroblasts. *Eur J Integr Med,* **40**.

Chandra S, Alam MT, *et al.* (2020) Healthy gut, healthy brain: the gut microbiome in neurodegenerative disorders. *Curr Top Med Chem*, **20**, 1142–1153.

Chatterjee S, Rudra T, *et al.* (2008) GST & CYP polymorphism related to tea drinking and oral pathology. *Int J Hum Genet*, 8, 295–299.

Chen D, Pamu S, *et al.* (2012) Novel epigallocatechin gallate (EGCG) analogs activate AMP-activated protein kinase pathway and target cancer stem cells. *Bioorg Med Chem*, **20**, 3031–3037.

Chen S-T, Kang L, *et al.* (2019) (-)-Epigallocatechin-3-Gallate Decreases Osteoclastogenesis via Modulation of RANKL and Osteoprotegrin. *Molecules*, **24**.

Chow HH, Cai Y, *et al.* (2001) Phase I pharmacokinetic study of tea polyphenols following single-dose administration of epigallocatechin gallate and polyphenon E. *Cancer Epidemiol Biomarkers Prev*, **10**, 53–58.

Chow HH, Cai Y, *et al.* (2003) Pharmacokinetics and safety of green tea polyphenols after multiple-dose administration of epigallocatechin gallate and polyphenon E in healthy individuals. *Clin Cancer Res*, 9, 3312–3319.

Chow HH, Hakim IA. (2011) Pharmacokinetic and chemoprevention studies on tea in humans. *Pharmacol Res*, **64**, 105–112.

Chow HHS, Hakim IA, *et al.* (2005) Effects of dosing condition on the oral bioavailability of green tea catechins after single-dose administration of polyphenon e in healthy individuals. *Clin Cancer Res*, **11**, 4627–4633.

Cialdella-Kam L, Ghosh S, *et al.* (2017) Quercetin and green tea extract supplementation downregulates genes related to tissue inflammatory responses to a 12-week high fat-diet in mice. *Nutrients*, **9**.

Ciesielski O, Biesiekierska M, *et al.* (2020) Epigallocatechin-3-gallate (EGCG) alters histone acetylation and methylation and impacts chromatin architecture profile in human endothelial cells. *Molecules*, **25**.

Collins QF, Liu HY, *et al*. (2007) Epigallocatechin-3-gallate (EGCG), a green tea polyphenol, suppresses hepatic gluconeogenesis through 5′-AMP-activated protein kinase. *J Biol Chem*, **282**, 30143–30149.

Cortés A, Solas M, *et al*. (2021) Expression of endothelial NOX5 Alters the integrity of the blood-brain barrier and causes loss of memory in aging mice. *Antioxidants (Basel)*, **10**.

Crew KD, Brown P, *et al*. (2012) Phase IB randomized, double-blinded, placebo-controlled, dose escalation study of polyphenon E in women with hormone receptor-negative breast cancer. *Cancer Prev Res*, **5**, 1144–1154.

Csaba G. (2019) Immunity and longevity. *Acta Microbiol Immunol Hung*, **66**, 1–17.

Daniel KG, Landis-Piwowar KR, *et al*. (2006) Methylation of green tea polyphenols affects their binding to and inhibitory poses of the proteasome beta5 subunit. *Int J Mol Med*, **18**, 625–632.

Deepashree S, Shivanandappa T, *et al*. (2022) Genetic repression of the antioxidant enzymes reduces the lifespan in Drosophila melanogaster. *J Comp Physiol B*, **192**, 1–13.

Del Rio D, Calani L, *et al*. (2010) Bioavailability of catechins from ready-to-drink tea. *Nutrition*, **26**, 52–533.

Didiano D, Shalaby T, *et al*. (2004) Telomere maintenance in childhood primitive neuroectodermal brain tumors. *Neuro-oncology*, **6**, 1–8.

Ding ML, Ma H, *et al*. (2017) Protective effects of a green tea polyphenol, epigallocatechin-3-gallate, against sevoflurane-induced neuronal apoptosis involve regulation of CREB/BDNF/TrkB and PI3K/Akt/mTOR signalling pathways in neonatal mice. *Can J Physiol Pharmacol*, **95**, 1396–1405.

Dobrzynska I, Szachowicz-Petelska B, *et al*. (2008) Effects of green tea on physico-chemical properties of liver cell membranes of rats intoxicated with ethanol. *Pol J Environ Stud*, **17**, 327–333.

Dong X, Wang D, *et al*. (2019) Efficacy of tea polyphenols (TP 50) against radiation-induced hematopoietic and biochemical alterations in beagle dogs. *J Tradit Chin Med*, **39**, 324–331.

Duan P, Zhang J, *et al*. (2020) Oolong tea drinking boosts calcaneus bone mineral density in postmenopausal women: a population-based study in southern China. *Arch Osteoporos*, **15**.

Duangjan C, Curran SP. (2022) Oolonghomobisflavans from Camellia sinensis increase Caenorhabditis elegans lifespan and healthspan. *Geroscience*, **44**, 533–545.

Dunnill C, Patton T, *et al*. (2017) Reactive oxygen species (ROS) and wound healing: the functional role of ROS and emerging ROS-modulating technologies for augmentation of the healing process. *Int Wound J*, **14**, 89–96.

Elbling L, Weiss RM, *et al.* (2005) Green tea extract and (-)-epigallocatechin-3-gallate, the major tea catechin, exert oxidant but lack antioxidant activities. *FASEB J*, **19**, 807–809.

Emoto Y, Yoshizawa K, *et al.* (2014) Green tea extract-induced acute hepatotoxicity in rats. *J Toxicol Pathol*, **27**, 163–174.

Enkhbat T, Nishi M, *et al.* (2018) Epigallocatechin-3-gallate enhances radiation sensitivity in colorectal cancer cells through nrf2 activation and autophagy. *Anticancer Res*, **38**, 6247–6252.

Feng WY. (2006) Metabolism of green tea catechins: an overview. *Curr Drug Metab*.

Fujiki H, Suganuma M, *et al.* (2002) Involvement of TNF-alpha changes in human cancer development, prevention and palliative care. *Mech Ageing Dev*, **123**, 1655–1663.

Gems D. (2022) The hyperfunction theory: an emerging paradigm for the biology of aging. *Ageing Res Rev*, **74**, 101557.

Gilardini L, Pasqualinotto L, *et al.* (2016) Effects of Greenselect Phytosome (R) on weight maintenance after weight loss in obese women: a randomized placebo-controlled study. *BMC Complement Altern Med*, **16**.

Golubev A, Hanson AD, *et al.* (2018) A tale of two concepts: harmonizing the free radical and antagonistic pleiotropy theories of aging. *Antioxid Redox Signal*, **29**, 1003–1017.

Gomez CR. (2021) Role of heat shock proteins in aging and chronic inflammatory diseases. *Geroscience*, **43**, 2515–2532.

Green S, Hillersdal L. (2021) Aging biomarkers and the measurement of health and risk. *Hist Philos Life Sci*, **43**, 28.

Guo T, Ho CT, *et al.* (2019) Oolong tea polyphenols ameliorate circadian rhythm of intestinal microbiome and liver clock genes in mouse model. *J Agric Food Chem*, **67**, 11969–11976.

Guoyuan Q, Yashi M, *et al.* (2017) Tea polyphenols ameliorates neural redox imbalance and mitochondrial dysfunction via mechanisms linking the key circadian regular Bmal1. *Food Chem Toxicol*, **110**, 189–199.

Gupta S, Hastak K, *et al.* (2001) Inhibition of prostate carcinogenesis in TRAMP mice by oral infusion of green tea polyphenols. *Proc Natl Acad Sci U S A*, **98**, 10350–10355.

Gusarov I, Shamovsky I, *et al.* (2021) Dietary thiols accelerate aging of C. elegans. *Nat Commun*, **12**, 4336.

Halder B, Das Gupta S, *et al.* (2012) Black tea polyphenols induce human leukemic cell cycle arrest by inhibiting Akt signaling: possible involvement of Hsp90, Wnt/β-catenin signaling and FOXO1. *FEBS J*, **279**, 2876–2891.

Hansen JM, Jones DP, *et al.* (2020) The Redox Theory of Development. *Antioxid Redox Signal*, **32**, 715–740.

Harman D. (1981) The aging process. *Proc Natl Acad Sci U S A*, **78**, 7124–7128.

Harman D. (2006) Free radical theory of aging: an update: increasing the functional life span. *Ann N Y Acad Sci*, 1067, 10–21.

Harper CE, Patel BB, *et al.* (2007) Epigallocatechin-3-gallate suppresses early stage, but not late stage prostate cancer in TRAMP mice: mechanisms of action. *Prostate*, **67**, 1576–1589.

He M, Zhao L, *et al.* (2009) Neuroprotective effects of (-)-epigallocatechin-3-gallate on aging mice induced by d-galactose. *Biol Pharm Bull*, **32**, 55–60.

Hecker L, Logsdon NJ, *et al.* (2014) Reversal of persistent fibrosis in aging by targeting Nox4-Nrf2 redox imbalance. *Sci Transl Med*, **6**.

Henning SM, Aronson W, *et al.* (2006) Tea polyphenols and theaflavins are present in prostate tissue of humans and mice after green and black tea consumption. *J Nutr*, **136**, 1839–1843.

Henning SM, Wang P, *et al.* (2013) Phenolic acid concentrations in plasma and urine from men consuming green or black tea and potential chemopreventive properties for colon cancer. *Mol Nutr Food Res*, **57**, 483–492.

Henning SM, Wang P, *et al.* (2015) Randomized clinical trial of brewed green and black tea in men with prostate cancer prior to prostatectomy. *Prostate*, **75**, 550–559.

Holczer M, Besze B, *et al.* (2018). Epigallocatechin-3-gallate (egcg) promotes autophagy-dependent survival via influencing the balance of mTOR-AMPK pathways upon endoplasmic reticulum stress. *Oxid Med Cell Longev*, **2018**, 6721530.

Hong M, Zhang R, *et al.* (2022) The interaction effect between tea polyphenols and intestinal microbiota: role in ameliorating neurological diseases. *J Food Biochem*, **46**, e13870.

Hsu S, Bollag WB, *et al.* (2003a) Green tea polyphenols induce differentiation and proliferation in epidermal keratinocytes. *J Pharm Exp Ther*, **306**, 29–34.

Hsu S, Lewis J, *et al.* (2003b) Green tea polyphenol targets the mitochondria in tumor cells inducing caspase 3-dependent apoptosis. *Anticancer Res*, **23**, 1533–1539.

Hsuan-Ti H, Tsung-Lin C, *et al.* (2020) Osteoprotective roles of green tea catechins. *Antioxidants*, **9**, 1136–1136.

Huang J-H, Huang C-C, *et al.* (2010). Protective effects of myricetin against ultraviolet-B-induced damage in human keratinocytes. *Toxicol in Vitro*, **24**, 21–28.

Huang Y, Liu L. (2017) Anti-aging and anti-osteoporosis effects of green teen polyphenol in a premature aging model of Bmi-1 knockout mice. *Int J Clin Exp Pathol*, **10**, 3765–3777.

Hwang JT, Ha J, *et al.* (2007) Apoptotic effect of EGCG in HT-29 colon cancer cells via AMPK signal pathway. *Cancer Lett*, **247**, 115–121.

Inoue H, Akiyama S, *et al.* (2011) High-dose green tea polyphenols induce nephrotoxicity in dextran sulfate sodium-induced colitis mice by down-regulation of antioxidant enzymes and heat-shock protein expressions. *Cell Stress Chaperones*, **16**, 653–662.

Inoue H, Maeda-Yamamoto M, *et al.* (2013) Low and medium but not high doses of green tea polyphenols ameliorated dextran sodium sulfate-induced hepatotoxicity and nephrotoxicity. *Biosci, Biotechnol, Biochem*, **77**, 1223–1228.

Ishii T, Mori T, *et al.* (2008) Covalent modification of proteins by green tea polyphenol (-)-epigallocatechin-3-gallate through autoxidation. *Free Radic Biol Med*, **45**, 1384–1394.

James KD, Kennett MJ, *et al.* (2018) Potential role of the mitochondria as a target for the hepatotoxic effects of (-)-epigallocatechin-3-gallate in mice. *Food Chem Toxicol*, **111**, 302–309.

Janssen RJ, Nijtmans LG, *et al.* (2006) Mitochondrial complex I: structure, function and pathology. *J Inherit Metab Dis*, **29**, 499–515.

Janssens GE, Grevendonk L, *et al.* (2022) Healthy aging and muscle function are positively associated with NAD+ abundance in humans. *Nat Aging*, **2**, 254–263.

Jatoi A, Ellison N, *et al.* (2003) A phase II trial of green tea in the treatment of patients with androgen independent metastatic prostate carcinoma. *Cancer*, **97**, 1442–1446.

Jia CM, Zhang FW, *et al.* (2021) Tea polyphenols prevent sepsis-induced lung injury via promoting translocation of DJ-1 to mitochondria. *Front Cell Dev Biol*, **9**, 622507.

Jones DP. (2015) Redox theory of aging. *Redox Biol*, **5**, 71–79.

Joyner PM. (2021) Protein adducts and protein oxidation as molecular mechanisms of flavonoid bioactivity. *Molecules*, **26**.

Justice JN, Ferrucci L, *et al.* (2018) A framework for selection of blood-based biomarkers for geroscience-guided clinical trials: report from the TAME Biomarkers Workgroup. *Geroscience*, **40**, 419–436.

Kammeyer A, Luiten RM. (2015) Oxidation events and skin aging. *Ageing Res Rev*, **21**, 16–29.

Katiyar SK, Afaq F, *et al.* (2001) Green tea polyphenol (-)-epigallocatechin-3-gallate treatment of human skin inhibits ultraviolet radiation-induced oxidative stress. *Carcinogenesis*, **22**, 287–294.

Kato Y, Higashiyama A, *et al.* (2021) Food phytochemicals, epigallocatechin gallate and myricetin, covalently bind to the active site of the coronavirus main protease in vitro. *Adv Redox Res*, **3**, 100021.

Kazi A, Wang Z, *et al.* (2004) Structure-activity relationships of synthetic analogs of (-)-epigallocatechin-3-gallate as proteasome inhibitors. *Anticancer Res*, **24**, 943–954.

Kim HS, Montana V, *et al.* (2013) Epigallocatechin gallate (EGCG) stimulates autophagy in vascular endothelial cells: a potential role for reducing lipid accumulation. *J Bbiol Chem*, **288**, 22693–22705.

Knaus UG. (2021) Oxidants in physiological processes. *Handb Exp Pharmacol*, **264**, 27–47.

Koju N, Qin ZH, *et al.* (2022) Reduced nicotinamide adenine dinucleotide phosphate in redox balance and diseases: a friend or foe? *Acta Pharmacol Sin*, **43**(8), 1889–1904.

Krishnan TR, Velusamy P, *et al.* (2014) Epigallocatechin-3-gallate restores the Bcl-2 expression in liver of young rats challenged with hypercholesterolemia but not in aged rats: an insight into its disparity of efficacy on advancing age. *Food Funct*, **5**, 916–926.

Kumar A, Yegla B, *et al.* (2018) Redox signaling in neurotransmission and cognition during aging. *Antioxid Redox Signal*, **28**, 1724–1745.

Kumazoe M, Nakamura Y, *et al.* (2017) Green tea polyphenol epigallocatechin-3-gallate suppresses toll-like receptor 4 expression via up-regulation of e3 ubiquitin-protein ligase RNF216. *J Biol Chem*, **292**, 4077–4088.

Lambert JD, Kennett MJ, *et al.* (2010). Hepatotoxicity of high oral dose (-)-epigallocatechin-3-gallate in mice. *Food Chem Toxicol*, **48**, 409–416.

Laurie SA, Miller VA, *et al.* (2005) Phase I study of green tea extract in patients with advanced lung cancer. *Cancer Chemother Pharmacol*, **55**, 33–38.

Le Bourg E. (2001) Oxidative stress, aging and longevity in Drosophila melanogaster. *FEBS Lett*, **498**, 183–186.

Lee AH, Su D, *et al.* (2013) Tea consumption reduces ovarian cancer risk. *Cancer Epidemiology*, **37**, 54–59.

Lee HS, Jun J-H, *et al.* (2014b) Epigalloccatechin-3-gallate inhibits ocular neovascularization and vascular permeability in human retinal pigment epithelial and human retinal microvascular endothelial cells via suppression of MMP-9 and VEGF activation. *Molecules*, **19**, 12150–12172.

Lee KO, Kim SN, *et al.* (2014a) Anti-wrinkle effects of water extracts of teas in hairless mouse. *Toxicol Res*, **30**, 283–289.

Lee S-H, Byeong-Gyun J, *et al.* (2018) Comparative analysis on anti-aging, anti-adipogenesis, and anti-tumor effects of green tea polyphenol epigallocatechin-3-gallate. *J Life Sci*, **28**, 1201–1211.

Lee S-J, Lee K-W. (2007) Protective effect of (-)-epigallocatechin gallate against advanced glycation endproducts-induced injury in neuronal cells. *Biol Pharm Bull*, **30**, 1369–1373.

Lee W-J, Cheng T-C, *et al.* (2021) Tea polyphenol epigallocatechin-3-gallate inhibits cell proliferation in a patient-derived triple-negative breast cancer xenograft mouse model via inhibition of proline-dehydrogenase-induced effects. *J Food Drug Anal*, **29**, 113–127.

Lefkimmiatis K, Grisan F, *et al.* (2021) Mitochondrial communication in the context of aging. *Aging Clin Exp Res*, **33**, 1367–1370.

Levy D, Reichert CO, *et al.* (2019) Paraoxonases activities and polymorphisms in elderly and old-age diseases: an overview. *Antioxidants (Basel)*, **8**.

Lewandowska U, Szewczyk K, *et al.* (2013) Overview of metabolism and bioavailability enhancement of polyphenols. *J Agric Food Chem*, **61**, 12183–12199.

Li B, Vik SB, *et al.* (2012) Theaflavins inhibit the ATP synthase and the respiratory chain without increasing superoxide production. *J Nutr Biochem*, **23**, 953–960.

Li HC, Yashiki S, *et al.* (2000) Green tea polyphenols induce apoptosis in vitro in peripheral blood T lymphocytes of adult T-cell leukemia patients. *Japanese J Cancer Res*, **91**, 34–40.

Li J, Chen C, *et al.* (2021c) Tea polyphenols regulate gut microbiota dysbiosis induced by antibiotic in mice. *Food Res Int (Ottawa, Ont.)*, **141**, 110153.

Li Q, Kakizaki M, *et al.* (2008) Green tea consumption and lung cancer risk: the Ohsaki study. *British J Cancer*, **99**, 1179–1184.

Li S, Wu H, *et al.* (2021b) Combined broccoli sprouts and green tea polyphenols contribute to the prevention of estrogen receptor-negative mammary cancer via cell cycle arrest and inducing apoptosis in HER2/neu mice. *J Nutr*, **151**, 73–84.

Li Y, Kračun D, *et al.* (2021a) Forestalling age-impaired angiogenesis and blood flow by targeting NOX: interplay of NOX1, IL-6, and SASP in propagating cell senescence. *Proc Natl Acad Sci U S A*, **118**.

Li Z, Wang S, *et al.* (2022) Advanced oxidative protein products drive trophoblast cells into senescence by inhibiting the autophagy: the potential implication of preeclampsia. *Front Cell Dev Biol*, **10**, 810282.

Liguori I, Russo G, *et al.* (2018) Oxidative stress, aging, and diseases. *Clin Interv Aging*, **13**, 757–772.

Lin IH, Ho M-L, *et al.* (2012) Smoking, green tea consumption, genetic polymorphisms in the insulin-like growth factors and lung cancer risk. *PLoS One*, **7**.

Lin J, Epel E. (2022) Stress and telomere shortening: insights from cellular mechanisms. *Ageing Res Rev*, **73**, 101507.

Lin Y, Huang J, *et al.* (2021) Preliminary study on hepatoprotective effect and mechanism of (-)-epigallocatechin-3-gallate against acetaminophen-induced liver injury in rats. *Iranian J Pharm Res*, **20**, 46–56.

Liu Y, Wu Z, *et al.* (2021) The role of the intestinal microbiota in the pathogenesis of host depression and mechanism of TPs relieving depression. *Food Funct*, **12**, 7651–7663.

Lorenz M. (2013) Cellular targets for the beneficial actions of tea polyphenols. *Am J Clin Nutr*, **98**, 1642S–1650S.

Louzada RA, Bouviere J, *et al.* (2020) Redox signaling in widespread health benefits of exercise. *Antioxid Redox Signal*, 10.1089/ars.2019.7949.

Lu Y, Wang Y, *et al.* (2021) Physiological dose of EGCG attenuates the health defects of high dose by regulating MEMO-1 in Caenorhabditis elegans. *Oxid Med Cell Longev*, **2021**, 5546493.

Luping Z, Mengqian S, *et al.* (2021) Tea polyphenols improve the memory in aging ovariectomized rats by regulating brain glucose metabolism in vivo and in vitro. *J Funct Foods*, **87**, 104856–104856.

Ma H, Zhang B, *et al.* (2019) Correlation analysis of intestinal redox state with the gut microbiota reveals the positive intervention of tea polyphenols on hyperlipidemia in high fat diet fed mice. *J Agric Food Chem*, **67**, 7325–7335.

Madreiter-Sokolowski CT, Sokolowski AA, *et al.* (2018) Targeting mitochondria to counteract age-related cellular dysfunction. *Genes (Basel)*, **9**.

Mah Y-J, Song JS, *et al.* (2014). The effect of epigallocatechin-3-gallate (EGCG) on human alveolar bone cells both in vitro and in vivo. *Arch Oral Biol*, **59**, 539–549.

Mähler A, Mandel S, *et al.* (2013) Epigallocatechin-3-gallate: a useful, effective and safe clinical approach for targeted prevention and individualised treatment of neurological diseases? *EPMA J*, **4**, 5.

Manach C, Williamson G, *et al.* (2005) Bioavailability and bioefficacy of polyphenols in humans. I. Review of 97 bioavailability studies. *Am J Clin Nutr*, **81**, 230s–242s.

Mantena SK, Roy AM, *et al.* (2005) Epigallocatechin-3-gallate inhibits photocarcinogenesis through inhibition of angiogenic factors and activation of CD8+ T cells in tumors. *Photochem Photobiol*, **81**, 1174–1179.

Marinovic MP, Morandi AC, *et al.* (2015) Green tea catechins alone or in combination alter functional parameters of human neutrophils via suppressing the activation of TLR-4/NFκB p65 signal pathway. *Toxicol in Vitro*, **29**, 1766–1778.

Martel J, Ojcius DM, *et al.* (2019) Hormetic effects of phytochemicals on health and longevity. *Trends Endocrinol Metab*, **30**, 335–346.

Martínez de Toda I, Vida C, *et al.* (2020) Redox parameters as markers of the rate of aging and predictors of life span. *J Gerontol A Biol Sci Med Sci*, **75**, 613–620.

Massudi H, Grant R, *et al.* (2012) Age-associated changes in oxidative stress and NAD+ metabolism in human tissue. *PLoS One*, **7**, e42357.

McLarty J, Bigelow RLH, *et al.* (2009). Tea polyphenols decrease serum levels of prostate-specific antigen, hepatocyte growth factor, and vascular endothelial growth factor in prostate cancer patients and inhibit production of hepatocyte growth factor and vascular endothelial growth factor in vitro. *Cancer Prev Res*, **2**, 673–682.

Menegazzi M, Campagnari R, *et al.* (2020). Protective effect of epigallocatechin-3-gallate (EGCG) in diseases with uncontrolled immune activation: could such a scenario be helpful to counteract COVID-19? *Int J Mol Sci*, **21**.

Meng J, Chen Y, *et al.* (2020) EGCG protects vascular endothelial cells from oxidative stress-induced damage by targeting the autophagy-dependent PI3K-AKT-mTOR pathway. *Ann Transl Med*, **8**, 200.

Mezera V, Endlicher R, *et al.* (2016) Effects of epigallocatechin gallate on tert-butyl hydroperoxide-induced mitochondrial dysfunction in rat liver mitochondria and hepatocytes. *Oxid Med Cell Longev*, **2016**, 7573131.

Miller RJ, Jackson KG, *et al.* (2012) A preliminary investigation of the impact of catechol-O-methyltransferase genotype on the absorption and metabolism of green tea catechins. *Eur J Nutr*, **51**, 47–55.

Mohan T, Narasimhan KKS, *et al.* (2020) Role of Nrf2 dysfunction in the pathogenesis of diabetic nephropathy: therapeutic prospect of epigallocatechin-3-gallate. *Free Radic Biol Med*, **160**, 227–238.

Murakami A. (2014) Dose-dependent functionality and toxicity of green tea polyphenols in experimental rodents. *Arch Biochem Biophys*, **557**, 3–10.

Murase T, Misawa K, *et al.* (2009) Catechin-induced activation of the LKB1/AMP-activated protein kinase pathway. *Biochem Pharmacol*, **78**, 78–84.

Mustata GT, Rosca M, *et al.* (2005) Paradoxical effects of green tea (Camellia sinensis) and antioxidant vitamins in diabetic rats: improved retinopathy and renal mitochondrial defects but deterioration of collagen matrix glycoxidation and cross-linking. *Diabetes*, **54**, 517–526.

Na H-K, Surh Y-J. (2008) Modulation of Nrf2-mediated antioxidant and detoxifying enzyme induction by the green tea polyphenol EGCG. *Food Chem Toxicol*, **46**, 1271–1278.

Nagao T, Hase T, *et al.* (2007) A green tea extract high in catechins reduces body fat and cardiovascular risks in humans. *Obesity*, **15**, 1473–1483.

Nagao T, Komine Y, *et al.* (2005) Ingestion of a tea rich in catechins leads to a reduction in body fat and malondialdehyde-modified LDL in men. *Am J Clin Nutr*, **81**, 122–129.

Nakachi K, Suemasu K, *et al.* (1998) Influence of drinking green tea on breast cancer malignancy among Japanese patients. *Japanese J Cancer Res: Gann*, **89**, 254–261.

Narumi K, Sonoda J-I, *et al.* (2014) Simultaneous detection of green tea catechins and gallic acid in human serum after ingestion of green tea tablets using ion-pair high-performance liquid chromatography with electrochemical detection. *J Chromatogr B-Anal Techn Biomed Life Sci*, **945**, 147–153.

Nguyen MM, Ahmann FR, *et al.* (2012) Randomized, double-blind, placebo-controlled trial of polyphenon E in prostate cancer patients before prostatectomy: evaluation of potential chemopreventive activities. *Cancer Prev Res*, **5**, 290–298.

Ohno S, Yu H, *et al.* (2012) Epigallocatechin-3-gallate modulates antioxidant and DNA repair-related proteins in exocrine glands of a primary Sjogren's syndrome mouse model prior to disease onset. *Autoimmunity*, **45**, 540–546.

Ohno T, Ohtani M, *et al.* (2016) Effect of green tea catechins on gastric mucosal dysplasia in insulin-gastrin mice. *Oncology Rep*, **35**, 3241–3247.

Okoro NO, Odiba AS, *et al.* (2021) Bioactive phytochemicals with anti-aging and lifespan extending potentials in Caenorhabditis elegans. *MoleculesI*, **26**.

Ortiz-Lopez L, Marquez-Valadez B, *et al.* (2016) green tea compound epigallo-catechin-3-gallate (egcg) increases neuronal survival in adult hippocampal neurogenesis in vivo and in vitro. *Neuroscience*, **322**, 208–220.

Pajonk F, Riedisser A, *et al.* (2006) The effects of tea extracts on proinflammatory signaling. *BMC Med*, **4**, 28.

Palliyaguru DL, Moats JM, *et al.* (2019) Frailty index as a biomarker of lifespan and healthspan: focus on pharmacological interventions. *Mech Ageing Dev*, **180**, 42–48.

Pandit AP, Joshi SR, *et al.* (2019) Curcumin as a permeability enhancer enhanced the antihyperlipidemic activity of dietary green tea extract. *BMC Complement Altern Med*, **19**, 129.

Patel R, Maru G. (2008) Polymeric black tea polyphenols induce phase II enzymes via Nrf2 in mouse liver and lungs. *Free Radic Biol Med*, **44**, 1897–1911.

Peng Y, Dai S, *et al.* (2021) Theanine improves high-dose epigallocatechin-3-gallate-induced lifespan reduction in Caenorhabditis elegans. *Foods*, **10**.

Pérez de la Cruz V, Korrapati SV, *et al.* (2014) Redox status and aging link in neurodegenerative diseases. *Oxid Med Cell Longev*, **2014**, 270291.

Pugh JN, Stretton C, *et al.* (2021) Exercise stress leads to an acute loss of mitochondrial proteins and disruption of redox control in skeletal muscle of older subjects: An underlying decrease in resilience with aging? *Free Radic Biol Med*, **177**, 88–99.

Ramlagan P, Rondeau P, *et al.* (2017) Comparative suppressing effects of black and green teas on the formation of advanced glycation end products (AGEs) and AGE-induced oxidative stress. *Food Funct*, **8**, 4194–4209.

Rasheed NOA, Ahmed LA, *et al.* (2018) Paradoxical cardiotoxicity of intraperitoneally-injected epigallocatechin gallate preparation in diabetic mice. *Sci Rep*, **8**, 7880.

Rehman H, Krishnasamy Y, *et al.* (2014) Green tea polyphenols stimulate mitochondrial biogenesis and improve renal function after chronic cyclosporin a treatment in rats. *PLoS One*, **8**, e65029.

Reiter CE, Kim JA, *et al.* (2010) Green tea polyphenol epigallocatechin gallate reduces endothelin-1 expression and secretion in vascular endothelial cells: roles for AMP-activated protein kinase, Akt, and FOXO1. *Endocrinology*, **151**, 103–114.

Robinson AR, Yousefzadeh MJ, *et al.* (2018). Spontaneous DNA damage to the nuclear genome promotes senescence, redox imbalance and aging. *Redox Biol*, **17**, 259–273.

Rocha A, Bolin AP, *et al.* (2016) Green tea extract activates AMPK and ameliorates white adipose tissue metabolic dysfunction induced by obesity. *Eur J Nutr*, **55**, 2231–2244.

Roth M, Timmermann BN, *et al.* (2011) Interactions of green tea catechins with organic anion-transporting polypeptides. *Drug Metab Dispos*, **39**, 920–926.

Sadowska-Bartosz I, Bartosz G. (2014) Effect of antioxidants supplementation on aging and longevity. *Biomed Res Int*, **2014**, 404680.

Saito E, Inoue M, *et al.* (2015a) Association of green tea consumption with mortality due to all causes and major causes of death in a Japanese population: the Japan Public Health Center-based Prospective Study (JPHC Study). *Ann Epidemiol*, **25**, 512–518.e513.

Saito K, Mori S, *et al.* (2015b) Epigallocatechin gallate stimulates the neuroreactive salivary secretomotor system in autoimmune sialadenitis of

MRL-Fas(lpr) mice via activation of cAMP-dependent protein kinase A and inactivation of nuclear factor kappa B. *Autoimmunity*, **48**, 379–388.

Salminen A, Kaarniranta K, *et al.* (2016) Age-related changes in AMPK activation: role for AMPK phosphatases and inhibitory phosphorylation by upstream signaling pathways. *Ageing Res Rev*, **28**, 15–26.

Sang S, Shao X, *et al.* (2007) Tea polyphenol (-)-epigallocatechin-3-gallate: a new trapping agent of reactive dicarbonyl species. *Chem Res Toxicol*, **20**, 1862–1870.

Santamarina AB, Carvalho-Silva M, *et al.* (2015) Decaffeinated green tea extract rich in epigallocatechin-3-gallate prevents fatty liver disease by increased activities of mitochondrial respiratory chain complexes in diet-induced obesity mice. *J Nutr Biochem*, **26**, 1348–1356.

Sarkar N, Sinha D. (2018) Epigallocatechin-3-gallate partially restored redox homeostasis in arsenite-stressed keratinocytes. *J Appl Toxicol*, **38**, 1071–1080.

Schiffers C, Reynaert NL, *et al.* (2021) Redox dysregulation in aging and COPD: role of NOX enzymes and implications for antioxidant strategies. *Antioxidants (Basel)*, **10**.

Scholl C, Lepper A, *et al.* (2018) Population nutrikinetics of green tea extract. *PLoS One*, **13**, e0193074.

Schumacher B, Pothof J, *et al.* (2021) The central role of DNA damage in the ageing process. *Nature*, **592**, 695–703.

Serrano JC, Gonzalo-Benito H, *et al.* (2013) Dietary intake of green tea polyphenols regulates insulin sensitivity with an increase in AMP-activated protein kinase α content and changes in mitochondrial respiratory complexes. *Mol Nutr Food Res*, **57**, 459–470.

Shanmugam T, Selvaraj M, *et al.* (2016) Epigallocatechin gallate potentially abrogates fluoride induced lung oxidative stress, inflammation via Nrf2/Keap1 signaling pathway in rats: An in-vivo and in-silico study. *Int Immunopharmacol*, **39**, 128–139.

Shen C-L, Chyu M-C, *et al.* (2010) Green tea polyphenols supplementation and Tai Chi exercise for postmenopausal osteopenic women: safety and quality of life report. *BMC Complement Altern Med*, **10**.

Shetty AK, Kodali M, *et al.* (2018) Emerging anti-aging strategies — scientific basis and efficacy. *Aging Dis*, **9**, 1165–1184.

Shields HJ, Traa A, *et al.* (2021) Beneficial and detrimental effects of reactive oxygen species on lifespan: a comprehensive review of comparative and experimental studies. *Front Cell Dev Biol*, **9**, 628157.

Shimizu K, Asakawa T, *et al.* (2014) Use of positron emission tomography for real-time imaging of biodistribution of green tea catechin. *PloS One*, **9**, e85520–e85520.

Song D, Ge J, *et al.* (2021) Tea polyphenol attenuates oxidative stress-induced degeneration of intervertebral discs by regulating the Keap1/Nrf2/ARE pathway. *Oxid Med Cell Longev*, **2021**, 6684147.

Srividhya R, Kalaiselvi P. (2013) Neuroprotective potential of epigallo catechin-3-gallate in PC-12 cells. *Neurochem Res*, **38**, 486–493.

Stanga S, Caretto A, *et al.* (2020) Mitochondrial dysfunctions: a red thread across neurodegenerative diseases. *Int J Mol Sci*, **21**.

Suh KS, Chon S, *et al.* (2010) Prooxidative effects of green tea polyphenol (-)-epigallocatechin-3-gallate on the HIT-T15 pancreatic beta cell line. *Cell Biol Toxicol*, **26**, 189–199.

Sun M, Shen Z, *et al.* (2019) Identification of the antiglycative components of Hong Dou Shan (Taxus chinensis) leaf tea. *Food Chem*, **297**.

Sun Q, Cheng L, *et al.* (2021) The interaction between tea polyphenols and host intestinal microorganisms: an effective way to prevent psychiatric disorders. *Food Funct*, **12**, 952–962.

Sun W, Liu X, *et al.* (2017) Epigallocatechin gallate upregulates NRF2 to prevent diabetic nephropathy via disabling KEAP1. *Free Radic Biol Med*, **108**, 840–857.

Sur S, Panda CK. (2017) Molecular aspects of cancer chemopreventive and therapeutic efficacies of tea and tea polyphenols. *Nutrition*, **43–44**, 8–15.

Syed DN, Afaq F, *et al.* (2007) Green tea polyphenol EGCG suppresses cigarette smoke condensate-induced NF-kappaB activation in normal human bronchial epithelial cells. *Oncogene*, **26**, 673–682.

Tang J, Zheng JS, *et al.* (2015) Tea consumption and mortality of all cancers, CVD and all causes: a meta-analysis of eighteen prospective cohort studies. *Br J Nutr*, **114**, 673–683.

Teichman EM, O'Riordan KJ, *et al.* (2020) When rhythms meet the blues: circadian interactions with the microbiota-gut-brain axis. *Cell Metab*, **31**, 448–471.

Tian J, Geiss C, *et al.* (2021) Green tea catechins EGCG and ECG enhance the fitness and lifespan of Caenorhabditis elegans by complex I inhibition. *Aging-US*, **13**, 22629–22648.

Tofolean IT, Ganea C, *et al.* (2016) Cellular determinants involving mitochondrial dysfunction, oxidative stress and apoptosis correlate with the synergic cytotoxicity of epigallocatechin-3-gallate and menadione in human leukemia Jurkat T cells. *Pharmacol Res*, **103**, 300–317.

Toolsee NA, Aruoma OI, *et al.* (2013) Effectiveness of green tea in a randomized human cohort: relevance to diabetes and its complications. *Biomed Research International*, **2013**.

Unno K, Pervin M, *et al.* (2017) Blood-brain barrier permeability of green tea catechin metabolites and their neuritogenic activity in human neuroblastoma SH-SY5Y cells. *Mol Nutr Food Res*, **61**.

Uysal U, Seremet S, *et al.* (2013) Consumption of polyphenol plants may slow aging and associated diseases. *Curr Pharm Des*, **19**, 6094–6111.

Valenti D, de Bari L, *et al.* (2013b) Negative modulation of mitochondrial oxidative phosphorylation by epigallocatechin-3 gallate leads to growth arrest and apoptosis in human malignant pleural mesothelioma cells. *Biochimica et biophysica acta*, **1832**, 2085–2096.

Valenti D, De Rasmo D, *et al.* (2013a) Epigallocatechin-3-gallate prevents oxidative phosphorylation deficit and promotes mitochondrial biogenesis in human cells from subjects with Down's syndrome. *Biochimica et biophysica acta*, **1832**, 542–552.

Valko M, Leibfritz D, *et al.* (2007) Free radicals and antioxidants in normal physiological functions and human disease. *The international journal of biochemistry & cell biology*, **39**, 44–84.

Van Amelsvoort JM, Van Hof KH, *et al.* (2001) Plasma concentrations of individual tea catechins after a single oral dose in humans. *Xenobiotica*, **31**, 891–901.

van Duynhoven J, Vaughan EE, *et al.* (2013) Interactions of black tea polyphenols with human gut microbiota: implications for gut and cardiovascular health. *Am J Clin Nutr*, **98**, 1631s–1641s.

van Woudenbergh GJ, Kuijsten A, *et al.* (2012) Tea consumption and incidence of type 2 diabetes in Europe: the EPIC-InterAct case-cohort study. *PLoS One*, **7**.

Victorelli S, Lagnado A, *et al.* (2019) Senescent human melanocytes drive skin ageing via paracrine telomere dysfunction. *EMBO J*, **38**, e101982.

Vidoni S, Zanna C, *et al.* (2013) Why mitochondria must fuse to maintain their genome integrity. *Antioxid Redox Signal*, **19**, 379–388.

Villegas L, Nørremølle A, *et al.* (2021) Nicotinamide adenine dinucleotide phosphate oxidases are everywhere in brain disease, but not in Huntington's disease? *Front Aging Neurosci*, **13**, 736734.

Viña J, Olaso-Gonzalez G, *et al.* (2020) Modulating oxidant levels to promote healthy aging. *Antioxid Redox Signal*, **33**, 570–579.

Vinereanu IV, Peride I, *et al.* (2021) The Relationship between advanced oxidation protein products, vascular calcifications and arterial stiffness in predialysis chronic kidney disease patients. *Medicina (Kaunas, Lithuania)*, 57.

Wang D, Wang T, *et al.* (2022) Green tea polyphenols upregulate the Nrf2 signaling pathway and suppress oxidative stress and inflammation markers in D-galactose-induced liver aging in mice. *Front Nutr*, **9**, 836112.

Wang D, Wang Y, *et al.* (2015) Green tea polyphenol (-)-epigallocatechin-3-gallate triggered hepatotoxicity in mice: responses of major antioxidant enzymes and the Nrf2 rescue pathway. *Toxicol Appl Pharmacol*, **283**, 65–74.

Wang D, Zhang M, *et al.* (2020) Green tea polyphenols mitigate the plant lectins-induced liver inflammation and immunological reaction in C57BL/6 mice via NLRP3 and Nrf2 signaling pathways. *Food Chem Toxicol*, **144**, 111576.

Wang G, Liu H, *et al.* (2014) Oolong tea drinking could help prevent bone loss in postmenopausal Han Chinese women (Retracted article. See vol. 76, pg. 325, 2018). *Cell Biochemistry And Biophysics*, **70**, 1289–1293.

Wang J-S, Luo H, *et al.* (2008) Validation of green tea polyphenol biomarkers in a phase II human intervention trial. *Food And Chemical Toxicology*, **46**, 232–240.

Wang J, Tang L, *et al.* (2018) Long-term treatment with green tea polyphenols modifies the gut microbiome of female sprague-dawley rats. *J Nutr Biochem*, **56**, 55–64.

Wang P, Heber D, *et al.* (2012) Quercetin increased bioavailability and decreased methylation of green tea polyphenols in vitro and in vivo. *Food Funct*, **3**, 635–642.

Wei R, Mao L, *et al.* (2018) Suppressing glucose metabolism with epigallocatechin-3-gallate (EGCG) reduces breast cancer cell growth in preclinical models. *Food Funct*, **9**, 5682–5696.

Wei Y, Xu J, *et al.* (2022) Recent advances in the utilization of tea active ingredients to regulate sleep through neuroendocrine pathway, immune system and intestinal microbiota. *Critical Rev Food Sci Nutr*, 1–29.

Weinreb O, Amit T, *et al.* (2007) A novel approach of proteomics and transcriptomics to study the mechanism of action of the antioxidant-iron chelator green tea polyphenol (-)-epigallocatechin-3-gallate. *Free Radic Biol Med*, **43**, 546–556.

Wen DT, Wang WQ, *et al.* (2020) Endurance exercise protects aging Drosophila from high-salt diet (HSD)-induced climbing capacity decline and lifespan decrease by enhancing antioxidant capacity. *Biol Open*, **9**.

Weng Z, Zhou P, *et al.* (2014) Green tea epigallocatechin gallate binds to and inhibits respiratory complexes in swelling but not normal rat hepatic mitochondria. *Biochem Biophys Res Commun*, **443**, 1097–1104.

Wu AH, Yu MC, *et al.* (2003) Green tea and risk of breast cancer in Asian Americans. *Int J Cancer*, **106**, 574–579.

Wu D. (2016) Green tea EGCG, T-cell function, and T-cell-mediated autoimmune encephalomyelitis. *J Investig Med*, **64**, 1213–1219.

Wu D, Wang J, *et al.* (2012) Green tea EGCG, T cells, and T cell-mediated autoimmune diseases. *Mol Aspects Med*, **33**, 107–118.

Xiao H, Jedrychowski MP, *et al.* (2020) A quantitative tissue-specific landscape of protein redox regulation during aging. *Cell*, **180**, 968.

Xie LW, Cai S, *et al.* (2020) Green tea derivative (-)-epigallocatechin-3-gallate (EGCG) confers protection against ionizing radiation-induced intestinal epithelial cell death both in vitro and in vivo. *Free Radic Biol Med*, **161**, 175–186.

Xiong LG, Chen YJ, *et al.* (2018) Epigallocatechin-3-gallate promotes healthy lifespan through mitohormesis during early-to-mid adulthood in Caenorhabditis elegans. *Redox Biol*, **14**, 305–315.

Xiong LG, Huang JA, *et al.* (2014). Black tea increased survival of Caenorhabditis elegans under stress. *J Agric Food Chem*, **62**, 11163–11169.

Xu J, Tu Y, *et al.* (2020) Prodrug of epigallocatechin-3-gallate alleviates choroidal neovascularization via down-regulating HIF-1 alpha/VEGF/VEGFR2 pathway and M1 type macrophage/microglia polarization. *Biomed Pharmacother*, 121.

Xu XX, Zheng G, *et al.* (2021) Theaflavin protects chondrocytes against apoptosis and senescence via regulating Nrf2 and ameliorates murine osteoarthritis. *Food Funct*, **12**, 1590–1602.

Yamamoto T, Staples J, *et al.* (2004) Protective effects of EGCG on salivary gland cells treated with gamma-radiation or cis-platinum(II)diammine dichloride. *Anticancer Res*, **24**, 3065–3073.

Yan J, Zhao Y, *et al.* (2013) Green tea catechins prevent obesity through modulation of peroxisome proliferator-activated receptors. *Sci China Life Sci*, **56**, 804–810.

Yan R, Ho CT, *et al.* (2020) Interaction between tea polyphenols and intestinal microbiota in host metabolic diseases from the perspective of the gut-brain axis. *Mol Nutr Food Res*, **64**, e2000187.

Yan Z, Zhong Y, *et al.* (2020) Antioxidant mechanism of tea polyphenols and its impact on health benefits. *Anim Nutr*, **6**, 115–123.

Yang CC, Wu CJ, *et al.* (2021) Green tea polyphenol catechins inhibit coronavirus replication and potentiate the adaptive immunity and autophagy-dependent protective mechanism to improve acute lung injury in mice. *Antioxidants (Basel)*, **10**.

Yang CS, Lee MJ, *et al.* (1999) Human salivary tea catechin levels and catechin esterase activities: implication in human cancer prevention studies. *Cancer Epidemiol Biomarkers Prev*, **8**, 83–89.

Yang CS, Wang X, *et al.* (2009) Cancer prevention by tea: animal studies, molecular mechanisms and human relevance. *Nat Rev Cancer*, **9**, 429–439.

Yang CS, Zhang J. (2019) Studies on the prevention of cancer and cardiometabolic diseases by tea: issues on mechanisms, effective doses, and toxicities. *J Agric Food Chem*, **67**, 5446–5456.

Yang CS, Zhang J, *et al.* (2016) Mechanisms of body weight reduction and metabolic syndrome alleviation by tea. *Mol Nutr Food Res*, **60**, 160–174.

Yang G, Shu X-O, *et al.* (2007) Prospective cohort study of green tea consumption and colorectal cancer risk in women. *Cancer Epidemiol Biomarkers Prev*, **16**, 1219–1223.

Yang J-R, Ren T-T, *et al.* (2020) Tea polyphenols attenuate staurosporine-induced cytotoxicity and apoptosis by modulating BDNF-TrkB/Akt and Erk1/2 signaling axis in hippocampal neurons. *IBRO reports*, **8**, 115–121.

Yao WL, Wen Q, *et al.* (2021) Different subsets of haematopoietic cells and immune cells in bone marrow between young and older donors. *Clin Exp Immunol*, **203**, 137–149.

Ye T, Zhen J, *et al.* (2015) Green tea polyphenol (-)-epigallocatechin-3-gallate restores Nrf2 activity and ameliorates crescentic glomerulonephritis. *PLoS One*, **10**, e0119543.

Yin JY, Duan SY, *et al.* (2017) Blood pressure is associated with tea consumption: A cross-sectional study in a rural, elderly population of Jiangsu China. *J Nutr Health Aging*, **21**, 1151–1159.

Yoon SO, Yun CH, *et al.* (2002) Dose effect of oxidative stress on signal transduction in aging. *Mech Ageing Dev*, **123**, 1597–1604.

Young ML, Franklin JL. (2019) The mitochondria-targeted antioxidant MitoQ inhibits memory loss, neuropathology, and extends lifespan in aged 3xTg-AD mice. *Mol Cell Neurosci*, **101**, 103409.

Zeng J, Xiao Q, *et al.* (2021) Advanced oxidation protein products aggravate age-related bone loss by increasing sclerostin expression in osteocytes via ROS-dependent downregulation of Sirt1. *Int J Mol Med*, **47**.

Zhang L, Jie G, *et al.* (2009) Significant longevity-extending effects of EGCG on Caenorhabditis elegans under stress. *Free Radic Biol Med*, **46**, 414–421.

Zhang X, Albanes D, *et al.* (2010) Risk of colon cancer and coffee, tea, and sugar-sweetened soft drink intake: pooled analysis of prospective cohort studies. *J Natl Cancer Inst*, **102**, 771–783.

Zhang YP, Yang XQ, *et al.* (2021a) Nrf2 signalling pathway and autophagy impact on the preventive effect of green tea extract against alcohol-induced liver injury. *J Pharm Pharmacol*, **73**, 986–995.

Zhang Z, Zhang Y, *et al.* (2021b) The neuroprotective effect of tea polyphenols on the regulation of intestinal flora. *Molecules*, **26**.

Zhao Y, Hu X, *et al.* (2017). ROS signaling under metabolic stress: cross-talk between AMPK and AKT pathway. *Mol Cancer*, **16**, 79.

Zheng X, Feng M, *et al.* (2021) Anti-damage effect of theaflavin-3′-gallate from black tea on UVB-irradiated HaCaT cells by photoprotection and maintaining cell homeostasis. *J Photochem Photobiol B-Biol*, **224**.

Zhou C, Zhang Y, *et al.* (2021) Association between serum advanced oxidation protein products and mortality risk in maintenance hemodialysis patients. *J Transl Med*, **19**, 284.

Zhou J, Tang L, *et al.* (2018) Green tea polyphenols modify gut-microbiota dependent metabolisms of energy, bile constituents and micronutrients in female Sprague-Dawley rats. *J Nutr Biochem*, **61**, 68–81.

Zhou J, Tang L, *et al.* (2020) Green tea polyphenols boost gut-microbiota-dependent mitochondrial TCA and urea cycles in Sprague-Dawley rats. *J Nutr Biochem*, **81**, 108395.

Zhu W, Tang H, *et al.* (2022) Epigallocatechin-3-o-gallate ameliorates oxidative stress-induced chondrocyte dysfunction and exerts chondroprotective effects via the Keap1/Nrf2/ARE signaling pathway. *Chem Biol Drug Des*, **100**(1), 108–120.

Zhu XH, Lu M, *et al.* (2015) In vivo NAD assay reveals the intracellular NAD contents and redox state in healthy human brain and their age dependences. *Proc Natl Acad Sci U S A*, **112**, 2876–2881.